Subcellular Biochemistry

Volume 23

Physicochemical Methods in the Study of Biomembranes

SUBCELLULAR BIOCHEMISTRY

SERIES EDITOR

J. R. HARRIS, Institute of Zoology, University of Mainz, Mainz, Germany

ASSISTANT EDITORS

H. J. HILDERSON, University of Antwerp, Antwerp, Belgium
D. A. WALL, SmithKline Beecham Pharmaceuticals, King of Prussia, Pennsylvania, U.S.A.

Recent Volumes in This Series:

A Continuation Order Plan is available for this series. A continuation order will bring delivery of each new volume immediately upon publication. Volumes are billed only upon actual shipment. For further information please contact the publisher.

Subcellular Biochemistry

Volume 23

Physicochemical Methods in the Study of Biomembranes

Edited by

Herwig J. Hilderson
University of Antwerp
Antwerp, Belgium

and

Gregory B. Ralston
University of Sydney
Sydney, Australia

SPRINGER SCIENCE+BUSINESS MEDIA, LLC

The Library of Congress cataloged the first volume of this title as follows:

Sub-cellular biochemistry.
London, New York, Plenum Press.
v. illus. 23 cm. quarterly.
Began with Sept. 1971 issue. Cf. New serial titles.
1. Cytochemistry—Periodicals. 2. Cell organelles—Periodicals.
QH611.S84 574.8'76   73-643479

ISBN 978-1-4613-5757-5   ISBN 978-1-4615-1863-1 (eBook)
DOI 10.1007/978-1-4615-1863-1

This series is a continuation of the journal *Sub-Cellular Biochemistry,*
Volumes 1 to 4 of which were published quarterly from 1972 to 1975

© 1994 Springer Science+Business Media New York
Originally published by Plenum Press, New York in 1994

Contributors

Rodney L. Biltonen   Departments of Biochemistry and Pharmacology, University of Virginia Health Sciences Center, Charlottesville, Virginia 22908

Véronique Cabiaux   Laboratoire de Chimie Physique des Macromolécules aux Interfaces, Université Libre de Bruxelles, B-1050 Brussels, Belgium

Rudy A. Demel   Centre for Biomembranes and Lipid Enzymology, Department of Biochemistry of Membranes, University of Utrecht, 3584 CH Utrecht, The Netherlands

Erik Goormaghtigh   Laboratoire de Chimie Physique des Macromolécules aux Interfaces, Université Libre de Bruxelles, B-1050 Brussels, Belgium

László I. Horváth   Institute of Biophysics, Biological Research Centre, H6701 Szeged, Hungary

Robert M. Johnson   Department of Biochemistry, Wayne State University, Detroit, Michigan 48201

Glenn F. King   Department of Biochemistry, University of Sydney, Sydney, NSW 2006, Australia

Kiaran Kirk   University Laboratory of Physiology, University of Oxford, OX1 3PT Oxford, England

Philip W. Kuchel   Department of Biochemistry, University of Sydney, Sydney, NSW 2006, Australia

Peter Laggner   Institute of Biophysics and X-ray Structure Research, Austrian Academy of Sciences, A-8010 Graz, Austria

Michael B. Morris   Department of Biochemistry, University of Sydney, Sydney, NSW 2006, Australia

Gregory B. Ralston   Department of Biochemistry, University of Sydney, Sydney, NSW 2006, Australia

Jean-Marie Ruysschaert   Laboratoire de Chimie Physique des Macro-molécules aux Interfaces, Université Libre de Bruxelles, B-1050 Brussels, Belgium

Ida L. van Genderen   Department of Cell Biology, Medical School, University of Utrecht, 3584 CX Utrecht, The Netherlands

Gerrit van Meer   Department of Cell Biology, Medical School, University of Utrecht, 3584 CX Utrecht, The Netherlands

Qiang Ye   Departments of Biochemistry and Pharmacology, University of Virginia Health Sciences Center, Charlottesville, Virginia 22908

Preface

In mammalian cells many physiological processes rely on the dynamics of the organization of lipids and proteins in biological membranes. The topics in this volume deal with physicochemical methods in the study of biomembranes. Some of them have a long and respectable history in the study of soluble proteins and have only recently been applied to the study of membranes. Some have traditionally been applied to studies of model systems of lipids of well-defined composition, as well as to intact membranes. Other methods, by their very nature, apply to organized bilayers comprised of both protein and lipid.

Van Meer and van Genderen provide us with an introduction to the field (Chapter 1). From their personal perspective regarding the distribution, transport, and sorting of membrane lipids, they formulate a number of biologically relevant questions and show that the physicochemical methods described in this book may contribute in great measure to solving these issues.

The methods of analytical ultracentrifugation have served faithfully for 60 years in the study of water-soluble proteins. The use of detergent extraction of membrane proteins, and the manipulation of density with H_2O/D_2O mixtures, has extended this technique to the study of proteins, and in particular their interactions, from biological membranes. As described by Morris and Ralston in Chapter 2, this technique can be used to determine a number of important properties of proteins.

Techniques such as monomolecular layers have had particular relevance to the structure of lipid bilayer membranes since the pioneering work of Gorter and Grendel in 1925, which led to the concept of lipid bilayers. The use of careful studies of the pressure–area characteristics of lipid monolayers, discussed in Chapter 3 by Demel, allows exploration of the interactions, both thermodynamic and kinetic, with the lipid layer of solutes dissolved in the aqueous subphase.

The physical properties and biological functions of biological membranes reflect both local and global organization of the lipids and proteins within the membrane. In Chapter 4, Ye and Biltonen describe the use of both differential scanning calorimetry and dynamic calorimetry for probing the organization and dynamics of lipids both in model membrane systems and in biological membranes.

Whole cells are more complicated than simple bilayers and often display interaction between the lipid bilayer and the underlying cytoskeleton. These interactions, only relevant in the intact membrane, are important for control of cell shape and deformability and may be probed through the technique of ektacytometry, as described by Johnson in Chapter 5.

The time scale of spin-label ESR spectroscopy is particularly suitable to resolving the motion of lipids and the restriction of such motion in the vicinity of membrane proteins, as discussed by Horváth in Chapter 6. The use of covalent ESR probes also provides a spatial dimension in probing the interaction between the protein and lipid moieties.

The use of NMR in the study of transport through membranes and into cells is based implicitly on the membrane limiting and defining the compartments inside and outside cells. The application of NMR to transport processes in biological membranes is addressed by Kuchel, Kirk, and King in Chapter 7. The authors demonstrate how this technique offers a number of advantages over more conventional techniques.

In Chapters 8, 9, and 10, Goormaghtigh, Cabiaux, and Ruysschaert, providing a very detailed account of the use of Fourier transform infrared spectroscopy, demonstrate that this method can lead to an understanding of the secondary structure of soluble proteins, and that it can be applied to the structure of proteins within a membrane as well. According to the authors, the method enables the simultaneous study of the structure of lipids and proteins in intact biological membranes without introduction of foreign perturbing probes. Chapter 8 is a summary of the basic knowledge accumulated on the different vibrations of interest for the study of proteins. In Chapter 9, the experimental problems related to recording spectra for proteins and the potentialities of the method for the study of the structure of amino acid side chains are examined. Chapter 10 deals more specifically with the recovery of protein secondary structures from the complex IR spectra.

X-ray diffraction is sensitive to regular spacing and orientation of the molecules within the lipid bilayer. In Chapter 11, Laggner describes recent developments in X-ray diffraction as applied to the lipid component of biological membranes, in particular the use of synchrotron radiation and fast detectors to overcome some of the earlier limitations of the method.

We are aware that this survey is far from complete. Many other techniques could have been included (e.g., electron microscopy) or are treated elsewhere (e.g., fluorospectroscopy in Volume 16 of this series). However, the range of approaches covered in the present volume is representative of the need for increasing sensitivity and resolution in attempting to understand the organization, interaction, and dynamics of both lipid and protein components within the biological membrane.

We thank Professor de Kruijff for his advice in compiling this book.

Herwig J. Hilderson
Antwerp, Belgium Gregory B. Ralston
Sydney, Australia

Contents

Chapter 3
Monomolecular Layers in the Study of Biomembranes
Rudy A. Demel

Chapter 4

**Differential Scanning and Dynamic Calorimetric Studies of
Cooperative Phase Transitions in Phospholipid Bilayer Membranes**
Qiang Ye and Rodney L. Biltonen

Chapter 5

Ektacytometry of Red Cells
Robert M. Johnson

Chapter 6

Spin-Label ESR Study of Molecular Dynamics of Lipid/Protein Association in Membranes

László I. Horváth

Chapter 7
NMR Methods for Measuring Membrane Transport
Philip W. Kuchel, Kiaran Kirk, and Glenn F. King

Chapter 8
**Determination of Soluble and Membrane Protein Structure
by Fourier Transform Infrared Spectroscopy: I. Assignments
and Model Compounds**
Erik Goormaghtigh, Véronique Cabiaux, and Jean-Marie Ruysschaert

Chapter 9
**Determination of Soluble and Membrane Protein Structure by
Fourier Transform Infrared Spectroscopy: II. Experimental
Aspects, Side Chain Structure, and H/D Exchange**
Erik Goormaghtigh, Véronique Cabiaux, and Jean-Marie Ruysschaert

Chapter 10
**Determination of Soluble and Membrane Protein Structure
by Fourier Transform Infrared Spectroscopy:
III. Secondary Structures**
Erik Goormaghtigh, Véronique Cabiaux, and Jean-Marie Ruysschaert

Chapter 11
**X-Ray Diffraction on Biomembranes with Emphasis
on Lipid Moiety**
Peter Laggner

Chapter 1

Intracellular Lipid Distribution, Transport, and Sorting

A Cell Biologist's Need for Physicochemical Information

Gerrit van Meer and Ida L. van Genderen

1. INTRODUCTION

Every scientist interested in the biological function of molecules or macro-molecular assemblies in cells sooner or later touches on questions concerning the interplay between the structure of these molecules and their physical environment. The present volume describes the application of physicochemical methods to study the structure, orientation, dynamics, and intermolecular interactions of molecules of biomembranes. These molecules include the membrane lipids and integral membrane proteins, but also proteins that are membrane-associated (some temporarily) such as cytoskeletal proteins, proteins responsible for the creation of lateral domains, and proteins involved in the various signal transduction cascades. The relevance of this field is obvious to anybody who has ever considered the multitude of biological processes that occur in cellular membranes, on the surface of cellular membranes, and across cellular membranes,

Gerrit van Meer and Ida L. van Genderen   Department of Cell Biology, Medical School, University of Utrecht, 3584 CX Utrecht, The Netherlands.

Subcellular Biochemistry, Volume 23: Physicochemical Methods in the Study of Biomembranes, edited by Herwig J. Hilderson and Gregory B. Ralston. Plenum Press, New York, 1994.

and who has tried to count the number of different molecules involved. Two general aims can be recognized: (1) to define general properties of the lipids and proteins in biomembranes; (2) to characterize the unique properties of the various components and relate these to specific functions.

One way to write an introduction to this volume would be to describe the membrane-associated processes essential for the functioning of living cells, to list the molecules involved, and to display the physicochemical questions that are being asked at the cutting edge of the field. We are afraid that, apart from the fact that we are not supposed to fill this volume all by ourselves, such a task goes far beyond our personal ability.

Instead, we shall describe our own interests in the dynamics of the organization of membrane lipids in mammalian cells to illustrate how crucial the application of physicochemical methods is for further progress in the field. This chapter should, therefore, be considered a very personal account of an area of membrane cell biology where the application of physicochemical methods is expected to have significant impact. It may be considered a plea to all scientists who possess the relevant expertise to consider the questions formulated. It is our task in this chapter to convince you of the biological relevance of these questions. This book is on the dynamics of biomembranes. Above all, what we hope is that it conveys the brilliance of the choreography, the intricacy of the dance, and the charm of the dancers (Figure 1).

2.   INTRACELLULAR LIPID TOPOLOGY: SOME PROBLEMS

2.1.   Lipid Composition of Membranes in Mammalian Cells

The major lipids in eukaryotic cell membranes are the glycerophospholipids phosphatidylcholine (PC) and phosphatidylethanolamine (PE), which account for 50–70% of the lipids in most membranes. Other abundant glycerophospholipids are phosphatidylserine (PS) and phosphatidylinositol (PI), whereas the di-and triphosphorylated forms of PI, PIP and PIP_2 occur in small quantities only. A second class of membrane lipids are the steroids, of which cholesterol is the main representative in mammalian cells. The third class, the sphingolipids, contain a ceramide backbone consisting of a sphingoid base with a fatty acyl chain hooked on to its 2-position by an amide linkage. The major sphingolipid is the phospholipid sphingomyelin (SM); this class also comprises the enormous variety of glycosphingolipids. The simplest glycosphingolipids are the monohexosylceramides glucosylceramide (GlcCer) and galactosylceramide (GalCer), while essentially all higher glycosphingolipids are derived from GlcCer by the subsequent addition of galactose, N-acetylgalactosamine, N-acetylglucosamine, sialic acid, and fucose units. Although this allows an almost limitless variation in gly-

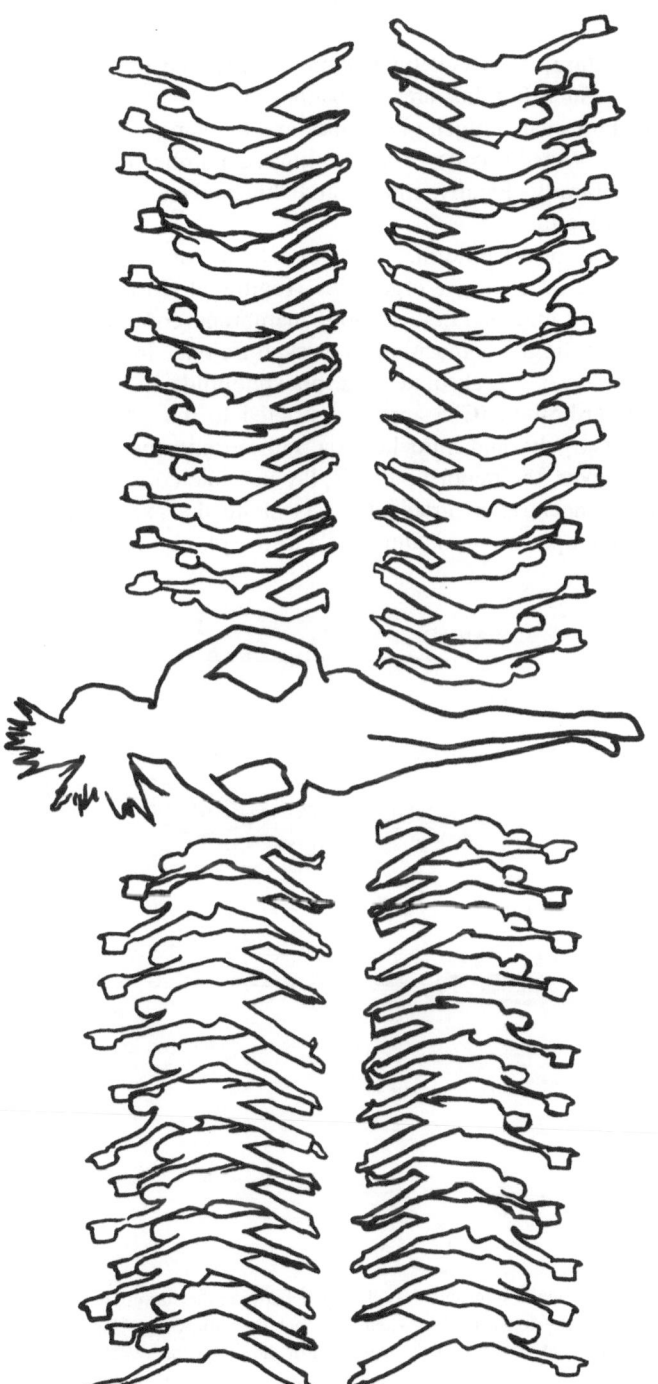

FIGURE 1. The fun of being a membrane lipid molecule. Stylish lipid dancers celebrate the extraordinary performance of a membrane protein molecule.

cosphingolipid structure, most cell types contain only one or two families of glycosphingolipids. Such families are for example the globo-series, leading to globoside and the Forssman antigen glycolipid, and the ganglio-series, comprising all gangliosides. Mostly, cells contain a limited number of glycosphingolipids of a certain family. Finally, mono- and dihexosylceramides containing galactose can be sulfated.

The vast majority of the lipid molecules in cells are organized in the membranes of the various intracellular organelles (Figure 2). Most organelles are thought to be separate entities, i.e., without membrane continuity with other organelles, and their lipid composition is very different. In some cases the compositional differences are absolute: The phospholipids lysobisphosphatidic acid (LBPA) and cardiolipin (CL) appear restricted to the lysosomes and the mitochondria, respectively, and mitochondria and peroxisomes do not contain sphingolipids. However, the major lipids display variable concentrations (Table I). An interesting concentration gradient exists along the exocytotic transport pathway from the endoplasmic reticulum (ER) to the plasma membrane, where

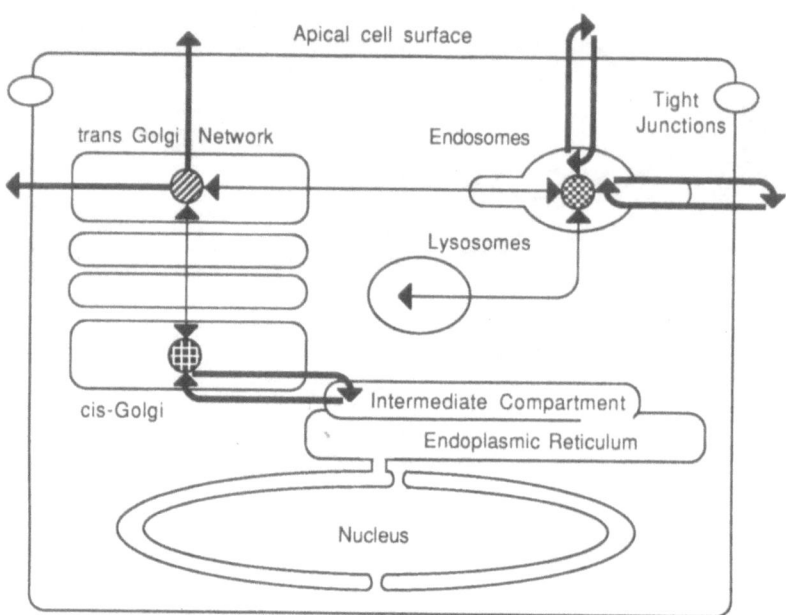

FIGURE 2. Pathways of vesicular transport. Vesicles cycle between the starting point, the ER, and the *cis*-Golgi (Hauri and Schweizer, 1992). Whether or not the intra-Golgi transport contains a unidirectional step is unclear at present. At the *trans* end of the Golgi the vesicle components are delivered to the plasmalemma and from it into the endocytic pathway. A shortcut connects the TGN via a prelysosomal compartment to the lysosomes; the latter present a point of no return for vesicular traffic.

Table I
Typical Lipid Composition of Some Intracellular Organelles of Rat Liver[a]

Phospholipids	Mitochondrial membrane	ER	Golgi membrane	Plasma membrane	Lysosomal membrane[b]
SM	0.5	2.5	7.6	16.0	20.3
PC	40.3	58.4	49.6	39.3	39.7
PI	4.6	10.1	12.2	7.7	4.5
PS	0.7	2.9	5.6	9.0	1.7
PE	34.6	21.8	19.6	23.3	14.1
CL	17.8	1.1	1.2	1.0	1.0
LBPA[b]	0.2				7.0
			Cholesterol/phospholipid (mol/mol)		
	0.03	0.08	0.16	0.38/0.76[c]	0.49[c]

[a]Data expressed as percentage of total phospholipid phosphorus from Zambrano et al. (1975). Similar data in Colbeau et al. (1971) and Brotherus and Renkonen (1977).
[b]Data from Wherrett and Huterer (1972).
[c]Data from Colbeau et al. (1971)

an enrichment is observed in SM, glycosphingolipids, and cholesterol at the cost of PC and PI. In addition, the plasma membrane is enriched in disaturated phospholipids (Keenan and Morré, 1970; Colbeau et al., 1971).

High concentrations of glycosphingolipids have been found in the plasma membrane. However, the general notion that the plasma membrane contains the bulk of the glycosphingolipids has been disproved by the observation of an intracellular accumulation of glycosphingolipids in various cell types (Symington et al., 1987; van Genderen et al., 1991). The relative sizes of the phospholipid pools have been calculated from cell fractionation data. In fibroblasts the plasma membrane contained 31% of the phospholipids (Brotherus and Renkonen, 1977), while Warnock et al. (1993) measured 50%. Although independent techniques seemed to confirm the value of 50% for the phospholipids and yielded 90% for cholesterol and SM (Lange et al., 1989; Lange, 1991), morphometric measurements on electron micrographs of the same cells led to the much lower value of 14% for the relative surface area of the plasma membrane, which should more or less equal the relative lipid content (Griffiths et al., 1989). These data are conflicting and the actual contribution of the plasma membrane lipids to the total cellular lipids remains an open question.

2.2. Transmembrane Lipid Asymmetry

2.2.1. Equilibrium Distribution

Evidence is accumulating to suggest that the plasma membrane of most cells is organized like the well-characterized asymmetric erythrocyte membrane (reviewed in (Devaux, 1991; 1993). Organelles late in the exocytotic transport pathway, the *trans*-Golgi network (TGN) and secretory granules, and organelles of the endocytotic pathway, endosomes and lysosomes, appear to have properties very similar to those of the plasma membrane. In the erythrocyte membrane most of the choline-phospholipids SM and PC are situated in the exoplasmic bilayer leaflet and most of the amino-phospholipids PS and PE are on the cytoplasmic surface. SM may be located in the exoplasmic leaflet exclusively. Also PS is essentially unilateral, whereas significant fractions of PC and PE are present in both bilayer leaflets. Glycosphingolipids in the plasma membrane have been found exclusively in the exoplasmic leaflet. However, the simple concept that all glycosphingolipids are organized in the exoplasmic, luminal leaflet of the membrane of organelles is challenged by recent discoveries concerning the siddeness of glycosphingolipid synthesis (see below).

Little is known about the transbilayer distribution of cholesterol. Whereas the preferential interaction between cholesterol and sphingolipids would predict an enrichment of cholesterol in the exoplasmic leaflet, a notion that seemed to be confirmed by X-ray studies on myelin (Caspar and Kirschner, 1971), recent data on fluorescent lipids in the erythrocyte membrane suggest the opposite (Schroeder *et al.*, 1991). This discrepancy is the more amazing as cholesterol constitutes more than 35 mol% of the plasma membrane lipids!

The transbilayer lipid distribution in the other intracellular membranes is also essentially unknown. It has even been questioned whether the ER membrane is a regular bilayer (see van Meer, 1986). Clearly, a solution to these questions is required to fully understand the structure of the membrane at the molecular level. More specifically, knowledge of the composition of each monolayer of the bilayer is needed to put into biological context experimental studies on lipid monolayers [Chapter 3 (Demel)], phospholipid phase transitions [Chapter 4 (Ye and Biltonen)], and lipid–protein interactions (Chapter 6 (Horvath)).

2.2.2. Transbilayer Translocation or Flipflop

The limiting membrane of eukaryotic cells, the plasma membrane, is a stable bilayer with low rates of transbilayer mobility (flip-flop). Glycosphingolipids do not translocate at all and this appears also to be true for SM (Devaux, 1991). In the case of PC, two pools exist in the plasma membrane that intermix with a half-time of flip-flop on the order of 10 hr or more at 37°C. Most likely this motion reflects spontaneous equilibration between the two monolayers. The process is slow on the time scale of intracellular transport steps, which display a typical half-time of minutes. In contrast, the amino-phospholipids are actively translocated to the cytoplasmic leaflet within seconds to minutes by an ATP-consuming amino-phospholipid translocator (Devaux, 1993). Does this enzyme correct a loss of lipid asymmetry during fusion and fission events? Is the lipid asymmetry of the plasma membrane ever scrambled, e.g., during signal transduction (Bratton, 1993; van Meer, 1993)? In such a case would the enzyme be sufficient to restore an asymmetric plasma membrane? What is the function of lipid asymmetry? Is it to differentiate the physicochemical properties of the two bilayer leaflets (El Hage Chahine et al., 1993), or does the translocator utilize ATP to provide the (so far unidentified) driving force for the formation of bilayer curvature in endocytosis? These are just a few of the fascinating questions concerning the transbilayer dynamics of the plasma membrane.

The major membrane lipid PC flips from its site of synthesis on the cytosolic surface of the ER to the luminal leaflet of the ER membrane and back in an energy-independent process, thereby allowing expansion of both sides of the membrane simultaneously. The stereospecificity of the flip-flop (Bishop and Bell, 1985) has suggested the involvement of a PC-specific flippase [a similar activity has been reported for the (specialized) bile-canalicular plasma membrane of hepatocytes (Berr et al., 1993), with a possible role in PC secretion into the bile]. Later studies observed a general translocation of phospholipids, including SM, across the ER bilayer with a half-time of 20 min at 37°C (Herrmann et al., 1990). An even more dramatic lipid translocation event precedes the N-glycosylation of proteins in the ER lumen. After synthesis on the cytosolic aspect of the ER, dolichol-pyrophospho-$(GlcNAc)_2$-$(Man)_5$ translocates across the membrane (Abeijon and Hirschberg, 1992). Also dolichol-phospho-mannose and dolichol-phospho-glucose required for further extension of the glyco-phospholipid tree translocate across the ER membrane. Finally, the simple glycosphingolipids GlcCer (van Meer, 1993) and GalCer (Burger et al., 1994) are synthesized on the cytosolic side of the Golgi and must translocate toward the Golgi lumen to be utilized for higher glycosphingolipid synthesis. The essential questions concerning the translocation of such a number of (very different) substrates are: What protein (complex) is responsible for each event, and how is the barrier function of the ER maintained in the process? Studies implicating the multidrug resistance proteins, which transport drugs across the plasma membrane, in lipid translocation (Higgins and Gottesman, 1992; Smit et al., 1993) suggest that methods to measure transport across membranes like those applied in Chapter 7 (Kuchel et al.) may contribute to solving these issues.

2.3. Lateral Distribution

Lipids rapidly diffuse in the plane of biomembranes (see Section 3.3.1). Still, approximately 50% of the cell types in the human body, the epithelial cells, display a stable lateral lipid heterogeneity in their plasma membrane. In these cells the plasma membrane is divided into a large-scale (>250 μm^2) apical and basolateral domain by the tight junctions, a zone of cell–cell contacts that surrounds the cell apex (Simons and van Meer, 1988). Many epithelial cells are organized like the simple epithelia of the intestine and kidney. Others, like hepatocytes, neurons, or myelin-forming oligodendrocytes and Schwann cells, display a more complex morphology. Still, all of these possess plasma membrane domains separated by tight junctions at the sites of contact with neighboring cells.

In the plasma membrane of intestinal cells, which has been characterized in most detail, the basolateral domain displays a typical plasma membrane composition, whereas the apical membrane which faces the external environment of the lumen of the gut is highly specialized. Compared with the basolateral membrane, it displays a fourfold enrichment in glycosphingolipids at the expense of, mainly, the phospholipid PC. The glycosphingolipids have an apical/basolateral polarity 17-fold higher than PC (Table II). The apical surface of intestinal cells turns out to be very special indeed. Its exoplasmic surface must be covered by a glycosphingolipid monolayer, with all phospholipids in the inner leaflet, a caveat being the unknown distribution of cholesterol.

The tight junctions maintain the lateral heterogeneity by constituting a diffusion barrier for lipids in the exoplasmic leaflet of the plasma membrane (Dragsten et al., 1981; Spiegel et al., 1985; van Meer and Simons, 1986). In the cytoplasmic bilayer leaflet, however, free lipid diffusion between the apical and basolateral domain was observed (van Meer and Simons, 1986). This could be explained by the lipid model of tight junction structure (Figure 3; Kachar and Reese, 1982; Pinto da Silva and Kachar, 1982). However, we have disproved the prediction by this model (in its simple form) that lipid molecules would be free to diffuse from one epithelial cell to the next through continuous exoplasmic leaflets of the apical and the basolateral membranes, both at the light and electron microscope level (van Meer et al., 1986; van Genderen et al., 1991). How a protein barrier can prevent lipid diffusion, and what the actual organization of the tight junction is, are questions asking for structural approaches like those to

Table II
Lipid Composition of the Plasma Membrane Domains of Mouse Intestinal Cells[a]

Lipid	Apical (mol%)		Basolateral (mol%)		Polarity (apical/basolateral)
Glycosphingolipid	33.3		7.9		4.22
Phospholipid	33.3		65.8		
PC		8.3		33.6	0.25
SM		2.8		6.5	0.43
Cholesterol	33.3		26.3		

[a]Calculated from cell fractionation data on the ICR strain of mouse, from Kawai et al. (1974).
  Very similar data for rat intestine in Brasitus and Schachter (1980), and for bovine urinary bladder in Stubbs et al. (1979). The latter tissue displayed a slight basolateral enrichment of SM (1.3×).

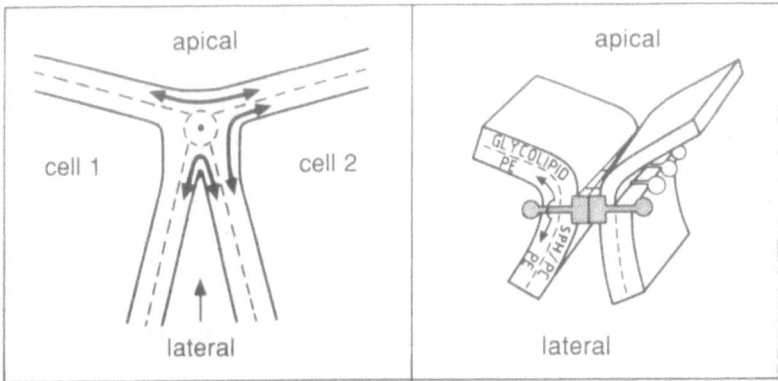

FIGURE 3. Models for the structure of the tight junction. (Left) In the lipid model, the core of the tight junction consists of a lipid cylinder surrounding the apex of the cell (Kachar and Reese, 1982; Pinto da Silva and Kachar, 1982). (Right) In protein models of the tight junction, proteins on the surface of one cell interact with proteins on the adjacent cell to restrict diffusion in between the cells and in the exoplasmic bilayer leaflet. Arrows indicate directions of free lipid diffusion. Adapted from van Meer and Simons (1986) by permission of *EMBO Journal*.

characterize protein structure described in Chapters 2 (Morris and Ralston) and 8–10 (Goormaghtigh *et al.*), potentially the methods to study lipid organization as in Chapter 11 (Laggner), and to describe the properties of the underlying cytoskeleton as in Chapter 5 (Johnson).

It is good to realize that a random distribution of the various lipid species in the plane of a biomembrane would not be a homogeneous distribution, but rather a fluctuating heterogeneity governed by diffusion and collision. Because each lipid displays characteristic possibilities to interact more or less intimately with each of the other types of lipid molecules, it has been an attractive idea that mixtures of lipids could spontaneously segregate into domains of molecules with preferential interactions (Glaser, 1993). This idea was fostered by model membrane studies, where simple lipid mixtures could be laterally separated into two types of environment by small changes in temperature or ionic conditions [Chapter 4 (Ye and Biltonen)]. In order to be of biological relevance the domains would have to interact specifically with certain proteins, whether integral membrane proteins or proteins on the surface. These interactions may be very short-lived [Chapter 6 (Horváth)], or be longer-lived but involve only a small number of lipid molecules, like the domains of as little as 100 glycosphingolipid molecules observed in erythrocytes (Thompson *et al.*, 1986), and still serve a biological purpose. However, physicochemical measurements in recent years have suggested the existence of substantially larger (μm) biological membrane domains as well (Jacobson and Vaz, 1992; Glaser, 1993), and interesting ideas concerning their potential functions in regulating bimolecular reactions have been generated from model membrane studies on domain connectivity (Vaz and Almeida, 1993). Cell biological studies over the last decade imply a function for lipid/protein domains in the sorting of both sphingolipids (van Meer et al., 1987; Kok et al., 1991) and of proteins anchored to the membrane by a glycosylphosphatidylinositol (GPI) tail (see Sargiacomo et al., 1993; Zurzolo et al., 1994) in the TGN. Independently, studies on caveolae, specialized areas in the plasma membrane, have led to the conclusion that similar GPI-protein/glycosphingolipid domains exist in the plasma membrane (Anderson, 1993; Dupree et al., 1993; Fiedler et al., 1993; Lisanti et al., 1993). Whether the lateral segregation of trans-membrane proteins observed during protein sorting in the TGN (Geuze et al., 1987) is caused solely by a one-to-one interaction of those proteins with cytosolic factors like adaptins and coat proteins (Pryer et al., 1992) or that transmembrane proteins aggregate into domains as well is an open question (see Section 3.3.3).

3. HOW DO LIPIDS GET WHERE THEY ARE?

3.1. Sites of Lipid Synthesis

Differences in lipid composition can in principle be caused by local synthesis. So it is important to define the site of synthesis for each lipid. The first way to approach this is to localize the enzyme responsible for the final step in the biosynthetic pathway. Sometimes an enzyme activity is located at two intracellular sites. The relative contribution of each enzyme pool to the synthesis of product will then depend on the relative amounts of substrate available at the two sites. Synthesis of the same lipid at two locations may even serve very different purposes (van Helvoort et al., 1994) and the reactions may be mediated by two different enzymes.

Whereas most steps of glycerophospholipid and cholesterol synthesis are thought to occur at the ER, sphingolipid synthesis is situated in the Golgi (Dennis and Vance, 1992). Unfortunately, the resolution of our knowledge is still very limited and many conflicting results on enzyme localization have been obtained (van Meer, 1993), mostly because the subcellular localization has essentially only been performed by cell fractionation techniques. Immunolocalization at the EM level would give definitive answers. However, immunolocalization of only a few enzymes has been reported and only by light microscopy. For example, although PC synthesis is generally assigned to the ER and Golgi (Dennis and Vance, 1992), the final steps of PC synthesis have been reported to occur at the nuclear membrane (Wang et al., 1993), in a mitochondrion-associated ER membrane (Vance, 1990) and in the transport vesicles mediating ER–Golgi transport (Slomiany et al., 1992). A similar situation exists for cholesterol synthesis, where the simple general view of cholesterol synthesis being completely localized in the ER has been challenged by the idea that the various steps are spread over a number of membranes including the plasma membrane, which would require shuttling intermediates back and forth (Lange et al., 1991). Also, the peroxisomes contain a number of enzymes required for cholesterol biosynthesis (Krisans, 1992). SM synthesis has been assigned to the cis-Golgi, with a limited contribution of the plasma membrane (Futerman et al., 1990; Jeckel et al., 1990), but resistance of the enzyme to relocation to the ER by brefeldin A would rather suggest localization in the TGN (van Meer and van't Hof, 1993). Some uncertainty also exists concerning the site of synthesis of GlcCer, precursor for essentially all higher glycosphingolipids. GlcCer synthesis has been assigned to the cis-Golgi and to another membrane fraction. The latter may be a pre-Golgi compartment (Futerman and Pagano, 1991), or trans-Golgi (Jeckel et al., 1992). In line with the latter suggestion, the activity in the unidentified membrane did not relocate to the ER in the presence of brefeldin A (Strous et al., 1993).

While glycerophospholipids are synthesized on the cytosolic aspect of cellular membranes, SM synthesis occurs in the luminal leaflet of the Golgi. Work over the last few years has firmly established that the synthesis of the simple glycosphingolipid GlcCer occurs on the cytosolic surface of the Golgi (Coste et al., 1986; Futerman and Pagano, 1991; Trinchera et al., 1991; Jeckel et al., 1992), whereas the next step, the addition of galactose by galactosyltransferase 2 to yield lactosylceramide (Gal-GlcCer), occurs in the lumen. Recent studies have revealed that synthesis of GalCer by galactosyltransferase 1, like GlcCer synthesis, is a cytosolic event, while also in that case the addition of the next galactose to yield Gal-GalCer occurs on the luminal side (Burger et al., 1994). Also the addition of sulfate to GalCer to yield sulfatide occurs at the luminal leaflet. Clearly, in order to become a substrate for galactosyltransferase 2 and sulfotransferase, GlcCer and GalCer must translocate across the Golgi membrane.

No single cellular membrane possesses all enzymes required for the synthesis of its full complement of lipids. In addition, lipids are not always synthesized in the membranes where their concentration is highest, e.g., cholesterol synthesis occurs in the ER but it is concentrated in the plasma membrane. So the conclusion must be that local synthesis alone cannot be the explanation for the specific lipid composition of organelles. Lipids must be transported from their organelle of synthesis to other organelles.

3.2. Sites of Lipid Degradation

An additional factor to explain the differences in lipid composition may be selective lipid hydrolysis. For example, the enrichment of cholesterol in the plasma membrane compared with the ER could be caused by selective hydrolysis of PC. For this, hydrolysis should be faster than equilibration of lipids between the two membranes. So, in order to judge whether local hydrolysis is a possible cause of differences in lipid composition, one should know the answers to the following questions: Where do the enzymes of lipid hydrolysis reside, and how fast is the reaction compared with lipid transport? We now know that phospholipases A and C are present not only in the lysosomal lumen but also in the plasma membrane, the nucleus, and the cytoplasm. This last location especially changes the importance of the question "where are the enzymes?" to "where do the enzymes act?," and even more importantly "when do they act? and how is their activity regulated?" Phospholipases A_2 and C, and sphingomyelinase at the plasma membrane can be activated during receptor-mediated signal transduction processes (Kolesnick, 1992; Nishizuka, 1992). Unfortunately, major uncertainties exist on what fraction of the plasma membrane lipids is degraded during a single round of activation and on the frequency of such events. Alarming numbers of 50% of a certain phospholipid hydrolyzed per hour have been published (see van Meer, 1993). Clearly, more careful studies of these phenomena are needed before we can reliably judge the influence of local hydrolysis for the equilibrium lipid distribution in cells.

3.3. Mechanisms of Lipid Transport

An explanation for the differences in lipid composition between the various intracellular membranes must be that the lipid transport between them displays selectivity. Transport of proteins between organelles and bulk transport of lipid molecules occurs through shuttling of membrane vesicles (Figure 2). Lipid molecules can also be transported between organelles by monomeric diffusion, in the plane of the membrane through possible membrane continuities or across the aqueous phase of the cytosol. Whether lipid diffusion is significant for intracellular lipid transport depends on the question whether organelles are connected by membrane continuities and, on the other hand, on the relative rate of lipid exchange processes compared with that of vesicular transport.

3.3.1. Lateral Diffusion via Membrane Continuities

Of the lipid transport processes, lipid diffusion in the plane of the membrane is the fastest process (on the order of $\mu m^2/sec$). It is relatively insensitive to lower temperatures, with a fourfold reduction between 37 and 10°C (Almeida *et al.*, 1992). It has been proposed that intracellular membranes are interconnected by membrane continuities that would allow free lipid diffusion between organelles to occur (Scow and Blanchette-Mackie, 1985). Such is the case for the nuclear membranes and the ER. However, the nuclear membranes are considered a domain of the ER membrane and not a separate organelle. The general belief is that intracellular organelles are separate entities and that lipid diffusion between them through (temporary) membrane continuities is insignificant. An exception to this general rule may be the contact sites between the inner and outer mitochondrial membrane, and diffusion via these structures of the phospholipids PS and PE has been proposed (see van Meer, 1993). Interestingly, these contact sites have been compared with the intercellular contacts at the tight junctions. The latter did not allow free lipid diffusion between the contacting cells (van Meer *et al.*, 1986). Membrane continuities between otherwise separate organelles have been proposed to be induced by the drug brefeldin A (Klausner *et al.*, 1992).

3.3.2. Monomeric Diffusion via the Aqueous Phase

Monomeric exchange between membranes is a slow process for most lipids with half-times of equilibration between lipid vesicles of 2 hr for cholesterol, days for phospholipids, and weeks for glycosphingolipids (Phillips *et al.*, 1987; Brown, 1992). Still, mitochondria and peroxisomes, organelles not connected to the vesicular transport pathways shown in Figure 2, can only acquire their phospholipids by exchange processes and do so with half-times of minutes (PC) to an hour (PE) (Yaffe and Kennedy, 1983). In fact, exchange between membranes in cells is considerably accelerated by two phenomena. First, the rate of spontaneous lipid transfer between cellular membranes is dramatically enhanced when they are closely apposed (Brown, 1992). This constellation has been found on the outside of the outer mitochondrial membrane where it interacts with a domain of the ER that has been termed the *mitochondrion-associated membrane fraction* (Vance, 1990; Cui *et al.*, 1993). The second factor accelerating lipid transfer is the presence in the cytosol of a group of lipid transfer proteins, which promote lipid exchange (Wirtz, 1991). These proteins, originally isolated from a cytosol fraction, can enhance the rate of lipid exchange *in vitro*. Some of them, however, were later found to be confined to the lumen of the lysosomes or peroxisomes, and their binding capacity is required for presenting a lipid substrate for an enzymatic reaction (glycosphingolipid catabolism and cholesterol synthesis, respectively; see van Meer, 1993). Some proteins, like the PC- and PI/PC transfer proteins, do appear to exert their function in the cytosol. It is unclear, after 25 years of discussing the role of such proteins, whether their function is the unidirectional, net transfer of lipids that is required for the growth of the membranes of mitochondria and peroxisomes, as they are always found carrying one lipid molecule. The same question applies to the idea that selective, monomeric retrieval of PC from the Golgi and transfer back to the ER could be the process responsible for enriching membrane proteins, secretory proteins, and lipids such as cholesterol (and sphingolipids) in the Golgi (Wieland *et al.*, 1987). An exciting recent observation is the requirement of the PI/PC transfer protein for vesicular traffic through the Golgi. In this case the protein is thought to probe the lipid composition and thereby act as a sensor. It is unclear whether its capability to transfer lipids is actually required for this function (Cleves et al., 1991; Dowhan, 1991). A function of the protein in replenishment of PI to the plasma membrane for inositol lipid signaling has been proposed (Van Paridon et al., 1987; Thomas et al., 1993). Finally, the recent discovery that GlcCer (and GalCer) has access to the cytoplasmic side of the Golgi membrane leaves open the possibility that a GlcCer transfer protein (Sasaki, 1990) shuttles GlcCer between organelles.

3.3.3. Carrier Vesicles

Only lipids present in the cytosolic leaflet of cellular membranes are available for exchange processes between organelles. Lipid molecules located in the lumen of organelles cannot equilibrate over all organelles and can only be transported by vesicles. For example, the complex glycosphingolipid Forssman antigen was found in the organelles along the vesicular transport pathways, and not in the mitochondria and the peroxisomes (van Genderen et al., 1991). The vesicles, of course, carry lipids in the cytosolic membrane leaflet as well. The vesicular transport pathways that connect the membranes of the vacuolar apparatus in the cell are shown in Figure 2. Two main cycles of vesicular transport can be discriminated. One involves the first steps of the secretory pathway, the cycle ER–intermediate compartment–cis-Golgi–ER (Wieland et al., 1987; Pelham, 1991), the other is the endocytotic cycle plasma membrane–endosome–(TGN)–plasma membrane (Griffiths et al., 1989; Kok et al., 1991; Watts and Marsh, 1992). Vesicular transport through these cycles is extensive. An area equivalent to 50% of the ER has been estimated to recycle through the cis-Golgi every 10 min (Wieland et al., 1987). A surface area equal to the plasma membrane recycles through the endosomes each hour (Griffiths et al., 1989). The two cycles are connected in at least one direction by the exocytotic pathway through the Golgi. Although evidence has been presented to suggest that retrograde transport through the Golgi stack occurs, it is not clear at present whether material can actually be transported from the TGN back into the Golgi stack. Each vesicular transport step involves a specific, complex set of proteins in budding, transport, docking, and fusion. Especially the various coat complexes required for budding (Schmid, 1993) and the proteins required for recognition and docking at the correct target membrane have been studied in detail (Bennett and Scheller, 1993). One question that is unanswered relates to the process that provides the energy for vesicle budding.

3.4. Specificity in Vesicular Transport

The various membranes along the vesicular transport pathways contain specific sets of membrane proteins. Essentially all of these have originally been inserted into the membrane of the ER after which they have been transported to the membrane(s) of their destination. For this purpose, each protein molecule carries structural targeting information. This information is read by a cellular sorting machinery that includes the molecule in, or excludes it from, forming transport vesicles (Pryer *et al.*, 1992). As sorting of proteins is thought to occur in virtually every organelle, multiple sorting signals may be present in each protein molecule, one for each transport step. The signals may be located at either side of the membrane. However, the proteins responsible for protein recognition and vesicle formation are on the cytoplasmic surface, organized in membrane coats (Robinson, 1992). If the signal is luminally oriented, as is the case for luminal soluble proteins but also for apical membrane proteins in epithelial cells, it must be transmitted to the cytosolic side, most likely by a transmembrane molecule. Exciting questions that have attracted much attention over the last years concern the nature of the sorting signals, their hierarchy, and the molecular interactions with proteins of the sorting machinery.

Because proteins serve very specific functions at specific locations it would seem necessary that proteins are sorted into the correct pathway and to the correct compartment with few mistakes. Highly efficient protein sorting is illustrated by the epithelial, kidney-derived MDCK cells, where newly synthesized apical proteins are transported from the TGN straight to the apical surface and basolateral proteins directly to the basolateral surface. Sorting of a similar efficiency has been observed in the transcytotic pathway (Mostov *et al.*, 1992). In hepatocytes, which appear to lack the direct apical pathway, all apical/basolateral sorting is performed by the transcytotic machinery (Hubbard *et al.*, 1989). A mixed situation has been observed in intestinal Caco-2 cells where newly synthesized apical proteins pass over the basolateral surface to various extents. Apparently, the sorting machinery in the TGN of Caco-2 cells has too low a capacity to handle all apical proteins, so that a fraction is missorted, which subsequently has to be corrected by a backup machinery in the transcytotic route (Louvard *et al.*, 1992). For lipids, the size of the vesicular transport pathway to the apical surface is not strongly reduced in Caco-2 cells as compared with MDCK (van't Hof and van Meer, 1990).

Membrane protein sorting implies that membrane proteins are continuously segregated from other protein molecules into lateral domains of proteins with the same destination. For proteins with sorting information in the cytoplasmic tail, the idea is that each protein interacts with a sorter molecule that in turn is part of a protein lattice or coat on the cytoplasmic surface of the membrane (Figure 4). An interesting alternative mechanism has been suggested for lateral segregation of transmembrane proteins in the Golgi. Accompanying the concentration of cholesterol along the exocytotic route through the Golgi the thickness of the membranes of the Golgi cisternae increases. Since Golgi resident proteins contain shorter membrane-spanning domains than plasma membrane proteins, only the latter would be able to partition into anterograde transport vesicles (Bretscher and Munro, 1993). Membrane lipids and GPI proteins lack a cytoplasmic tail and do not span the membrane. Recent studies have led to the concept that sorting of these components is not governed by a one-to-one relation between the molecule to be sorted and a sorter, but rather that the sorter recognizes a domain or cluster of these molecules.

Over the years, we have observed that, after synthesis, a variety of glucosylceramide analogs is preferentially delivered to the apical surface of epithelial cells in culture. For kidney-derived MDCK cells, the recorded apical/ basolateral polarity of delivery was ≥ 2, whereas in intestinal Caco-2 cells values were 2–9. In contrast, SM arrived at the surface with polarities ≤ 1 (van Meer *et al.*, 1987; van't Hof and van Meer, 1990; van't Hof *et al.*, 1992; van Meer and van't Hof, 1993). Recent work has demonstrated that in MDCK cells short-chain analogs of the simple glycosphingolipid lactosylceramide are preferentially transported to the apical domain, while GalCer, which differs from glucosylceramide only in the orientation of a single hydroxyl group, sulfatide and GalGalCer are preferentially transported basolaterally (van der Bijl *et al.*, 1994). Clearly, sorting of lipids is far less absolute (or efficient) than sorting of proteins. However, the sorting between GlcCer and SM expressed as the difference in polarity of delivery in well-differentiated Caco-2 cells with a value of about 7 (van Genderen and van Meer, 1993), was very similar to the apical enrichment of glycosphingolipids over SM in intestinal tissue (Table II).

As many of the sphingolipids are synthesized in the Golgi lumen, and since there is growing evidence that also GlcCer and GalCer are transported from the TGN to the plasma membrane on the luminal side of carrier vesicles (van Meer, 1993; van Meer and van't Hof, 1993), the simplest mechanism to generate the different polarities of delivery would be a sorting event driven by glycosphingolipid microdomain formation in the luminal bilayer leaflet of the TGN, followed by vesicular traffic to either plasma membrane domain (van Meer *et al.*, 1987; Simons and van Meer, 1988). Whereas the glycosphingolipids based on GlcCer appear to partition into apical precursor domains, glycosphingolipids based on GalCer seem to aggregate into a basolateral domain, which should also be the case for the major lipid of the basolateral cell surface PC. Sphingolipids and not glycerolipids have a tendency to form hydrogen bonds (Pascher, 1976). It is a fascinating question as to what physicochemical difference between glucose- and galactose-based lipids drives their segregation. Methods like those described in Chapters 3 and 4 may contribute to elucidation of the molecular mechanism of lipid sorting.

Although in our studies on lipid transport, so far, sorting has been expressed as the enrichment of one lipid over another, GlcCer over SM, a more accurate way to describe sorting would be to compare the surface density of a lipid in the various pathways from the TGN. However, the actual pathway sizes are not known. Only the application of a nonspecific marker for transport, either a membrane marker or a volume marker, will allow a correct assessment of the (relative) size of the various vesicular transport pathways (van Genderen and van Meer, 1993).

Over the last 5 years evidence has been accumulating to suggest that GPI proteins partition into the GlcCer domain at the luminal surface of the TGN. In MDCK cells and in Caco-2 cells (Lisanti and Rodriguez-Boulan, 1990) all GPI proteins were found to be apical, while a reversal of lipid sorting in Fischer rat thyroid (FRT) cells was accompanied by a reversal in polarity of GPI proteins (Zurzolo et al., 1994). Clustering of GPI proteins in the exocytotic pathway of MDCK cells was independently demonstrated by the use of fluorescent techniques (Hannan et al., 1993).

Consistent with the notion of a glycosphingolipid/GPI protein aggregate, both GPI proteins and sphingolipids displayed a selective resistance against extraction by Triton X-100 in the cold (Brown and Rose, 1992). However, especially SM, a basolaterally enriched lipid, was resistant to detergent extraction, suggesting that this behavior may not directly be related to apical sorting. Still, the method has served as an elegant procedure to enrich proteins that are involved in GPI protein clustering (Kurzchalia et al., 1992; Fiedler et al., 1993; Lisanti et al., 1993; Sargiacomo et al., 1993). One protein isolated in this way, VIP21, turned out to be caveolin (Glenney, 1992; Kurzchalia et al., 1992; Dupree et al., 1993), a protein component of caveolae. Caveolae are specialized microdomains at the plasma membrane, which were found to be enriched in GPI-linked proteins and glycosphingolipids (Montesano et al., 1982; Parton, 1994). Although the caveolae were first thought to fulfill a function in endocytosis, this is now questioned (van Deurs et al., 1993). Instead, a function in signal transduction involving tyrosine kinases has been suggested (Anderson, 1993; Lisanti et al., 1993). Although caveolin has been found in both apical and basolateral TGN-derived transport vesicles, and on both the apical and basolateral cell surface, it is still an important candidate for being a subunit of the epithelial sorting machinery: FRT cells which display reversed sorting of glycosphingolipids and GPI-linked proteins do not express caveolin (Sargiacomo et al., 1993; Zurzolo et al., 1994)!

While glycosphingolipids may cluster spontaneously, evidence has been presented to suggest that the self-binding of the GPI proteins was induced by low pH and ionic conditions (Fukuoka et al., 1992; Hannan et al., 1993), very similar to the selective aggregation of secretory and membrane proteins in the TGN preceding the formation of secretory granules (Bauerfeind and Huttner, 1993). If caveolin were part of the sorting machinery, it should in some way interact with the luminal microdomain and with cytosolic elements. It has been suggested that caveolin is a transmembrane protein (Lisanti et al., 1993; Sargiacomo et al., 1993), but evidence has also been provided that both the N- and C-terminus of the molecule are oriented toward the cytosolic surface (Dupree et

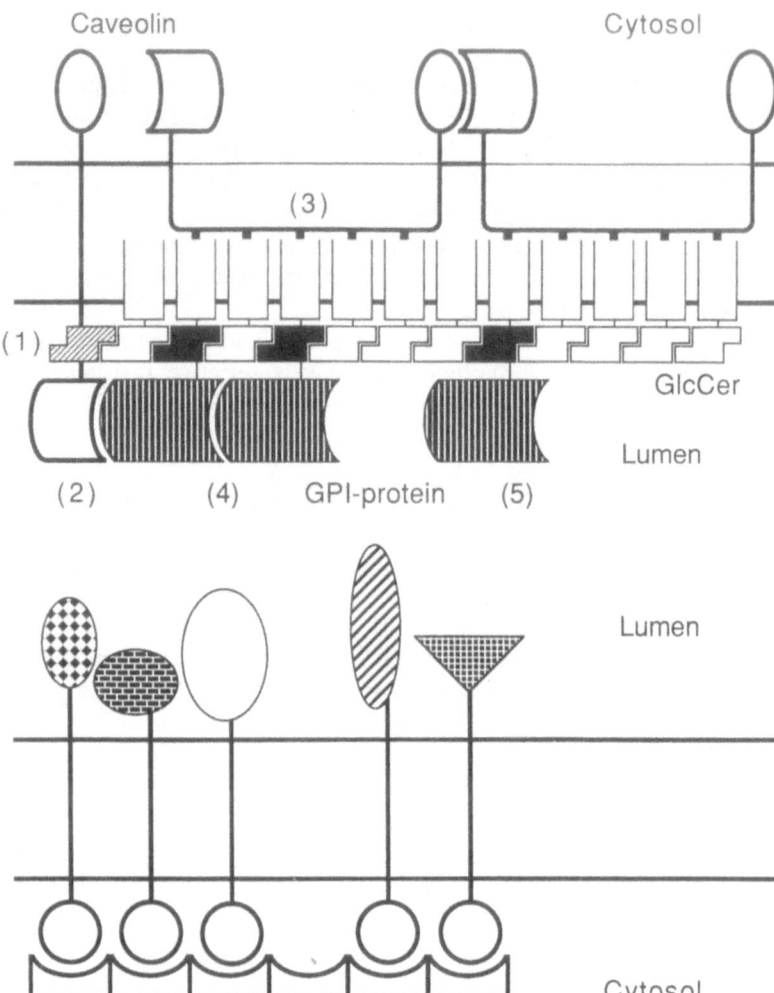

FIGURE 4. Microdomains and lipid sorting in the TGN. (Top) Self-aggregating GPI proteins and glycosphingolipids (GlcCer) may be recognized by caveolin to form an apical precursor microdomain. Caveolin, in association with other proteins, may penetrate the membrane and recognize the luminal microdomain by interacting with the lipid moieties (1) or the protein domain of GPI proteins (2). Alternatively, caveolin may penetrate the membrane only partially, and interact with the microdomain in the hydrophobic membrane interior (3). The GPI proteins may self-aggregate (4) or associate with the glycosphingolipids (5). (Bottom) Simplified view of sorting of a variety of basolateral proteins containing targeting information in their cytoplasmic tail by a cytosolic protein scaffold or coat.

al., 1993). Both topologies have interesting consequences for interactions with the exoplasmic microdomain of glycosphingolipids and GPI proteins (Figure 4). If caveolin has a luminal domain, this domain could interact with the GPI proteins and the glycosphingolipids directly. Even more interesting would be the possibility that it is exposed on the cytosolic side of the membrane only and recognizes the luminal microdomain by an interaction in the membrane interior (Figure 4).

While in all cells the glycosphingolipid/GPI protein domains appear to be included in anterograde transport vesicles, in epithelial cells they are recognized as having an apical or basolateral destination depending on the cell type (Sargiacomo et al., 1993; Zurzolo et al., 1993). Analogous to the proposed self-aggregation of glycosphingolipids in the TGN of epithelial cells, we have suggested that microdomain formation of the sphingolipid SM with cholesterol plays a part in the sorting event on the cis side of the Golgi (van Meer, 1993). A case of sphingolipid sorting has also been reported in endocytosis (Kok et al., 1991). After uptake into endosomes, GlcCer in the luminal leaflet was observed to be transported to the TGN, whereas luminal SM recycled to the cell surface. The combined evidence suggests that sphingolipid domains play a role in sorting lipids and GPI-linked proteins at potentially three locations in the cell (Figure 2). To our surprise, so far, we have not observed sphingolipid sorting along the transcytotic pathway in epithelial cells (van Genderen and van Meer, 1994).

Major questions regarding the function of caveolae at the plasma membrane and the role of microdomains in epithelial sorting are now: What is the nature of the lipid/protein interactions in the microdomains, what are the dynamic properties of the microdomains, and how are they recognized by the cytosolic sorting machinery? The physicochemical techniques presented in this volume can be applied so as to increase our insight in the structure and dynamic properties of membrane components. A number of them may be applicable to elucidate some of the many aspects of these questions.

4. FINAL REMARKS

It is the aim of the scientific enterprise to gather knowledge, insight, and understanding. An essential role is reserved for the data collected in the process. However, although we must appreciate and respect the actual state of our knowledge, it is the story of how this knowledge was and is obtained that reveals the real marvel of scientific discovery. The present volume provides detailed accounts of this quest for knowledge. It allows appreciation of the elegant experimental approaches, and it makes the reader experience the exhilaration of the scientific breakthroughs. May the reader enjoy it. In the end, it is one of the major forces driving the progress of science.

20   Gerrit van Meer and Ida L. van Genderen

5. REFERENCES

Abeijon, C., and Hirschberg, C. B., 1992, Topography of glycosylation reactions in the endoplasmic reticulum, *Trends Biochem. Sci.* **17**:32–36.
Almeida, P.F.F., Vaz, W.L.C., and Thompson, T. E., 1992, Lateral diffusion in the liquid phases of dimyristoylphosphatidylcholine/cholesterol lipid bilayers: A free volume analysis, *Biochemistry* **31**:6739–6747.
Anderson, R.G.W., 1993, Plasmalemmal caveolae and GPI-anchored membrane proteins, *Curr. Opin. Cell Biol.* **5**:647–652.
Bauerfeind, R., and Huttner, W. B., 1993, Biogenesis of constitutive secretory vesicles, secretory granules and synaptic vesicles, *Curr. Opin. Cell Biol.* **5**:628–635.
Bennett, M. K., and Scheller, R. H., 1993, The molecular machinery for secretion is conserved from yeast to neurons, *Proc. Natl. Acad. Sci. USA* **90**:2559–2563.
Berr, F., Meier, P. J., and Stieger, B., 1993, Evidence for the presence of a phosphatidylcholine translocator in isolated rat liver canalicular plasma membrane vesicles, *J. Biol. Chem.* **268**:3976–3979.
Bishop, W. R., and Bell, R. M., 1985, Assembly of the endoplasmic reticulum phospholipid bilayer: The phosphatidylcholine transporter, *Cell* **42**:51–60.
Brasitus, T. A., and Schachter, D., 1980, Lipid dynamics and lipid–protein interactions in the rat enterocyte basolateral and microvillus membranes, *Biochemistry* **19**:2763–2769.
Bratton, D. L., 1993, Release of platelet activation factor from activated neutrophils. Transglutaminase-dependent enhancement of transbilayer movement across the plasma membrane, *J. Biol. Chem.* **268**:3364–3373.
Bretscher, M. S., and Munro, S., 1993, Cholesterol and the Golgi apparatus, *Science* **261**:1280–1281.
Brotherus, J., and Renkonen, O., 1977, Phospholipids of subcellular organelles isolated from cultured BHK cells, *Biochim. Biophys. Acta* **486**:243–253.
Brown, D. A., and Rose, J. K., 1992, Sorting of GPI-anchored proteins to glycolipid-enriched membrane subdomains during transport to the apical cell surface, *Cell* **68**:533–544.
Brown, R. E., 1992, Spontaneous lipid transfer between organized lipid assemblies, *Biochim. Biophys. Acta* **1113**:375–389.
Burger, K.N.J., van der Bijl, P., and van Meer, G., 1994, A Golgi lipid galactosyltransferase-1-activity is oriented towards the cytosol, while two galactosyltransferase-2 activities face the Golgi lumen, submitted.
Caspar, D.L.D., and Kirschner, D. A., 1971, Myelin membrane structure at 10 Å resolution, *Nature New Biol.* **231**:46–52.
Cleves, A. E., McGee, T. P., Whitters, E. A., Champion, K. M., Aitken, J. R., Dowhan, W., Goebl, M., and Bankaitis, V. A., 1991, Mutations in the CDP-choline pathway for phospholipid biosynthesis bypass the requirement for an essential phospholipid transfer protein, *Cell* **64**:789–800.
Colbeau, A., Nachbaur, J., and Vignais, P. M., 1971, Enzymic characterization and lipid composition of rat liver subcellular membranes, *Biochim. Biophys. Acta* **249**:462–492.
Coste, H., Martel, M. B., and Got, R., 1986, Topology of glucosylceramide synthesis in Golgi membranes from porcine submaxillary glands, *Biochim. Biophys. Acta* **858**:6–12.
Cui, Z., Vance, J. E., Chen, M. H., Voelker, D. R., and Vance, D. E., 1993, Cloning and expression of a novel phosphatidylethanolamine N-methyltransferase. A specific biochemical and cytological marker for a unique membrane fraction in rat liver, *J. Biol. Chem.* **268**:16655–16663.
Dennis, E. A., and Vance, D. E., eds., 1992, *Phospholipid Biosynthesis, Methods Enzymol.* **209**:1–544.
Devaux, P. F., 1991, Static and dynamic lipid asymmetry in cell membranes, *Biochemistry* **30**:1163–1173.

Devaux, P. F., 1993, Lipid transmembrane asymmetry and flip-flop in biological membranes and in lipid bilayers, *Curr. Opin. Struct. Biol.* **3**:489–494.

Dowhan, W., 1991, Phospholipid-transfer proteins, *Curr. Opin. Cell Biol.* **3**:621–625.

Dragsten, P. R., Blumenthal, R., and Handler, J. S., 1981, Membrane asymmetry in epithelia: Is the tight junction a barrier to diffusion in the plasma membrane? *Nature* **294**:718–722.

Dupree, P., Parton, R. G., Raposo, G., Kurzchalia, T. V., and Simons, K., 1993, Caveolae and sorting in the *trans*-Golgi network of epithelial cells, *EMBO J.* **12**:1597–1605.

El Hage Chahine, J. M., Cribier, S., and Devaux, P. F., 1993, Phospholipid transmembrane domains and lateral diffusion in fibroblasts, *Proc. Natl. Acad. Sci. USA* **90**:447–451.

Fiedler, K., Kobayashi, T., Kurzchalia, T. V., and Simons, K., 1993, Glycosphingolipid-enriched, detergent-insoluble complexes in protein sorting in epithelial cells, *Biochemistry* **32**:6365–6373.

Fukuoka, S.-I., Freedman, S. D., Yu, H., Sukhatme, V. P., and Scheele, G. A., 1992, GP-2/THP gene family encodes self-binding glycosylphosphatidylinositol-anchored proteins in apical secretory compartments of pancreas and kidney, *Proc. Natl. Acad. Sci. USA* **89**:1189–1193.

Futerman, A. H., and Pagano, R. E., 1991, Determination of the intracellular sites and topology of glucosylceramide synthesis in rat liver, *Biochem. J.* **280**:295–302.

Futerman, A. H., Stieger, B., Hubbard, A. L., and Pagano, R. E., 1990, Sphingomyelin synthesis in rat liver occurs predominantly at the cis and medial cisternae of the Golgi apparatus, *J. Biol. Chem.* **265**:8650–8657.

Geuze, H. J., Slot, J. W., and Schwartz, A. L., 1987, Membranes of sorting organelles display lateral heterogeneity in receptor distribution, *J. Cell Biol.* **104**:1715–1723.

Glaser, M., 1993, Lipid domains in biological membranes, *Curr. Opin. Struct. Biol.* **3**:475–481.

Glenney, J. R., Jr., 1992, The sequence of human caveolin reveals identity with VIP21, a component of transport vesicles, *FEBS Lett.* **314**:45–48.

Griffiths, G., Back, R., and Marsh, M., 1989, A quantitative analysis of the endocytic pathway in baby hamster kidney cells, *J. Cell Biol.* **109**:2703–2720.

Hannan, L. A., Lisanti, M. P., Rodriguez-Boulan, E., and Edidin, M., 1993, Correctly sorted molecules of a GPI-anchored protein are clustered and immobile when they arrive at the apical surface of MDCK cells, *J. Cell Biol.* **120**:353–358.

Hauri, H.-P., and Schweizer, A., 1992, The endoplasmic reticulum–Golgi intermediate compartment, *Curr. Opin. Cell Biol.* **4**:600–608.

Herrmann, A., Zachowski, A., and Devaux, P. F., 1990, Protein-mediated phospholipid translocation in the endoplasmic reticulum with a low lipid specificity, *Biochemistry* **29**:2023–2027.

Higgins, C. F., and Gottesman, M. M., 1992, Is the multidrug transporter a flippase? *Trends Biochem. Sci.* **17**:18–21.

Hubbard, A. L., Stieger, B., and Bartles, J. R., 1989, Biogenesis of endogenous plasma membrane proteins in epithelial cells, *Annu. Rev. Physiol.* **51**:755–770.

Jacobson, K., and Vaz, W.L.C., eds., 1992, *Domains in Biological Membranes*, *Commun. Mol. Cell. Biophys.* **8**:1–114.

Jeckel, D., Karrenbauer, A., Birk, R., Schmidt, R. R., and Wieland, F., 1990, Sphingomyelin is synthesized in the cis Golgi, *FEBS Lett.* **261**:155–157.

Jeckel, D., Karrenbauer, A., Burger, K.N.J., van Meer, G., and Wieland, F., 1992, Glucosylceramide is synthesized at the cytosolic surface of various Golgi subfractions, *J. Cell Biol.* **117**:259–267.

Kachar, B., and Reese, T. S., 1982, Evidence for the lipidic nature of tight junction strands, *Nature* **296**:464–466.

Kawai, K., Fujita, M., and Nakao, M., 1974, Lipid components of two different regions of an intestinal epithelial cell membrane of mouse, *Biochim. Biophys. Acta* **369**:222–233.

Keenan, T. W., and Morré, D. J., 1970, Phospholipid class and fatty acid composition of Golgi apparatus isolated from rat liver and comparison with other cell fractions, *Biochemistry* **9:**19–25.

Klausner, R. D., Donaldson, J. G., and Lippincott-Schwartz, J., 1992, Brefeldin A: Insights into the control of membrane traffic and organelle structure, *J. Cell Biol.* **116:**1071–1080.

Kok, J. W., Babia, T., and Hoekstra, D., 1991, Sorting of sphingolipids in the endocytic pathway of HT29 cells, *J. Cell Biol.* **114:**231–239.

Kolesnick, R., 1992, Ceramide: A novel second messenger, *Trends Cell Biol.* **2:**232–236.

Krisans, S. K., 1992, The role of peroxisomes in cholesterol metabolism, *Am. J. Respir. Cell Mol. Biol.* **7:**358–364.

Kurzchalia, T. V., Dupree, P., Parton, R. G., Kellner, R., Virta, H., Lehnert, M., and Simons, K., 1992, VIP21, a 21-kD membrane protein, is an integral component of *trans*-Golgi-network-derived transport vesicles, *J. Cell Biol.* **118:**1003–1014.

Lange, Y., 1991, Disposition of intracellular cholesterol in human fibroblasts, *J. Lipid Res.* **32:**329–339.

Lange, Y., Swaisgood, M. H., Ramos, B. V., and Steck, T. L., 1989, Plasma membranes contain half the phospholipid and 90% of the cholesterol and sphingomyelin in cultured human fibroblasts, *J. Biol. Chem.* **264:**3786–3793.

Lange, Y., Echevarria, F., and Steck, T. L., 1991, Movement of zymosterol, a precursor of cholesterol, among three membranes in human fibroblasts, *J. Biol. Chem.* **266:**21439–21443.

Lisanti, M. P., and Rodriguez-Boulan, E., 1990, Glycophospholipid membrane anchoring provides clues to the mechanism of protein sorting in polarized epithelial cells, *Trends Biochem. Sci.* **15:**113–118.

Lisanti, M. P., Tang, Z., and Sargiacomo, M., 1993, Caveolin forms a hetero-oligomeric protein complex that interacts with an apical GPI-linked protein: Implications for the biogenesis of caveolae, *J. Cell Biol.* **123:**595–604.

Louvard, D., Kedinger, M., and Hauri, H. P., 1992, The differentiating intestinal epithelial cell: Establishment and maintenance of functions through interactions between cellular structures, *Annu. Rev. Cell Biol.* **8:**157–195.

Montesano, R., Roth, J., Robert, A., and Orci, L., 1982, Non-coated membrane invaginations are involved in binding and internalization of cholera and tetanus toxins, *Nature* **296:**651–653.

Mostov, K., Apodaca, G., Aroeti, B., and Okamoto, C., 1992, Plasma membrane protein sorting in polarized epithelial cells, *J. Cell Biol.* **116:**577–583.

Nishizuka, Y., 1992, Intracellular signaling by hydrolysis of phospholipids and activation of protein kinase C, *Science* **258:**607–614.

Parton, R. G., 1994, Ultrastructural localization of gangliosides: GM_1 is concentrated in caveolae. *J. Histochem. Cytochem.* **42:**155–166.

Pascher, I., 1976, Molecular arrangements in sphingolipids. Conformation and hydrogen bonding of ceramide and their implication on membrane stability and permeability, *Biochim. Biophys. Acta* **455:**433–451.

Pelham, H.R.B., 1991, Recycling of proteins between the endoplasmic reticulum and Golgi complex, *Curr. Opin. Cell Biol.* **3:**585–591.

Phillips, M. C., Johnson, W. J., and Rothblat, G. H., 1987, Mechanism and consequences of cellular cholesterol exchange and transfer, *Biochim. Biophys. Acta* **906:**223–276.

Pinto da Silva, P., and Kachar, B., 1982, On tight-junction structure, *Cell* **28:**441–450.

Pryer, N. K., Wuestehube, L. J., and Schekman, R., 1992, Vesicle-mediated protein sorting, *Annu. Rev. Biochem.* **61:**471–516.

Robinson, M. S., 1992, Adaptins, *Trends Cell Biol.* **2:**293–297.

Sargiacomo, M., Sudol, M., Tang, Z. L., and Lisanti, M. P., 1993, Signal transducing molecules and glycosyl-phosphatidylinositol-linked proteins form a caveolin-rich insoluble complex in MDCK cells, *J. Cell Biol.* **122:**789–807.

Sasaki, T., 1990, Glycolipid transfer protein and intracellular traffic of glucosylceramide, *Experientia* **46**:611–616.

Schmid, S. L., 1993, Biochemical requirements for the formation of clathrin- and COP-coated transport vesicles, *Curr. Opin. Cell Biol.* **5**:621–627.

Schroeder, F., Nemecz, G., Wood, W. G., Joiner, C., Morrot, G., Ayraut-Jarrier, M., and Devaux, P. F., 1991, Transmembrane distribution of sterol in the human erythrocyte, *Biochim. Biophys. Acta* **1066**:183–192.

Scow, R. O., and Blanchette-Mackie, E. J., 1985, Why fatty acids flow in cell membranes, *Prog. Lipid Res.* **24**:197–241.

Simons, K., and van Meer, G., 1988, Lipid sorting in epithelial cells, *Biochemistry* **27**:6197–6202.

Slomiany, A., Grzelinska, E., Kasinathan, C., Yamaki, K.-i., Palecz, D., Slomiany, B. A., and Slomiany, B. L., 1992, Biogenesis of endoplasmic reticulum transport vesicles transferring gastric apomucin from ER to Golgi, *Exp. Cell Res.* **201**:321–329.

Smit, J.J.M., Schinkel, A. H., Oude Elferink, R.P.J., Groen, A. K., Wagenaar, E., van Deemter, L., Mol, C.A.A.M., Ottenhoff, R., van der Lugt, N.M.T., van Roon, M.A., van der Valk, M. A., Offerhaus, G.J.A., Berns, A.J.M., and Borst, P., 1993, Homozygous disruption of the murine *mdr2* P-glycoprotein gene leads to a complete absence of phospholipid from bile and to liver disease, *Cell* **75**:451–462.

Spiegel, S., Blumenthal, R., Fishman, P. H., and Handler, J. S., 1985, Gangliosides do not move from apical to basolateral plasma membrane in cultured epithelial cells, *Biochim. Biophys. Acta* **821**:310–318.

Strous, G. J., van Kerkhof, P., van Meer, G., Rijnboutt, S., and Stoorvogel, W., 1993, Differential effects of brefeldin A on transport of secretory and lysosomal proteins, *J. Biol. Chem.* **268**:2341–2347.

Stubbs, C. D., Ketterer, B., and Hicks, R. M., 1979, The isolation and analysis of the luminal plasma membrane of calf urinary bladder epithelium, *Biochim. Biophys. Acta* **558**:58–72.

Symington, F. W., Murray, W. A., Bearman, S. I., and Hakomori, S.-i., 1987, Intracellular localization of lactosylceramide, the major human neutrophil glycosphingolipid, *J. Biol. Chem.* **262**:11356–11363.

Thomas, G.M.H., Cunningham, E., Fensome, A., Ball, A., Totty, N. F., Truong, O., Hsuan, J. J., and Cockcroft, S., 1993, An essential role for phosphatidylinositol transfer protein in phospholipase C-mediated inositol lipid signaling, *Cell* **74**:919–928.

Thompson, T. E., Barenholz, Y., Brown, R. E., Correa-Freire, M., Young, W. W., Jr., and Tillack, T. W., 1986, Molecular organization of glycosphingolipids in phosphatidylcholine bilayers and biological membranes, in: *Enzymes of Lipid Metabolism* (L. Freysz, ed.,), pp. 387–396, Plenum Press, New York.

Trinchera, M., Fabbri, M., and Ghidoni, R., 1991, Topography of glycosyltranferases involved in the initial glycosylations of gangliosides, *J. Biol. Chem.* **266**:20907–20912.

Vance, J. E., 1990, Phospholipid synthesis in a membrane fraction associated with mitochondria, *J. Biol. Chem.* **265**:7248–7256.

van der Bijl, P., Lopes-Cardozo, M., and van Meer, G., 1994, Glycosphingolipids with a glucosylceramide backbone display an apical preference in epithelial MDCK and Caco-2 cells, whereas a galactosylceramide backbone acts as a basolateral signal, submitted.

van Deurs, B., Holm, P. K., Sandvig, K., and Hansen, S. H., 1993, Are caveolae involved in clathrin-independent endocytosis? *Trends Cell Biol.* **3**:249–251.

van Genderen, I. L., and van Meer, G., 1993, Lipid sorting—Measurement and interpretation, *Biochem. Soc. Trans.* **21**:235–239.

van Genderen, I. L., van Meer, G., Slot, J. W., Geuze, H. J., and Voorhout, W. F., 1991,

Subcellular localization of Forssman glycolipid in epithelial MDCK cells by immuno-electronmicroscopy after freeze-substitution, *J. Cell Biol.* **115**:1009–1019.

Van Genderen, I., and van Meer, G., 1994, Glucosylceramide and galactosylceramide analogs are sorted after biosynthesis but not during transcytosis in MDCK cells, submitted.

van Helvoort, A., van't Hof, W., Ritsema, T., Sandra, A., and van Meer, G., 1994, Conversion of diacylglycerol to phosphatidylcholine on the basolateral surface of epithelial (Madin–Darby canine kidney) cells. Evidence for the reverse action of the sphingomyelin synthase, *J. Biol. Chem.*, **269**:1763–1769.

van Meer, G., 1986, The lipid bilayer of the ER, *Trends Biochem. Sci.* **11**:194–195, 401.

van Meer, G., 1993, Transport and sorting of membrane lipids, *Curr. Opin. Cell Biol.* **5**:661–673.

van Meer, G., and Simons, K., 1986, The function of tight junctions in maintaining differences in lipid composition between the apical and the basolateral cell surface domains of MDCK cells, *EMBO J.* **5**:1455–1464.

van Meer, G., and van't Hof, W., 1993, Epithelial sphingolipid sorting is insensitive to reorganization of the Golgi by nocodazole, but is abolished by monensin in MDCK cells and by brefeldin A in Caco-2 cells, *J. Cell Sci.* **104**:833–842.

van Meer, G., Gumbiner, B., and Simons, K., 1986, The tight junction does not allow lipid molecules to diffuse from one epithelial cell to the next, *Nature* **322**:639–641.

van Meer, G., Stelzer, E.H.K., Wijnaendts-van-Resandt, R. W., and Simons, K., 1987, Sorting of sphingolipids in epithelial (Madin–Darby canine kidney) cells, *J. Cell Biol.* **105**:1623–1635.

Van Paridon, P. A., Gadella, T.W.J., Jr., Somerharju, P. J., and Wirtz, K.W.A., 1987, On the relationship between the dual specificity of the bovine brain phosphatidylinositol transfer protein and membrane phosphatidylinositol levels, *Biochim. Biophys. Acta* **903**:68–77.

van't Hof, W., and van Meer, G., 1990, Generation of lipid polarity in intestinal epithelial (Caco-2) cells: Sphingolipid synthesis in the Golgi complex and sorting before vesicular traffic to the plasma membrane, *J. Cell Biol.* **111**:977–986.

van't Hof, W., Silvius, J., Wieland, F., and van Meer, G., 1992, Epithelial sphingolipid sorting allows for extensive variation of the fatty acyl chain and the sphingosine backbone, *Biochem. J.* **283**:913–917.

Vaz, W.L.C., and Almeida, P.F.F., 1993, Phase topology and percolation in multi-phase lipid bilayers: Is the biological membrane a domain mosaic? *Curr. Opin. Struct. Biol.* **3**:482–488.

Wang, Y., Sweitzer, T. D., Weinhold, P. A., and Kent, C., 1993, Nuclear localization of soluble CTP-phosphocholine cytidylyltransferase, *J. Biol. Chem.* **268**:5899–5904.

Warnock, D. E., Roberts, C., Lutz, M. S., Blackburn, W. A., Young, W. W., Jr., and Baenziger, J. U., 1993, Determination of plasma membrane lipid mass and composition in cultured Chinese hamster ovary cells using high gradient magnetic affinity chromatography, *J. Biol. Chem.* **268**:10145–10153.

Watts, C., and Marsh, M., 1992, Endocytosis: What goes in and how? *J. Cell Sci.* **103**:1–8.

Wherrett, J. R., and Huterer, S., 1972, Enrichment of bis-(monoacylglyceryl) phosphate in lyso-somes from rat liver, *J. Biol. Chem.* **247**:4114–4120.

Wieland, F. T., Gleason, M. L., Serafini, T. A., and Rothman, J. E., 1987, The rate of bulk flow from the endoplasmic reticulum to the cell surface, *Cell* **50**:289–300.

Wirtz, K.W.A., 1991, Phospholipid transfer proteins, *Annu. Rev. Biochem.* **60**:73–99.

Yaffe, M. P., and Kennedy, E. P., 1983, Intracellular phospholipid movement and the role of phospholipid transfer proteins in animal cells, *Biochemistry* **22**:1497–1507.

Zambrano, F., Fleischer, S., and Fleischer, B., 1975, Lipid composition of the Golgi apparatus of rat kidney and liver in comparison with other subcellular organelles, *Biochim. Biophys. Acta* **380**:357–369.

Zurzolo, C., van't Hof, W., van Meer, G., and Rodriguez-Boulan, E., 1994, VIP21/caveolin, glycosphingolipid clusters, and the sorting of glycosylphosphatidylinositol-anchored proteins in epithelial cells, *EMBO J.*, **13**:42–53.

Chapter 2

Biophysical Characterization of Membrane and Cytoskeletal Proteins by Sedimentation Analysis

Michael B. Morris and Gregory B. Ralston

1. INTRODUCTION

Analytical ultracentrifugation can be used to determine a number of important properties of proteins including molecular weight, the thermodynamic parameters governing self-associations and heterogeneous interactions, and nonideality. Hydrodynamic properties including the sedimentation, diffusion, and frictional coefficients, and molecular shape can also be determined. There are three fundamental experiments from which information can be obtained: (1) sedimentation velocity, in which the rate of transport of a sedimenting protein boundary is measured, (2) sedimentation equilibrium, in which the concentration distribution of protein is measured in the absence of net flow, and (3) diffusion experiments, in which the rate of spreading of a protein boundary is determined. Usually the data are obtained using one or more of the optical systems available on an analytical ultracentrifuge (Schachman, 1959; Svedberg and Pederson, 1940; Van Holde, 1971). However, the preparative ultracentrifuge, in combination with precise fractionation techniques, can provide information of similar accuracy in many cases (Attri and Minton, 1983, 1984, 1986; Howlett, 1987; Minton, 1989).

Michael B. Morris and Gregory B. Ralston   Department of Biochemistry, The University of Sydney, Sydney, NSW 2006, Australia.

Subcellular Biochemistry, Volume 23: Physicochemical Methods in the Study of Biomembranes, edited by Herwig J. Hilderson and Gregory B. Ralston. Plenum Press, New York, 1994.

1.1. Sedimentation Equilibrium

Molecular weights from as small as 350 up to several million can be determined from the concentration distribution of a protein or protein complex at sedimentation equilibrium (Teller, 1973; Yphantis, 1964). Since the method is based firmly on thermodynamic principles, molecular weights can be obtained without the use of calibration markers, often to an accuracy and precision of 1% or better. The molecular weights of the heterodimer and tetramer of the red-cell cytoskeletal protein, spectrin, have been determined using sedimentation equilibrium in the analytical ultracentrifuge (Dunbar and Ralston, 1981) and that of the detergent-solubilized human insulin receptor from experiments in the preparative ultracentrifuge (Pollet *et al.*, 1981; Ullrich *et al.*, 1985).

Sedimentation equilibrium is still the method of choice for analyzing self-associations. The method has a distinct advantage over many others in that data can be obtained at chemical equilibrium without perturbing the equilibrium, and small amounts of heterogeneity, which can seriously affect estimates of thermodynamic parameters, can be readily detected (Johnson *et al.*, 1981; Milthorpe *et al.*, 1975; Ralston and Morris, 1992; Teller, 1973; Yphantis, 1964). Furthermore, the technique allows interactions that are both very weak (equilibrium constants $\approx 10 \text{ M}^{-1}$) and very strong ($> 10^7 \text{ M}^{-1}$) to be quantified (Attri and Minton, 1986; Muramatsu and Minton, 1989). Analysis of an association reaction over a range of temperature, pH, and salt concentration allows a determination of the enthalpy, entropy, and heat capacity of the association, and an indication of the types of bonds involved (Aune and Timasheff, 1971; Aune *et al.*, 1971; Ralston, 1991).

Ligand binding to proteins, and the effect of ligands on the polymerization or enzymatic activity of an acceptor protein, can also be examined. The ligand can be a small molecule such as an allosteric effector (Holzenburg *et al.*, 1989; Steer *et al.*, 1990), an enzyme substrate (Hudson *et al.*, 1983; Lindenthal and Schubert, 1991), or another protein (Howlett, 1987, 1992a; Mulzer *et al.*, 1990). The exoplasmic portion of type 1 tumor necrosis factor (TNF) receptor polymerizes on binding of TNF-α (Pennica *et al.*, 1992). This work exploits the use of molecular biological techniques to express recombinant domains of proteins whose functions can then be assessed independently of the rest of the protein. An additional advantage of the use of recombinant techniques for membrane proteins is that soluble domains can be purified and analyzed in the absence of detergents.

Sedimentation equilibrium can be used to quantify the thermodynamic nonideality of a protein resulting from its charge and size. The inclusion of nonideality allows the correct molecular weight and the values of other thermodynamic parameters to be determined (Adams and Fujita, 1963; Adams *et al.*, 1978; Johnson *et al.*, 1981; Milthorpe *et al.*, 1975; Morris and Ralston, 1985).

Protein interactions can be examined under conditions approximating the

highly nonideal solutions found inside cells. The use of inert, space-filling molecules, such as dextran, to "crowd" solutions has begun to revolutionize our
understanding of cellular events (Minton, 1983; Tellam *et al.*, 1983). Although
little use has been made to date of sedimentation equilibrium analysis for investigating the effects of solution crowding on association behavior of proteins, the
method has been shown to have potential in this type of study (Wills and Winzor,
1992).

1.2. Sedimentation Velocity

The sedimentation coefficient, s, of a protein or protein complex can be
obtained by analyzing the rate of movement of the solute boundary in a sedimentation velocity experiment (Chervenka, 1973; Schachman, 1959; Van Holde,
1971). The value of s is useful for characterizing new proteins and can be used in
combination with other data to obtain the molecular weight of a protein (Chervenka, 1973; Schachman, 1959; Van Holde, 1971). The molecular weights of the
heterodimer and tetramer of spectrin were determined from a combination
of sedimentation velocity and diffusion experiments in the analytical ultracentrifuge. The values obtained (470 and 990 kDa) agreed with those obtained
from sedimentation equilibrium experiments (Dunbar and Ralston, 1981). The
rate of formation and the size of small oligomers of actin, at concentrations
above and below the critical concentration for the formation of F-actin, have also
been examined using sedimentation velocity. The results have been correlated
with the ability of these oligomers to participate in the polymerization of F-actin
(Attri *et al.*, 1991).

The preparative ultracentrifuge can be especially useful for crude mixtures
of proteins, since advantage can be taken of rapid and sensitive fractionation and
detection systems that selectively monitor the sedimentation of the protein of
interest (Attri and Minton, 1986; Flörke *et al.*, 1990; Howlett, 1987; Jacques *et
al.*, 1990; Minton, 1989).

Sedimentation velocity data can be used to assess the frictional properties of
a protein (Schachman, 1959; Van Holde, 1971). In combination with data from
other techniques (such as small-angle X-ray scattering, viscosity, and light scattering), it is possible to make reasonable estimates of the contribution to the
frictional properties resulting from the size and shape of the protein, its degree of
hydration, and the roughness of the protein surface (Creeth and Knight, 1965;
Kumosinski and Pessen, 1985; Nichol and Winzor, 1985; Pilz *et al.*, 1979). The
hydrodynamic behavior of proteins whose shapes cannot be approximated by
simple models such as spheres, ellipsoids, or rods can be modeled by assemblies
of spheres (García de la Torre, 1992; García de la Torre and Bloomfield, 1978).

Changes in the value of s for a protein with changes in solution conditions or
on the binding of ligand can reflect conformational changes. The s value of the

$(\alpha\beta)_2$ form of the rat insulin receptor in Triton X-100 increases by 15% on the binding of insulin, apparently reflecting a structural change that is important for tyrosine autophosphorylation and consequent signal transduction (Flörke *et al.*, 1990). For the $\alpha\beta$ form of the receptor, there is no detectable change in s following insulin binding nor is there any stimulation of autophosphorylation (Flörke *et al.*, 1990).

Sedimentation velocity experiments can be used to quantify the free energy changes associated with self-associating proteins. The method is dependent on several simplifying assumptions; nevertheless it offers a rapid alternative to sedimentation equilibrium for obtaining data on such systems. The dimerization of a mixture of β-lactoglobulin A and B has been studied using this method and a value of 0.5 liter/g obtained for the equilibrium constant (Gilbert and Gilbert, 1973).

In summary, a series of ultracentrifuge experiments can provide a variety of biophysical information that characterize a protein and its ability to interact with itself or other proteins. In many instances the methodology is superior to, and less equivocal than, other techniques that attempt to provide the same information. On the other hand, results obtained from the ultracentrifuge frequently complement data obtained using other techniques.

With the renewed interest in protein chemistry it has again become obvious that the analytical ultracentrifuge is an essential tool for protein characterization. However, the difficulty of operating the older instruments and a lack of standardization of computer interfacing and software have, until recently, held back the ultracentrifuge in this resurgence. Recent developments in instrumentation and computerization should quickly correct this situation.

We will deal here with the basic concepts and uses of the sedimentation velocity and sedimentation equilibrium experiments, the instrumentation, sample preparation, and some of the ways for analyzing data. Examples from the literature will be given that highlight the use of the analytical and preparative ultracentrifuges for water-soluble membrane proteins, such as cytoskeletal components, as well as for membrane proteins that can be kept in a soluble state only in the presence of detergents or lipids.

2. BASIC CONCEPTS

When a solution containing a single, ideal, nonassociating protein is centrifuged in a sector-shaped cell at sufficiently high speed, a boundary forms at the meniscus of the solution column and begins to move away from it (Figure 1a). If we could observe the concentration of solute across the sedimenting boundary at different times using, for example, an absorbance optical system, we would note the following: The boundary will be symmetrical with the midpoint being the point of inflection, and the boundary will continually spread with time as it sediments as a result of diffusion (Chervenka, 1973; Van Holde, 1971). The initial height of the boundary will depend on the loading concentration of the protein but the height will decrease with time since the molecules sedimenting radially through the sector-shaped cell will result in a dilution effect (Chervenka, 1973; Svedberg and Pederson, 1940). The concentration distribution curve can be differentiated with respect to distance to give the concentration *gradient* across the boundary. Alternatively, the concentration gradient can be observed directly using the schlieren optical system (Figure 1b). The boundary will now appear Gaussian in shape with the midpoint corresponding to the maximum concentration gradient (Chervenka, 1973). In this differential representation, the area under the Gaussian curve reflects the concentration of the sedimenting solute (Schachman, 1959). A third option is to use the Rayleigh interference optical system, which, like the schlieren system, relies on changes in refractive index as the concentration of the protein increases across the boundary (Figure 1c). Increments in the refractive index cause the interference fringes to curve up and the number of fringes crossed is a measure of the change in concentration across the boundary (Babul and Stellwagen, 1969; Chervenka, 1973).

Regular measurements of the movement of the boundary are the basis for the sedimentation velocity experiment. We can continue to watch the sedimenting boundary until all of the protein piles up at the base of the solution column. Eventually, all net movement of protein molecules will cease as the effects of sedimentation and diffusion cancel exactly at all points in the solution column. This situation is known as sedimentation equilibrium, and potentially a great deal of information can be obtained from the distribution of concentration as a function of radial distance. At the speeds used for sedimentation velocity, the concentration gradient is so steep that the data cannot be obtained reliably. However, if the speed selected is reduced substantially compared with that of the sedimentation velocity run, a more gentle concentration gradient will be established and the concentration distribution can be obtained with great precision and accuracy. As with sedimentation velocity experiments, any of the three optical systems might be used. The choice of optical system is not trivial and depends, among other things, on the nature of the sample and the operating conditions used.

3. SAMPLE PREPARATION

Apart from maintaining constant pH there are several important considerations when preparing protein samples for the ultracentrifuge. Nonideality resulting from the presence of a charged macromolecule can be greatly reduced by adding salt to a concentration of 0.1 M or greater, or by working close to the protein's isoelectric point (Pederson, 1958). The added salt can itself give rise to

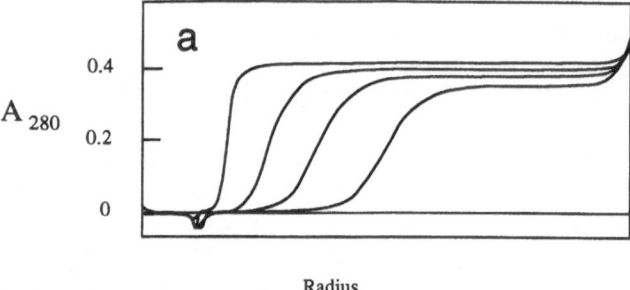

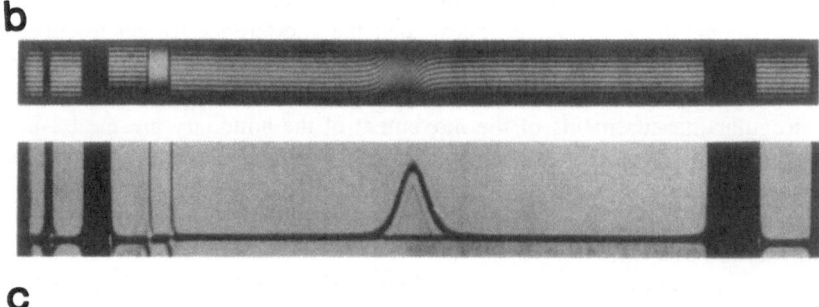

FIGURE 1. (a) Sedimentation velocity profiles of a sedimenting protein boundary observed at various times using absorbance optics. The speed selected must be sufficient to rapidly deplete the meniscus region of a measurable amount of solute and form a sharp boundary, as seen in the trace on the far left. In successive traces the boundary broadens with time because of diffusion. The initial height of the boundary, as determined by the absorbance in the plateau region, depends on the loading concentration of protein and also decreases with time because of radial dilution. The sharp upturn in absorbance at the base of the cell is caused by piling up of the protein. (b) Sedimenting boundary observed with schlieren optics which measure the concentration *gradient*. The schlieren pattern is Gaussian in shape for a single, ideal solute, with the maximum corresponding to the point of inflection in a corresponding absorbance trace. The schlieren pattern of the *solvent* provides a baseline and allows accurate measurements of the solute boundary to be made (see Sections 3 and 4.1), particularly when there is substantial skewing of the boundary because of sedimentation of the solvent components (Schachman, 1959). (c) Interference pattern of a sedimenting protein boundary using Rayleigh optics. The fringes are flat where there is no concentration gradient. The vertical displacement of the fringes in the boundary, or the number of fringes crossed with increasing radial distance, is a measure of the relative change in concentration of the protein. (d) Interference pattern of a protein solution at sedimentation equilibrium. The greatly reduced speed of sedimentation equilibrium experiments with respect to sedimentation velocity experiments means the concentration gradient is much more gentle. Measurements of fringe displacement can be made which provide information on the relative changes in protein concentration with radial distance (see Section 4.3). (b and c reprinted from Schachman, 1959.)

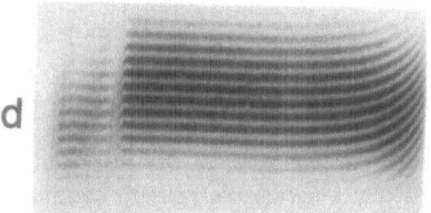

FIGURE 1. (*Continued*)

a "secondary charge effect" resulting from differential sedimentation or diffusion of the components of the added salt (Svedberg and Pederson, 1940). Thus, it is generally desirable to use a salt whose anion and cation components are not markedly different in sedimentation or diffusion. NaCl or KCl are most frequently used to avoid these problems (Schachman, 1959).

Much of the theory relating to the ultracentrifugation of molecules presupposes the presence of only two components: solvent and solute. This is an especially important consideration when optical systems are used and each solution channel has a matching channel containing the solvent. The two-component system can be closely approximated by dialyzing the sample against a large volume of the solvent containing the desired buffer and salts (Casassa and Eisenberg, 1964). The diffusate is then used in the solvent channel.

When absorbance optics are used, it is generally desirable to employ solvent systems that do not contain components that absorb strongly at the wavelength of interest. Carboxyl groups and halides begin to absorb significantly below 230 nm and for work at wavelengths below 200 nm, NaF is usually used in place of NaCl.

Immiscible layering oils are sometimes used in ultracentrifuge experiments; e.g., to raise the solution column off the base of the channel (Chervenka, 1973) or to fill the spherical portion at the base of a preparative ultracentrifuge tube (Attri and Minton, 1986). These oils can cause denaturation and aggregation of some proteins such as spectrin (Adams and Lewis, 1968; Morris and Ralston, 1984).

For membrane proteins, the choice of detergent and the concentration used to solubilize the protein are often complicated by the need to retain biological activity (Tanford and Reynolds, 1976). Testing with a variety of different detergents may be required to determine which ones maximize solubilization of the protein in the active form. For a study on erythropoietin receptor, Triton X-100 proved to be suitable in these respects (Mayeux *et al.*, 1990). Subtle problems with detergents can occur: For rhodopsin, the native spectrum of 11-*cis*-retinal is retained after solubilization of the protein in both digitonin and $C_{12}Me_2NO$. After bleaching by light, the native spectrum can be recovered in digitonin but not in $C_{12}Me_2NO$ (Sardet *et al.*, 1976; Yeager *et al.*, 1976). If membrane proteins, such as cytochrome P450, are purified in a highly delipidated state, care should be taken to add sufficient detergent to overcome artificial clustering of the protein (Behlke, 1992).

The critical micelle concentration of the detergent also may prove to be important since the protein must be dialyzed to equilibrium against its buffer before being used in the analytical ultracentrifuge (Casassa and Eisenberg, 1964), or if the solution column is to be optically scanned following sedimentation in the preparative ultracentrifuge (Attri and Minton, 1983, 1984). Very low critical concentrations may mean impractically long dialysis times.

Ideally, phospholipid vesicles would be used to simulate the endogenous environment of the solubilized protein. However, the lipid/protein ratio in such vesicles is usually too large for effective measurement of the protein moiety in the ultracentrifuge and there is the danger of a significant percentage of the vesicles containing more than one copy of the protein (Tanford and Reynolds, 1976). In some cases, these problems may be overcome: Purified band 3 monomers have been reconstituted into egg phospatidylcholine proteoliposomes and subsequently transformed into small vesicles using a French Press (Lindenthal and Schubert, 1991). After fractionation of the vesicles, sedimentation equilibrium studies showed that a single copy of the monomer was present in each of the vesicles containing protein (Lindenthal and Schubert, 1991).

If absorbance optics are to be used, then the optical properties of the detergent must be considered. Detergents that contain benzene rings, such as the Triton range, are unsuitable if the concentration distribution of a protein is to be measured at 280 nm. The use of detergents for ultracentrifuge work also relies on the partial specific volume of the protein-bound detergent being equal to that of the unbound detergent. For this reason it is probably best to avoid detergents such as Lubrol WX that contain a mixture of detergents of different chain length and different partial specific volumes. For such detergents there is the likelihood of preferential binding to the protein of detergent molecules with higher critical micelle concentrations (Reynolds and Tanford, 1976). An error of 12% in the molecular weight of the A1 polypeptide of the high-density lipoprotein has been attributed largely to this effect (Reynolds and McCaslin, 1985; Reynolds and Tanford, 1976).

Denaturants such as sodium dodecyl sulfate (SDS) and guanidine hydrochloride are sometimes used to dissociate protein complexes so that the molecular weight of the constituent polypeptide chains can be determined by sedimentation equilibrium (Tanford and Reynolds, 1976). However, guanidine hydrochloride frequently fails to completely dissociate integral-membrane protein complexes (Tanford and Reynolds, 1976). SDS is more reliable in this

Table I
Loading Concentrations of Protein for Sedimentation Equilibrium and Sedimentation Velocity Experiments in the Analytical Ultracentrifuge Using Various Optical Systems

Experiment	Optical system	Loading concentration (g/liter)
Sedimentation equilibrium	Interference	0.2–5
	Schlieren	1–70
	Absorbance (200 nm)[a]	0.005–0.1[a]
	Absorbance (280 nm)	0.1–2[b]
Sedimentation velocity	Interference	0.5–10
	Schlieren	1–20
	Absorbance (200 nm)[a]	0.005–0.1[a]
	Absorbance (280 nm)	0.1–2[b]
	Fluorescence	$> 10^{-5}$

[a]XL-A ultracentrifuge only.
[b]Values are for the XL-A. The maximum value would be halved for the Model E.

regard, but, for example, the coat protein of filamentous bacteriophage f1 remains dimeric in SDS (Makino *et al.*, 1975) and appears to retain the structure it adopts when inserted in the cell membrane of its host, *Escherichia coli* (Nozaki *et al.*, 1976a).

The amount of material required for an ultracentrifuge experiment depends, in large part, on the means by which the concentration distribution is to be measured. Table I indicates the range of loading concentrations that can used in the analytical ultracentrifuge when the concentration distribution is to be measured by various methods. Table II indicates the volumes of solution required and the lengths of the solution columns for various types of experiment.

Table II
Volumes of Solution and Column Lengths for Sedimentation Equilibrium and Sedimentation Velocity Experiments Performed in the Analytical Ultracentrifuge

Experiment	Volume (μl)	Column length (mm)
Sedimentation equilibrium		
Intermediate speed/high speed/low speed	3×130[a]	3[a]
Low speed (short column)	4×15[b]	1[b]
Sedimentation velocity	450	12

[a]Using a six-channel centerpiece (Yphantis, 1964) or three separate double-sector centerpieces.
[b]Using an eight-channel centerpiece (Correia and Yphantis, 1992; Yphantis, 1960).

4. INSTRUMENTATION

Recently, Beckman launched the XL-A analytical ultracentrifuge, which represents a long overdue modernization of the analytical ultracentrifuge, particularly in the areas of computer interfacing, software, electronics and miniaturization. It is a radical departure from, and is in most ways superior to, the Beckman Model E and other instruments of the 1950s, 1960s, and 1970s.

The Model E, in turn, is the descendant of the pioneering machines developed by Nobel laureate Thé Svedberg in the 1920s and 1930s. Svedberg and his colleagues not only built the first machines but developed much of the theory behind their use and laid the foundations for the practical versatility of the analytical ultracentrifuge (Svedberg and Pederson, 1940). It was on Svedberg's machine in Uppsala in 1924 that Svedberg and Fåhraeus obtained the first evidence that proteins (in this case hemoglobin) were large, discrete molecules with well-defined molecular weights rather than polydisperse colloidal suspensions (Svedberg and Fåhraeus, 1926). Later these ideas were extended to include the then remarkable suggestion that proteins could be composed of subunits.

For the analysis of proteins, an ultracentrifuge must have the following capabilities:

1. It must be able to spin protein samples at selected speeds up to 60,000 rpm ($\sim$250,000g) with minimal variation in the selected speed and minimum precession of the spinning rotor. Significant variations in speed lead to convection, which can erode the concentration gradient of the sedimenting protein. Thus, some form of electronic speed control is required. Precession (or rotor wobble) also results in convection, which will erode concentration gradients. Rotor wobble is probably the most common instrumental anomaly with the Beckman Model E. The XL-A has an induction motor and dynamic damping, which has eliminated the problem of rotor wobble.

2. There must be a good temperature control. Temperature variations, especially temperature gradients, also lead to convection. Temperature control in the XL-A is markedly improved over the Model E and relies on thermoelectric cooling of the chamber.

3. For an analytical ultracentrifuge, an optical system is required to measure the concentration or concentration gradient of the protein sample during the run. This means that the protein samples must be loaded into specialized cells, which are transparent to the light. In addition, there must be a means of recording the data. This can be done either photographically (in the Model E) or photoelectrically (in the XL-A, and with the scanner attachment for the Model E). The XL-A has fully computerized data storage capabilities (Giebeler, 1992), while some researchers have fitted on-line data acquisition systems to the Model E (Laue, 1992; Lewis, 1992; Schmidt *et al.*, 1990).

4.1. Rotors and Cells

Figure 2 shows the basic setup of the XL-A. The instrument is essentially a normal floor ultracentrifuge (based on the Beckman Optima XL preparative instrument) with some special features incorporated. The titanium rotor contains two pairs of holes for accepting the cells in which the samples are loaded. Some

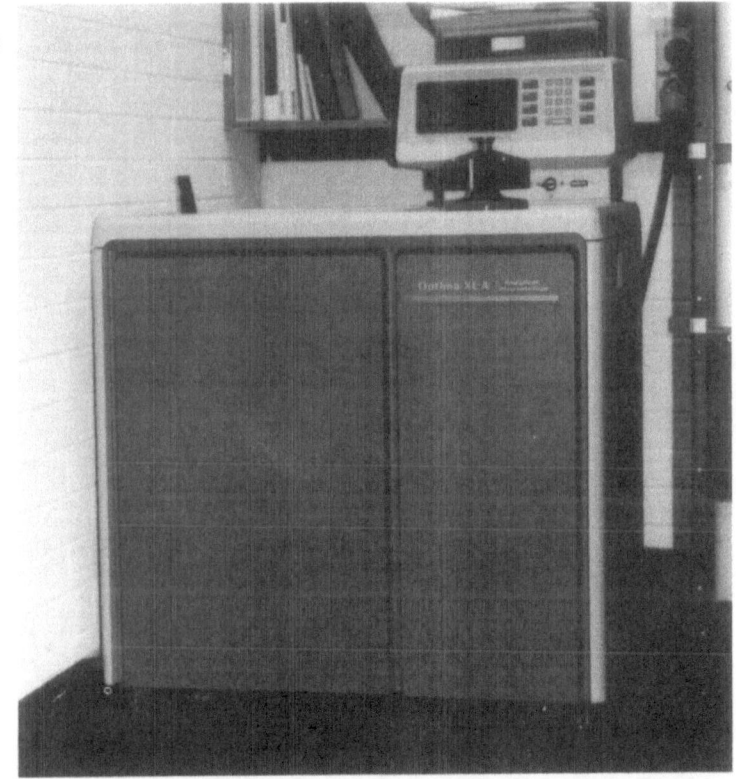

FIGURE 2. (a) The Beckman XL-A analytical ultracentrifuge. The instrument is based on the Beckman Optima XL preparative machine. (b) The XL-A contains a xenon flash lamp and a photomultiplier tube mounted in the bottom of the centrifuge. A removable housing contains the diffraction grating, reflectors for absorbing stray light, and an incident detector to aid in measuring pulse-to-pulse changes in light intensity. The firing of the flash lamp is timed so that light can pass through a selected channel in a selected cell sitting in the rotor. (c) Components of a typical cell, including a double-sector centerpiece. Solute is loaded into one of the channels while solvent, which has been dialyzed to equilibrium against the solution, is loaded into the other channel (see Sections 3 and 4.1). Both the centerpiece and the barrel have filling holes allowing the channels to be filled by syringe once the cell has been assembled.

b

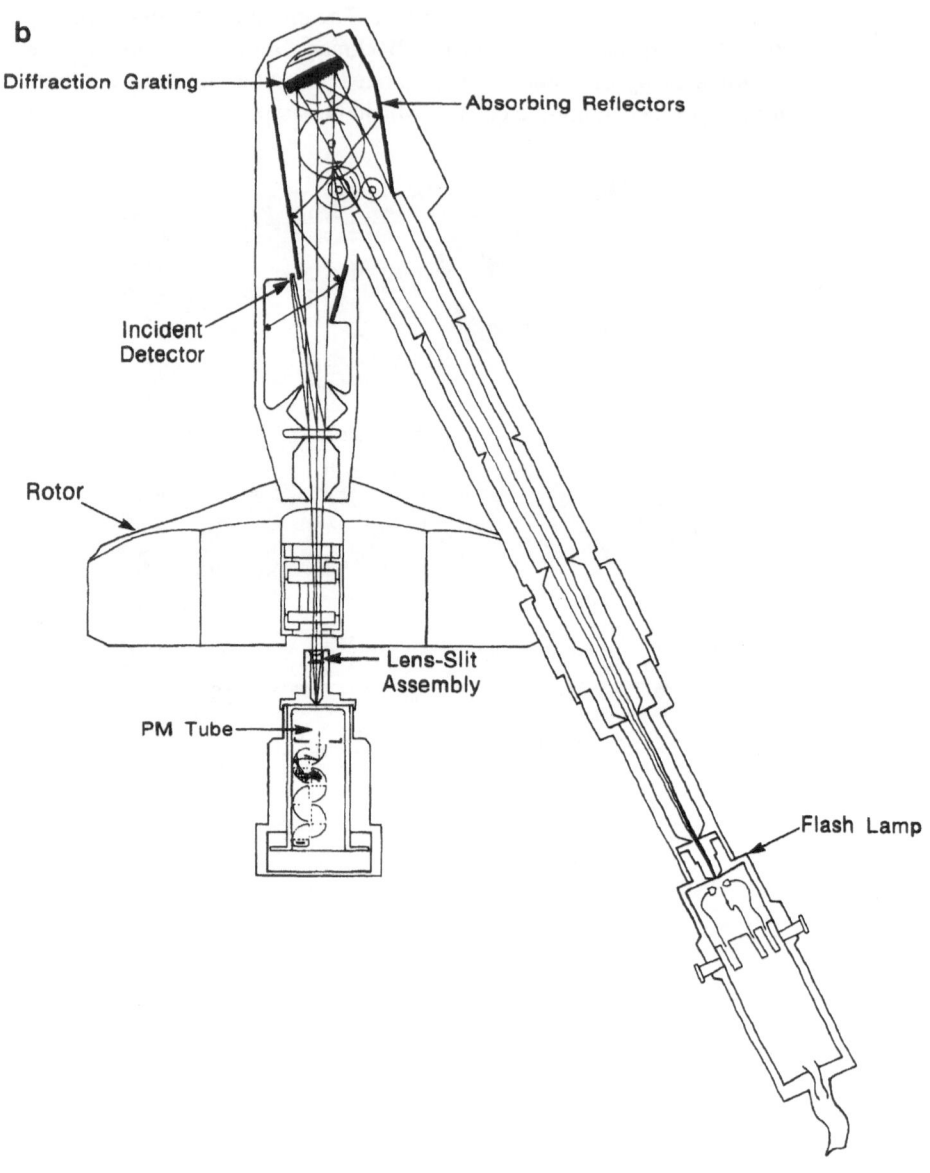

Diffraction Grating

Absorbing Reflectors

Incident
Detector

Rotor

Lens-Slit
Assembly

PM Tube

Flash Lamp

FIGURE 2. (*Continued*)

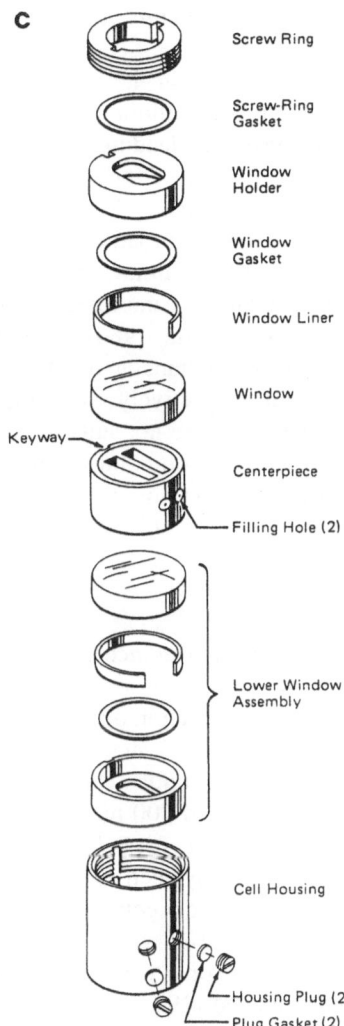

C

Screw Ring

Screw-Ring
Gasket

Window
Holder

Window
Gasket

Window Liner

Window

Keyway

Centerpiece

Filling Hole (2)

Lower Window
Assembly

Cell Housing

Housing Plug (2)

Plug Gasket (2)

FIGURE 2. (*Continued*)

rotors for the Model E have four pairs of holes (Mächtle, 1992). A counter-balance must occupy one of the holes and provides distance calibration. The cells themselves are the only pieces that can be transferred to the new machine from the Model E. A typical cell consists of a centerpiece containing two sector-shaped holes, or channels, into which are loaded the protein solution and the matching solvent, the diffusate, respectively. The centerpiece is sandwiched between quartz or sapphire windows and this assembly, along with gaskets and a screw ring, is sheathed by the barrel.

The sector shape of the channels in the double-sector centerpiece allows protein molecules to sediment without colliding with the walls of the centerpiece. Such collisions would create convection leading to erosion of the protein gradient in sedimentation velocity experiments. The length of the solution column is measured along the direction of the centrifugal force; that is, from the narrow end of the sector (the top of the cell) to the broad end of the sector (the base of the cell). For the double-sector cell, the maximum solution column length is 12 mm (Chervenka, 1973). Other specialized centerpieces have six (Ansevin *et al.*, 1970; Yphantis, 1964) or eight (Yphantis, 1960) channels allowing three or four samples to be run in the same cell simultaneously. The maximum column lengths with these cells are 3 and 1 mm, respectively. Since the individual channels are not sector-shaped, these centerpieces are not suitable for sedimentation velocity experiments. However, the short solution columns make them especially useful for sedimentation equilibrium since the time to equilibrium is inversely proportional to the square of the length of the solution column (Teller, 1973).

4.2. Optical Systems

The Model E is fitted with interference and schlieren optics. Some researchers have replaced the mercury light source with a laser light source and added multiplexing so that the concentration distributions in several sample cells can be recorded in a single experiment (Laue, 1992; Paul and Yphantis, 1972; Yphantis, 1980). Scanner absorbance optics can also be fitted to the Model E allowing wavelengths between 250 and 700 nm to be selected and absorbances up to 1.0 to be measured (Chervenka, 1973; Schachman and Edelstein, 1973). A fluorescence optical system has recently been developed for the Model E offering the potential to detect extremely low concentrations of protein (Table I; Schmidt *et al.*, 1990).

The XL-A is currently available only with scanner absorbance optics but the wavelength range has been extended (190–800 nm) and the maximum measurable absorbance increased to beyond 2.0. The stability and photometric accuracy of the XL-A absorbance system is superior to that of the Model E giving rise to greater precision and reproducibility of the absorbance scans. With the XL-A, wavelengths around 200 nm can be selected to take advantage of the high extinction coefficients of proteins here compared to 280 nm (Table I).

None of the optical systems can be used for the full range of applications of the analytical ultracentrifuge. Absorbance optics are very sensitive (Table I) and a protein can be detected selectively in a mixture if it is bound to a chromophore whose absorbance maximum is shifted to wavelengths longer than ~350 nm (Mulzer *et al.*, 1990; Muramatsu and Minton, 1989). However, the accuracy of the data ultimately depends on the accuracy of the extinction coefficient. Further complications arise if the absorbance of the protein does not obey Beer's law, if polymers of a self-associating protein have different extinction coefficients, or if there is turbidity or adsorption of the protein to the cell windows (Laue, 1992). In addition, the solvent components must not absorb strongly at the wavelengths of interest nor must there by any differential changes in absorbance of the solvent components in the solution compared with the diffusate following dialysis. An example of the last problem occurs with dithiothreitol (DTT), which is frequently used to keep proteins reduced. The presence of protein accelerates the formation of oxidized DTT relative to the diffusate. Oxidized DTT absorbs maximally at ~280 nm thereby giving the impression of increased protein concentration.

The two refractometric optical systems are more robust with respect to changes in solvent conditions and solute aggregation. Though these two systems are less sensitive than absorbance optics, very high concentrations of protein can be examined (Table I) and the radial resolution is superior (Laue, 1992; Rowe *et al.*, 1992). The precision and accuracy of the interference system, in particular, is superior to absorbance optics.

4.3. Data Storage and Manipulation

Schlieren and interference patterns obtained from experiments in the Model E are normally recorded on photographic plates. Transcribing the photographic record of the interference fringes from sedimentation equilibrium experiments to obtain concentration versus distance data is a slow task (Teller, 1973). Automated plate readers have been developed to alleviate this problem (DeRosier *et al.*, 1972; Richards and Richards, 1974; Rowe *et al.*, 1992; Teller, 1967) and an on-line system has been developed (Laue, 1992). An automated plate reading and analysis system for schlieren patterns obtained at sedimentation equilibrium has also been developed (Rowe *et al.*, 1992). Similarly, it is possible to modify the absorbance system of the Model E so that data can be transferred directly to a computer rather than just to a chart recorder (Sackett *et al.*, 1989). On the other hand, the absorbance system of the XL-A comes fully computerized with respect to data collection and storage.

Many ultracentrifuge experiments can be analyzed without the aid of computer programs although such programs clearly facilitate the analysis (Laue *et al.*, 1992). Other experiments, such as sedimentation equilibrium experiments, require very extensive manipulation and analysis of the concentration versus distance data, including model fitting by nonlinear regression techniques (Johnson *et al.*, 1981; Kim *et al.*, 1977; Morris and Ralston, 1985; Teller, 1973). In these cases, effective analysis requires the use of a computer and programs are available from certain authors (Harding *et al.*, 1992; Johnson *et al.*, 1981; Laue *et al.*, 1992; Lewis, 1992).

5.   SEDIMENTATION VELOCITY

5.1.   The Sedimentation and Frictional Coefficients

The sedimentation coefficient, s, can be determined by measuring the movement of a protein boundary in a sedimentation velocity experiment (Chervenka, 1973; Van Holde, 1971):

$$s \equiv \mathrm{d}\ln x / \omega^2 \mathrm{d}t = M(1 - \bar{v}\rho)/Nf \tag{1}$$

where s is usually quoted in Svedberg units ($1\ S \equiv 10^{-13}$ sec), x is the radial position of the midpoint of the boundary at time, t, and ω is the angular velocity of the rotor in radians per second. M and f are the molar weight and the frictional coefficient of the protein, respectively, $\bar{v}$ is the partial specific volume of the protein (approximately equal to the volume of solvent, in milliliters, displaced by each gram of the protein), ρ is the density of the solvent, and N is Avogadro's number.

For a single, nonassociating solute, s can be calculated from the slope of the linear plot of $\log x$ versus t, usually to a precision and accuracy of 0.5% or better (Chervenka, 1973). A small, compact protein will sediment more slowly than a large, compact protein and therefore will have a smaller value of s. On the other hand, a compact protein will sediment more quickly than an asymmetrical or expanded protein of the same mass since the compact shape will result in less friction. Solvation and rugosity (the roughness of the particle surface) will also affect the frictional properties of the protein. From this it is clear that s is not directly proportional to M since f depends on the hydrodynamic radius of the protein. For globular proteins with similar solvation and partial specific volume, $f \propto M^{1/3}$ and so, in principle, $s \propto M^{2/3}$.

The value of s, and hence f, is dependent on the solvent composition and temperature even if there is no change in the shape, solvation, or molecular weight of the protein with a change in conditions. Thus, a particular value of s is not especially helpful unless it is related to some set of standard conditions, usually taken to be water at 20°C:

$$s_{20,w} = s[(1 - \bar{v}\,\rho)_{20,w}\eta_{T,b}]/[(1 - \bar{v}\,\rho)_{T,b}\eta_{20,w}] \tag{2}$$

where η is viscosity and the subscripts b and T refer to the buffer composition and the experimental temperature, respectively.

5.2.   Concentration Dependence

For a single, nonassociating solute the value of the sedimentation coefficient decreases with increasing protein concentration. This occurs because as the concentration of a solute more dense than the solvent increases across the sedimenting boundary, the viscosity and density of the solution increase. There is also backflow of solvent to make room for the sedimenting molecules and to fill the space formerly occupied by the sedimenting molecules. Since sedimentation is measured relative to the cell's frame of reference and not to the flow of solvent, this also leads to a decrease in s (Schachman, 1959). Finally, there is an electric field set up as the charged macromolecules separate from their small counterions. The macromolecule and its counterions are forced to move together. This effect is reduced, but not eliminated, by the addition of salt to the buffer (Pederson, 1958).

In order to eliminate the effects of concentration dependence on the sedimentation coefficient, a plot of s versus concentration must be extrapolated to zero to obtain s^0. The following equations can be used to guide the extrapolation:

$$s \approx s^0(1 - k_s c_o) \tag{3}$$

$$1/s = (1 + k_s c_0)/s^0 \tag{4}$$

where c_0 is the initial loading concentration of the protein in the grams per liter scale and k_s is a measure of the concentration dependence.

Equation (3) can be used if the concentration dependence of s appears to be linear, which is usually the case for small, compact proteins over a wide concentration range or for large, expanded or asymmetrical proteins providing the concentration range is limited. An example of the latter is the spectrin tetramer ($M = 960$ kg/mol), a highly expanded protein with moderate asymmetry, where s versus c_0 remains approximately linear only up to about 5 g/liter (Figure 3a). A plot of $1/s$ versus c_0 using Equation (4) is linear over a wider concentration range than a plot of s versus c_0 and can be used, for example, to guide the extrapolation for very large, highly asymmetrical molecules such as synthetic myosin filaments ($M \approx 1.8 \times 10^3$ kg/mol; Persechini and Rowe, 1984). Note that in Figure 3b, a plot of s versus c_0 for this solute is curvilinear.

A natural consequence of the concentration dependence of sedimentation is self-sharpening of the boundary which occurs because concentration, and hence the concentration dependence, increases through the boundary region. Molecules in the leading edge of the boundary experience greater frictional drag and tend to lag back into the boundary while those in the dilute solution at the trailing edge of the boundary move more rapidly. Thus, while k_s is a quantitative measure of the concentration dependence, self-sharpening of the sedimenting boundary is a qualitative measure. When viewed by schlieren optics, self-sharpened boundaries appear spikelike rather than Gaussian in form. Grossly nonideal solutes give rise to razor-sharp boundaries during sedimentation (Schachman, 1951) and

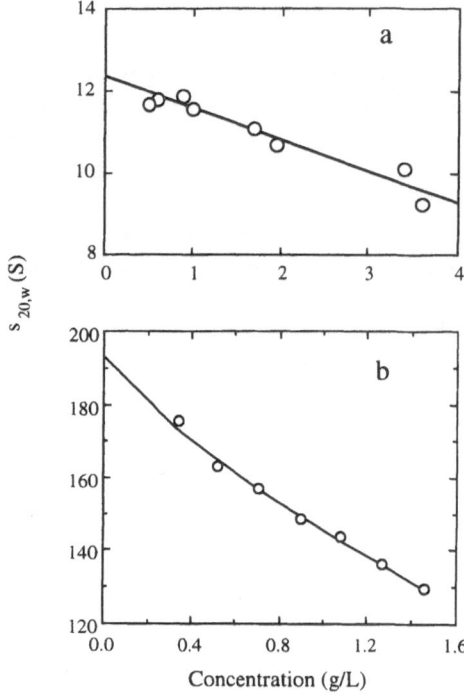

$s_{20,w}$ (S)

Concentration (g/L)

FIGURE 3. (a) Plot of the sedimentation coefficient, s, against the loading concentration for the tetramer of the red cell cytoskeletal protein, spectrin ($M = 960$ kg/mol). The concentration dependence of sedimentation leads to a decrease in s with increasing concentration (see Section 5.2). The dependence is approximately linear over the concentration range. Equation (3) can be used to guide the extrapolation to zero concentration to obtain a value of the sedimentation coefficient, s^0, which is independent of the effects of concentration. (b) Plot of s versus concentration for synthetic myosin filaments ($M \approx 1.8 \times 10^3$ kg/mol). The plot is curvilinear for this very large, highly asymmetric complex, indicating a large concentration dependence. A plot of $1/s$ versus concentration, however, is linear (see Figure 3, Persechini and Rowe, 1984) and Equation (4) can be used to guide the extrapolation to zero concentration to obtain s^0. The value of s^0 obtained from the plot of $1/s$ versus concentration has been used to guide the freehand curve in the plot of s versus concentration. (b redrawn from Figure 3 in Persechini and Rowe, 1984.)

even the spectrin heterodimer shows obvious signs of self-sharpening (Ungewickell and Gratzer, 1978).

5.3. Determination of Molecular Weight

Molecular weight can be determined from sedimentation velocity experiments using several methods of varying degrees of rigor.

5.3.1. Using Calibration Standards

The simplest approach is to assume that the protein is compact and globular. In this case, an apparent molecular weight can be estimated crudely from the value of $s^0_{20,w}$ and a calibration curve of standard, globular proteins for which the known values of $s^0_{20,w}$ are plotted against $M^{2/3}$ (Van Holde, 1975). Clearly, if the protein is expanded or asymmetrical the apparent molecular weight will be smaller than the true molecular weight. Similarly, if the protein has an unusual partial specific volume because, for example, it is conjugated to lipid or carbohydrate, or if the degree of hydration is especially large or small this will give an apparent molecular weight substantially different from the true molecular weight.

5.3.2. Using the Concentration Dependence of Sedimentation

Molecular weight can also be determined from the concentration dependence of the sedimentation coefficient, k_s, using the following equation (Persechini and Rowe, 1984; Rowe, 1977, 1992):

$$M = (6\pi\eta_{20,w}s^0_{20,w})^{3/2}\{(3\ \bar{v}/4\pi)[(k_s/2\ \bar{v}) - (\bar{v}_s/\bar{v})]\}^{1/2} \tag{5}$$

where $\bar{v}_s$ is the partial specific volume of the hydrated protein.

Equation (5) is derived from Stokes's law for the movement of spheres through a fluid, and a definition for k_s based on a moving frame of reference within the solution column as well as the conventional cell frame of reference (Lavrenko et al., 1992; Rowe, 1977, 1992). In Equation (5), all terms can be determined experimentally including the ratio $\bar{v}_s/\bar{v}$, which can be calculated from the concentration dependence of the reduced viscosity and sedimentation coefficient (Rowe, 1977, 1992). In any event, the equation is very insensitive to changes in $\bar{v}_s/\bar{v}$ since this quantity is small in relation to $k_s/2\bar{v}$, particularly for large, highly asymmetrical proteins.

The method is especially applicable to the determination of molecular weights greater than 10^6, particularly since other methods may prove problematic for these larger proteins. For example, light scattering may suffer from clarification problems, and a breakdown of the Rayleigh–Gans approximation for highly asymmetric proteins. Sedimentation equilibrium would only be attained very slowly and very low speeds would be required. These low speeds may not be attainable without significant rotor wobble unless the XL-A or a large rotor in the Model E is available. If the protein is susceptible to damage with time, sedimentation equilibrium may also be unacceptable. Sedimentation equilibrium using the method of Yphantis with very short solution columns (Correia and Yphantis, 1992; Yphantis 1960; see Section 7.3.3) may be an acceptable alternative but the limitations of this method should be borne in mind: small amounts of noninteracting particles (such as irreversible aggregates) which are large relative to the protein of interest will skew the value of the molecular weight. Sedimentation velocity will be more robust in this respect since such relatively large particles will rapidly sediment away from the boundary of interest.

5.3.3. Using a Combination of Sedimentation and Diffusion Coefficients

The most common approach for obtaining molecular weights from sedimentation velocity experiments is based on the following expression (Van Holde, 1971):

$$M = (s^0RT)/[D^0(1 - \bar{v}\ \rho)] \tag{6}$$

where R is the universal gas constant, T is the absolute temperature, and D^0 (cm^2/sec) is the diffusion coefficient at infinite dilution. Equation (6) is a rearrangement of Equation (1) where (Tanford, 1961)

$$D^0 = RT/Nf \qquad (7)$$

D^0 can be obtained from an analysis of the sedimenting boundary or from measuring the time-dependent spreading of a stationary boundary in either the analytical or preparative ultracentrifuges (see Section 6). Alternatively, D^0 can be obtained from quasielastic light scattering (Bloomfield and Lim, 1978).

There are several points to note about obtaining molecular weights using the second and third methods outlined above. First, the molecular weight is that of the unhydrated species and is obtained without assumptions about the shape of the protein and independent of any calibration markers. The last point is particularly noteworthy when these methods are compared with gel electrophoresis under denaturing conditions. Though gel electrophoresis is rapid and can be used routinely and diagnostically, it has no sound theoretical basis and ultimately relies on calibration. For these reasons, errors in molecular weights obtained from denaturing gels frequently occur and are often large. Commonly, these errors result from differential binding of the denaturant to the protein moiety compared with binding to the calibration proteins, or because the protein (such as a glycoprotein) has intrinsically unusual mobility. Values can even vary substantially depending on the denaturant used.

Second, the molecular weight estimate from sedimentation analysis is for the protein as it exists in solution, whether this is a single polypeptide chain or a complex held together by covalent or noncovalent interactions. If the protein is a complex, there is the potential to determine the stoichiometry by using sedimentation equilibrium experiments in the presence of denaturant (Lee and Timasheff, 1979).

Third, to obtain values of the molecular weight, $\bar{v}$ and ρ must be known. The value of ρ will depend on the choice of solvent (generally H_2O or D_2O), temperature, and the presence of other components such as buffer salts (but *excluding* the solute of interest; Wills *et al.*, 1993). ρ can be determined using densimetry, from tables, or from empirical equations (Laue *et al.*, 1992). The value of $\bar{v}$ can be determined experimentally using densimetry (which requires large amounts of protein at high concentration) or from sedimentation velocity experiments in both H_2O and D_2O (Edelstein and Schachman, 1967; see Section 5.3.5). Sedimentation equilibrium can also be used to determine $\bar{v}$ (see Section 7.3.5). Generally, however, $\bar{v}$ is calculated from the amino acid composition using a simple additive procedure based on the partial specific volumes and molecular weights of the residues (Cohn and Edsall, 1943; Durchschlag, 1986). The calculated value is usually accurate to within 1% for proteins if the correct composition is used. A 1% error in $\bar{v}$ leads to an error of ~3% in the term $(1 - \bar{v}\rho)$ and hence in M [Equation (6)]. If the composition and relative weight of conjugated carbohydrate of a glycoprotein is known, this can be easily incorporated into the calculation of $\bar{v}$ (Durchschlag, 1986). However, even without this correction, the value of $\bar{v}$ is usually accurate to within 3% (Laue et al., 1992).

5.3.4. Density Gradient Centrifugation in the Preparative Ultracentrifuge

Two techniques have been developed that allow sedimentation coefficients to be obtained from the preparative ultracentrifuge. In the more recent development, a small amount of solute, such as 0.25% sucrose, is used to form a small, self-generating concentration gradient that is just sufficient to stabilize the *boundary* of the sedimenting protein during deceleration of the rotor and subsequent manipulation of the tube (Attri and Minton, 1983, 1984, 1986). Measurement of the boundary involves either the use of absorbance optics to scan the tube following centrifugation (Attri and Minton, 1983, 1984) or fractionation of the tube's contents in aliquots as small as 1 μl which can then be quantified with respect to a protein-bound probe such as a radiolabel (Attri and Minton, 1986). Both processes are computer controlled, are rapid enough to avoid significant spreading of the boundary (Attri and Minton, 1986), and provide data with an accuracy and precision similar to that obtained from the analytical ultracentrifuge. In addition, the sensitivity using the fractionation method and radiolabeled protein can exceed that of absorbance optics (Attri and Minton, 1986) and the use of radiolabeled ligand or a sensitive assay allows specific proteins in crude mixtures to be analyzed. Once the sedimentation coefficient has been obtained by these methods, the procedures outlined above can be used to calculate the molecular weight.

When optical scanning or sophisticated fractionation techniques are not available, large, linear density gradients of sucrose (e.g., 5–30%) are used to stabilize the protein boundary (or, more usually, a narrow zone or "band" of protein, initially layered over the sucrose gradient) during deceleration and subsequent manipulation.

Determination of the sedimentation coefficient, $s^0_{20,w}$, requires that the density and viscosity throughout the sucrose gradient be known. By combining Equations (1) and (2) (Siegel and Monty, 1966):

$$s^0_{20,w} = (1 - \bar{v}\rho_{20,w})/(\omega^2 t \eta_{20,w}) \int_{x_m}^{x_t} \eta_x/[x(1 - \bar{v}\rho_x)] \cdot dx \qquad (8)$$

where t is the time of centrifugation, x_m and x are the distance from the center of the rotor to the meniscus and to the center of the solute band, respectively, and ρ_x

and η_x are the density and viscosity of the solvent, respectively, at the temperature of the experiment and at radial position x.

Following fractionation of the solution column at the completion of the experiment, the density of the sucrose in the fractions can be determined refractiometrically and the viscosities calculated from the density and the temperature (Smigel and Fleischer, 1977). The integral in Equation (8) can then be solved numerically to obtain the value of $s_{20,w}$ for the protein of interest (Siegel and Monty, 1966; Smigel and Fleischer, 1977), provided the value of $\bar{v}$ for the protein is known or can be estimated (see Section 5.3.3). If the protein is not pure, the value of $\bar{v}$ can be measured by performing sedimentation velocity experiments in H_2O and D_2O buffers (see Section 5.3.5).

Calculation of viscosities of the fractions based on the measured densities can lead to error because viscosity is highly temperature dependent and there is considerable uncertainty about the temperature in the swinging buckets of the rotor (Smigel and Fleischer, 1977). This error can be reduced by running a series of standard proteins with known sedimentation coefficients and making an empirical adjustment to the temperature, and hence to the viscosity, until the standard closest to that of the sample protein assumes its correct value of s (Hutchison and Fox, 1989; Smigel and Fleischer, 1977). The revised viscosities are then used in Equation (8).

Once the sedimentation coefficient has been determined, the second and third methods outlined above (Sections 5.3.2 and 5.3.3) can be used to calculate the molecular weight. Pure protein is required if the second method is to be used and usually is required to determine the value of D^0 for the third method. Generally, however, with density gradient centrifugation, molecular weight is calculated from a combination of s^0 and f [Equation (1)], where f is determined from the Stokes radius, R_s (Van Holde, 1971):

$$f = 6\pi\eta R_s \tag{9}$$

It is common for a value of R_s to be estimated from gel filtration: A standard curve is constructed from the Stokes radii of compact, globular proteins against $(-\log K_{av})^{1/2}$, where K_{av} is a measure of the fractional elution volume (Siegel and Monty, 1966).

Determination of molecular weights using a combination of density gradient centrifugation and gel filtration is common practice, largely because it can be used for impure samples and only basic equipment and computation are required. There are, however, several theoretical and practical problems: In particular, preferential hydration of proteins in the presence of sucrose will occur (Lee et al., 1979) and there is, as yet, no rigorous theoretical basis for the use of gel filtration to obtain Stokes radii (Tanford and Reynolds, 1976). While gel filtration appears to be adequate for globular proteins, it fails for large, asymmetric proteins such as myosin (Nozaki et al., 1976b).

5.3.5. Detergent-Solubilized Proteins

The molecular weight of protein solubilized by detergents also can be determined using sedimentation velocity. For a protein–detergent complex in a crude mixture, band sedimentation through sucrose density gradients in the preparative ultracentrifuge is often the method used. The most important additional factor to be determined is the amount of bound detergent. This can be measured directly using, for example, radiolabeled detergent (Clarke, 1975; Garrigos et al., 1993; Kuchel et al., 1978). Normally, however, the amount of bound detergent is determined indirectly through measurement of the partial specific volume for the protein–detergent complex, $\bar{v}_c$. Two methods are commonly used (Clarke, 1975; Smigel and Fleischer, 1977). Both depend on sedimenting the solute under identical conditions in H_2O and D_2O (or $D_2{}^{18}O$), and both allow the value of $s_{20,w}$ and $\bar{v}_c$ for the protein–detergent complex to be solved simultaneously. Using the method of Clarke (1975), the value of $\bar{v}_c$ can be calculated by

$$\bar{v}_c = [s_H\eta/(s_D\eta_D) - 1]/[\rho_D s_H \eta_H/(s_D\eta_D) - \rho_H] \tag{10}$$

where the subscripts H and D refer to samples centrifuged in H_2O and D_2O, respectively, and the values of η and ρ are those at x_{av}, taken to be the radial position halfway between the meniscus and the sedimenting band (Clarke, 1975). The values of s_H and s_D are also calculated at x_{av}, in accordance with Equation (1) and assuming that the velocity of the band is constant for a given angular velocity. Having determined the value of $\bar{v}_c$, the value of $s_{20,w}$ can be calculated from the following form of Equation (2) (Clarke, 1975):

$$s_{20,w} = s_H(\eta_H/\eta_{20,w})(1 - \bar{v}_c\rho_{20,w})/(1 - \bar{v}_c\rho_H) \tag{11}$$

The uncertainty in the value of η at x_{av} (Section 5.3.4) can be minimized by calculating the viscosities at the x_{av} positions for a set of standard proteins with known values of $s_{20,w}$ and $\bar{v}$ (Clarke, 1975). The viscosity of x_{av} for the protein of interest is then obtained by interpolation. Clarke applied this method to a range of integral membrane proteins and other detergent-solubilized proteins. Using the method of Smigel and Fleischer (1977), a range of values of $\bar{v}_c$ are used to determine a range of values of $s_{20,w}$ using Equation (8). The intersection of the two curves generated for H_2O and D_2O gives the values of $\bar{v}_c$ and $s_{20,w}$ for the protein–detergent complex. Smigel and Fleischer used this method for the prostaglandin E-binding protein from rat liver membranes (Figure 4).

For both methods, the Stokes radius, R_s, of the protein–detergent complex is again estimated from gel filtration using standard proteins for calibration (see Section 5.3.4; Siegel and Monty, 1966) and the molecular weight of the complex is calculated using Equations (9) and (1). That fraction of the molecular weight of

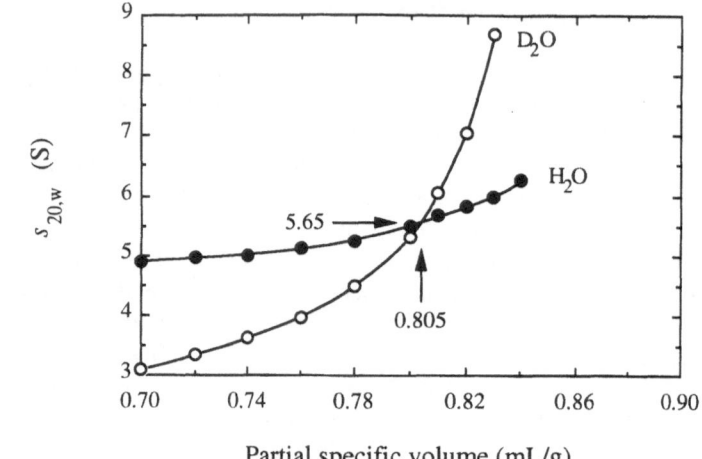

FIGURE 4. Simultaneous determination of $s_{20,w}$ (the sedimentation coefficient corrected to water at 20°C) and partial specific volume, $\bar{v}_c$, for the detergent-solubilized prostaglandin E-binding protein. Rat liver plasma membranes were labeled with tritiated prostaglandin E_1 and a crude mixture containing the binding protein was extracted from the membranes using Triton X-100. Aliquots of the extract were layered over 10–34% (w/v) sucrose gradients made up in buffer containing either H_2O or D_2O to alter the solvent density. Following centrifugation and fractionation, the position of the binding protein band in both H_2O and D_2O buffers was determined by scintillation counting. Values of $s_{20,w}$ for protein in H_2O and D_2O were calculated for a range of values of $\bar{v}_c$ using Equation (8). The intersection of the two curves gives the value of $s_{20,w}$ and $\bar{v}_c$ for the protein–detergent complex. (Adapted from Smigel and Fleischer, 1977.)

the complex present as protein, X_p, can be calculated simply from the values of $\bar{v}$ for the complex ($\bar{v}_c$), the detergent ($\bar{v}_{Det}$), and the protein moiety itself ($\bar{v}_P$):

$$X_p = (\bar{v}_c - \bar{v}_{Det})/(\bar{v}_P - \bar{v}_{Det}) \qquad (12)$$

Partial specific volumes for most detergents are known (Durchschlag, 1986), while that for the protein would have to be measured or estimated from the composition. Note that in calculating the molecular weight of the protein–detergent complex, the values for s, ρ, and η used in Equations (1) and (9) are those corresponding to water at 20°C as the solvent. For the adenosine A_2-like binding site of the placental membrane (Hutchison and Fox, 1989) it would appear that the viscosity of the sucrose solution has been used to calculate a molecular mass of 230 kDa for the complex. In fact, with the correct value for viscosity, the value for the molecular mass is ~180 kDa indicating that the protein is free of bound detergent.

All of the precautions and limitations of sucrose density gradient centrifuga-tion outlined in Section 5.3.4 apply when detergent is present. Other assumptions and restrictions also apply (Smigel and Fleischer, 1977; Tanford and Reynolds, 1976). In particular, the standard proteins must not bind the detergent and the assumption is made that sucrose does not affect the binding of the detergent to the protein nor the state of aggregation of the protein. There is evidence suggesting that the effects of sucrose are not negligible in either respect (Lee et al., 1979; Simons et al., 1973).

Clearly there is the potential for some of the assumptions of sucrose gradient centrifugation to fail in practice leading to significant inaccuracy in the molecular weight. The precision of measurement is also relatively low. These factors make an unambiguous determination of the stoichiometry of a protein complex difficult, if not impossible, in most cases. Nevertheless, for unpurified protein, and if sophisticated fractionation equipment is not available (Attri and Minton, 1986), density gradient centrifugation usually offers the only means of obtaining an initial estimate of the mass and stoichiometry of a protein complex.

5.4. Hydrodynamic Analysis of Proteins: Particle Size and Shape

It is possible to obtain estimates of the shape and degree of hydration of a protein from sedimentation velocity experiments and, in combination with data obtained from other types of experiment, these estimates can be refined.

From Equation (1) the frictional coefficient, f, can be determined if the molecular weight and $s_{20,w}^0$ are known. In addition, the value of f_0 can be calculated using Equation (9), where f_0 is the frictional coefficient for the protein if it were an unhydrated, incompressible sphere of equivalent molecular weight and partial specific volume. That is, f_0 represents the minimum possible frictional coefficient and the frictional ratio f/f_0 represents the degree of deviation from this minimum. As a guide, a value of f/f_0 less than ~1.2 indicates that the protein is compact and globular, the value differing significantly from 1.0 due, primarily, to hydration and rugosity of the surface of the molecule. Values progressively larger than 1.2 arise as the protein becomes, in addition, more expanded or asymmetrical. However, it is not possible unambiguously to assign the contribution from each of these effects from sedimentation velocity studies alone.

In calculating f_0 from Equation (9), a value for the Stokes radius must be used. The Stokes radius can be estimated in a number of ways (Laue et al., 1992; Teller, 1976), the most common employing the following equation (Laue et al., 1992; Schachman, 1959):

$$R_s \approx (3M \, \bar{v}/4\pi N)^{1/3} \tag{13}$$

The relationship is approximate because $\bar{v}$ includes contributions from solute–solvent interactions in addition to the volume of solvent occupied by the protein (Laue et al., 1992).

The shape of the protein can be modeled using the value of f/f_0. Generally, if no other data are available, the model selected is a prolate or oblate ellipsoid from which the axial ratio of the protein can be obtained (Van Holde, 1971). An unambiguous choice cannot be made between prolate and oblate ellipsoids based on the frictional coefficient alone. Some resolution of this ambiguity of shape can be achieved using a combination of the frictional ratio with other information, such as a measure of the thermodynamic nonideality of the protein (Nichol and Winzor, 1985). Formulas for other shapes such as cylinders (García de la Torre, 1992; Tirado and García de la Torre, 1979) and triaxial ellipsoids (Harding and Rowe, 1983, 1984) are also available.

Values for the axial ratio obtained with these models are the *maximum* values since the effects of hydration, rugosity, or expansion of the molecule are not incorporated. Thus, the values are intended only as a guide. Unfortunately, these axial ratios, in isolation, are commonly used for further computation or as definitive evidence for the asymmetry of the protein without regard to the contribution from other effects (Rogers and Sykes, 1990; Smigel and Fleischer, 1977; Smith and Agre, 1991).

It is possible to measure or estimate the contribution of hydration to the frictional ratio and to assume that the remaining contribution is related to the shape of the molecule (García de la Torre, 1992; Laue *et al.*, 1992):

$$(f/f_0)_{hyd} = (\delta\rho_{20,w}/\bar{v} + 1)^{1/3} \qquad (14)$$

$$(f/f_0)_{shp} = (f/f_0)/(f/f_0)_{hyd} \qquad (15)$$

where δ is the degree of hydration in grams H_2O per gram protein and the subscripts "shp" and "hyd" refer to the contributions to the frictional ratio assigned to the shape of the protein and its hydration, respectively. The difficulty is that the value of δ can vary from ~0.2 to 0.6 depending on the protein (Pessen and Kumosinski, 1985). Estimates of the degree of hydration can be obtained from the amino acid composition (Kuntz, 1971; Lee and Timasheff, 1979) or can be measured using various techniques such as small-angle X-ray scattering (Kumosinski and Pessen, 1985). Caution should also be exercised here since the various techniques measure different combinations of properties of water and therefore will provide different results. It is not known to what degree each of the measured properties of water contribute to the friction of the sedimenting protein. On the other hand, the degree of hydration can be measured from the frictional ratio if the shape of the protein is known, for example, from electron microscopy.

Additional experimental information can prove useful in elucidating the shape of a protein. In particular, the ratio of the concentration dependence of the sedimentation coefficient, k_s, to the intrinsic viscosity, $[\eta]$, of the protein can be used to assess whether the protein is expanded or asymmetric (Creeth and

Knight, 1965). A value of $k_s/[\eta]$ near 1.6 indicates a spherical molecule regardless of the degree of expansion, while a decrease in the ratio indicates asymmetry. Myosin, for example, has a $k_s/[\eta]$ value of 0.3 indicating gross asymmetry (Creeth and Knight, 1965). The spectrin heterodimer and tetramer have values of 1.15 and 1.37, respectively (Dunbar and Ralston, 1981). The axial ratios of these oligomers can be calculated from the $k_s/[\eta]$ values using a prolate ellipsoid model (Creeth and Knight, 1965). The axial ratios are modest, being 7.2 for the heterodimer and 4.0 for the tetramer (Dunbar and Ralston, 1981). On the other hand, the frictional ratios are 2.3 for the heterodimer and 2.8 for the tetramer (Dunbar and Ralston, 1981) and on this basis alone spectrin would appear to be highly asymmetric. However, the $k_s/[\eta]$ data strongly suggest that the very high values of the frictional coefficient for spectrin are related, in large part, to expansion.

The radius of gyration, R_G, measured by light, neutron, or small-angle X-ray scattering, can also be used to assess the shape of the molecule and assist in the interpretation of the frictional ratios (Cantor and Schimmel, 1980; Kumosinski and Pessen, 1985; Pilz et al., 1979). The shape of a protein or protein complex can be modeled, a theoretical value for R_G calculated, and the theoretical value compared with the measured value for R_G to indicate if the shape is appropriate (Cantor and Schimmel, 1980; Pilz et al., 1979). In particular, small-angle X-ray scattering has been used in combination with sedimentation velocity data to assess the shape, hydration, and rugosity of globular proteins (Kumosinski and Pessen, 1985).

When sucrose density gradient centrifugation has been used, the frictional ratio relating to the shape of the protein is often calculated from a combination of Equations (9), (13), (14), and (15) (Smigel and Fleischer, 1977):

$$(f/f_o)_{shp} = R_s\{4\pi N/[3M(\bar{v} + \delta/\rho_{20,w})]\}^{1/3} \tag{16}$$

In these cases, the values of R_s has usually been obtained from gel filtration and the value of δ is often chosen to be 0.2 g H_2O/g protein, with very little justification (Smigel and Fleischer, 1977; Smith and Agre, 1991).

For detergent-solubilized proteins, the values of R_s, M, and $\bar{v}$ are those for the protein–detergent complex and not the protein moiety alone. In the case of the 28-kDa channel-like protein from the erythrocyte membrane, use of Equation (16) has yielded a value of 1.63 for $(f/f_o)_{shp}$, which the authors took to indicate that this oligomeric protein is very asymmetric. However, the value for M was chosen as that of the protein moiety, not the protein–detergent complex. Recalculation with the molecular weight determined for the complex reduces the value of $(f/f_0)_{shp}$ to 1.45. In addition, the value of δ was selected, apparently arbitrarily, as 0.2 g H_2O)/g protein (Smith and Agre, 1991), but if this value were doubled the frictional ratio would be reduced further to 1.36. Remembering that the rugosity and any expansion would contribute to the ratio of 1.36, the protein–detergent complex may actually be close to spherical in shape. In addition, inferences with respect to the shape of the protein moiety alone generally rely on the assumption that the bound detergent is evenly distributed over the protein surface, which may not be true (Yeager *et al.*, 1976). Clearly, this is a case where further data are required before the shape of the protein can be more confidently assessed.

5.5. Conformational Changes

Large changes in the value of $s^0_{20,w}$ under different solution conditions will reflect, for the most part, changes in the shape or swelling of the protein, provided that the molecular weight remains constant. For example, the concentration dependence of the sedimentation coefficient has been used to show that the conformation of synthetic myosin filaments is sensitive to the concentration of Mg^{2+} (Persechini and Rowe, 1984). A change in conformation was reflected by a large increase in $s^0_{20,w}$ and a large decrease in k_s as the concentration of Mg^{2+} increased from 0.2 mM to 3 mM. At the same time molecular weight remained constant (Persechini and Rowe, 1984). These data indicate that the filaments adopt a more compact conformation in the presence of Mg^{2+}.

Chondrus crispus flavodoxin undergoes a large change in the frictional ratio when it is converted to apoflavodoxin on removal of the flavin analogue (Rogers and Sykes, 1990). For flavodoxin, the value of the frictional ratio is 1.3, indicating that the protein is compact and nearly spherical. X-ray crystallography shows that the protein has a maximum axial ratio of ~1.4 (Rogers and Sykes, 1990). For apoflavodoxin, the frictional ratio increases to 1.9. The molecular weight of both forms of the protein are the same based on the values of $s^0_{20,w}$ and $D^0_{20,w}$. The axial ratio of apoflavodoxin was calculated to be 20 or 30, depending on the use of prolate or oblate ellipsoid model, respectively (Rogers and Sykes, 1990). However, these values are the *maximum* possible values and clearly are gross overestimates since the substantial contributions from hydration or rugosity are neglected.

Precise determination of changes in the sedimentation coefficient of less than 1% can be obtained using difference sedimentation velocity methods (Kirschner and Schachman, 1971; Newell and Schachman, 1990; Richards and Schachman, 1959; Smith and Schachman, 1973). Small differences of this kind might be brought about, for example, by the addition of a ligand or substrate to the protein that alters the protein's conformation. Along with data from other methods, such as densimetry (to assess changes in preferential hydration; Kratky *et al.*, 1973) and small-angle X-ray scattering (Kumosinski and Pessen, 1985), difference sedimentation provides a potentially powerful method for investigating small conformational changes in proteins.

5.6. Heterogeneity

By means of qualitative and quantitative analysis of sedimentation velocity profiles, one can analyze sample purity and the properties of polydisperse and paucidisperse systems. Heterogeneity may take several forms from the presence of one or more contaminating species to true polydispersity. A pure solute in two or more conformational states that do not interconvert (or that interconvert only slowly during the experiment) may also be detected as heterogeneity in a sedimentation velocity experiment. A special case of apparent heterogeneity occurs when a pure solute self-associates (see Section 5.7).

A sedimenting boundary that is obviously asymmetrical is indicative of heterogeneity (although hypersharpening arising from a strong degree of concentration dependence can also lead to boundary asymmetry; Fujita, 1975). However, heterogeneity is often ignored because the boundary appears to be Gaussian in shape on visual inspection. Only careful measurement can verify if the boundary is truly Gaussian in shape. If the boundary is asymmetrical, then the square root of the second moment of the concentration distribution can be used to obtain an "equivalent boundary" position (Chervenka, 1973; Schachman, 1959):

$$x_e^2 = x_p^2 - (2/c_p) \int_{x_m}^{x_p} cr \cdot dr \qquad (17)$$

where x_e is the equivalent boundary position and x_p is a position in the plateau region with concentration c_p. Values of x_e can then be used in Equation (1) to calculate the weight-average sedimentation coefficient of the solute.

A useful means of detecting small amounts of heterogeneity is to monitor the concentration of protein in the sedimenting boundary. The concentration will decline with time because of radial dilution (Chervenka, 1973; Svedberg and Pederson, 1940):

$$c_0 = c_t(x/x_m)^2 \qquad (18)$$

where c_t is the measured concentration in the plateau region at time t when the boundary is at x. For a homogeneous solute, the corrected concentration should equal c_0 at all times. A decrease in the corrected concentration indicates that minor components are sedimenting away from the boundary of the main component.

Individual boundaries may become resolved if noninteracting species with sufficiently different sedimentation rates are present. In these paucidisperse systems, the concentrations of the individual species may be readily calculated. An additional (and frequently ignored) consideration in these cases is the Johnston–Ogston effect, which arises because of the concentration dependence of the sedimenting species (Johnston and Ogston, 1946; Schachman, 1959; Van Holde,

1971). The net result of this effect for a pair of sedimenting boundaries is to reduce the true concentration of the rapidly sedimenting boundary and correspondingly to increase the concentration of the slower boundary. As with all cases involving concentration dependence, the Johnston–Ogston effect is minimized by working at low concentrations. The effect is progressively more severe as the concentration dependence of the more rapidly sedimenting species increases or as the sedimentation coefficients of the species match progressively more closely (Schachman, 1959). In extreme cases, the faster-moving boundary may be completely eliminated (Trautman, 1964).

Self-sharpening of boundaries (Section 5.2) can readily obscure visual signs of heterogeneity and claims of homogeneity on the basis of visual inspection of spikelike boundaries should always be treated with caution. For example, spectrin–actin–protein 4.1 junctional complexes isolated from the red cell cytoskeleton appear to migrate as a symmetrical, spikelike boundary in the ultracentrifuge with a sedimentation coefficient of ~20 S (Morris and Kaufman, 1989). The main source of heterogeneity in these large, highly expanded complexes is the number of spectrin molecules tethered to the actin–protein 4.1 core. According to electron microscopy studies this number may vary from five to eight (Liu *et al.*, 1987). Because of self-sharpening, it is very unlikely that any of these species would be resolved by sedimentation velocity despite claims to the contrary (Morris and Kaufman, 1989).

A detailed analysis of the sedimenting boundary, however, can provide quantitative information on the heterogeneity even in the presence of self-sharpening (Baldwin, 1957a; Mächtle, 1988, 1992; Schachman, 1959; Stafford, 1992a,b; Williams and Saunders, 1954). The situation is complicated by the process of diffusion, which, like heterogeneity, tends to broaden the boundary with time. With very large particles, however, diffusion may be negligible during the time of the experiment, and boundary-broadening may be considered only in terms of heterogeneity. This approach was taken with neurofilament–microtubule association (Runge *et al.*, 1981), which leads to the formation of very large particles with small diffusion coefficients.

5.7. Self-Associating Systems

The thermodynamics of a self-associating solute can also be analyzed using sedimentation velocity under favorable circumstances. For a reversible self-association the exact shape and number of sedimenting boundaries will depend on the rates of the forward and reverse chemical reactions, the magnitude of the centrifugal field, the frictional (and, therefore, the diffusional) coefficient of each species, and the concentration dependence of each species (Fujita, 1975; Schachman, 1959).

In an extreme case, the rates of forward and reverse chemical reactions may be so slow that no significant interconversion of oligomers occurs during the sedimentation velocity experiment. For spectrin, the distribution of oligomers established by incubation at temperatures above 30°C is apparently kinetically trapped at 4°C, allowing the boundaries from individual oligomers to be separated by sedimentation velocity (Ungewickell and Gratzer, 1978). Determination of the concentration changes across the heterodimer and tetramer boundaries allowed a reasonably accurate estimate of the equilibrium constant for tetramer formation to be calculated ($\sim 1 \times 10^6$ M^{-1}; Ungewickell and Gratzer, 1978). On the other hand, the use of relatively insensitive schlieren optics together with single-sector cells (which require the experimenter to estimate the position of the baseline) prevented detection of larger oligomers of spectrin (Ungewickell and Gratzer, 1978). The use of double-sector cells with schlieren or absorbance optics clearly showed the ability of spectrin to self-associate extensively beyond the tetramer (Morris and Ralston, 1984).

Most of the interest in the study of self-associations using sedimentation velocity has centered on reactions that are so rapid with respect to the rate of

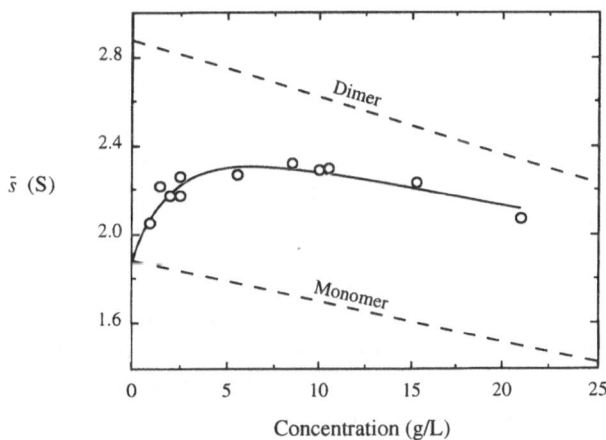

FIGURE 5. Plot of weight-average sedimentation coefficients, $\bar{s}_{20,w}$, as function of total solute concentration obtained from sedimentation velocity experiments performed on β-lactoglobulin (Townend et al., 1960). The coefficients were calculated according to Equations (1), (2), and (17). The dashed lines represent the concentration dependence of sedimentation coefficients of the monomer and dimer of β-lactoglobulin. Note that the initial rise in the values of $\bar{s}_{20,w}$ caused by dimerization is eventually overwhelmed by this concentration dependence resulting in a maximum in the plot. The solid line represents the best fit of a monomer–dimer model using nonlinear regression (Gilbert and Gilbert, 1973). The returned value of the equilibrium constant was 0.5 liter/g. (Redrawn from Gilbert and Gilbert, 1973).

sedimentation that the system is close to equilibrium in all parts of the solution column throughout the experiment (Adams, 1992; Cann and Goad, 1973; Gilbert and Gilbert, 1973; Holloway and Cox, 1974). In these cases, the weight-average sedimentation coefficient [determined through the application of Equations (17) and (1)] will increase with increasing loading concentration of solute. In the case of the dimerization of β-lactoglobulin, a maximum appears in the plot of s versus c_0 because of the competing effect of concentration dependence (Figure 5; Gilbert and Gilbert, 1973).

In general, keeping the parameters involved in analyzing even simple self-associations to a manageable number requires several assumptions to be made. In particular, a good description of the effects of diffusion on the self-association during sedimentation is lacking and, unless independent information on the frictional coefficient of each species is known, a single frictional coefficient is usually assumed (Gilbert and Gilbert, 1973; Holloway and Cox, 1974). Nevertheless, even with these simplifying assumptions, good estimates of the equilibrium constants of monomer–dimer and monomer–n-mer associations are possible (Gilbert and Gilbert, 1973; Holloway and Cox, 1974; Luther et al., 1986; Winzor et al., 1977).

6. DIFFUSION

The diffusion coefficient, D, of a protein can be determined by measuring the time-dependent spreading of the boundary. These measurements are usually made on a stationary boundary formed by gently layering solvent over a solution of the solute at low rotor speeds with the use of a "synthetic boundary" centerpiece (Chervenka, 1973). Layering takes place through a small pore connecting the solvent and solute channels. As the rotor speed increases, pressure forces the extra fluid in the solvent channel to flow through the pore and over the solution. A sharp boundary is formed and the rate of spreading can then be recorded (Chervenka, 1973). It is then a simple process to determine the value of D (in units of cm^2/sec) from the slope of a plot of the square of the effective width of the boundary versus time (Chervenka, 1973).

As with the sedimentation coefficient (Section 5.1), D is affected by absolute temperature, T, solvent composition, and the concentration of protein. The value of the diffusion coefficient at infinite dilution, $D°$, can be obtained by extrapolating values from a series of experiments to infinite dilution. $D_{20,w}$ can be calculated from D in analogous fashion to $s_{20,w}$ (Chervenka, 1973):

$$D_{20,w} = D(293.2/T)(\eta_{T,b}/\eta_{20,w}) \tag{19}$$

The value of the diffusion coefficient, along with that of the sedimentation coefficient, can be used to calculate the molecular weight of the protein [Equation (6)]. In addition, the diffusion coefficient can be used to calculate the frictional coefficient [Equation (7)] thereby allowing an appraisal of the hydrodynamic properties of the protein (see Section 5.4).

Diffusion coefficients can, in principle, also be obtained from the spreading of the sedimenting boundary. Thus, s and D can be obtained from a single experiment using either the analytical (Baldwin, 1957b; Svedberg and Pederson, 1940) or preparative (Minton, 1989; Muramatsu and Minton, 1987) ultracentrifuges. However, corrections must be applied if the values of s and D, over the concentration range used, display significant concentration dependence (Baldwin, 1957b; Fujita, 1975). The diffusion coefficient is very sensitive to heterogeneity; for a heterogeneous solute, the apparent value of the coefficient determined from the spreading of a sedimenting boundary will be larger than that obtained from a stationary boundary (Schachman, 1959).

The extinction coefficient of a protein can be determined from a stationary diffusing boundary using a combination of refractometric optics and absorbance. The refractive increment of most proteins falls into a narrow range so that, using interference optics, every 4.1 fringes crossed in the diffusing boundary is equivalent to 1 g/liter of protein (Babul and Stellwagen, 1969). The absorbance of the protein can be determined either prior to the ultracentrifuge experiment using a spectrophotometer or during the experiment using the absorbance optical system.

7. SEDIMENTATION EQUILIBRIUM

Sedimentation equilibrium allows anhydrous molecular weights to be obtained over an extremely wide range (350 to several million) without recourse to calibration or assumptions about the shape of the solute (Haschemeyer and Bowers, 1970; Schachman, 1959; Teller, 1973; Van Holde, 1967, 1971). It is still the preferred method for quantifying the thermodynamic properties of self-associations. The advantages of the method are that a wide concentration range can be examined in a single experiment and data can be obtained at chemical equilibrium without perturbing the equilibrium. Small amounts of heterogeneity, which can seriously affect the values of thermodynamic parameters, can be detected at very low levels (Johnson *et al.*, 1981; Ralston and Morris, 1992; Teller, 1973). More recently, sedimentation equilibrium has been used with great success to analyze heterogeneous interactions (Howlett, 1987, 1992a; Lewis and Youle, 1986; Minton, 1989, 1990).

7.1. Theory

The equations describing a single, ideal solute at sedimentation equilibrium are (Haschemeyer and Bowers, 1970; Van Holde, 1967)

$$d\ln[c(r)]/dr^2 = M(1 - \bar{v}\rho)\omega^2/2RT \tag{20}$$

$$c(r) = c(r_m)\exp[M(1 - \bar{v}\rho)\omega^2(r^2 - r_m^2)/2RT] \tag{21}$$

Equation (21) is the integrated form of Equation (20) between a reference position r_m (usually taken as the meniscus) and any radial position r.

If more than one ideal, nonassociating component is present, then Equations (20) and (21) become (Haschemeyer and Bowers, 1970)

$$d\ln[c_i(r)]/(dr^2) = M_i(1 - \bar{v}\rho)\omega^2/2RT \tag{22}$$

$$c(r) = \sum_{i=1}^{n} c_i(r) = \sum_{i=1}^{n} c_i(r_m)\exp[M_i(1 - \bar{v}\rho)\omega^2(r^2 - r_m^2)/2RT] \tag{23}$$

where the subscript i refers to one of the n components. Equation (23) shows that the distribution of total concentration is simply the sum of exponentials of the individual components.

Rearranging and integrating Equation (22) between the meniscus, r_m, and the base, r_b, of the cell, and taking into account the conservation of mass gives (Schachman, 1959; Van Holde, 1971)

$$[c(r_b) - c(r_m)]/c_0 = M_w[(1 - \bar{v}\rho)\omega^2(r_b^2 - r_m^2)/2RT] \tag{24}$$

where

$$M_w = \sum_{i=1}^{n} c_i M_i \bigg/ \sum_{i=1}^{n} c_i \tag{25}$$

The initial loading concentration, c_0, is the sum of the loading concentrations of the individual components, c_i. For simplicity, the value of $\bar{v}$ is taken to be the same for all components. M_w defines the weight-average molecular weight obtained for all components at sedimentation equilibrium (Van Holde, 1971). By this treatment the value of M_w that is determined is that for the original loading concentration.

Similar treatments also allow the z-average molecular weight, M_z, to be determined (especially with the use of schlieren optics), where M_z is defined as (Van Holde, 1971)

$$M_z = \sum_{i=1}^{n} c_i M_i^2 \bigg/ \sum_{i=1}^{n} c_i M_i \tag{26}$$

The reader is referred to Creeth and Knight (1965) for a detailed discussion on the use of different optical systems for the determination of particular average molecular weights.

7.2. Meniscus Depletion, Intermediate-Speed and Low-Speed Sedimentation Equilibrium

Since the data required for analysis of a sedimentation equilibrium experiment are $c(r)$ versus r, absorbance and interference optics are used more commonly than schlieren optics. With absorbance measurements, concentrations can be determined directly through the use of the extinction coefficient for the protein. However, for interference optics only the concentration relative to a reference point can be measured. Therefore, the absolute concentration at some convenient point in the cell must be determined (see Creeth and Pain, 1967). The simplest way is to sediment the solution at a speed high enough to deplete the meniscus region of solute to a level below the limits of measurement. In this way, all concentrations are measured relative to the zero concentration at the meniscus. This is known as the meniscus depletion or high-speed method (Yphantis, 1964).

There are several advantages to this method including the fact that concentrations close to zero can be measured (Teller, 1973; Yphantis, 1964). The use of short solution columns (1–3 mm) and relatively high speeds means that the time to sedimentation equilibrium is greatly reduced (Correia and Yphantis, 1992; Yphantis, 1960, 1964), and specially constructed centerpieces allow three or four different loading concentrations of solute to be centrifuged simultaneously (Ansevin et al., 1970; Yphantis, 1960, 1964) using only a small amount of material (Table I).

An alternative to the meniscus depletion method is the intermediate-speed method in which the centrifuge is run at a slightly lower speed so that a maximum concentration of 0.04 mg/ml of protein is present at the meniscus. If interference optics are used, the exact concentration can be determined by an iterative process (Teller, 1973). The advantage of this method is that the lower speed reduces the steepness of the concentration gradient near the bottom of the cell thereby facilitating the accuracy and precision of measurements in this region.

For the low-speed method, the speed of the centrifuge is again reduced and a more gentle concentration gradient is now formed throughout the solution column (Chervenka, 1973; Schachman, 1959; Teller, 1973). With this method a high concentration of protein may be present at the meniscus. This may exclude the use of absorbance optics but the method is well suited to interference optics provided a suitable method is available to determine the meniscus concentration (LaBar, 1965; Richards and Schachman, 1959). Alternatively, the meniscus concentration need not be measured at all: Instead, Equation (24) can be used and the value of $[c(r_b) - c(r_m)]$ measured by interference optics, since it is simply the difference in concentration between the ends of the solution column. The initial loading concentration, c_0, can be measured from a synthetic boundary experiment (Section 6).

Sedimentation equilibrium experiments can also be performed in the preparative ultracentrifuge using either fixed angle or swinging bucket rotors (Howlett, 1987, 1992a; Minton, 1989; Pollet, 1985). The $c(r)$ versus r data can be obtained by optically scanning the tubes (Attri and Minton, 1983, 1984, 1986) or by fractionating the solution column using various methods (Howlett, 1992a; Minton, 1989). The use of preparative ultracentrifuge requires that a small amount of inert solute (<5 mg/ml) be added, which stabilizes the concentration gradient of the protein. If optical scanning is to be used, sucrose (Attri and Minton, 1983) or dextran (Howlett, 1992a) can be used for this purpose. If the solution column is to be fractionated and the concentration gradient determined from a radiolabel or some specific biochemical assay, then metrizamide (Ralston *et al.*, 1989) or even another protein, such as serum albumin, can be used to stabilize the gradient (Howlett, 1987, 1992a). In all cases the added solute must not interact with the protein or affect its behavior. This may not always be the case for albumin (Howlett, 1987, 1992a).

7.3. Calculating Molecular Weight

Molecular weight can be calculated in a number of ways from sedimentation equilibrium data. The accuracy and precision of the results is generally superior to that obtained from sedimentation velocity since the thermodynamic basis of sedimentation equilibrium means that few, if any, assumptions are needed to analyze the data. Thus, it is preferable to use sedimentation equilibrium where possible.

7.3.1. Using $\ln[c(r)]$ versus r^2

The molecular weight of a single, ideal protein can be calculated from the slope of a plot of $\ln[c(r)]$ versus r^2, provided $\bar{v}$ and ρ are known [Equation (20)]. However, the plot will curve upwards if more than one sedimenting component is present or if there is protein–protein association. It is frequently difficult to detect this curvature, particularly when the density of data points is low, yet the calculated value of the molecular weight can be affected significantly (Munk and Cox, 1972). Downward curvature in the $\ln[c(r)]$ versus r^2 plot resulting from solution nonideality (see Section 7.6) may also be difficult to observe and may also lead to inaccurate values for molecular weight (Munk and Cox, 1972).

7.3.2. Using Weight-Average Molecular Weights

Differentiation of the $\ln[c(r)]$ versus r^2 data at each radial position yields the apparent weight-average molecular weight, M_w^{app} corresponding to total concentration $c(r)$ (Kim *et al.*, 1977; Teller, 1973). In this way a plot of point-average M_w^{app} values versus $c(r)$ can be constructed. For a single, ideal component, the data points in a plot of M_w^{app} versus $c(r)$ will be distributed around a straight line with a slope of zero. If the solution is nonideal, values of M_w^{app} will decrease with increasing concentration of protein. In this case, extrapolation of the data to zero concentration will yield a value of the molecular weight independent of nonideality (Teller, 1973). If more than one species is present, M_w^{app} will increase with total concentration and, in the meniscus depletion experiments, the molecular weight of the smallest species present can be obtained by extrapolating to zero concentration (Teller, 1973; Yphantis, 1964).

M_w^{app} versus $c(r)$ curves often allow a determination of the molecular weight of a pure solute to an accuracy and precision of 1–2% (Correia and Yphantis, 1992; Van Holde, 1971; Yphantis, 1964). In addition, qualitative analysis of the plot of M_w^{app} versus $c(r)$ can provide useful information on heterogeneity, nonideality, and associations (see below).

7.3.3. Using Very Short Solution Columns

Rapid determination of molecular weights can be obtained by means of low-speed experiments with very short (<1 mm) solution columns (Correia and Yphantis, 1992; Yphantis, 1960). Using these short columns, a value of M_z^{app} can be obtained which pertains to the initial loading concentration of solute, c_0 (Correia and Yphantis, 1992; Yphantis, 1960). If the centrifuge speed is sufficiently low such that the value of $M(1 - \bar{v}\rho)\omega^2/RT < 1.04$ cm^{-2} for all species present, then a value for M_w^{app} can be calculated. Again, this value refers to the initial loading concentration of protein (Correia and Yphantis, 1992; Yphantis, 1960).

The short-column method was used to determine the molecular weight of ribonuclease A to within 1% and bovine serum albumin to within about 3% (Correia and Yphantis, 1992). A distinct advantage of this method is the extremely short times required to reach equilibrium: For example, a solution of *E. coli* ribosomes (molecular weight $> 1.4 \times 10^6$) reaches equilibrium in a 0.75-mm column in 4–5 hr instead of 70 hr in a 3-mm column. However, the accuracy and precision of the method, compared with the meniscus depletion method in 3-mm solution columns, is compromised by the need to perform a double (numerical) differentiation to obtain M_z^{app} and by the need to determine the loading concentration of the protein to calculate M_w^{app} (Correia and Yphantis, 1992; Yphantis, 1960). In addition, values of M_z^{app} are very sensitive to the presence of small amounts of high-molecular-weight contaminants [Equation (26)].

The above methods are commonly used for calculating the molecular weights of proteins from sedimentation equilibrium data. However, the differentiation required to obtain M_w^{app} and M_z^{app} values often imposes sinusoidal error on the data (e.g., Johnson and Yphantis, 1978; Laue *et al.*, 1984). Attempts to minimize this error have relied on various methods of smoothing and differentiation (Horbett and Teller, 1972; Roark and Yphantis, 1969; Teller, 1973). Smoothing has been claimed to be essential for obtaining accurate point-average molecular weights when large numbers of closely spaced data points are measured (Teller, 1973). However, there is no guarantee that systematic error will be minimized effectively. In addition, the Gaussian error on the data is largely destroyed. Conversion of the $c(r)$ data to a logarithmic function also distorts the original Gaussian error.

7.3.4. Direct Fitting of the $c(r)$ versus r^2 Data

The reduced accuracy and precision associated with the use of logarithms, differentiation, and smoothing can be eliminated by directly fitting Equation (21) or (23) (Johnson *et al.*, 1981). The advantage of using these equations is that no manipulation of the concentration values is required. The exponential nature of these functions requires that a nonlinear least-squares curve-fitting program be used to obtain an estimate of the molecular weight. Several nonlinear regression programs have been specifically tailored to this purpose and are available from certain authors (Johnson *et al.*, 1981; Lewis, 1992).

7.3.5. Detergent-Solubilized Proteins

The molecular weight of detergent-solubilized proteins can be determined without knowledge of the amount of detergent bound to the protein moiety (Reynolds and McCaslin, 1985; Reynolds and Tanford, 1976). The method relies on an expansion of the expression for the buoyant molecular weight of the protein–detergent complex, $M_C(1 - \phi'\rho)$ (Casassa and Eisenberg, 1964; Reynolds and Tanford, 1976):

$$M_c(1 - \phi'\rho) = M_P(1 - \bar{v}_P\rho) + M_{Det}(1 - \bar{v}_{Det}\rho)$$
$$= M_P[(1 - \bar{v}_P\rho) + \delta_{Det}(1 - \bar{v}_{Det}\rho)] \qquad (27)$$

where M_P is the molecular weight of the protein moiety, M_{Det} is the sum of the molecular weights of the bound detergent molecules, $\bar{v}_P$ and $\bar{v}_{Det}$ are the partial specific volumes of the protein moiety and bound detergent, respectively, and δ_{Det} is the number of grams of detergent bound per gram of protein moiety. Since the value of $\bar{v}_{Det}$ is known for most detergents, D_2O can be used to adjust ρ to $1/\bar{v}_{Det}$. The term $\delta_{Det}(1 - \bar{v}_{Det}\rho)$ then equals zero irrespective of the amount of bound detergent, thereby allowing the molecular weight of the protein moiety alone to be determined by any one of the methods described above.

There are two variants of this procedure. The first relies on measuring δ_{Det} directly by, for example, equilibrium dialysis or chromatographic procedures using radiolabeled detergents (Garrigos *et al.*, 1993; Robinson and Tanford, 1975; Tanford and Reynolds, 1976). This method was used to determine the molecular weight of the Thy-1 glycoprotein solubilized from rat thymocyte membranes (Kuchel *et al.*, 1978). The value agreed with that obtained when the contribution from detergent was masked by adjusting the solution density to $1/\bar{v}_{Det}$ (Kuchel *et al.*, 1978).

The second variation relies on determining ϕ' from sedimentation velocity experiments performed in buffer prepared in both H_2O and D_2O (or $D_2^{18}O$) (Edelstein and Schachman, 1967; Howlett, 1987, 1992b):

$$\phi' = (k - P)/(\rho_D - \rho_H P) \tag{28}$$

where P is the experimentally determined ratio σ_D/σ_H; $\sigma = M_C(1 - \phi')\omega^2/2RT$. k is the ratio of the molecular weight of the protein–detergent complex in the deuterated solvent (where exchangeable protons are replaced by deuterium) compared with that in H_2O. The value of δ_{Det} and M_P can then be calculated using (Howlett, 1987, 1992b)

$$\phi' = (\bar{v}_P + \delta_{Det}\bar{v}_{Det})/(1 + \delta_{Det}) \tag{29}$$

$$M_P = M_C/(1 + \delta_{Det}) \tag{30}$$

The value of k will be close to $1.0155 \times$ mole fraction of D_2O for the protein moiety alone (Edelstein and Schachman, 1967) but the presence of detergent will alter it slightly. A simple iterative procedure can be used to obtain k for the protein–detergent complex (Howlett, 1987). This method has been used to determine a molecular mass of 175 kDa for the protein–detergent complex of γ-glutamyltransferase solubilized from rat kidney microsomes using Triton X-100. A value of 79 kDa was then calculated for the protein moiety alone (Howlett *et al.*, 1982a). A similar approach was used to determine a molecular mass of 310 kDa for the insulin receptor solubilized from cultured human lymphoblastoid cells (Pollet *et al.*, 1981).

7.4.   Tests for Heterogeneity

Heterogeneity arises from many sources including the presence of contaminating proteins, proteolytic breakdown products, and irreversibly formed aggregates. Heterogeneity may be generated during the course of the sedimentation equilibrium experiment, an important consideration for unstable solutes. The use of the meniscus depletion method and short solution columns can reduce the time to equilibrium substantially (see Sections 7.2 and 7.3).

The presence of even very small levels of heterogeneity can significantly affect estimates of the values of thermodynamic parameters irrespective of the technique used. Unlike other techniques, sedimentation equilibrium allows heterogeneity to be detected at very low levels for the time at which data were collected (Johnson et al., 1981; Munk and Cox, 1972; Ralston and Morris, 1992; Teller, 1973).

Equation (24) shows that a measure of the concentration distribution of a species at sedimentation equilibrium is given by $[c_i(r_b) - c_i(r_m)]$. The value of this expression will depend on the loading concentration of the species and the operating conditions of the experiment; i.e., the angular velocity, the distance of the meniscus from the axis of rotation, and length of the solution column ($r_b - r_m$). As an example, take a solution containing two noninteracting species. If two experiments are performed in which the loading concentrations or operating conditions differ, the change in $[c_i(r_b) - c_i(r_m)]$ between the two experiments will be greater for the species with the larger buoyant molecular weight. Furthermore, a selected total concentration of protein that is common to both experiments will contain different proportions of the two species. Thus, the value of M_w^{app} or M_z^{app} at the selected total concentration in the two experiments will be different and plots of M_w^{app} versus $c(r)$ for the two experiments will diverge. In some cases, the type of heterogeneity can be determined by inspection of the plots (Teller, 1973).

The use of M_w^{app} versus $c(r)$ plots to test for heterogeneity is somewhat compromised by the data manipulation required to obtain the point-average values of M_w^{app}. Figure 6a shows M_w^{app} versus $c(r)$ curves for three loading concentrations of spectrin centrifuged to sedimentation equilibrium in a meniscus depletion experiment. The curves do not overlap, suggesting a small amount of heterogeneity but this may be a result of the error imposed on the M_w^{app} values by the differentiation process.

Two other methods are available for testing for the presence of heterogeneity which do not impose further error on the original $c(r)$ versus r data. The first of these is the omega function, $\Omega(r)$ (Milthorpe et al., 1975):

$$\Omega(r) = c(r)\exp[\phi M(r_F^2 - r^2)]/c(r_F) \tag{31}$$

where $c(r_F)$ is a reference concentration and $\phi = (1 - \bar{v}\rho)\omega^2/2RT$. For a sample centrifuged at different loading concentrations or under different operating conditions, $c(r_F)$ will necessarily occur at a different radial position, r_F, in the various solution columns. By definition, $\Omega(r) = 1$ at $c(r_F)$. If the sample is heterogeneous, the data from the various channels will not overlap over their common concentration range when a common value of $c(r_F)$ is used (Milthorpe et al., 1975). As with plots of M_w^{app} versus $c(r)$, inspection of the diverging curves can, in some cases, reveal the type of heterogeneity present (Ralston and Morris,

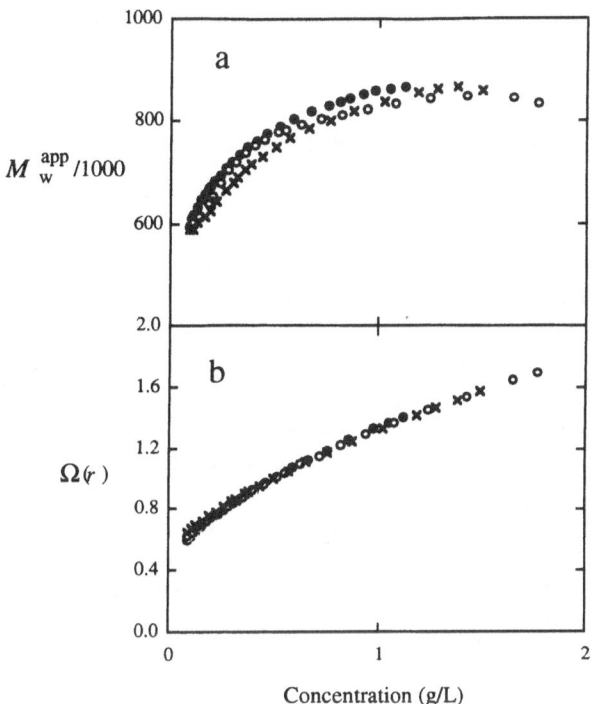

FIGURE 6. (a) Plot of apparent weight-average molecular weight, M_w^{app}, against total concentration for three loading concentrations of the red cell cytoskeletal protein, spectrin, centrifuged to sedimentation equilibrium in a meniscus depletion experiment (see Section 7.2). The protein self-associates in an indefinite manner leading to a sharp initial increase in M_w^{app} values with increasing concentration (Section 7.5). Strong nonideality resulting from the expansion and charge of the protein eventually overwhelms the effects of self-association resulting in a maximum in the data (Section 7.6). The data for the three loading concentrations of protein do not fully overlap over their common concentration range suggesting heterogeneity in the sample. (b) Data from the same experiment plotted as the Omega function, $\Omega(r)$, against total concentration. Data from the three channels overlap very closely, indicating that the sample is homogeneous. The lack of overlap seen in (a) is probably related to error introduced in the differentiation process required to obtain M_w^{app} values. [See Equation (20) and Section 7.3.1.] The three loading concentrations used were 0.2, 0.7, and 2.0 g/liter.

1992). The definition of the omega function implies that the value of M must be known, but an incorrect value of M will affect the data from all channels equally. Thus, any reasonable value of M will be adequate for testing heterogeneity.

Figure 6 shows that the apparent heterogeneity observed for a spectrin sample when the data from three channels are plotted as M_w^{app} versus $c(r)$ is not detected when the data are replotted as $\Omega(r)$ versus $c(r)$, indicating that the diverging $M_w{}^{app}$ distributions result from the differentiation process used to obtain the point-average $M_w{}^{app}$ values.

Both the molecular weight and omega distributions can be assessed for heterogeneity in a more objective fashion than visual inspection: Nonlinear regression can be used to fit a suitable model to the data obtained from each channel. (The model may incorporate nonideality and self-association; see Sections 7.5 to 7.7.) If the values of the parameters returned from each channel are the same, within error, then the system is taken to be homogeneous, but if the values are significantly different, then the system is heterogeneous. This method has been used on $\Omega(r)$ and $M_w{}^{app}$ distributions to check heterogeneity in spectrin samples (Morris and Ralston, 1984, 1989). The limitation in this approach is that the errors quoted for the parameters are the standard errors which are true reflections of the symmetrical error space of linear models but are only estimates of the asymmetrical error space associated with nonlinear models (Johnson *et al.*, 1981). This limitation has been overcome by Johnson *et al.* (1981) who fit suitable models to individual data sets directly through Equation (23). The fits include an estimation of the asymmetry of the error space associated with each parameter of the nonlinear model. The method does have several limitations (Johnson *et al.*, 1981; Morris and Ralston, 1985) but it nevertheless provides an objective test for detecting very small levels of heterogeneity (Correia and Yphantis, 1992; Johnson *et al.*, 1981). A similar approach could be used for omega distributions, but not for $M_w{}^{app}$ distributions since the error space for $M_w{}^{app}$ values has already been distorted.

7.5. A Description of Protein Self-Association

Self-associations can be described in the molar scale either by sequential reactions or parallel reactions.

Sequential:

$$A_1 + A_i \rightleftharpoons A_{i,i+1} \qquad K_{i,i+1} = [A_{i+1}]/([A_i][A_1]) \qquad (32)$$

Parallel:

$$iA_1 \rightleftharpoons A_i \qquad K_i = [A_i]/[A_1]^i \qquad (33)$$

The two descriptions are equivalent since

$$K_i = K_{1,2}K_{2,3} \ldots K_{i-1,i} \qquad (34)$$

A_1 is the protomer or effective monomer, that is, the smallest species participating in the self-association. For sedimentation equilibrium experiments on spectrin in 100 mM NaCl using interference optics, the protomer is the heterodimer, not its component monomer chains, since there is no detectable dissociation of the heterodimer over the concentration range that can be measured with these optics (Kam et al., 1977; Morris and Ralston, 1984, 1985). At higher concentrations of salt, however, detectable dissociation of the heterodimer occurs so that the protomers become the individual monomer chains (Cole and Ralston, 1992).

Self-associations can also be described in the g/liter scale by replacing the sequential molar equilibrium constants, $K_{i-1,i}$, with the g/liter analogues, $k_{i-i,i}$. Conversion between the two sets of units is as follows:

$$K_{i-1,i} = k_{i-1,i}(i - 1)M_1/i \qquad (35)$$

$$k_i = k_{1,2}k_{2,3}\ldots k_{i-1,i} \qquad (36)$$

where M_1 is the molar weight of the protomer. The g/liter sequential equilibrium constants, $k_{i-1,i}$, should not be confused with the molar equilibrium constants expressed in the g/liter scale, $K'_{i-1,i}$. The latter are obtained by dividing the $K_{i-1,i}$ by the molar weight of the protomer:

$$k'_{i-1,i} = K_{i-1,i}/M_1 \qquad (37)$$

For an ideal self-associating system at chemical equilibrium, the total concentration, c_T, is given by

$$c_T = \Sigma k_i c_1^i \qquad (38)$$

Substitution of Equation (38) into Equation (23) shows that for a self-associating system at sedimentation equilibrium:

$$c(r) = \sum_{i=1}^{n} k_i[c_1(r_m)]^i \exp[iM_1(1 - \bar{v}\rho)\omega^2(r^2 - r_m^2)/2RT] \qquad (39)$$

where the approximation is made that $\bar{v}$ describes the partial specific volume of all oligomeric species. Note that sedimentation equilibrium will not be achieved until chemical equilibrium has been reached at every point in the solution column (Adams, 1969).

Equations (38) and (39) show that for a solute capable of participating in reversible self-association reactions, the individual oligomeric species do not act independently but as a single, thermodynamic component. The concentration of each oligomeric species is dependent *only* on the total solute concentration and not on the loading concentration of protein or the operating conditions of the centrifuge. Thus, plots of M_w^{app} versus $c(r)$ or $\Omega(r)$ versus $c(r)$ from individual channels will overlap over their common concentration range (Figure 6). If the system is heterogeneous, including the presence of irreversibly formed aggregaters and protomer incapable of participating in self-association reactions, the curves from individual channels will diverge.

7.6. Nonideality

Nonideality results from the fact that real molecules occupy space; each macromolecule occupies a volume in solution from which other macromolecules are excluded. The excluded volume is effectively increased if the macromolecules carry an electrostatic charge which increases the mutual separation of the molecules through electrostatic repulsion (Kim et al., 1977; Roark and Yphantis, 1971; Teller, 1973; Van Holde, 1971; Wills and Winzor, 1992). Nonideality is also greater for expanded or asymmetric molecules, for which the excluded volume is increased. Although there has been a tendency simply to ignore the effects of nonideality, in general its effects cannot be ignored, even for globular proteins at low concentrations (Correia and Yphantis, 1992).

The nonideality can be expressed in terms of the thermodynamic activity, a, which depends on the concentration, c, and an "activity coefficient," y:

$$a = cy \tag{40}$$

The value of y, and hence of a, depends on the concentration of the macromolecule and a measure of the mutual exclusion between molecules. For a homogeneous solution of a single type of macromolecule, the activity coefficient can be expressed as a Taylor's series in the concentration (Fujita, 1975):

$$\ln y = BMc + CM^2c^2 + \cdots \tag{41}$$

where B and C are known respectively as the second and third "virial coefficients," and have a strict meaning in statistical mechanics (Wills and Winzor, 1992). When the degree of nonideality is relatively small, it is convenient to truncate the above series at the first term:

$$\ln y \approx BMc \tag{42}$$

The effect of nonideality in sedimentation experiments is to reduce the apparent molecular weight as calculated through application of Equation (20):

$$M^{\text{app}} = M/(1 + c\, d\ln y/dc) \approx M/(1 + BMc) \tag{43}$$

In the case of a macromolecular solute capable of self-association, nonideality has two consequences: First, the sedimentation of each of the oligomeric species will be affected by nonideality. This leads to a reduction in the apparent weight-average molecular weight compared with the ideal weight-average molecular weight. Second, the association reaction itself will be influenced; the *apparent* equilibrium constant, k_i^{app}, based on the concentrations of the various species, will be enhanced with increased "crowding" of the solution (Minton, 1983).

The *thermodynamic* equilibrium constant, k_i, can be written in terms of the activities of the various species:

$$k_i = c_i y_i / (c_1 y_1)^i = (c_i/c_1^i)(y_i/y_1^i) = k_i^{app} (y_i/y_1^i) \qquad (44)$$

In general, the ratio (y_i/y_1^i) will not be unity, and in these cases quantitative assessment of nonideality requires knowledge of the various activity coefficients, y_i. The values of y_i, in turn, depend on the concentrations of all species, as well as on the volumes that each excludes from all of the others. This problem is intractable without knowledge of the detailed particle geometries, and even then iterative methods are required (Wills *et al.*, 1980).

Adams and Fujita (1963) provided a simple means of overcoming this problem in many cases: Over a limited range of concentration, where neglect of all but the second virial coefficient was justifiable, the assumption could be made that the second virial coefficient of each species in a self-association reaction had the same value (in units of liters mol g^{-2}). With this assumption, only a single value of B is required to characterize the association reaction, making the problem tractable in a computational sense. The measured, apparent molecular weights and the true molecular weights can then be interconverted (Adams and Fujita, 1963; Fujita, 1975; Kim *et al.*, 1977):

$$M_w^{app} = M_w/(1 + BM_w c_T) \qquad (45)$$

While association reactions result in an increase in weight-average molecular weight with concentration, the effect of nonideality is to reduce the apparent molecular weight. Thus, a decrease in the measured, or apparent, molecular weight for a pure, nonassociating solute, or a maximum in the molecular weight distribution for a self-associating solute is indicative of nonideality (Figure 6a).

In addition, implicit in the Adams–Fujita approximation is that the equilibrium constant is unaffected by nonideality, since

$$y_i = \exp(BM_i c) = \exp(iBM_1 c) \qquad (46)$$

whence

$$y_i/y_1^i = \exp(iBM_1 c)/[\exp(BM_1 c)]^i = 1 \qquad (47)$$

The Adams–Fujita approximation has been shown to be most closely approached for end-to-end association of rodlike particles or for highly charged globular particles (Ogston and Winzor, 1975).

It must be stressed that the Adams–Fujita approximation is only valid over a limited concentration range; the approximation progressively fails with increasing total solute concentration. Furthermore, at higher concentrations, the higher virial coefficients become increasingly important (Roark and Yphantis, 1971).

Chatelier and Minton (1987) have examined nonideal effects in the self-association of globular proteins, and have suggested an empirical relationship that appears to be valid over a wide concentration range:

$$M_w = M_w^{app} \exp (7.86 f_v) \tag{48}$$

where f_v is the volume fraction of solution that is occupied by solute.

Wills and Winzor (1992) have described a method in which the logarithm of the monomer activity can be expressed as a polynomial in the *total* concentration of self-associating solute [a relationship reminiscent of Equation (41)]. The coefficients of this polynomial can in turn be expressed as various combinations of the equilibrium constants and excluded volume terms. This approach avoids the iterative calculations normally required to determine the distributions of concentration and activity of the various oligomers throughout the solution column.

7.7. Analysis of Self-Associating Systems

A variety of methods are available for analyzing self-association systems at sedimentation equilibrium (Adams *et al.*, 1978; Johnson *et al.*, 1981; Kim *et al.*, 1977; Milthorpe *et al.*, 1975; Morris and Ralston, 1985; Teller, 1973). Each method requires that a model be selected. For example, the model selected might be a discrete self-association such as a monomer–dimer or a monomer–n-mer self-association. Or the model may be some form of indefinite self-association in which sequential addition of protomers to an oligomer can continue without limit (Adams *et al.*, 1978; Johnson *et al.*, 1981). Fitting the models depends on expressing the concentration of protomer in terms of the total concentration of protein and the equilibrium constant(s) describing each step in the reaction. Nonideality should be included in the form of one or more parameters that can be quantified in the fitting process (see Section 7.6). A summary of some of the methods follows.

7.7.1. Using the Molecular Weight Distribution

Self-association models have traditionally been fitted to M_w versus $c(r)$ distributions based on the following general equation (Adams *et al.*, 1978; Jeffrey, 1966; Kim *et al.*, 1977):

$$M_1/M_w = (c/c_1)(dc_1/dc) \qquad (49)$$

For example, for a monomer–dimer reaction, the following equations apply:

$$c_1 = (-1 + \alpha)/2k_2 \qquad (50)$$

$$M_w = 2M_1\alpha/(1 + \alpha) \qquad (51)$$

where $\alpha = (1 + 4k_2c_T)^{1/2}$. A fit of this model to a molecular weight distribution using nonlinear regression techniques would return a value for the equilibrium constant, k_2. The appropriateness of the model would then be checked by examining the size of the standard error in k_2, the value of the sum of squares of the residuals to the fit, and the distribution of the residuals (Cleland, 1967; Teller, 1973). Nonideality can be readily incorporated through the use of Equations (45) or (48). This method has been used to *reject* the nonideal monomer–dimer model as being a suitable description of the self-association of spectrin (Morris and Ralston, 1984).

The method requires knowledge of M_1. If M_1 is not known it can be estimated by extrapolating the molecular weight distribution to zero concentration or else M_1 can be incorporated as an additional parameter in the fitting process. Alternatively, a series of fits using different values of M_1 can be made to the molecular weight distribution. The best value of M_1 would be that which minimized the sums of squares of residuals of the fit (Teller, 1973).

The use of molecular weight distributions has the advantage that some models can be rejected on the basis of inspection. For example, if the values of M_w rise above the molecular weight of the dimer, then a monomer–dimer model can be rejected.

Several methods rely on combinations of weight-average, z-average, and number-average molecular weights. These methods allow the choice of models to be directed and the values of the parameters to be calculated by simple graphical procedures (Adams *et al.*, 1978; Beckerdite *et al.*, 1980; Hoagland and Teller, 1969; Kim *et al.*, 1977; Roark and Yphantis, 1969). In some cases, the manipulation of the data is so extensive that the correct model for describing the self-association can be rejected, or estimates of the values of the parameters can be significantly in error (Morris, 1985). These linearization procedures may provide a general guide to the choice of self-association model but otherwise are best avoided.

7.7.2. Using the Omega Distribution

$\Omega(r)$ versus $c(r)$ distributions can also be used to estimate the values of the thermodynamic parameters of a self-associating solute (Milthorpe *et al.*, 1975; Morris and Ralston, 1985; Ralston and Morris, 1992). In the original formulation of the method, the omega distribution [Equation (31)], extrapolated to zero concentration, allowed the activity of the monomer, a_1 (r_F), at the reference to be determined (Milthorpe et al., 1975):

$$\lim_{c(r) \to 0} \Omega(r) \equiv \Omega_0 = a_1(r_F)/c(r_F) \tag{52}$$

From the value of $a_1(r_F)$, the activity of the monomer at each radial position for which the total concentration has been measured can be calculated:

$$a_1(r) = a_1(r_F)\exp[\phi M_1(r^2 - r_F^2)] \tag{53}$$

A plot of $a_1(r)$ versus $c(r)$ could then be constructed and fitted with various reaction models (Milthorpe et al., 1975). The extrapolation does, however, become progressively more difficult as the strength of the self-association increases and, in addition, the plot if $a_1(r)$ versus $c(r)$ is an artificially smoothed distribution (Morris and Ralston, 1985). A better alternative is to directly employ the omega function written in terms of the thermodynamic activity of the protomer (Milthorpe et al., 1975):

$$\Omega(r) = a_1(r_F)c(r)/[a_1(r)c(r_F)] \tag{54}$$

Equation (54) shows that the omega function itself is uniquely dependent on the total solute concentration and can be written in terms of the self-association parameters and nonideality. Therefore, a direct fit of a self-association model to the omega function avoids extrapolation and its associated problems (Morris and Ralston, 1985).

Direct fits to the omega function have been used to determine the mode of self-association of spectrin at 30°C in 100 mM NaCl (Morris and Ralston, 1985, 1989). The precision of the data allowed several discrete and indefinite self-association models to be rejected, including models developed on the basis of results obtained from quasithermodynamic techniques such as native gel electrophoresis (Shahbakhti and Gratzer, 1986). The results indicated that spectrin self-associates in an indefinite fashion by the sequential addition of the heterodimers either by the so-called SEK III mechanism (in which all but the first molar equilibrium constant have the same value) or an AK I mechanism (in which the $K_{i-1,1}$ progressively decrease) (Adams et al., 1978). Analysis of sedimentation equilibrium experiments over a range of pH, temperature, and concentration showed that the SEK III model provided the better description (Ralston, 1991). These experiments also allowed the enthalpy, entropy, and heat capacity of the self-association to be calculated and comments to be made on the type of bonds involved (Cole and Ralston, 1992; Ralston, 1991).

7.7.3. Using the Concentration Distribution

Self-association models can be fitted directly to Equation (23). The general equation for discrete self-associations is given by Johnson $et\ al.$ (1981). The functional form, F, for a monomer–n-mer self-association, including the Adams–Fujita approximation for nonideality, is (Johnson $et\ al.$, 1981):

$$Y_i \approx F = \sum_{i=1}^{n} c_1(r_m)^i \exp[(\ln k_i)/i + \sigma_1(r^2 - r_m^2)/2 - B(F - \delta c)]^i + \delta c \qquad (55)$$

where Y_i is the measured absorbance or fringe displacement at radial position r, $\sigma_1 = 2M_1\phi$, and δc is an additive constant that allows for systematic error in the evaluation of absolute concentrations from the relative concentrations obtained from interference optics. The values to be returned for this model are the reaction model parameters (k_n, B, σ_1) and the additional parameters δc and $c_1(r_m)$. When several sets of data are analyzed simultaneously, the nonlinear, least-squares regression analysis will return a set of values, with confidence limits, for the reaction model parameters appropriate for the combined sets of data, but separate values of δc and $c_1(r_m)$ for each channel.

7.8. Heterogeneous Associations

Associations between dissimilar molecules are of wider biological relevance than self-association reactions. However, heterogeneous associations have been more difficult to study. Nevertheless, thermodynamic information on such systems can still be obtained through the use of sedimentation equilibrium. In these cases, all species cannot be assumed to have the same partial specific volume, absorbance coefficient, or refractive index increment. The assessment of nonideality in such cases is also more difficult, since the composition of the solution (and hence the activity coefficients) varies with radial distance, and it has been common to restrict analysis to conditions that approximate ideal behavior; i.e., where associations are relatively strong, and consequently the concentrations of individual species can be kept relatively low.

Under these pseudoideal conditions, each species may be considered to be distributed as

$$c_i = c_{i,m} \exp[M_i(1 - \bar{v}\rho)\omega^2(r^2 - r_m^2)/2RT] \qquad (56)$$

where $c_{i,m}$ is the meniscus concentration of species i. The relevant apparent equilibrium constants may be estimated from the concentrations. This type of analysis has been applied to the association of the subunits of ricin (Lewis and Youle, 1986).

It is also possible to analyze heterogeneous associations through the derived data of apparent weight-average molecular weight of each thermodynamic component as a function of the concentration of that component, provided that it is possible to distinguish between the individual components (Minton, 1990). For example, an interaction between oxyhemoglobin and the tetramer of band 3 was determined by selectively measuring the concentration distribution of oxy-hemoglobin at 505 nm (Schuck and Schubert, 1991). The binding of ankyrin to band 3 tetramers was facilitated by labeling ankyrin with fluorescein isothiocyanate, allowing its concentration distribution to be measured at 495 nm (Mulzer *et al.*, 1990).

The preparative ultracentrifuge can also be used to great advantage in analyzing heterogeneous interactions if individual components are, for example, radiolabeled (Howlett, 1987, 1992a; Howlett *et al.*, 1982b, 1983; Minton, 1989). This approach has been used to quantify the strength of the interaction between calmodulin and spectrin (Husain *et al.*, 1984).

In spite of difficulties in the analysis of nonideal heterogeneous systems, it has proven possible to incorporate nonideality through iterative processes that take account of the change in composition with radial distance (Nichol and Winzor, 1976). A variation of the omega analysis has been used to analyze the nonideal interaction between lysozyme and ovalbumin (Jeffrey *et al.*, 1979; Nichol *et al.*, 1976).

8. REFERENCES

Adams, E. T., 1969, Chemically reacting systems of the type A + B ↔ AB. I. Sedimentation equilibrium of ideal solutions, *Ann. N.Y. Acad. Sci.* **164**:226–244.

Adams, E. T., 1992, Sedimentation coefficients of self-associating species. Analysis of monomer-dimer-n-mer associations and some indefinite self-associations, in: *Analytical Ultracentrifugation in Biochemistry and Polymer Science* (S. E. Harding, A. J. Rowe, and J. C. Horton, eds.), pp. 407–427, Royal Society of Chemistry, Cambridge.

Adams, E. T., and Fujita, H., 1963, Sedimentation equilibrium in reacting systems, in: *Ultracentrifugation in Theory and Experiment* (J. W. Williams, ed.), pp. 119–129, Academic Press, New York.

Adams, E. T., and Lewis, M. S., 1968, Sedimentation equilibrium in reacting systems. VI. Studies with β-lactoglobulin A, *Biochemistry* **7**:1044–1053.

Adams, E. T., Tang, L.-H., Sarquis, J. L., Barlow, G. H., and Norman, W. M., 1978, Self-association in protein solutions, in: *Physical Aspects of Protein Interactions* (N. Catsimpoolas, ed.), pp. 1–55, Elsevier, Amsterdam.

Ansevin, A. T., Roark, D. E., and Yphantis, D. A., 1970, Improved ultracentrifuge cells for high-speed sedimentation equilibrium studies with interference optics, *Anal. Biochem.* **34**:237–261.

Attri, A. K., and Minton, A. P., 1983, An automated method for determination of the molecular weight of macromolecules via sedimentation equilibrium in a preparative ultracentrifuge, *Anal. Biochem.* **133**:142–152.

Attri, A. K., and Minton, A. P., 1984, An automated method for the determination of the sedimentation coefficient of macromolecules using a preparative ultracentrifuge, *Anal. Biochem.* **136:**407–415.

Attri, A. K., and Minton, A. P., 1986, Technique and apparatus for automated fractionation of the contents of small centrifuge tubes: Application to analytical ultracentrifugation, *Anal. Biochem.* **152:**319–328.

Attri, A. K., Lewis, M. S., and Korn, E. D., 1991, The formation of actin oligomers studied by analytical ultracentrifugation, *J. Biol. Chem.* **266:**6815–6824.

Aune, K. C., and Timasheff, S. N., 1971, Dimerization of α-chymotrypsin. I. pH dependence in the acid region, *Biochemistry* **10:**1609–1617.

Aune, K. C., Goldsmith, L. C., and Timasheff, S. N., 1971, Dimerization of α-chymotrypsin. II. Ionic strength and temperature dependence, *Biochemistry* **10:**1617–1622.

Babul, J., and Stellwagen, E., 1969, Measurement of protein concentration with interference optics, *Anal. Biochem.* **28:**216–221.

Baldwin, R. L., 1957a, Boundary spreading in sedimentation velocity experiments. 4. Measurement of the standard deviation of a sedimentation coefficient distribution: Application to bovine albumin and β-lactoglobulin, *Biochem. J.* **65:**490–502.

Baldwin, R. L., 1957b, Boundary spreading in sedimentation velocity experiments. 5. Measurement of the diffusion coefficient of bovine albumin by Fujita's equation, *Biochem. J.* **65:**503–512.

Beckerdite, J. M., Wan, C. C., and Adams, E. T., 1980, Analysis of various indefinite self-associations of the AK type, *Biophys. Chem.* **12:**199–214.

Behlke, J., 1992, Disaggregation of the membrane protein P450 by detergents, in: *Analytical Ultracentrifugation in Biochemistry and Polymer Science* (S. E. Harding, A. J. Rowe, and J. C. Horton, eds.), pp. 484–494, Royal Society of Chemistry, Cambridge.

Bloomfield, V. A., and Lim, T. K., 1978, Quasi-elastic laser light scattering, *Methods Enzymol.* **48:**415–494.

Cann, J. R., and Goad, W. B., 1973, Measurements of protein interactions mediated by small molecules using sedimentation velocity, *Methods Enzymol.* **27:**296–306.

Cantor, C. R., and Schimmel, P. R., 1980, *Biophysical Chemistry,* Freeman, San Francisco.

Casassa, E. F., and Eisenberg, H., 1964, Thermodynamic analysis of multicomponent systems, *Adv. Protein Chem.* **19:**287–395.

Chatelier, R. C., and Minton, A. P., 1987, Sedimentation equilibrium in macromolecular solutions of arbitrary concentration. I. Self-associating proteins, *Biopolymers* **26:**507 524.

Chervenka, C. H., 1973, *A Manual of Methods for the Analytical Ultracentrifuge,* Spinco Division, Beckman Instruments, Palo Alto.

Clarke, S., 1975, The size and detergent binding of membrane proteins, *J. Biol. Chem.* **250:**5459–5469.

Cleland, W. W., 1967, The statistical analysis of enzyme kinetic data, *Adv. Enzymol. Relat. Areas Mol. Biol.* **21:**1–32.

Cohn, E. J., and Edsall, J. T., 1943, *Proteins, Amino Acids and Peptides as Ions and Dipolar Ions,* pp. 370–381, Reinhold, New York.

Cole, N., and Ralston, G. B., 1992, The effects of ionic strength on the self-association of human spectrin, *Biochim. Biophys. Acta* **1121:**23–30.

Correia, J. J., and Yphantis, D. A., 1992, Equilibrium sedimentation in short solution columns, in: *Analytical Ultracentrifugation in Biochemistry and Polymer Science* (S. E. Harding, A. J. Rowe, and J. C. Horton, eds.), pp. 231–252, Royal Society of Chemistry, Cambridge.

Creeth, J. M., and Knight, C. G., 1965, On the estimation of the shape of macromolecules from sedimentation and viscosity measurements, *Biochim. Biophys. Acta* **102:**549–558.

Creeth, J. M., and Pain, R. H., 1967, The determination of molecular weights of biological macromolecules by ultracentrifuge methods, *Prog. Biophys. Mol. Biol.* **17:**217–287.

DeRosier, D. J., Munk, P., and Cox, D. J., 1972, Automatic measurement of interference photographs from the ultracentrifuge, *Anal. Biochem.* **50**:139–153.

Dunbar, J. C., and Ralston, G. B., 1981, Hydrodynamic characterization of the heterodimer of spectrin, *Biochim. Biophys. Acta* **667**:177–184.

Durchschlag, H., 1986, Specific volumes of biological macromolecules and some other molecules of biological interest, in: *Thermodynamic Data for Biochemistry and Biotechnology* (H.-J., Hinz, ed.), pp. 45–128, Springer-Verlag, Berlin.

Edelstein, S. J., and Schachman, H. K., 1967, The simultaneous determination of partial specific volumes and molecular weights with microgram quantities, *J. Biol. Chem.* **242**:306–311.

Flörke, R.-R., Klein, H. W., and Rienauer, H., 1990, Structural requirements for signal transduction of the insulin receptor, *Eur. J. Biochem.* **191**:473–482.

Fujita, H., 1975, *Foundations of Ultracentrifugal Analysis*, Wiley, New York.

García de la Torre, J., 1992, Sedimentation coefficients of complex biological particles, in: *Analytical Ultracentrifugation in Biochemistry and Polymer Science* (S. E. Harding, A. J. Rowe, and J. C. Horton, eds.), pp. 333–345, Royal Society of Chemistry, Cambridge.

García de la Torre, J., and Bloomfield, V. A., 1978, Hydrodynamic properties of macromolecular complexes. IV. Intrinsic viscosity theory, with applications to once-broken rods and multisubunit proteins, *Biopolymers* **17**:1605–1627.

Garrigos, M., Centeno, F., Deschamps, S., Moller, J. V., and le Maire, M., 1993, Sedimentation equilibrium of detergent-solubilized membrane proteins in the preparative ultracentrifuge, *Anal. Biochem.* **208**:306–310.

Giebeler, R., 1992, The Optima XL-A: A new analytical ultracentrifuge with a novel precision absorption optical system, in: *Analytical Ultracentrifugation in Biochemistry and Polymer Science* (S. E. Harding, A. J. Rowe, and J. C. Horton, eds.), pp. 16–25, Royal Society of Chemistry, Cambridge.

Gilbert, L. M., and Gilbert, G. A., 1973, Sedimentation velocity measurement of protein association, *Methods Enzymol.* **27**:273–296.

Harding, S. E., and Rowe, A. J., 1983, Modeling biological macromolecules in solution. II. The general tri-axial ellipsoid, *Biopolymers* **22**:1813–1829.

Harding, S. E., and Rowe, A. J., 1984, Erratum. Modeling biological macromolecules in solution. II. The general tri-axial ellipsoid, *Biopolymers* **23**:843.

Harding, S. E., Horton, J. C., and Morgan, P. J., 1992, A FORTRAN program for the model independent molecular weight analysis of macromolecules using low speed or high speed sedimentation equilibrium, in: *Analytical Ultracentrifugation in Biochemistry and Polymer Science* (S. E. Harding, A. J. Rowe, and J. C. Horton, eds.), pp. 275–294, Royal Society of Chemistry, Cambridge.

Haschemeyer, R. H., and Bowers, W. F., 1970, Exponential analysis of concentration or concentration difference data for discrete molecular weight distribution in sedimentation equilibrium, *Biochemistry* **9**:435–445.

Hoagland, V. D., and Teller, D. C., 1969, Influence of substrates on the dissociation of rabbit muscle D-glyceraldehyde 3-phosphate dehydrogenase, *Biochemistry* **8**:594–602.

Holloway, R. R., and Cox, D. J., 1974, Computer simulation of sedimentation in the ultracentrifuge. VII. Solutes undergoing indefinite self-association, *Arch. Biochem. Biophys.* **160**:595–602.

Holzenburg, A., Engel, A., Kessler, R., Manz, H. J., Lustig, A., and Aebi, U., 1989, Rapid isolation of OmpF porin–LPS complexes suitable for structure–function studies, *Biochemistry* **28**:4187–4193.

Horbett, T. A., and Teller, D. C., 1972, An experimental study of baseline reproducibility and its effect on high-speed sedimentation equilibrium data, *Anal. Biochem.* **45**:86–99.

Howlett, G. J., 1987, Air-driven ultracentrifuge for sedimentation equilibrium and binding studies, *Methods Enzymol.* **150**:447–463.

Howlett, G. J., 1992a, The preparative ultracentrifuge as an analytical tool, in: *Analytical Ultracentrifugation in Biochemistry and Polymer Science* (S. E. Harding, A. J. Rowe, and J. C. Horton, eds.), pp. 32–48, Royal Society of Chemistry, Cambridge.

Howlett, G. J., 1992b, Sedimentation analysis of membrane proteins, in: *Analytical Ultracentrifugation in Biochemistry and Polymer Science* (S. E. Harding, A. J. Rowe, and J. C. Horton, eds.), pp. 470–483, Royal Society of Chemistry, Cambridge.

Howlett, G. J., Birch, H., Dickson, P. W., and Schreiber, G., 1982a, Determination of the molecular weight of detergent-solubilized enzymes by sedimentation equilibrium in an air-driven ultracentrifuge, *Biochem. Biophys. Res. Commun.* 105:895–901.

Howlett, G. J., Dickson, P. W., Birch, H., and Schreiber, G., 1982b, Studies on ^{125}I-labelled proteins in rat plasma using an air-driven ultracentrifuge: Protein–protein interactions and nonideality, *Arch. Biochem. Biophys.* 215:309–318.

Howlett, G. J., Roche, P. J., and Schreiber, G., 1983, Protein–protein interactions: Analysis of the interaction of concanavalin A with serum glycoproteins by sedimentation equilibrium using an air-driven ultracentrifuge, *Arch. Biochem. Biophys.* 224:178–185.

Hudson, G. S., Howlett, G. J., and Davidson, B. E., 1983, The binding of tyrosine and NAD$^+$ to chorismate mutase/prephenate dehydrogenase from *Escherichia coli* K12 and the effects of these ligands on the activity and self-association of the enzyme. Analysis in terms of a model, *J. Biol. Chem.* 258:3114–3210.

Husain, A., Howlett, G. J., and Sawyer, W. H., 1984, The interaction of calmodulin with human and avian spectrin, *Biochem. Biophys. Res. Commun.* 122:1194–1200.

Hutchison, K. A., and Fox, I. H., 1989, Purification and characterization of the adenosine A$_2$-like binding site from human placental membranes, *J. Biol. Chem.* 264:19898–19903.

Jacques, Y., Le Mauff, B., Godard, A., Naulet, J., Concino, M., Marsh, H., Ip, S., and Soulillou, J.-P., 1990, Biochemical study of a recombinant soluble interleukin-2 receptor. Evidence for a homodimer structure, *J. Biol. Chem.* 265:20252–20258.

Jeffrey, P. D., 1966, An equilibrium ultracentrifuge study of the self-association of bovine insulin, *Biochemistry* 5:489–498.

Jeffrey, P. D., Nichol, L. W., and Teasdale, R. D., 1979, Studies of macromolecular heterogeneous associations involving cross-linking: A re-examination of the ovalbumin–lysozyme system, *Biophys. Chem.* 10:379–387.

Johnson, M. L., and Yphantis, D. A., 1978, Subunit association and heterogeneity of *Limulus polyphemus* hemocyanin, *Biochemistry* 17:1448–1455.

Johnson, M. L., Correia, J. J., Yphantis, D. A., and Halvorson, H. R., 1981, Analysis of data from the analytical ultracentrifuge by nonlinear least-squares techniques, *Biophys. J.* 36:575–588.

Johnston, J. P., and Ogston, A. G., 1946, A boundary anomaly found in the ultracentrifugal sedimentation of mixtures, *Trans. Faraday Soc.* 42:789–799.

Kam, Z., Josephs, R., Eisenberg, H., and Gratzer, W. B., 1977, Structural study of spectrin from human erythrocyte membrane, *Biochemistry* 16:5568–5572.

Kim, H., Deonier, R. C., and Williams, J. W., 1977, The investigation of self-association reactions by equilibrium ultracentrifugation, *Chem. Rev.* 77:659–690.

Kirschner, M. W., and Schachman, H. K., 1971, Conformational changes in proteins as measured by difference sedimentation studies. II. Effect of stereospecific ligands on the catalytic subunit of aspartate transcarbamylase, *Biochemistry* 10:1919–1926.

Kratky, O., Leopold, H., and Stabinger, H., 1973, The determination of the partial specific volume of proteins by the mechanical oscillator technique, *Methods Enzymol.* 27:98–110.

Kuchel, P. W., Campbell, D. G., Barclay, A. N., and Williams, A. F., 1978, Molecular weights of the Thy-1 glycoproteins from rat thymus and brain in the presence and absence of deoxycholate, *Biochem. J.* 169:411–417.

Kumosinski, T. F., and Pessen, H., 1985, Structural determination of hydrodynamic measurements of proteins in solution through correlations with X-ray data, *Methods Enzymol.* **117:**154–182.

Kuntz, I. D., 1971, Hydration of macromolecules. III. Hydration of polypeptides, *J. Am. Chem. Soc.* **93:**514–516.

LaBar, F. E., 1965, A procedure for molecular weight measurements: Application to chymotrypsinogen A, *Proc. Natl. Acad. Sci. USA* **54:**31–36.

Laue, T. M., 1992, On-line data acquisition and analysis from the Rayleigh interferometer, in: *Analytical Ultracentrifugation in Biochemistry and Polymer Science* (S. E. Harding, A. J. Rowe, and J. C. Horton, eds.), pp. 63–89, Royal Society of Chemistry, Cambridge.

Laue, T. M., Johnson, A. E., Esmon, C. T., and Yphantis, D. A., 1984, Structure of bovine blood coagulation factor Va. Determination of the subunit associations, molecular weights, and asymmetries by analytical ultracentrifugation, *Biochemistry* **23:**1339–1348.

Laue, T. M., Shah, B. D., Ridgeway, T. M., and Pelletier, S. L., 1992, Computer-aided interpretation of analytical sedimentation data for proteins, in: *Analytical Ultracentrifugation in Biochemistry and Polymer Science* (S. E. Harding, A. J. Rowe, and J. C. Horton, eds.), pp. 90–125, Royal Society of Chemistry, Cambridge.

Lavrenko, P. N., Linow, K. J., and Gornitz, E., 1992, The concentration dependence of the sedimentation coefficient of some polysaccharides in very dilute solution, in: *Analytical Ultracentrifugation in Biochemistry and Polymer Science* (S. E. Harding, A. J. Rowe, and J. C. Horton, eds.), pp. 517–531, Royal Society of Chemistry, Cambridge.

Lee, J. C., and Timasheff, S. N., 1979, The calculation of partial specific volumes of proteins in 6M guanidine hydrochloride, *Methods Enzymol.* **61:**49–57.

Lee, J. C., Gekko, K., and Timasheff, S. N., 1979, Measurement of preferential solvent interactions by densimetric techniques, *Methods Enzymol.* **61:**26–49.

Lewis, M. S., 1992, Data acquisition and analysis systems for the absorption optical system of the analytical ultracentrifuge, in: *Analytical Ultracentrifugation in Biochemistry and Polymer Science* (S. E. Harding, A. J. Rowe, and J. C. Horton, eds.), pp. 126–137, Royal Society of Chemistry, Cambridge.

Lewis, M. S., and Youle, R. J., 1986, Ricin subunit association, *J. Biol. Chem.* **261:**11571–11577.

Lindenthal, S., and Schubert, D., 1991, Monomeric erythrocyte band 3 transports anions, *Proc. Natl. Acad. Sci. USA* **88:**6540–6544.

Liu, S.-C., Derrick, L. H., and Palek, J., 1987, Visualization of the hexagonal lattice in the erythrocyte membrane skeleton, *J. Cell Biol.* **104:**527–536.

Luther, M. A., Cai, G.-Z., and Lee, J. C., 1986, Thermodynamics of dimer and tetramer formations in rabbit muscle phosphofructokinase, *Biochemistry* **25:**7931–7937.

Mächtle, W., 1988, Coupling particle size distribution technique, *Angew. Makromol. Chem.* **162:**35–52.

Mächtle, W., 1992, Analysis of polymer dispersions with an eight-cell-AUC-multiplexer: High resolution particle size distribution and density gradient techniques, in: *Analytical Ultracentrifugation in Biochemistry and Polymer Science* (S. E. Harding, A. J. Rowe, and J. C. Horton, eds.), pp. 147–175, Royal Society of Chemistry, Cambridge.

Makino, S., Woolford, J. L., Tanford, C., and Webster, R. E., 1975, Interaction of deoxycholate and of detergents with the coat protein of bacteriophage f1, *J. Biol. Chem.* **250:**4327–4332.

Mayeux, P., Casadevall, N., Lacombe, C., Muller, O., and Tambourin, P., 1990, Solubilization and hydrodynamic characteristics of the erythropoietin receptor. Evidence for a multimeric complex, *Eur. J. Biochem.* **194:**271–278.

Milthorpe, B. K., Jeffrey, P. D., and Nichol, L. W., 1975, The direct analysis of sedimentation equilibrium results obtained with polymerizing systems, *Biophys. Chem.* **3:**169–176.

Minton, A. P., 1983, The effect of volume occupancy upon the thermodynamic activity of proteins: Some biochemical consequences, *Mol. Cell. Biochem.* **55:**119–140.

Minton, A. P., 1989, Analytical ultracentrifugation with preparative ultracentrifuges, *Anal. Biochem.* **176**:209-216.

Minton, A. P., 1990, Quantitative characterization of reversible molecular associations via analytical centrifugation, *Anal. Biochem.* **190**:1-6.

Morris, M., 1985, The self-association of human spectrin, Ph.D. thesis, University of Sydney.

Morris, M., and Ralston, G. B., 1984, A reappraisal of the self-association of human spectrin, *Biochim. Biophys. Acta* **788**:132-137.

Morris, M., and Ralston, G. B., 1985, Determination of the parameters of self-association by direct fitting of the omega function, *Biophys. Chem.* **23**:49-61.

Morris, M., and Ralston, G. B., 1989, A thermodynamic model for the self-association of human spectrin, *Biochemistry* **28**:8561-8567.

Morris, S. A., and Kaufman, M., 1989, Ultracentrifugal analysis of the junction complexes of the red cell membrane cytoskeletal network: Application to hereditary spherocytosis and metabolically depleted cells. *Blut* **59**:385-389.

Mulzer, K., Kampmann, L., Petrasch, P., and Schubert, D., 1990, Complex associations between protein analyzed by analytical ultracentrifugation: Studies on the erythrocyte membrane proteins band 3 and ankyrin, *Colloid Polym. Sci.* **268**:60-64.

Munk, P., and Cox, D. J., 1972, Sedimentation equilibrium of protein solutions in concentrated guanidinium chloride. Thermodynamic nonideality and protein heterogeneity, *Biochemistry* **11**:687-697.

Muramatsu, N., and Minton, A. P., 1987, An automated method for rapid determination of diffusion coefficients via measurements of boundary spreading, *Anal. Biochem.* **168**:345-351.

Muramatsu, N., and Minton, A. P., 1989, Hidden self-association of proteins, *J. Mol. Recog.* **1**:166-171.

Newell, J. O., and Schachman, H. K., 1990, Amino acid substitutions which stabilize aspartate transcarbamylase in the R state disrupt both homotropic and heterotropic effects, *Biophys. Chem.* **37**:183-196.

Nichol, L. W., and Winzor, D. J., 1976, Allowance for composition dependence of activity coefficients in the analysis of sedimentation equilibrium results obtained with heterogeneously associating systems, *J. Phys. Chem.* **80**:1980-1983.

Nichol, L. W., and Winzor, D. J., 1985, The use of covolume in the estimation of protein axial ratio, *Methods Enzymol.* **117**:182-198.

Nichol, L. W., Jeffrey, P. D., and Milthorpe, B. K., 1976, The sedimentation equilibrium of heterogeneously associating systems and mixtures of noninteracting solutes: Analysis without determination of molecular weight averages, *Biophys. Chem.* **4**:259-267.

Nozaki, Y., Chamberlain, B. K., Webster, R. E., and Tanford, C., 1976a, Evidence for a major conformational change of coat protein in assembly of f1 bacteriophage, *Nature* **259**:335-337.

Nozaki, Y., Schechter, N. M., Reynolds, J. A., and Tanford, C., 1976b, Use of gel chromatography for the determination of the Stokes radii of proteins in the presence and absence of detergents. A reexamination, *Biochemistry* **15**:3884-3890.

Ogston, A. G., and Winzor. D. J., 1975, Treatment of thermodynamic nonideality in equilibrium studies on associating systems, *J. Phys. Chem.* **79**:2496-2500.

Paul, C. H., and Yphantis, D. A., 1972, Pulsed laser interferometry (PLI) in the analytical ultracentrifuge: I. System design, *Anal. Biochem.* **48**:588-604.

Pederson, K. O., 1958, On charge and specific ion effects on sedimentation in the ultracentrifuge, *J. Phys. Chem.* **62**:1282-1290.

Pennica, D., Kohr, W. J., Fendly, B. M., Shire, S. J., Raab, H. E., Borchardt, P. E., Lewis, M., and Goeddel, D. V., 1992, Characterization of a recombinant extracellular domain of the type I tumor necrosis receptor: Evidence for the tumor necrosis factor-α induced receptor aggregation, *Biochemistry* **31**:1134-1141.

Persechini, A., and Rowe, A. J., 1984, Modulation of myosin filament conformation by physiological levels of divalent cations, *J. Mol. Biol.* **172**:23–39.

Pessen, H., and Kumosinki, T. F., 1985, Measurement of protein hydration by various techniques,
Methods Enzymol. **117**:219–255.

Pilz, I., Glatter, O., and Kratky, O., 1979, Small-angle X-ray scattering, *Methods Enzymol.* **61**:148–
249.

Pollet, R. J., 1985, Characterization of macromolecules by sedimentation equilibrium in the air-
turbine ultracentrifuge, *Methods Enzymol.* **117**:3–27.

Pollet, R. J., Haase, B. A., and Standaert, M. L., 1981, Characterization of detergent-solubilized
membrane proteins. Hydrodynamic and sedimentation equilibrium properties of the insulin
receptor of the cultured human lymphoblastoid cell, *J. Biol. Chem.* **256**:12118–12126.

Ralston, G. B., 1991, Temperature and pH-dependence of the self-association of human spectrin,
Biochemistry **30**:4179–4186.

Ralston, G. B., and Morris, M. B., 1992, The use of the omega function for sedimentation equilibri-
um analysis, in: *Analytical Ultracentrifugation in Biochemistry and Polymer Science* (S. E.
Harding, A. J. Rowe, and J. C. Horton, eds.), pp. 253–274, Royal Society of Chemistry,
Cambridge.

Ralston, G. B., Teller, D. C., and Bukowski, T., 1989, The use of metrizamide for stabilizing against
convection in sedimentation equilibrium, *Anal. Biochem.* **178**:198–201.

Reynolds, J. A., and McCaslin, D. R., 1985, Determination of protein molecular weight in complexes with detergent without knowledge of binding, *Methods Enzymol.* **117**:41–53.

Reynolds, J. A., and Tanford, C., 1976, Determination of molecular weight of the protein moiety on
the protein–detergent complex without direct knowledge of detergent binding, *Proc. Natl.
Acad. Sci. USA* **73**:4467–4470.

Richards, E. G., and Schachman, H. K., 1959, Ultracentrifuge studies with Rayleigh interference
optics. I. General applications, *J. Phys. Chem.* **63**:1578–1591.

Richards, J. H., and Richards, E. G., 1974, Light-difference detector for reading Rayleigh fringe
patterns from the ultracentrifuge, *Anal. Biochem.* **62**:523–530.

Roark, D. E., and Yphantis, D. A., 1969, Studies of self-associating systems by equilibrium
ultracentrifugation, *Ann. N.Y. Acad. Sci.* **164**:245–278.

Roark, D. E., and Yphantis, D. A., 1971, Equilibrium centrifugation of nonideal systems. The
Donnan effect in self-associating systems, *Biochemistry* **10**:3241–3249.

Robinson, N. C., and Tanford, C., 1975, The binding of deoxycholate, Triton X-100, sodium
dodecyl sulfate, and phosphatidylcholine vesicles to cytochrome b5, *Biochemistry* **14**:369–378.

Rogers, L. J., and Sykes, G. A., 1990, Conformational changes in *Chondrus crispus* flavodoxin on
dissociation of FMN and reconstitution with flavin analogues, *Biochem. J.* **272**:775–779.

Rowe, A. J., 1977, The concentration dependence of transport processes: A general description
applicable to sedimentation, translational diffusion, and viscosity coefficients of macromolecular solutes, *Biopolymers* **16**:2595–2611.

Rowe, A. J., 1992, The concentration dependence of sedimentation, in: *Analytical Ultracentrifugation in Biochemistry and Polymer Science* (S. E. Harding, A. J. Rowe, and J. C. Horton, eds.),
pp. 394–406, Royal Society of Chemistry, Cambridge.

Rowe, A. J., Jones, S. W., Thomas, D. G., and Harding, S. E., 1992, Methods for off-line analysis
of sedimentation velocity and sedimentation equilibrium patterns, in: *Analytical Ultracentrifugation in Biochemistry and Polymer Science* (S. E. Harding, A. J. Rowe, and J. C.
Horton, eds.), pp. 49–62, Royal Society of Chemistry, Cambridge.

Runge, M. S., Laue, T. M., Yphantis, D. A., Lifsics, M. R., Saito, A., Altin, M., Reinke, K., and
Williams, R. C., 1981, ATP-induced formation of an associated complex between microtubules
and neurofilaments, *Proc. Natl. Acad. Sci. USA* **78**:1431–1435.

Sackett, D. L., Lippoldt, R. E., Gibson, C., and Lewis, M. S., 1989, Easily assembled digital data acquisition system for the analytical ultracentrifuge, *Anal. Biochem.* **180**:319–325.

Sardet, C., Tardieu, A., and Luzzati, V., 1976, Shape and size of bovine rhodopsin: A small-angle X-ray scattering study of a rhodopsin–detergent complex, *J. Mol. Biol.* **105**:383–407.

Schachman, H. K., 1951, Ultracentrifuge studies of tobacco mosaic virus, *J. Am. Chem. Soc.* **73**:4808–4811.

Schachman, H. K., 1959, *Ultracentrifugation in Biochemistry*, Academic Press, New York.

Schachman, H. K., and Edelstein, S. J., 1973, Ultracentrifugal studies with absorption optics and a split-beam photoelectric scanner, *Methods Enzymol.* **27**:3–59.

Schmidt, B., Rappold, W., Rosenbaum, V., Fischer, R., and Riesner, D., 1990, A fluorescence detection system for the analytical ultracentrifuge and its application to proteins, nucleic acids, and viruses, *Colloid Polym. Sci.* **268**:45–54.

Schuck, P., and Schubert, D., 1991, Band 3–haemoglobin associations. The band 3 tetramer is the oxyhaemoglobin binding site, *FEBS Lett.* **293**:81–84.

Shahbakhti, F., and Gratzer, W. B., 1986, Analysis of the self-association of human red cell spectrin, *Biochemistry* **25**:5969–5975.

Siegel, L. M., and Monty, K. J., 1966, Determination of molecular weights and frictional ratios of proteins in impure systems by use of gel filtration and density gradient centrifugation. Application to crude preparations of sulfite and hydroxylamine reductases, *Biochim. Biophys. Acta* **112**:346–362.

Simons, K., Helenius, A., and Garoff, H., 1973, Solubilization of the membrane proteins from Semliki Forest virus with Triton X100, *J. Mol. Biol.* **80**:119–133.

Smigel, M., and Fleischer, S., 1977, Characterization of Triton X-100-solubilized prostaglandin E binding protein of rat liver plasma membrane, *J. Biol. Chem.* **252**:3689–3696.

Smith, B. L., and Agre, P., 1991, Erythrocyte M_r 28,000 transmembrane protein exists as a multisubunit oligomer similar to channel proteins, *J. Biol. Chem.* **266**:6407–6415.

Smith, G. D., and Schachman, H. K., 1973, Effect of D_2O and NAD on the sedimentation properties and structure of glyceraldehyde phosphate dehydrogenase, *Biochemistry* **12**:3789–3801.

Stafford, W. F., 1992a, Methods for obtaining sedimentation coefficient distributions, in: *Analytical Ultracentrifugation in Biochemistry and Polymer Science* (S. E. Harding, A. J. Rowe, and J. C. Horton, eds.), pp. 359–393, Royal Society of Chemistry, Cambridge.

Stafford, W. F., 1992b, Boundary analysis in sedimentation transport experiments: A procedure for obtaining sedimentation coefficient distributions using the time derivative of the concentration profile, *Anal. Biochem.* **203**:295–301.

Steer, C. J., Osborne, J. C., and Kempner, E. S., 1990, Functional and physical molecular size of the chicken hepatic lectin determined by radiation inactivation and sedimentation equilibrium analysis, *J. Biol. Chem.* **265**:3744–3749.

Svedberg, T., and Fåhraeus, R., 1926, A new method for the determination of the molecular weight of the proteins, *J. Am. Chem. Soc.* **48**:430–438.

Svedberg, T., and Pederson, K. O., 1940, *The Ultracentrifuge*, Oxford University Press, London.

Tanford, C., 1961, *Physical Chemistry of Macromolecules*, Wiley, New York.

Tanford, C., and Reynolds, J. A., 1976, Characterization of membrane proteins in detergent solutions, *Biochim. Biophys. Acta* **457**:133–170.

Tellam, R. L., Sculley, M. J., and Nichol, L. W., 1983, The influence of poly(ethylene glycol) 6000 on the properties of skeletal-muscle actin, *Biochem. J.* **213**:651–659.

Teller, D. C., 1967, Modification of the Nikon microcomparator for accurate reading of Rayleigh interference patterns, *Anal. Biochem.* **19**:256–264.

Teller, D. C., 1973, Characterization of proteins by sedimentation equilibrium in the analytical ultracentrifuge, *Methods Enzymol.* **27**:346–441.

Teller, D. C., 1976, Accessible area, packing volumes and interaction surfaces of globular proteins, *Nature* **260**:729–731.

Tirado, M. M., and García de la Torre, J., 1979, Translational friction coefficients of rigid, symmetric top macromolecules. Application to circular cylinders, *J. Chem. Phys.* **71**:2581–2587.

Townend, R., Weinberger, L., and Timasheff, S. N., 1960, Molecular interactions in β-lactoglobulin. IV. The dissociation of β-lactoglobulin below pH 3.5, *J. Am. Chem. Soc.* **82**:3175–3179.

Trautman, R., 1964, Ultracentrifugation, in: *Instrumental Methods of Experimental Biology* (D. W. Newman, ed.), pp. 211–297, Macmillan Co., New York.

Ullrich, A., Bell, J. R., Chen, E. Y., Herrera, R., Petruzzelli, L. M., Dull, T. J., Gray, A., Coussens, L., Liao, Y.-C., Tsubokawa, M., Mason, A., Seeburg, P. H., Grunfield, C., Rosen, O. M., and Ramachandran, J., 1985, Human insulin receptor and its relationship to the tyrosine kinase family of oncogenes, *Nature* **313**:756–761.

Ungewickell, E., and Gratzer, W., 1978, Self-association of human spectrin. A thermodynamic and kinetic study, *Eur. J. Biochem.* **88**:379–385.

Van Holde, K. E., 1967, Sedimentation equilibrium, in: *Fractions,* Vol. 1, pp. 1–10, Beckman Instruments, Palo Alto.

Van Holde, K. E., 1971, *Physical Biochemistry,* Prentice-Hall, Englewood Cliffs, N.J.

Van Holde, K. E., 1975, Sedimentation analysis of proteins, in: *The Proteins,* 3rd ed., Vol. 1 (H. Neurath and R. L. Hill, eds.), pp. 225–291, Academic Press, New York.

Williams, J. W., and Saunders, W. M., 1954, Size distribution analysis in plasma extender systems. II. Dextran, *J. Phys. Chem.* **58**:854–859.

Wills, P. R., and Winzor, D. J., 1992, Thermodynamic non-ideality and sedimentation equilibrium, in: *Analytical Ultracentrifugation in Biochemistry and Polymer Science* (S. E. Harding, A. J. Rowe, and J. C. Horton, eds.), pp. 311–330, Royal Society of Chemistry, Cambridge.

Wills, P. R., Nichol, L. W., and Siezen, R. J., 1980, The indefinite self-association of lysozyme: Consideration of composition-dependent activity coefficients, *Biophys. Chem.* **11**:71–82.

Wills, P. R., Comper, W. D., and Winzor, D. J., 1993, Thermodynamic nonideality in macromolecular solutions: Interpretations of virial coefficients, *Arch. Biochem. Biophys.* **300**:206–212.

Winzor, D. J., Tellam, R., and Nichol, L. W., 1977, Determination of the asymptotic shapes of sedimentation velocity patterns for reversibly polymerizing solutes, *Arch. Biochem. Biophys.* **178**:327–332.

Yeager, M., Schoenborn, B., Engelman, D., Moore, P., and Stryer, L., 1976, Neutron scattering analysis of the shape, molecular weight and amphipathic structure of an intrinsic membrane protein: Rhodopsin, *Biophys. J.* **16**:36a.

Yphantis, D. A., 1960, Rapid determination of molecular weights of peptides and proteins, *Ann. N.Y. Acad. Sci.* **88**:586–601.

Yphantis, D. A., 1964, Equilibrium ultracentrifugation in dilute solutions, *Biochemistry* **3**:297–317.

Yphantis, D. A., 1980, Pulsed laser interferometry in the ultracentrifuge, *Methods Enzymol.* **61**:3–12.

Chapter 3

Monomolecular Layers in the Study of Biomembranes

Rudy A. Demel

1. INTRODUCTION

Monomolecular layers have been of great value in the development of concepts of biomembranes. Phospholipids and other amphipathic membrane compounds with a balanced ration between the hydrophobic and hydrophilic part form an oriented monolayer with the polar portion in contact with the aqueous phase and the hydrocarbons extended above. Membrane lipids form insoluble monolayers since the concentration of lipid in the aqueous phase is essentially negligible.

Early monolayer studies led Gorter and Grendel (1925) to the concept of a phospholipid bilayer as the basic structure of biological membranes. They determined that the area occupied by the lipids spread as a monolayer at the air–water interface was twice the area of the extracted erythrocytes. Within the membrane a large number of different types of lipid molecules are present and each class of lipid is comprised of molecules containing a variety of hydrocarbon chains of different length and degrees of unsaturation (Rouser, 1983). Compression of the monolayer allows an accurate measurement of the molecular area. Proteins can be associated with the membrane in different ways. Membrane proteins can be

Rudy A. Demel   Centre for Biomembranes and Lipid Enzymology, Department of Biochemistry of Membranes, University of Utrecht, 3584 CH Utrecht, The Netherlands.

Subcellular Biochemistry, Volume 23: Physicochemical Methods in the Study of Biomembranes, edited by Herwig J. Hilderson and Gregory B. Ralston. Plenum Press, New York, 1994.

studied in monolayers in pure form as well as in their interaction with lipids. A special advantage of monolayers is the ability to vary the density and pressure of the monolayer which is especially useful in studying lipid-protein interactions and enzymatic reactions.

2. EXPERIMENTAL TECHNIQUES FOR THE STUDY OF MONOMOLECULAR FILMS

2.1. Measurement of Surface Pressure

Since membrane constituents generally form insoluble monolayers, the value of Γ (surface concentration) is simply the amount of material on the surface per unit area. The film pressure π is defined as the difference between the surface tension of the aqueous phase (γ_0) and that of film-covered surface (γ_f):

$$\pi = \gamma_0 - \gamma_f$$

The Wilhelmy slide method which measures the change in γ rather than π directly is capable of measuring low values with a precision of 0.01 mN/m. The plate is partly immersed and the change in tension at constant position of the plate is measured:

$$\pi = - \Delta\gamma = -\Delta F/2w$$

ΔF is the change in force and w is the width of the plate (the thickness of the plate is normally neglected) (Gaines, 1966). As slide materials, glass, platinum, mica, and filter paper have been used. Standard dimensions are a width of 1.96 cm and height of 1 cm. A zero contact angle has to be maintained. Surface roughness decreases the contact angle and increases the reversibility when the contact angle is advanced or receded from its original position. Monolayer deposition can be of influence as deposition is the basis of the Langmuir–Blodgett technique. As a measuring device an electrobalance or even a force transducer can be used. As a spreading solvent, chloroform (containing up to 30 v/v percent methanol) is satisfactory for most lipids.

The film pressure can be directly measured by the horizontal force on a float separating the film from a clean solvent surface, a method originally described by Langmuir in 1917 (Gaines, 1966; Demel, 1974). The advantage of this method is that there is no interference of the contact angle. On the other hand the measuring device is difficult to install in small dishes. When compounds are injected into the subphase the clean solvent surface can become contaminated.

The determination of pressure–area (π–A) isotherms is the most common measurement that is performed on monolayers and from these measurements most of our knowledge of the phase behavior of monolayers has been derived. For this type of measurement convenient trough dimensions are $\sim 30 \times 15 \times 1$ cm. After spreading an exact amount of surface-active agent (for lipid, normally 50 nmol), the film is compressed at a constant rate (e.g., 0.25 nm^2/mol min) and the change in surface pressure is recorded.

Proteins at a concentration of less than 0.06 mg/ml in the appropriate buffer solution can be spread as a film down a water-wetted 5-mm glass rod or roughened glass plate into the interface at a flow rate of less than 0.2 ml/min (Trurnit, 1960). The glass rod is acid-cleaned and positioned in the interface whereas the point of delivery of the protein solution is approximately 0.5 cm above. Spreading of dilute protein solutions will diminish possible aggregation. Retention at the interface can be promoted by spreading onto a high-ionic-strength subphase solution. The spreading efficiency can be examined by using radiolabeled protein which is recovered from the interface.

2.2. Interaction of Solutes Added to the Subphase (Proteins, Toxins, Drugs) with Lipid Monolayers

2.2.1. Changes in Surface Pressure at Constant Surface Area

Peptide incorporation into various lipid monolayers at constant surface area after formation of a phospholipid monolayer at a given initial surface pressure is reflected in an increase in the surface pressure as represented in Figure 1. Peptides that interact with the head groups of phospholipid monolayers—e.g., polylysine and prothrombin (Mayer *et al.*, 1983)–cause only a small (1–3 mN/m) increase in surface pressure. Peptides that insert also into the acyl chain region cause a larger change in monolayer surface pressure. For this type of experiment, even very small dishes can be used (e.g., 2 cm in diameter and a volume of 2 ml). The width of the Wilhelmy slide is then reduced to 0.98 cm. The pressure change caused by the penetrating solute will finally terminate the penetration. Additionally absorbed layers can be formed which can be detected by surface radioactivity only (Pilon *et al.*, 1987).

2.2.2. Changes in Surface Area at Constant Surface Pressure

The penetration of peptide in the lipid monolayer can be compensated by increasing the surface area. The surface pressure should be well above the equilibrium spreading pressure of the protein, otherwise protein films will be formed without possible lipid interaction. Since this system is operated by a computer-controlled movable barrier, it is necessary to ensure that there is no film leakage over longer time periods. The relative differences for different lipid

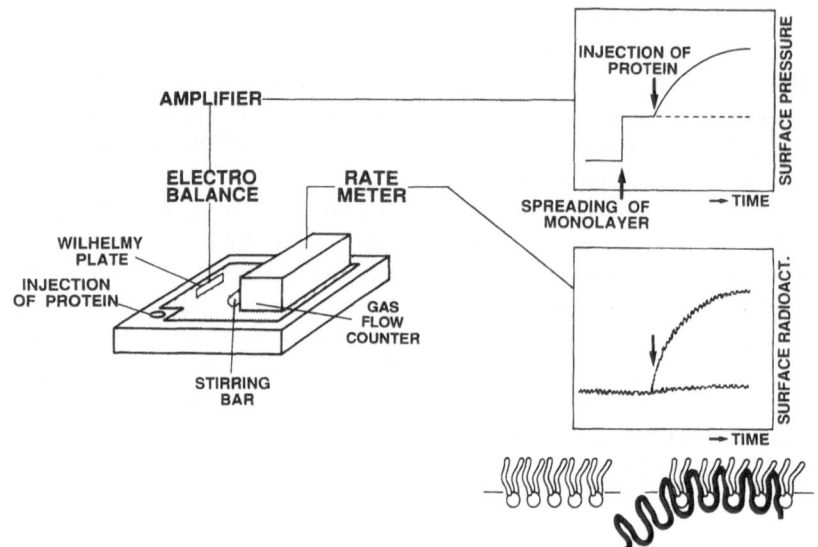

FIGURE 1. Measurement of surface pressure and surface radioactivity at constant surface area, demonstrating the interaction of lipid and protein at the interface.

films are comparable for changes in surface pressure at constant are and changes in surface area at constant surface pressure. However, in the latter case a saturating concentration is not easily obtained (Tamm, 1986; Pilon *et al.*, 1987; Seelig and MacDonald, 1989). With respect to the physiological situation, neither of the two systems can be considered as ideal. A biological membrane will not expand without limit. On the other hand, pressure fluctuation will probably be limited and occurs only locally.

2.3. Surface Radioactivity

The determination of surface radioactivity can be very useful to quantify the amount of solute interacting with the monomolecular layer or material disappearing from the interface. Peptides can be made radioactive either by incorporation of radioactive amino acid residues or by labeling an amino acid chain. ^{14}C-labeling of cysteinyl residues (Tamm, 1986) or lysine residues (Jentoft and Dearborn, 1979), gives specific activities which can be detected easily by means of an atomic gas flow detector with a thin polyethylene window (Demel, 1982). For ^{3}H labels direct detection is difficult and sampling of the subphase or collecting of the monolayer is more appropriate. Surface radioactivity is a very sensitive method for studying lipolytic reactions. Small amounts of radiolabeled substrate lipids are mixed with matrix lipids which are nonhydrolyzable lipids. In this way the sensitivity can be increased since hydrolysis of picomoles of substrate will lead to a drastic change in surface radioactivity. Desportion of the product from the interface could be rate-limiting for free fatty acids. Fatty acid-free BSA added to the subphase at a concentration of 0.1 μM will enhance the desorption of [^{14}C]oleic acid, which follows first-order kinetics with a half time of 3 min. This is at least ten times faster than the rate of triolein hydrolysis by hepatic lipase (Laboda et al., 1986) or lipoprotein lipase (Demel et al., 1982a).

The rate of hydrolysis can be calculated from the slope of the decrease in surface radioactivity. The range of soft β radiation emitted by the ^{14}C-labeled nucleus is approximately 300 μm in water so that ^{14}C-labeled protein near the surface is detected including protein molecules that penetrate into the lipid phase and protein molecules that absorb to the lipid interface by ionic interaction (Pilon et al., 1987). At the same time vesicle binding to the interface can be determined (Demel et al., 1989; Rojo et al., 1991). Background radioactivity can be determined by counting the surface radioactivity when the subphase contains the same total radioactivity of a soluble non-surface-active molecule (e.g., sucrose).

2.4. Surface Potential

Surface potential gives information regarding the orientation of the monolayer constituents as well as subphase interaction. Surface potential measurements are normally performed in conjunction with force–area measurements. The procedures involve the use of an α-emitter. The air ionization leads to sufficient conductivity to measure potential differences by means of a high-impedance dc voltmeter that serves as a zero indicator in a standard potentiometer circuit. The potential of the film-converted surface is normally compared with the film-free surface.

A method of monolayer surface potential measurement described by Mozaffary (1991) makes use of a specially designed vibrating element without the need of a trough electrode (with the subphase only being grounded with a gold wire). In this way a sudden change of -350 mV is observed at the gas–liquid expanded phase transition of various phosphatidylcholine films. A gradual surface potential change which is a function of the fatty acid composition (consequently the physical state and compressibility) of the phosphatidylcholine films can reach a maximum value of -650 mV.

The surface potential of a monolayer is defined as

$$\Delta v = \frac{12\ \pi}{A} \cdot \mu\perp$$

Δv is obviously dependent on the area A of the film. Δv values are frequently converted to μ, the surface dipole moment:

$$\mu\perp = \mu_1 + \mu_2 + \mu_3$$

where μ_1 = dipole moment of the polar groups, μ_2 = contribution of the reorientation of the water molecules, μ_3 = dipole moment of the terminal methyl groups of the fatty acid chains (Gaines, 1966).

The surface potential will reflect changes in film pressure and substrate composition. Considering the different contributions to the surface potential, so far only a qualitative molecular assignment is possible (Adamson, 1990; Taylor et al., 1990).

2.5. Surface Rheology

Visoelastic properties are apparent for different protein monolayers and of lipids in a condensed state. The surface viscosity can be measured by a canal-type viscometer by determining the flow of a film through a narrow canal under a two-dimensional pressure difference. Therefore, this measurement cannot be made at a single surface pressure nor is the shear rate constant (Adamson, 1990). When a liquid–air interface, initially at rest, is disturbed by changing the magnitude of surface area, this disturbance is propagated along the interface by longitudinal waves (Van den Tempel and Lucassen-Reynders, 1983). This is expressed by the following equation:

$$\epsilon = \frac{d\gamma}{d\ln A}$$

where A = surface area, ϵ = dilational elasticity, and γ = surface tension. This equation can be interpreted as follows: the interface with its monolayer possesses a resistance against a change in surface area. In the case of harmonic variations, ϵ can be treated as a complex quantity,

$$\epsilon = |\epsilon| \, e^{j\phi_\epsilon}$$

in which $|\epsilon|$ is the modulus of the dilational elasticity. ϕ_ϵ can be considered as the loss angle which reflects a time lag between surface area deformation and surface tension. Using the asymmetric method of area change at 20°C the $|\epsilon|$ versus π plots for lecithins of increasing chain length show interesting differences correlated with the thermodynamic state of the monolayer (Figure 2). Dibehenylphosphatidylcholine and distearoylphosphatidycholine are in a condensed state and show a high elasticity at low pressures, which increases to $|\epsilon|$ values beyond 300 mN/m at compression. However, at a surface pressure of 55 mN/m there is a sharp drop in $|\epsilon|$ probably indicating collapse and the formation of clusters. A similar behavior is noted for dipalmitoylphosphatidylcholine, except that this lipid is in a liquid state at pressures below 10 mN/m, reflected by a low elasticity. This is also the case for didecanoylphosphatidylcholine and dimyr-

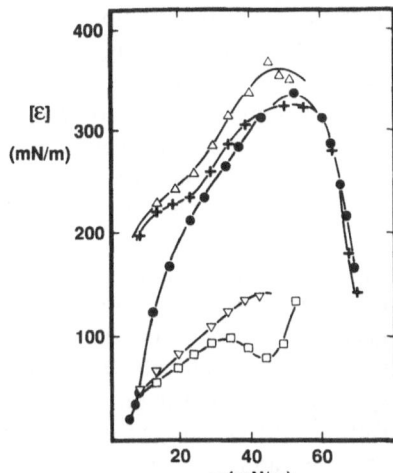

FIGURE 2. Elasticity modulus · ϵ | versus surface pressure (π) of saturated phosphatidylcholines of increasing chain length at 20°C, ω = 0.63 rad/sec. $\triangle$, dibehenoyl (C_{22}); +, distearoyl (C_{18}); ●, dipalmitoyl (C_{16}); □, dimyristoyl (C_{14}); ∇, didecanoyl (C_{10}). (Gieles, P. M. C., Thesis Technical University Eindhoven, The Netherlands, 1987.)

istoylphosphatidylcholine. The latter, which has a phase transition in the bulk phase at 23°C, shows a minimum |ϵ| at a surface pressure of 40 mN/m, reflecting the liquid condensed–liquid expanded phase transition. The phase angles remain quite small indicating little viscoelastic behavior (Gieles, 1987). Also for proteins, measurement of the surface elasticity can provide information on phase transitions and the physical state of the protein monolayer (Tournois *et al.*, 1989). Adsorbed films of β-casein which are flexible show frequency-dependent dilational moduli related to relaxation of the film. In contrast to β-casein, the films of the globular proteins BSA and lysozyme are essentially rigid and have moduli that are independent of frequency (Graham and Phillips, 1980).

2.6. Langmuir–Blodgett Monolayer Films

Many optical properties of monomolecular layers can be studied more easily after transfer to a solid support. This method for transferring monolayers is referred to as the Langmuir–Blodgett (LB) film technique. The solid support (e.g., quartz or glass) is dipped vertically while the surface pressure is kept constant. The decrease in area of the interface divided by that of the substrate covered with the monolayer for each stroke is called the deposition ratio and should ideally be 1. Films can be deposited by downward movement only (x-type), by upward movement only (z-type), or by upward and downward movement (y-type) (Sugi, 1985). The type of deposition depends on the monolayer-forming material and on the conditions of deposition. For lipid and protein monolayers a z-type LB film of one layer is readily formed. CD and FTIR spectra demonstrate the secondary structure of proteins (Oosterlaken-Dijksterhuis *et al.*,

1991b) or the effect of lipid–protein interaction (Cornell *et al.*, 1989; Demel *et al.*, 1990). By comparing the hydrolysis by pancreatic and snake venom phospholipase A_2 of fluorescent monolayers of pyrene-labeled phosphatidylglycerol on solid support and monolayers at the air–water interface, it has been demonstrated that the pressure dependency of the enzymatic reaction is the same in both systems. This means that the surface pressure of the lipid layer is preserved after transfer to the solid support (Thuren *et al.*, 1990).

The LB film technique has been successfully applied to the construction of stable and photoactivatable films of chromatophore membranes and isolated reaction centers from photosynthetic bacteria. Z-type multilayers deposited on glass slides were analyzed by spectrophotometric and redox potentiometric techniques. The results obtained indicate that the in vivo properties of the photosynthetic apparatus in the deposited films are essentially unchanged (Alegria and Dutton, 1991). This means that activity and orientation of integral membrane proteins can be preserved after spreading at the air–water interface and transfer to a solid support. A horizontal lifting method can be used instead of the usual Blodgett technique by lowering a hydrophobically precoated plate in a horizontal position until it touches the monolayer on the water surface. In this way *x*-type layers can be formed (Fukuda *et al.*, 1976; Sugi, 1985). It is suggested that molecular packing and orientation are better preserved by this method of transfer to a solid surface. Horizontal transfer onto hydrophobic carbon-coated grids viewed by high-contrast dark-field electron microscopy exhibits a microporous structure which is related to an incomplete liquid expanded–liquid condensed phase transition of dipalmitoylphosphatidylcholine (Reinhardt-Schlegel *et al.*, 1991).

Monolayers have been used to crystallize proteins on lipid films as was shown for IgE anti-2,6-dinitrophenyl antibody. When a hydrophobic surface (carbon-shadowed electron microscope grid) was passed through the monolayer, a phospholipid monolayer was transferred to this surface with the hydrocarbon tails of the lipid next to the solid surface. This monolayer was subsequently exposed to a solution containing the antibody (Uzgiris, 1987).

The transfer of monolayers to a solid support can also be used for electron microscopy studies as described by Ries *et al.* (1975). A collodion-covered grid is mounted 2 mm beneath the interface. The surface level is rapidly reduced after compression of the monolayer to the collapse pressure. The monolayer deposited on the grid shadowed with Pt/C shows horizontally stacked layers in the collapse phase of condensed monolayers of dipalmitoylphosphatidylcholine and long vertical ridges in mixed monolayers that contain up to 20% of a second component. Fluid monolayers show no detectable structures at collapse (Van Liempd *et al.*, 1987).

A technique to form supported phospholipid bilayers by fusion of small unilamellar vesicles to supported phospholipid monolayers on quartz is described by Kalb *et al.* (1992). This method could have advantages for the introduction of proteins. Monolayers that adhered to alkylated quartz slides by lowering through the monolayer have to remain covered by an aqueous phase (Subramaniam *et al.*, 1986).

2.7. Epifluorescence Microscopy of Monolayers

Information regarding the lateral distribution and organization of membrane constituents is still scarce. Fluorescence microscopy of monolayers doped with a low concentration of fluorescent lipid offers a new possibility of characterizing the static and dynamic properties of coexisting phases. Some unexpected morphologies have been discovered that provide information about the nature of monolayer phases and have connections to pattern formation in other systems (Knobler, 1990).

A monolayer containing a small amount of a fluorescent probe is excited with a laser or an arc lamp and an epifluorescence microscope with a long-distance objective mounted on an x–y stage is used to scan the monolayer. Images of the monolayer are detected with a high-sensitivity television camera and recorded on videotape for subsequent analysis. A description of a fluorescence film balance is given by Heyn *et al.* (1991). Different fluorescent probes have been used, as lipids containing the N-(7-nitro-2,1,3-benzoxadiol-4-yl)moiety (NBD) attached to different positions (Weis, 1991) or a pyrene label (Eklund *et al.*, 1988). In the transition region from fluid to condensed state, the solid domains are regularly distributed in a fluid phase, indicating mutual repulsion of the domains.

The coexistence region of continuously compressed monolayers is illustrated in Figure 3, which is a series of epifluorescence micrographs of a dye containing dipalmitoylphosphatidylcholine monolayers at increasing surface pressures, as reproduced from Weis (1991). In the initial stages of the phase transition, small nucleations of the solid phase appear as black spots from which the fluorescent probe is excluded. The size of the domain structures was dependent on the compression rate. Larger domains with a greater diversity of sizes were seen at a slow compression rate and smaller and more irregular in shape at rapid compression (Nag *et al.*, 1991).

At increasing pressure the solid phase domains grow in size but not in number. These periodic patterns are also obtained for supported lipid monolayers on alkylated glass (McConnell *et al.*, 1984). Using the appropriate filters and fluid and condensed phase can be visualized separately. After minimization of surface flow, Peters and Beck (1983) were able to measure the translational diffusion coefficient (D) using fluorescence microphotolysis. In monolayers of dipalmitoylphosphatidylcholine, D decreases by more than three orders of magnitude by passing from the fluid to the crystalline state. Absolute D values are

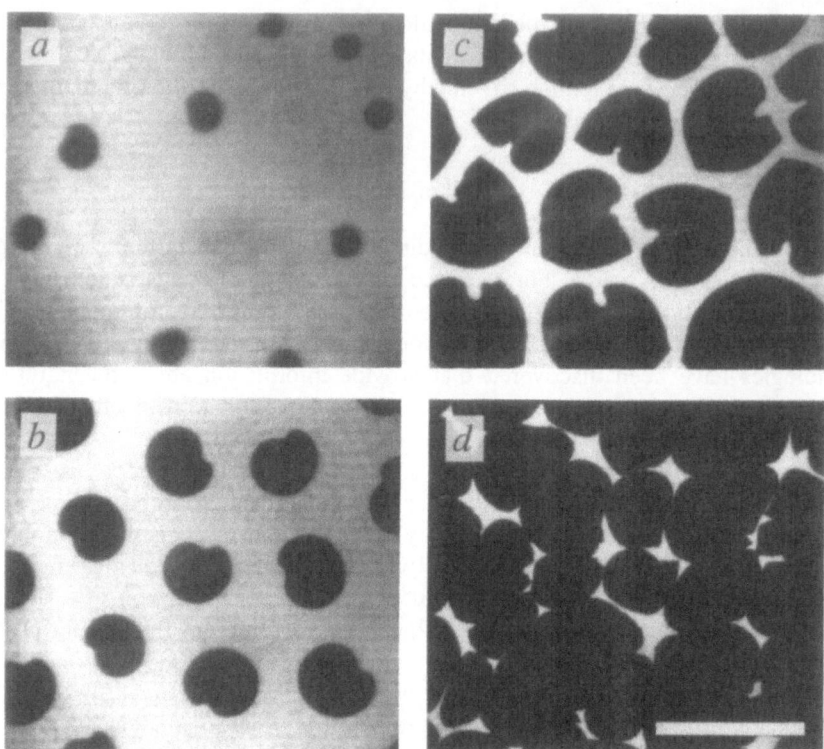

FIGURE 3. Fluorescence micrographs of a monolayer consisting of dipalmitoyl-phosphatidyl-choline and 2 mol% NBD-phosphatidylcholine on a distilled water subphase. The photographs record the appearance of the monolayer at various points in the solid–fluid coexistence region, with values of A and π, respectively, of: (a) 0.814 nm², 2.4 mN/m; (b) 0.674 nm², 4.7 mN/m; (c) 0.515 nm², 12.4 mN/m; (d) 0.415 nm², 24.1 mN/m. Subphase temperature 19.5°C; sale bar = 50 μm (Weis, 1991).

somewhat smaller in bilayers than in monolayers. This is made plausible by interdigitation, or shear forces between the two monolayers and the reduction by integral membrane proteins (Peters and Beck, 1983).

Lateral phase separation of lipid mixtures was studied by Eklund *et al.* (1988) and observed for dipalmitoylphosphatidylcholine-pyrene-labeled phosphatidic acid in the presence of Ca^{2+} by the increased pyrene excimer intensity of the separated phosphatidic acid-enriched domains. In films containing 50 mol% cholesterol, no phase separation could be detected at the resolution available. In monomolecular layers of phospholipids containing the antenna protein B800–850, a homogeneous distribution of a functional protein was achieved, whereas reorientation of the protein by increasing pressure led to a drastic change in fluorescence yield. Incorporation of the reaction center of photosynthetic bacterium *Rhodopseudomonas sphaernoides* into phospholipid monolayers containing the antenna protein gave a highly efficient energy transfer proving the reconstitution of both proteins in the monolayer (Heckl *et al.*, 1985).

2.8. Optical Methods to Determine Monolayer Thickness

The thickness of lipid monolayers and the effect of lipid—protein interaction was studied by specular reflection of neutrons (Johnson *et al.*, 1991). For pure phospholipids the thickness of the hydrocarbon region of dimyristoylphosphatidylcholine monolayers increased from 1.14 ± 0.15 nm in the expanded state to 1.58 ± 0.15 nm in the condensed state whereas the head group region is ~ 1.0 nm in both phases. Lipid–spectrin coupling is enhanced by electrostatic interaction as the volume fraction of spectrin in the head group region increases to 0.22 in a mixed monolayer of dimyristoylphosphatidylcholine and negatively charged dimyristoylphosphatidylglycerol in the condensed state. Consistent with spectrin penetration of the monolayer, the area per molecule increases by $\sim 20\%$ but does not significantly affect the thickness of the head group or tail region Polylysine, on the other hand, does not penetrate the head group region but forms a layer electrostatically absorbed to the charged head group. Ellipsometry has been shown to be a powerful method for deriving information on membrane thickness and orientation of molecules in monolayers at the air–water interface. This method uses the principle that the state of polarization of polarized light changes on reflection at an interface. For dioleoylphosphatidylcholine and rod outer segment phospholipids, a thickness of 1.89–2.38 nm was determined (Ducharme *et al.*, 1985). To orient rhodopsin at the nitrogen–water interface comparable to its natural transmembrane orientation required a lateral pressure of 38 mN/m (Salesse *et al.*, 1990). This value is, however, determined in the absence of membrane lipids.

A completely new possibility in monolayer studies was introduced by Helm *et al.* (1991) using X-ray diffraction and reflection. Compression of the laterally isotropic fluid led to a change in thickness of the hydrophobic moiety. The main transition involved a 10% increase in electron density of the hydrophobic region, a dehydration of the head group, and an increased positional ordering.

3. EXAMPLES OF THE USE OF MONOMOLECULAR LAYERS

3.1. Pressure–Area Characteristics of Lipids

Pressure–area curves are characterized by different states which are more or less analogous to the three-dimensional solids, liquids, and gases. Thus, in condensed films the molecules are closely packed and well-aligned. At the collapse point the molecules are at their maximum density and the molecular area is minimal. In gaseous films the molecules probably lie flat and are widely separated. Discontinuities mark apparent phase transitions. When there are two phases present, a monolayer has only one degree of freedom and fixing the temperature fixes π. In the liquid expanded state (LE) the hydrocarbon chains are lifted from the surface but remain largely disordered. In the intermediate region the film consists of clusters of crystallized molecules that are immiscible with the uncrystallized molecules (Phillips, 1972). This is very elegantly visualized by fluorescence microscopy (Knobler, 1990; Weis, 1991). The long relaxation times of 0.1 to 1.2 sec (Träuble, 1971) for the cooperative transition liquid expanded $\rightarrow$ crystalline phase explain why the region of the phase transition in the π–A curve is sensitive to the rate of compression of the monolayer. The sensitivity of the phase transition to temperature leads to a variety of isotherms, shifting the phase transition to higher pressures with increasing temperature (Phillips, 1972; Blume, 1979; Albrecht et al., 1981). The heat of the phase transition for phosphatidylcholine in the monolayer and bilayer cases are similar, confirming the correspondence of the two systems (Phillips and Chapman, 1968).

For a mixed-chain fatty acid derivative the mean molecular area is about 0.6 nm^2 (at 30 mN/m) which means that the surface area of the saturated fatty acid chain has increased (Demel et al., 1967, 1972a). This is in line with the observation of Phillips (1970) that mixtures of unsaturated phosphatidylcholine (dioleoylphosphatidylcholine) and saturated phosphatidylcholine (distearoyl- or dipalmitoylphosphatidylcholine) show an increase in molecular area as long as mixing occurs. This is not the case for dioleoylphosphatidylcholine and dibehenoylphosphatidylcholine. For most membrane lipids except dipalmitoylphosphatidylcholine, collapse pressures range from 43 to 46 mN/m (Mayer et al., 1983).

The increase in molecular area of phospholipid monolayers with increasing unsaturation (ranging from 0.45 nm^2 for dipalmitoylphosphatidylcholine to 0.76 nm^2 for dilinoleoylphosphatidylcholine, at 30 mN/m) or decreasing chain length (dimyristoyl- or dilauroylphosphatidylcholine) correspond well with the increases in permeability of the lipid bilayer (Demel et al., 1972a; De Kruijff et al., 1972). Monolayers are very useful for determining effects of pH and ion concentration on the molecular packing of membrane lipids; anionic lipids are particularly sensitive to such effects (Mattai et al., 1989; Demel et al., 1992a).

It is apparent that phosphatidylcholine monolayers are more expanded than phosphatidylethanolamine monolayers. Dipalmitoylphosphatidylcholine undergoes a transition from liquid-expanded to a condensed state at 10 m/Nm and 22°C, whereas dimyristoylphosphatidylethanolamine behaves similarly at the same temperature (Phillips, 1972). The introduction of fluorescent groups can affect the surface properties significantly (Pisarchick and Thompson, 1990).

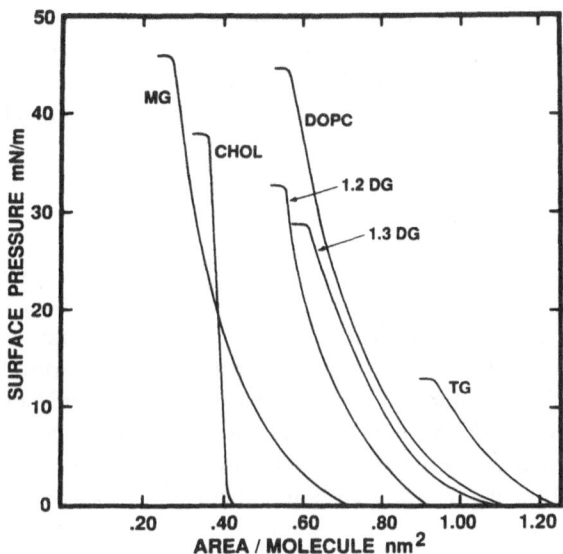

FIGURE 4. Pressure–area curves for monooleoylglycerol (MG); 1,2-dioleoylglycerol (1.2 DG); 1,3-dioleoylglycerol (1.3 DG); 1,2-dioleoylphosphatidylcholine (DOPC); trioleoylglycerol (TG); and cholesterol (CHOL) at the air–10 mM Tris (pH 7.4) interface at 22°C.

Small impurities that are not even detected by HPTLC can greatly affect the surface properties of lipids as was demonstrated for gangliosides (Fidelio *et al.*, 1991).

Some of the principal qualities of (membrane) lipids are demonstrated in Figure 4. At a surface pressure of 30 mN/m the molecular area of dioleoylphosphatidylcholine is 0.63 nm². Conversion into diolcoylglycerol will lead to a change in molecular area particularly in the 1,2 derivative. The reductions in collapse pressure indicate the reduced interfacial stability of dioleoylglycerols. It is clear that trioleoylglycerol is not a membrane-forming lipid; it lacks the amphipathic character and shows a very low collapse pressure. Monooleoylglycerol, on the other hand, forms a stable monolayer. The pressure–area curve of cholesterol demonstrates its particular character with low compressibility.

The force–area curves at the *n*-heptane–water interface are quite different from those at the air–water interface (Phillips, 1972). Distearoylphosphatidylethanolamine and distearoylphosphatidylcholine undergo two-dimensional phase transitions at 0.80 and 1.00 nm²/molecule whereas dipalmitoylphosphatidylcholine does not show a phase transition. The molecular expansion follows because the heptane molecules can enter between the adjacent lipid chains. Only at high surface pressures depending on chain length and polar head group of the phospholipid, is the molecular area identical to that found at the air–water interface. Cholesterol cannot be spread as a monolayer at the alkane–water interface since it will dissolve in the organic phase (Demel, unpublished data). It can be concluded that the oil–water phase is not an ideal model for biological membranes.

3.2. Phospholipid–Sterol Interactions

Monolayer studies have largely contributed to the understanding of the membrane properties of cholesterol. Cholesterol is a main component of mammalian biological membranes and abundant in most (Demel and De Kruijff, 1976). Although cholesterol forms a very condensed monolayer, it is not crystalline but rather liquid condensed with a collapse point at 38 mN/m and a molecular area of 0.38 nm^2 (Pethica, 1955). The molecular area, and consequently the interfacial orientation, is greatly affected by the polar moiety which is, for cholesterol and for membrane sterols in general, a 3β oriented enolic hydroxy group. Keto derivatives or autoxidation products of cholesterol show an increased molecular area (Theunissen et al., 1986). The monolayer method can also simply discriminate between an orientation perpendicular or parallel to the interface when the polar groups are distributed over the extremities of the molecule as in 25-hydroxycholesterol. In this case the hydrophobic stretch prevents a horizontal orientation and the molecule is still oriented perpendicular although the collapse pressure is strongly reduced (Theunissen et al., 1986). These properties can explain the increased membrane permeability and cytotoxic effects of 25-hydroxycholesterol. Hydroxylation at position 22, however, allows a horizontal orientation of 22,R-hydroxycholesterol at low surface pressures but not of 22,S-hydroxycholesterol (Gallay et al., 1984). The striking modification of the orientation of 22,R-hydroxycholesterol compared with cholesterol can explain the high activation energy of the 22,R-hydroxylation step in steroidogenesis catalyzed by the membrane-bound cytochrome P-450.

Since biological membranes are composed of complex mixtures of lipids, it is relevant to study mutual interactions in detail. Monomolecular layers offer a number of advantages such as a wide range of compositions (without changes in the curvature of the membrane caused by particle size) and molecular packing (changing the surface pressure).

The two-dimensional miscibility can be indicated by the deviation of the average molecular composition isobars from the additivity rule (Gaines, 1966). Miscibility is also revealed by the collapse pressure. Variation of the collapse pressure between the limits of the single components as a function of composition indicates their miscibility (Crisp, 1949; Joos and Demel, 1969). More direct evidence may also be seen by fluorescence microscopy. Molecules with very small differences in chain length show cocrystallization (dipalmitoyl- and distearoylphosphatidylcholine). For distearoyl- and dimyristoylphosphatidylcholine, the difference in chain length is already too great for cocrystallization.

Of special interest are the interactions of phospholipids with sterols, particularly cholesterol. The first evidence for such interactions came from monolayer studies of De Bernard (1958) using mixtures of egg phosphatidylcholine and cholesterol. Since it can be assumed that cholesterol will occupy essentially the same area in both pure and mixed films, it was concluded that cholesterol reduces the molecular area of egg phosphatidylcholine. The phospholipid–sterol interaction is governed by hydrophobic forces and therefore will be influenced by the acyl composition of the phosopholipid. Saturated phosphatidylcholines of intermediate chain length such as dimyristoyl- and dilauroylphosphatidylcholine are condensed by cholesterol (Demel et al., 1967; Standish and Pethica, 1967), whereas short-chain derivatives such as dicapryl- and dinonanoylphosphatidylcholine showed little or no condensation (Joos and Demel, 1969). With the exception of dilinoleoyl- and dilinolenoylphosphatidylcholine, unsaturated phosphatidylcholine derivatives condensed by cholesterol (Demel et al., 1972a; Gosh et al., 1973). These findings are completely in line with studies using lipid bilayers demonstrating reduced permeability in the presence of cholesterol, or with NMR or ESR studies determining chain mobilities (Demel and De Kruijff, 1976). The above effects are detectable at 5–50 mol% cholesterol. Interestingly enough, using fluorescence microscopy of monomolecular layers, it was possible to determine effects at very low cholesterol concentrations (Weis and McConnell, 1985; Meller, 1985). At concentrations as low as 0.1 mol% the domain geometry was altered probably by cholesterol binding to the interface between liquid and crystal regions. Fluorescence microphotolysis revealed that the translocational diffusion in phospholipid monolayers, which is of the same order as in the bilayer, is strongly reduced in the gel phase (Peters and Beck, 1983). In the presence of cholesterol the translational diffusion is reduced in the liquid crystalline phase but increased in the gel phase (Meller, 1985). This is in line with differential scanning calorimetry measurements demonstrating the liquefying effect of cholesterol in the gel phase (Demel and De Kruijff, 1976). The structure and orientation of the polar moiety of cholesterol are of critical importance (Demel et al., 1972b). However, a free hydroxyl group is not a prerequisite. The cholesteryl-methyl ether is nearly as effective as cholesterol although its interfacial stability is already reduced (Demel et al., 1984b). This proves that there is no direct interaction of the polar group involved. In phospholipid bilayers, cholesterol tends to increase the extent of hydration of the polar groups, which is also in support of the idea that cholesterol does not associate directly with the polar group of phospholipids (Lundberg et al., 1978).

The possible preferential affinity of cholesterol for different lipid classes was also investigated. The interaction of cholesterol with sphingomyelin was demonstrated by Grönberg et al. (1991), who showed that the condensing effect of cholesterol was larger on sphingomyelin than that observed with phosphatidylcholine. Also the enzyme-catalyzed cholesterol oxidation in monolayers containing 3-hydroxy-substituted N-stearoylsphingomyelin was lower than the rate measured for dipalmitoylphosphatidylcholine.

These results correlate well with the molecular dynamics of sphingomyelin on cholesterol in biological membranes as it retards the rate of cholesterol desorption from lipid bilayers (Gold and Phillips, 1990). Early studies with differential scanning calorimetry of bilayer systems consisting of mixtures of sphingomyelin and phosphatidylcholine, which gave phase separation, demonstrated a preferential interaction of cholesterol with sphingomyelin (Demel et al., 1977a).

3.3. Relationship between Monolayer Surface Pressure and Lateral Pressure in a Lipid Bilayer

A difficulty in using the monolayer as a model membrane has been deciding what surface pressure is the most significant. The bilayer concept of Gorter and Grendel (1925) was based on compression of a monolayer of lipid extracted from red cells to an arbitrary low surface pressure. The question of the appropriate compression and hence the relationship between molecular packing in monolayers and that in bilayers was attacked in different ways.

MacDonald and Simon (1987) proposed that the most condensed state of the monolayer (at collapse) formed by continuous addition of phospholipid to the air–water interface of a constant-area trough exists in equilibrium with unstressed liposomes.

Gruen and Wolfe (1982) reported a theoretical value of 50 mN/m based on the incorrect assumption that a monolayer at the alkane–water interface would be a closer analogy of the lipid bilayer than a monolayer at an air–water interface. This value would exceed the collapse pressure of most membrane lipids. Examining the area change at the monolayer and bilayer phase transition the closest agreement between the two systems was found at a surface pressure of approximately 30 mN/m (Blume, 1979). By comparison of the action of purified phospholipases on monolayer films at various interfacial pressures with the action on erythrocyte membranes (Demel et al., 1975), it was concluded that the packing of the lipids in the outer layer of the erythrocyte is comparable to a lateral film pressure of 31–35 mN/m.

Comparative studies on the binding of local anesthetics to monolayers and bilayers led to the conclusion that the packing of phospholipids in a bilayer may be similar to the packing in monolayers at an equivalence pressure of about 32–33 mN/m (Seelig, 1987). Another approach measured the lateral pressure required to maintain the transmembrane orientation of rhodopsin. The orientation of rhodopsin at the nitrogen–water interface was determined by ellipsometrie measurement of the film thickness, based on a surface pressure of 38 mN/m. On the other hand, pressures of 45 mN/m would lead to the formation of multilayers (Salesse et al., 1990). This study was performed in the absence of lipid, thus neglecting their possible contribution to protein orientation. The molecular packing and consequently the lateral surface pressure are highly dependent on the curvature of the membrane surface. There is strong evidence that the surface pressure of serum lipoproteins will depend on their size and may range from 20 to 25 mN/m for these particles (Jackson et al., 1986).

The value for the lateral surface pressure of the erythrocyte membrane was also of consequence for the calculations made by Gorter and Grendel (1925) on the presence of a membrane bilayer. They determined the surface area of the extracted lipids at a very low surface pressure (approximately 5 mN/m). On the other hand, they did not account for the presence of integral membrane proteins. Fortunately, these two errors canceled each other. At a surface pressure of 30–35 mN/m the ratio of the surface area of the extracted lipids to the area of the extracted erythrocytes of 1.5 (implying that 75% of the membrane is covered by a lipid bilayer) accounts for the fact that 25% of the erythrocyte surface area is occupied by integral proteins (Zwaal et al., 1976).

3.4. Lipid–Protein Interactions

Most proteins will partition between interface and aqueous phase. The peptide equilibrium surface pressure at saturation (taken as the peptide collapse pressure) is determined by injecting increasing amounts of peptide into the subphase in the absence of lipid. The aqueous buffer is stirred while the peptide monolayer is formed at a period of 20–40 min (Tamm et al., 1989).

Compression curves of proteins and the derived area per residue can give indications for the interfacial orientation of the protein. The area per residue for a completely unfolded protein at the surface assuming β-form conformation is 0.15–0.17 nm²/residue (Birdi, 1973). Spread monolayers of homopolypeptides that are α-helical assume areas of 0.13–0.19 nm²/residue (Malcolm, 1968). At saturation, monolayer coverage of the molecular areas for apolipoprotein A-I and A-II are 0.15 and 0.13 nm²/residue, consistent with monolayers consisting largely of α-helical protein molecules lying with their long axes parallel to the interface (Krebs et al., 1988). It is concluded that at all π values, more segments of apolipoprotein A-II than of A-I are out of the plane of the interface in loops or tails. For the very hydrophobic lung surfactant proteins (SP-B and SP-C), at surface pressures of 35 mN/m, areas of 0.16 and 0.17 nm² were measured corresponding to a high α-helix content mainly in the plane of the monolayer (Oosterlaken-Dijksterhuis et al., 1991b). The area of the 25-residue signal peptide of subunit IV of yeast cytochrome oxidase (COXIV-25) measured by the area increase at constant pressure of 25 mN/m and quantification of the interfacial peptide concentration revealed a molecular are of 5.60 ± 1.70 nm². Since

COXIV-25 is up to 50% α-helical in membranes that contain phosphatidyl-glycerol, it was concluded that the α-helical segment is probably aligned parallel to the plane of the membrane (Tamm, 1986).

In a perpendicular orientation of the transmembrane protein rhodopsin, the molecular area was found to be 23.00 nm^2, which corresponds to 0.06 nm^2 per amino acid residue (Salesse *et al.*, 1990). The monolayer technique offers excellent possibilities for determining specific affinities in lipid–protein interactions. Monolayer penetration is monotonically reduced with increasing surface pressure pointing to the dependence of hydrophobic protein–lipid interaction on hydrocarbon chain density. Using radiolabeled protein the average surface radioactivity and protein surface concentration are generally directly proportional. Desorbtion is very slow when the film containing the protein is transferred to a fresh subphase containing no protein (Quinn and Dawson, 1970; Bazzi and Nelsestuen, 1988).

Plasma apolipoproteins showed an interaction with different membrane-forming lipids (Jackson, et al., 1979; Phillips and Krebs, 1986). While in the mammalian system triglycerides are transported by serum lipoproteins, in insects such as the migratory locust which use lipid to fuel flight; lipid transport is in the form of diacylglycerol-rich low-density lipoproteins. Contrary to triglycerides, which are mostly present in the core of the particle, diglycerides are present in the surface layer (compare their properties as presented in Figure 4). It could be shown that apolipophorin III interacts preferentially with diacylglycerol monolayers and therefore will play an essential role in the formation of low-density lipophorin (Demel *et al.*, 1992b). The high affinity of the acetylcholine receptor protein for cholesterol was first established in monolayer experiments (Popot *et al.*, 1978) before it was recognized in reconstitution experiments (Toyoshima and Unwin, 1988).

The spectrin membrane skeleton that lines the cytoplasmic surface of erythorocytes may be involved in maintaining transbilayer membrane asymmetry of phospholipids. This is supported by the finding that erythrocyte protein 4.1 interacts more readily with the lipids of the inner than of the outer leaflet of the membrane (Shiffer *et al.*, 1988).

The interaction of cytochrome b_5 with lipid monolayers was not charge dependent whereas cytochrome c interaction was (Heckl *et al.*, 1987). Both proteins preferentially partition into the fluid phase. Preferential adsorption of cytochrome c to negatively charged monolayers was visualized by reflection spectroscopy (Kozarac *et al.*, 1988). A detailed comparison of apocytochrome c and cytochrome c demonstrated remarkable differences in lipid specificities. Like the majority of mitochondrial proteins, apocytochrome c is synthesized in the cytosol on free ribosomes as a precursor that is subsequently imported and converted into the holoprotein by attachment of the heme group. Apocytochrome c and cytochrome c have identical polypeptide chains but differ in secondary and tertiary structure, that is, apocytochrome c is exceptional in that it does not have a cleavable signal sequence. The initial electrostatic interaction of apocytochrome c with anionic phospholipids is followed by penetration of the protein in between the acyl chains. It shows a similar interaction for all anionic lipids. In strong contrast, the holoprotein discriminates enormously between cardiolipin, for which it has a high affinity, and phosphatidylserine and phosphatidylinositol, for which it has a much lower affinity. For these latter lipids the interaction is primarily electrostatic. Interactions of cytochrome c with cardiolipin monolayers at high surface pressure suggest the formation of a specific complex of the two molecules (Demel et al., 1989). Earlier indications were obtained by Quinn and Dawson (1969). These results are in accordance with the final localization of cytochrome c.

The signal peptide is believed to function in envelope targeting of preproteins. Many researchers support a model in which both binding and translocation are protein-mediated. On the other hand, it has been suggested that signal sequences may facilitate membrane insertion and translocation by a direct interaction with phospholipids. A major driving force is the high hydrophobic nature of the signal sequence. Binding to the lipid interface will eventually reduce diffusion to a two-dimensional lattice.

Amphiphilic segments are predicted within practically all mitochondrial signal sequences. The mitochondrial signal sequences are predominantly positively charged whereas these peptides also exhibit a pronounced conformational polymorphism in different lipid systems. The amphiphilic signal sequences share this property with the hydrophobic signal peptides (Jones et al., 1990; Hoyt et al., 1991).

The isolated signal peptide can insert into phospholipid monolayers up to high limiting pressures. The incorporation of the positively charged peptide is strongly enhanced by the presence of negatively charged phospholipids (Cornell et al., 1989). It is suggested that parallel with mitochondrial targeting, the induction of N-terminal amphiphilic structures is important for initiation of membrane insertion of bacterial phosphoenolpyruvate-dependent phosphotransferase system permeases. The induction of α-helical secondary structure after lipid association amounts to 65% (Tamm et al., 1989). However, the export-defective mutant peptide in LamB (A13D) is not significantly reduced in α-helix content. The ability to form an α-helix structure is not sufficient to ensure proper signal function. Also the length of the hydrophobic stretch is of critical importance (Jones et al., 1990). The higher affinity of the PhoE signal peptide for anionic lipids is in line with the in vivo as well as the cell-free translocation systems which showed that translocation is less effective when phosphatidylglycerol biosynthesis is reduced (De Vrije et al., 1988). Monolayer data have shown that the interaction with the PhoE signal peptide is maximal at 30—40 mol% anionic lipid (Demel et al., 1990) and coincides with the adoption of α-helix structure (Keller *et al.*, 1992). This concentration of anionic lipids corresponds to that found in wild-type *Escherichia coli* inner membranes. Quantification of the interacting peptide, and measurement of the occupied surface area derived from the induced pressure change, allow calculation of the molecular area. In this way it was found that the molecular area of the PhoE signal peptide right after interaction was about 3.70 nm². After further interaction this value was reduced to 1.60 nm² (Bastenburg *et al.*, 1988a,b; Demel *et al.*, 1990). These results were interpreted as an unlooping of the signal peptide and as a possible first step in protein translocation.

The affinities for the lipid monolayer are reflected in the concentrations of the peptide required for half-maximal surface pressure change (the apparent k_D's). For LamB signal peptides it has been shown that peptides corresponding to export-defective sequences either do not insert into the hydrocarbon region of the lipid monolayer or do so at much higher concentrations than for the export-active peptides (McKnight *et al.*, 1989). Direct evidence for the involvement of the signal sequence and its specificity for the target membrane was obtained for preferredoxin. Interaction with monolayers of chloroplast lipids was measured for preferredoxin and the chemically prepared transit sequence but neither for the ferredoxin apoprotein nor the holoprotein. There was little affinity for the membrane lipids of mitochondria by any of these proteins. The monogalactosyl-diglyceride, characteristic for chloroplast membranes, was largely responsible for the interaction of preferredoxin with lipid monolayers (Van't Hof *et al.*, 1993).

The involvement of other proteins in the translocation of precursor proteins such as SecA, has been studied in monolayer model systems as well. It was established that SecA insertion was greatly enhanced by negatively charged lipids such as phosphatidylglycerol and cardiolipin, which is remarkable since SecA has an overall net negative charge. ATP decreased both insertion and binding of SecA to phosphatidylglycerol monolayers. ADP and phosphate decreased SecA insertion to the same extent as ATP, but the binding of SecA was only slightly reduced. On the basis of the data, a cycle of SecA binding, insertion, and dissociation from the membrane was postulated (Breukink *et al.*, 1992). The monolayer data are in line with SecA binding to liposomes of negatively charged lipids resulting in a large increase in SecA ATPase activity (Lill *et al.*, 1990).

3.5. Lipolytic Reactions at the Air–Water Interface

Monolayers at the air–water interface are an excellent model system to study lipolytic reactions of membrane lipids and to follow even picomoles of substrate conversion. Interfacial lipolytic reactions have been shown to be surface pressure-dependent. Above a characteristic critical packing density of the substrate molecules the enzyme cannot penetrate. This property was used to compare the action of different phospholipases on monolayers and intact erythrocytes to determine the lateral pressure of the latter (Demel et al., 1975).

The product desorption from the interface can be measured as a change in surface pressure and/or surface radioactivity, or at constant surface pressure as a change in surface area. This barostat method (Verger and De Haas, 1976) has been applied specially for short-chain lipid derivatives and requires a high surface concentration of the substrate in order to measure a change in surface area. The use of a trough consisting of two compartments connected by a narrow surface channel gives linear kinetics in contrast to the nonlinear plot obtained with a one-compartment trough. This method was also used to study the action of phospholipase C in a two-step reaction; enzymatic hydrolysis of didodecanoyl-phosphatidylcholine (di C14:0) by phospholipase C, generating 1,2-diacylglycerol, followed by the action of pancreatic lipase to give rise to fatty acid and 2-monoacylglycerol, which desorbs rapidly. In this way the use of radiolabeled substrates, and diacylglycerol accumulation in the film, can be avoided (Moreau et al., 1988). By measuring the time of enzyme activity and the amount of radioactive enzyme in excess at the interface, it was shown (Pattus et al., 1979) that the lag periods in lipolysis are related to a slow penetration of the enzyme into the interface. By measuring the influence of surface pressure on the induction time of different phospholipases A_2, the enzymes could be classified according to their penetration power.

The action of phospholipase A_2 was characterized optically after labeling the enzyme with a fluorescent marker. Enzyme domain formation was seen on hydrolysis in the phase transition region (Grainger et al., 1990). For long-chain derivatives an excess of bovine serum albumin (BSA) in the subphase is necessary to solubilize lysophospholipid and fatty acid such that product desorption is not the rate-limiting step. The tensioactivity of BSA prevents measurements below 22–23 mN/m.

Serum lipoproteins show a large variation in composition and size, and therefore in molecular packing. The surface monolayer consists of phospholipids, cholesterol, and amphipathic apolipoproteins surrounding a core of nonpolar lipids. Triglycerides and cholesteryl esters have a very low interfacial stability and are consequently present in minor amounts in the surface layer. This small quantity is of considerable biological importance since the enzymatic reactions involving these lipids and water-soluble enzymes are interfacial. Using radiolabeled substrate the hydrolysis of triglycerides in monomolecular layers can be followed while present at a concentration less than 5 mol% (McLean et al., 1986).

Lipase-catalyzed hydrolysis of a triglyceride shows the complexity of the kinetics in lipolytic reactions where the first hydrolysis product is a substrate in a consecutive lipolytic reaction. Lipoprotein lipase (LpL) and hepatic triglyceride lipase (HL) have the capacity to hydrolyze triglycerides as well as phospholipids; however, they have different preferences for lipoprotein substrates. LpL is found in extrahepatic tissues and utilizes chylomicrons and very low density lipoprotein (VLDL) as physiological substrates and requires apolipoprotein C-II as an activator. In LpL the triglyceridase activity exceeds the phospholipase A_1 activity. HL, which is involved in processing lipoproteins and high-density lipoproteins, is an active triglyceridase as well as phospholipase A_1. Varying the composition and concentration of lipids on the monolayer as well as varying the surface pressure revealed that the two enzymes are quite different (Demel et al., 1982a, 1984a; Demel and Jackson, 1985; Jackson et al., 1986; Laboda et al., 1986).

Laboda et al. (1988) have shown that the rate of hydrolysis of triolein by HL is affected by the phospholipid matrix and the presence of cholesterol. A number of human apolipoproteins (A-I, A-II, C-II, and C-III ≥ 0.1 nM) were found to inhibit triolein hydrolysis by HL. At these concentrations apolipoproteins did not affect the lipid surface pressure (π_i 28 mN/m), and therefore a competition between apolipoproteins and HL for binding to the lipid surface is suggested. Apolipoprotein E activated HL action on phosphatidylethanolamine at surface pressures below 15 mN/m. At high surface pressures of 25 mN/m, all apolipoproteins were inhibitory (Thuren et al., 1991).

Most studies with lipolytic enzymes have been done primarily in vitro using pure lipids as substrates. In vivo, however, interfaces are composed of a complex mixture of lipids and proteins. Lipolytic activities involve the simultaneous participation of several classes of lipids. The hydrolysis of phosphatidylinositol monolayers by phosphatidylinositol phosphodiesterase showed a limiting pressure of 33 mN/m. The addition of phosphatidic acid or phosphatidylglycerol increased the limiting pressure (Hirasawa et al., 1981). Conversely, addition of phosphatidylcholine causes a marked fall in the pressure threshold for film hydrolysis. Unsaturated amphiphiles such as oleic acid could be additive or synergistic in their interaction with some phospholipids. A possible self-amplifying potential for phosphatidylinositol hydrolysis could be of great importance in vivo in the enhanced breakdown of phosphatidylinositol after agonist–receptor interaction.

In studies on the enzymatic synthesis-hydrolysis of cholesteryloleate in mixed surface films, phosphatidylcholine, when present in a molar excess relative to the sum of free and esterified cholesterol, is inhibitory. Inhibition is associated with segregation of the substrate into reactive and unreactive pools which are not exchangeable. It is suggested that in the presence of phosphatidylcholine the reactivity or availability of substrate is also affected. Bile salts and other surfactants reverse the inhibition presumably by disrupting the unreactive phosphatidylcholine—substrate complex (Bhat and Brockman, 1981). Using initial enzyme adsorption flux the role of oleic acid in the regulation of the hydrolysis of cholesteryloleate in lipid films was demonstrated. The strong adsorption of the enzyme to the hydrolysis product, oleic acid, constitutes a kind of product activation (Bhat and Brockman, 1982).

The presence of cholesteryloleate affects the LpL-catalyzed hydrolysis of trioleoylglycerol in mixed monolayers containing egg phosphatidylcholine (Demel and Jackson, 1985). In phospholipid monolayers containing 5.0 or 7.5 mol% trioleoylglycerol the further addition of cholesteryloleate caused a decrease in LpL activity. In contrast, addition of cholesteryloleate to phospholipid monolayers containing 1–5 mol% trioleoylglycerol enhanced enzyme activity (Figure 5).

Based on force–area measurements the cholesteryl ester-mediated decrease in LpL activity observed at high substrate concentrations may be explained by displacement of trioleoylglycerol from the monolayer thereby reducing the interfacial trioleoylglycerol concentration available for enzyme catalysis. Based on these monolayer studies it is suggested that the relative amount of cholesteryl ester in plasma triaclyglycerol-rich lipoproteins plays a regulatory role in determining the rate at which triaclyglycerols are cleared from circulation.

The action of combined phospholipases can lead to a stimulation of one reaction by product formation of the other reaction. Phospholipase C (*Bacillus cereus*) does not act on intact erythrocytes and is not capable of hydrolyzing phosphatidylcholine in a single or mixed phosphatidylcholine film at pressures above 32 mN/m. However, when sphingomyelin in the film is degraded either first or simultaneously by sphingomyelinase from *S. aureus* thereby producing ceramides, it appears that phospholipase C is at once capable of hydrolyzing phosphatidylcholine even at a film pressure of 38 mN/m (Demel et al., 1975). A comparable action can be seen for phospholipase C from *C. welchii* which can degrade phosphatidylcholine as well as sphingomyelin and is active by itself on intact erythrocytes. In monolayers the single phosphatidylcholines alone are not hydrolyzed at pressures above 28 mN/m but sphingomyelin monolayers were found to be attacked at the highest surface pressure attainable. Mixtures of phosphatidylcholine and sphingomyelin appeared to facilitate the hydrolysis of phosphatidylcholine by phospholipase from *C. welchii* and allowed hydrolysis at high surface pressures (Demel et al., 1975).

Using enzymatic reactions of monomolecular layers can be a very sensitive method to sense such conditions as molecular packing and lateral distribution, which are very important for the functioning of biological membranes. Enzyme-catalyzed oxidation of cholesterol in mixed phospholipid monolayers revealed the stoichiometry at which free cholesterol clusters disappear. Cholesterol–phospholipid stoichiometry is twice as high for sphingomyelin as for phosphatidylcholine (Slotte, 1992) presenting new evidence for the role of sphingomyelin in membranes with high cholesterol concentrations. In mixed monolayers that show phase separation, the increase in lipolytic activity can be explained by a larger

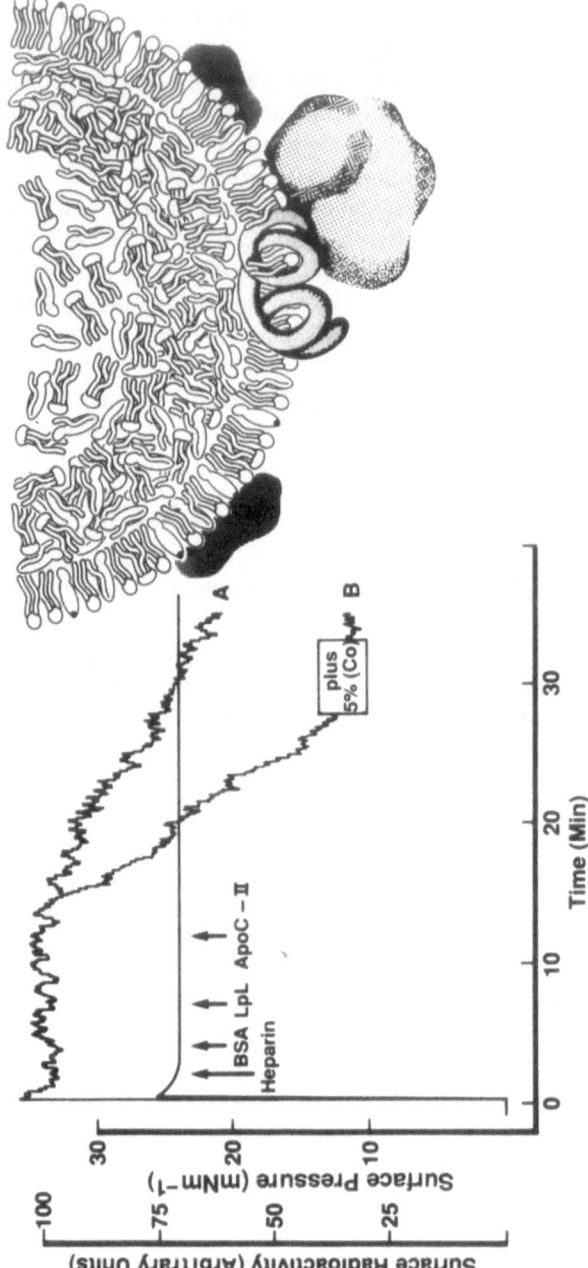

FIGURE 5. Effect of 5 mol% cholesteryl ester on lipoprotein lipase-catalyzed hydrolysis of 2.5 mol% [^{14}C]trioleoylglycerol (A) in the absence and (B) in the presence of cholesteryloleate (Demel and Jackson, 1985). Triacylglycerol substrates and cholesteryl esters reside primarily in the core of the lipoprotein, with a small percentage in the surface monolayer. Apo C-II residing in the monolayer surface of the lipoprotein particle is a specific activator for lipoprotein lipase. The surface of the triglyceride-rich particle is mimicked as monolayers at thte air-water interface.

amount of enzyme that penetrates the interface. Packing defects that occur in the membrane during phase transition and phase separation are believed to be responsible (Yedgar et al., 1982).

3.6. Vesicle Spreading and Vesicle Monolayer Interaction

Vesicles of lipids or membranes can be spread to a monolayer at the interface by the Trurnit (1960) method or by self-assembly from a vesicle suspension that was added to the subphase (Heyn et al., 1990). Surface films generated from lipid vesicles are organized as a monolayer (Schürholz and Schindler, 1991). This conclusion is based on the fact that lipid-to-protein rations of surface layers match those of the vesicles used. Monolayer organization was also concluded from ellipsometric studies (Salesse et al., 1987) where a film thickness of 2.0 ± 0.1 nm was measured for spreading of vesicles and monolayers spread from an organic solvent. Analysis of the surface layer spread from erythrocytes revealed that all integral components of the membrane were present in the surface layer and with the same stoichiometrics as in the erythrocyte membrane. Nonspanning proteins such as NADH oxidase and acetylcholinesterase and the membrane-spanning protein $Ca^{2+}-ATPase$ retained activity after spreading (Pattus et al., 1981).

The experimental procedure involves the distribution of vesicles (2–5 μg/ml) at the surface of a sandblasted glass slide (2 × 3 cm^2) which is cleaned with Na-dichromate-sulfuric acid and double-distilled water and inserted at one end of the trough. Vesicles adsorbed to the monolayer can be removed by a glass roller (Schürholz and Schindler, 1991) or a wet bridge of filter paper (Heyn et al., 1990). Surface denaturation of membrane proteins is prevented when the film generated at low surface pressures is compressed to pressures ≥ 15 mN/m within a minute (Pattus et al., 1981). Expansion to lower surface pressures (5 mN/m) was found to result in irreversible area increases caused by unfolding a membrane proteins (Schürholz and Schindler, 1991).

The spreading of membrane vesicles under controlled conditions offers the possibility of forming monolayers with retention of the biological activity of the proteins although uncertainties still remain with respect to their structure and conformation. The freeze-fracture face of human erythrocyte membrane films suggests that the monolayers are formed (Pattus et al., 1981). Isolated reaction centers from photosynthetic bacteria solubilized in octyl-β-glucopyranoside could be spread at the air–water interface under controlled conditions without a significant loss in electron transfer capability (Alegria and Dutton, 1991). Optimal protein densities and orientations were obtained at pressures of 25–30 mN/m. Higher pressures produced a loss of orientation. This study does not provide evidence for the structure of the lipid layer formed in the presence of bilayer-spanning proteins. It is suggested that multilamellar lipid domains might be present. The combination of monolayer and vesicle systems creates the possibility of studying membrane interactions on line without the need for separations. Membrane fusion or the formation of transient associations between monolayers and vesicles can be monitored by labeling the vesicles. Release of monolayer compounds can also be followed by the change in surface radioactivity. Simultaneous measurement of surface pressure can indicate changes in surface concentration.

The interaction of mitochondrial creatine kinase with spread mitochondrial membranes (from both inner and outer membranes) showed that the octameric form is more effective than the dimeric. Intermembrane contact formation was demonstrated by measuring the change in surface radioactivity after the injection of radiolabeled vesicles into the subphase. Other enzymes of the intermembrane space, as well as cytosolic isoenzymes of creatine kinase, failed to induce contact site formation (Rojo et al., 1991). The interaction with two opposing membranes would enable mitochondrial creatine kinase to be held in direct contact with the inner membrane adenine nucleotide translocator and outer membrane porin. These enzymes transport the creatine kinase substrates across the inner (ATP, ADP) and outer (creatine, phosphocreatine) membranes of mitochondira. Whereas mitochondrial creatine kinase does not lead to a fusion of vesicles with the monolayer (the surface pressure is hardly affected by the vesicle injection) poly-L-lysine does, indicated by the sharp increase in pressure (Rojo et al., 1991). Poly-L-lysine itself hardly penetrates the lipid layer. Another example of possible contact site formation between inner and outer mitochondrial membranes was provided by monolayers and vesicles formed of both membrane lipid constituents in the presence of apo- and holocytochrome c (Demel et al., 1989). Only apocytochrome c was able to induce close contacts as is illustrated in Figure 6. The differences in interaction between the two proteins are mainly related to differences in their tertiary structure and not the presence of the heme group itself. It is the initial unfolded structure of apocytochrome c that is thought to be responsible for the high penetrative power of the protein and its ability to induce close membrane contact, whereas the folded structure of cytochrome c is responsible for a specific interaction with cardiolipin.

Monomolecular layers at the air–water interface are an ideal system to study the properties of lung surfactant which is aligned at the alveolar surface and is responsible for the decrease in surface tension (Van Golde et al., 1988). The special properties of dipalmitoylphosphatidylcholine, the major compound in lung surfactant, has attracted the attention of many investigators. Lung surfactant is only made in the later stages of gestation and premature birth is likely to be accompanied by surfactant deficiency which will result in respiratory distress syndrome (RDS). Recently, a class of hydrophobic proteins have been described that are probably involved in the formation of the lung surfactant monolayer (Hawgood et al., 1987; Possmayer, 1988). It has been shown that the presence of

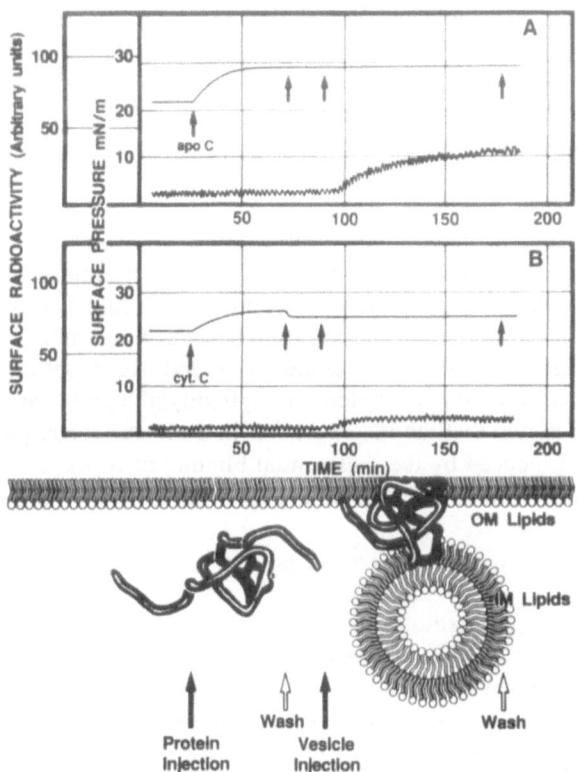

FIGURE 6. The effect of apo- (A) and holocytochrome *c* (B) on contact formation between a monolayer of mitochondrial outer membrane lipids and ^{14}C-labeled vesicles of mitochondrial inner membrane lipids (Demel *et al.*, 1989). The procedure of monolayer formation, protein injection, washing to eliminate unbound protein, vesicle injection, and washing to eliminate unbound vesicles is indicated.

surfactant protein-B (SP-B) in phospholipid vesicles enhances the adsorption of phospholipids to an air–water interface as revealed by an increase in surface pressure. The rate of pressure increase is even enhanced when the proteins were present in a performed phospholipid monolayer after triggering with Ca^{2+} (Oosterlaken-Dijksterhuis *et al.*, 1991a). The hydrophobic proteins with molecular masses of 18 and 5 kDa show an extremely high interfacial stability with collapse pressures of 37 mN/m which is in the range of membrane lipids. Circular dichroism measurements of the collected monolayers showed a very high α-helix content (Oosterlaken-Dijksterhuis *et al.*, 1991b). The protein at the interface is responsible for the vesicle binding and the amount of lipid bound is proportional to the monolayer protein concentration. Surprisingly, mixed monolayers of surfactant protein and phospholipid led to an increase in mean molecular area (Oosterlaken-Dijksterhuis *et al.*, 1991b). This is in agreement with fluorescence measurements in vesicle systems showing, on the one hand, an ordering effect of the bilayer surface and, on the other hand, disruption of acyl chain packing (Horowitz *et al.*, 1992). It can be argued that the first effect is required for vesicle binding and a dehydration effect of Ca^{2+} will induce a monolayer vesicle fusion reaction. The second effect of decreased acyl chain packing will facilitate the insertion of lipids from the vesicles into the monolayer.

Since dipalmitoylphosphatidylcholine, the major lipid of lung surfactant, has a higher collapse pressure than other membrane lipids, its enrichment was ascribed to a selective squeezing out at high surface pressures. However, using radiolabeled lipids no evidence could be found for a selective loss of dioleoylphosphatidylcholine or dioleoylphosphatidylglycerol (Van Liempd *et al.*, 1987). It is possible that the incorporation into the surfactant monolayer is not random and is induced by the preferential binding of SP-A to dipalmitoylphosphatidylcholine (Kuroki and Akino, 1991).

3.7. Protein-Mediated Lipid Transfer

For dynamic properties of membranes such as the transfer and exchange of lipids, which is a common process and of great importance to the functioning of cells, monomolecular layers are an interesting model system. In the case where the monolayer serves as a donor membrane, transfer to vesicles in the subphase or eventually to a separated monolayer, can be measured by following the change in surface radioactivity.

Specific transfer proteins can be isolated from the sytosolic fraction of mammalian tissues, plants, yeast, and bacteria (Wirtz, 1991). The specificity and mechanism of the bovine liver phosphatidylcholine transfer protein were elucidated using the monolayer model system. Although the surface radioactivity of the [^{14}C]phosphatidylcholine monolayer decreases rapidly in the presence of acceptor membranes and transfer protein, the surface pressure is not affected, showing that it acts as an exchange protein rather than as a transfer protein. Demonstration of the exchange between two separate monolayers and the binding of lipid to the protein present in the subphase proves a carrier mechanism (Demel *et al.*, 1973). A similar mechanism is apparent for the phosphatidylinositol transfer protein which as a high affinity for phosphatidylinositol and a low affinity for phosphatidylcholine. Although the protein mediated net transfer of phosphatidylinositol in model systems, there was no net mass transfer (Demel *et al.*, 1977b). It is interesting to observe the effect of membrane compounds not involved in the reaction. Phosphatidylcholine and sphingomyelin show a close resemblance of the head group, although their ratio is known to be membrane-specific (Van Deenen and De Gier, 1964). Sphingomyelin is not transferred by the phosphatidylinositol transfer protein but the transfer of phosphatidylinositol is inhibited by sphingomyelin (Demel et al., 1982b). Net mass transfer of galactosylceramide is catalyzed by the glycolipid transfer protein from pig brain (Sasaki and Demel, 1985). This can be rationalized by the fact that the glycolipid is bound to less than 50% of the transfer protein. Glycolipid concentrations in biological membranes regulate many essential functions. Therefore, it is interesting to see that the galactosylceramide transfer from a mixed galactosylceramide—phosphatidylcholine monolayer is already maximal at a concentration of 4 mol%. This could mean that the lateral distribution of the glycolipid is affected by phosphatidylcholine, which by itself is not transferred. The bovine nonspecific lipid transfer protein (SCP2) is more surface active and does not display any distinct specificity in lipid transfer. In membrane studies transfer proteins have proven to be a very useful tool in altering the lipid composition and symmetry of vesicles and biological membranes. The nonspecific lipid transfer protein does not act as a lipid carrier since no significant lipid binding was apparent. It is proposed that this protein may facilitate the transfer of lipids by being part of a transient collisional complex between donor and acceptor membrane (Van Amerongen et al., 1989). The exact function of fatty acid-binding proteins is still unclear. A possible involvement in fatty acid transfer was studied in the monolayer system. Oleic acid was withdrawn from monolayers at a high rate and the results indicate that the fatty acid-binding protein–fatty acid complex may function as an intermediate in the transfer of fatty acids between membranes (Peeters et al., 1989).

3.8. Interaction of Toxins and Drugs with Monolayers

Toxins and drugs either interact with the membrane or have to be transferred over the membrane. Their efforts on membranes can be studied in model membranes as monomolecular layers. Polyene antibiotics are widely used for their fungicidic activity. It was shown by the use of monomolecular layers that sterols are the site of action (Demel et al., 1965). The higher sensitivity of fungi for polyene antibiotics corresponds with the higher affinity for ergosterol (the fungal sterol) over cholesterol. This was demonstrated by permeability experiments with liposomes (Teerlink et al., 1980). for amphotericin B there is conclusive evidence that the polyene antibiotic–sterol complex is organized in a pore structure. Two such complexes will generate a hydrophilic channel which traverses the membrane.

Structural reorientation of toxin in the membrane as a cause of lytic activity is suggested for cardiotoxin III (Bougis et al., 1981). The apparent molecular area of cardiotoxin III during insertion into negatively charged phospholipid monolayers is dependent on the surface pressure and cardiotoxin III may exist in two different conformations—"flat" (14.0 nm^2) or edgewise (4.2 nm^2). The

surface activity of cardiotoxin is strongly increased in the presence of phospholipids. Many antitumor drugs such as doxorubicin and celiptium are positively
charged and share an important electrostatic interaction with negatively charged
phosphates of cardiolipin (Nicolay *et al.*, 1988). Recent evidence for a specific
interaction of doxorubicin with polyphosphoinostides in monomolecular layers
(De Wolf *et al.*, 1991) supports findings that doxorubicin can affect phospho-
inositide metabolism in cancer cells.

Substance P, a positively charged neurotransmitter peptide, caused an expansion of the monolayer surface area attributed to the insertion in lipid films of
particularly anionic lipids (Seelig and MacDonald, 1989). The biphasic surface
area increase suggested that two binding states are involved. Based on monolayer
and vesicle experiments a perpendicular insertion of the peptide is rejected
(Duplaa *et al.*, 1992). The hydrophobic binding between egg phosphatidylcholine and the fusion peptide of influenza virus hemagglutinin occur on acidifica-
tion of the media (Brunner, 1989) mimicking the low pH of the endosomal
compartment. From monolayer studies it is concluded that the low-pH-induced
activation of hemagglutinin of influenza virus results from a conformation
change. The surface activity of isolated hemagglutinin appears mainly to be
caused by penetration into the lipid monolayer of protein domains other than the
amino-terminus of subunit 2 of hemagglutinin; domains of subunit 1 could be
involved (Burger *et al.*, 1991). Lipid specificities of toxin interaction were
determined for diphertia toxin. The low-pH-driven lipid interaction of this toxin
is favored by the presence of acidic phospholipids without an apparent requirement for a particular class of anionic lipids (Demel *et al.*, 1991). Tetanus toxin
penetration into lipid monolayers is facilitated by acidic phospholipds and glycosphingolipid sulfates (Schiavo *et al.*, 1991). Electrostatic interactions play an
important role and are required as a first step in the process of penetration of
colicin A domains into negatively charged dilauroylphosphatidylglycerol monolayers (Frenette *et al.*, 1989).

4. PERSPECTIVE

The results obtained with monomolecular layers at the air–water interface
correlated well with those obtained with lipid bilayers. There is conclusive
evidence, reported in many papers, that monomolecular layers are a relevant
model system for biological membranes. This model system offers the possibility
for a more direct interpretation of penetration, adsorption, desorption, and molecular parameters resulting from interfacial interactions. Total internal reflection
fluorescence microscopy can be used to investigate the association of fluores-
cently labeled ligands at specific sites on supported planar model membranes.
Promising approaches for the visualization of monomolecular layers by scanning

tunneling microscopy will provide more detailed information on organization and distribution of membrane compounds. Monomolecular layers can also be used as a two-dimensional lattice for protein crystallization in order to elucidate secondary and tertiary structure (Darst *et al.*, 1991). The use of supported monolayers as biochemical sensors is a feasible perspective.

5. REFERENCES

Adamson, A. W., 1990, *Physical Chemistry of Surfaces*, Wiley–Interscience, New York.

Albrecht, O., Gruler, H., and Sackmann, E., 1981, Pressure–composition phase diagrams of cholesterol–lecithin, cholesterol–phosphatidic acid and lecithin–phosphatidic acid mixed monolayers: A Langmuir film balance study, *J. Colloid Interface Sci.* **79**:319–338.

Alegria, G., and Dutton, P. L. 1991, Langmuir–Blodgett monolayer films of bacterial photosynthetic membranes and isolated reaction centers: Preparation, spectrophotometric and electrochemical characterization, *Biochim, Biophys. Acta* **1057**:239–257.

Batenburg, A. M., Brasseur, R., Ruysschaert, J. M., Van Scharrenburg, G. J. M., Slotboom, A. J., Demel, R. A., and De Kruijff, B., 1988a, Characterization of the interfacial behavior and structure of the signal sequence of *Escherichia coli* outer membrane pore protein, *J. Biol. Chem.* **263**:4202–4207.

Batenburg, A. M., Demel, R. A., Verkleij, A. J., and De Kruijff, B., 1988b, Penetration of the signal sequences of *Escherichia coli* PhoE protein into phospholipid model membranes leads to lipid-specific changes in signal peptide structure and alterations of lipid organization, *Biochemistry* **27**:5678–5685.

Bazzi, M. D., and Nelsestuen, G. L., 1988, Association of protein kinase C with phospholipid monolayers: Two stage irreversible binding, *Biochemistry* **27**:6776–6783.

Bhat, S. G., and Brockman, H. L., 1981, Enzymatic synthesis—hydrolysis of cholesteryl oleate in surface films, *J. Biol. Chem.* **256**:3017–3023.

Bhat, S. G., and Brockman, H. L. 1982, Lipid hydrolysis catalyzed by pancreatic cholesterol esterase. Regulation by substrate and product phase distribution and packing density, *Biochemistry* **21**:1547–1552.

Birdi, K. S., 1973, Spread monolayer films of protein at the air–water interface. *J. Colloid Interface Sci.* **43**:545–547.

Blume, A., 1979, A comparative study of the phase transitions of phospholipid bilayers and monolayers, *Biochim, Biophys. Acta* **557**:32–44.

Bougis, P., Rochat, H., Piéroni, G., and Verger, R., 1981, Penetration of phospholipid monolayers by cardiotoxins, *Biochemistry* **20**:4915–4920.

Breukink, E., Demel, R. A., De Korte-Kool, G., and De Kruijff, B., 1992, SecA insertion into phosopholipids is stimulated by negatively charged lipids and inhibited by ATP. A monolayer study, *Biochemistry* **31**:1119–1124.

Brunner, J., 1989, Testing topological models for the membrane penetration of the fusion peptide of influenza virus hemagglutinin, *FEBS Lett.* **257**:369–372.

Burger, K.N.J., Wharton, S. A., Demel, R. A., and Verkleij, A. J., 1991, Interaction of influenza virus hemagglutinin with a lipid monolayer. A comparison of the surface activities of intact virions isolated hemagglutinins and synthetic fusion peptide, *Biochemistry* **30**:11173–11180.

Cornell, D. G., Dluhy, R. A., Briggs, M. S., McKnight, J., and Gierasch, L. M., 1989, Conformation and orientations of a signal peptide interacting with phospholipid monolayers, *Biochemistry* **28**:2789–2797.

Crisp, D. J., 1949, *Surface Chemistry*, pp. 17–22, Butterworths, London.

Darst, S. A., Ahlers, M., Meller, P. H., Kubalek, E. W., Blankenburg, R., Ribi, H. O., Ringsdorf, H., and Kornberg, R. D. 1991, Two dimensional crystals of streptavidin on biotinylated lipid layers and their interactions with biotinylated macromolecules, *Biophys. J.* **59**:387–396.

De Bernard, L., 1958, Associations moléculaires entre les lipides. II Lécithine et cholesterol, *Bull. Soc. Chim. Biol.* **40**:161–168.

De Kruijff, B., Demel, R. A., and Van Deenen, L.L.M. 1972, The effect of cholesterol and epicholesterol incorporation on the permeability and phase transition of intact *Acholeplasma laidlawii* cell membranes and derived liposomes, *Biochim. Biophys. Acta* **255**:331–347.

Demel, R. A., 1974, Model membrane monolayers—Description of use and interaction, *Methods Enzymol.* **32**:539–545.

Demel, R. A., 1982, Lipid–protein interactions in monomolecular layers, *in: Current Topics in Membranes and Transport* (A. Martonosi, ed.), pp. 159–164, Plenum Press, New York.

Demel, R. A., and De Kruijff, B., 1976, The function of sterols in membranes. *Biochim. Biophys. Acta.* **457**:109–132.

Demel, R. A., and Jackson, R. L., 1985, Lipoprotein lipase hydrolysis of trioleoylglycerol in a phospholipid interface, effect of cholestryl-oleate on catalysis. *J. Biol. Chem.* **260**:9589–9592.

Demel, R. A., Van Deenen, L.L.M., and Kinsky, S. C., 1965, Penetration of lipid monolayers by polyene antibiotic, *J. Biol. Chem.* **240**:2749–2753.

Demel, R. A., Van Deenen, L.L.M., and Pethica, B. A., 1967, Monolayer interactions of phospholipids and cholesterol, *Biochim, Biophys. Acta* **135**:11–19.

Demel, R. A., Geurts van Kessel, W.S.M., and Van Deenen, L.L.M., 1972a, The properties of polyunsaturated lecithins in monolayers and liposomes and the interactions of these lecithins with cholesterol, *Biochim. Biophys. Acta* **266**:26–40.

Demel, R. A., Bruckdorfer, K. R., and Van Deenen, L.L.M., 1972b, Structural requirements of sterols for the interaction with lecithin at the air–water interface, *Biochim. Biophys. Acta* **255**:311–320.

Demel, R. A., Wirtz, K.W.A., Kamp, H. H., Geurts van Kessel, W.S.M., and Van Deenen, L.L.M., 1973, Phosphatidylcholine exchange protein from beef liver, *Nature* **246**:102–105.

Demel, R. A., Geurts van Kessel, W.S.M., Zwaal, R.F.A., Roelofsen, B., and Van Deenen, L.L.M., 1975, Relation between various phospholipase actions on human red cell membranes and the interfacial phospholipid pressure in monolayers, *Biochim. Biophys. Acta* **406**:97–107.

Demel, R. A., Jansen, J.W.C.M., Van Dijck, P.W.M., and Van Deenen, L.L.M., 1977a, The preferential interaction of cholesterol with different classes of phospholipids, *Biochim, Biophys. Acta* **465**:1–10.

Demel, R. A., Kalsbeek, R., Wirtz, K.W.A., and Van Deenen, L.L.M., 1977b, The protein-mediated net transfer of phosphatidylinositol in model systems, *Biochim. Biophys. Acta* **466**:10–22.

Demel, R. A., Shirai, K., and Jackson, R. L., 1982a, Lipoprotein lipase-catalyzed hydrolysis of tri[14C] oleoylglycerol in a phospholipid interface, *Biochim. Biophys. Acta* **713**:629–637.

Demel, R. A., Van Bergen, B.G.M., Van den Eeeden, A.L.G., Zborowski, J., and Defize, L.H.K., 1982b, Transfer properties of the bovine brain phospholipid transfer protein specificity towards phosphatidyl choline analogs and the inhibitory effect of sphingomyelin, *Biochim. Biophys. Acta* **710**:264–270.

Demel, R. A., Dings, P. J., and Jackson, R. L., 1984a, Effect of monolayer lipid structure and composition on the lipoprotein lipase-catalyzed hydrolysis of triacyl-glycerol, *Biochim. Biophys. Acta* **793**:399–408.

Demel, R. A., Lala, A. K., Kumari, S. N., and Van Deenen, L.L.M., 1984b, The effect of sterol oxygen function on the interaction with phospholipids, *Biochim. Biophys. Acta* **771**:142–150.

Demel, R. A., Jordi, W., Lambrechts, H., Van Damme, H., Hovius, R., and De Kruijff, B.,

1989, Differential interactions of apo- and holocytochrome c with acidic membrane lipids in model systems and the implications for their import into mitochondria, *J. Biol. Chem.* **264**:3988–3997.

Demel, R. A., Goormaghtigh, E., and De Kruijff, B., 1990, Lipid and peptide specificities in signal peptide–lipid interaction in model membranes, *Biochim. Biophys. Acta* **1027**:155–162.

Demel, R. A., Schiavo, G., De Kruijff, B., and Montecucco, C., 1991, Lipid interaction of diphtheria toxin and mutants. A study with phospholipid and protein monolayer, *Eur. J. Biochem.* **197**:481–486.

Demel, R. A., Yin, C. C., Lin, B. Z., and Hauser, H., 1992a, Monolayer characteristics and thermal behavior of natural and synthetic phosphatidic acids, *Chem. Phys. Lipids* **60**:209–223.

Demel, R. A., Van Doorn, J. M., and Van Der Horst, D. J., 1992b, Insect apolipophorin III. Interaction of locust apolipophorin III with diacylglycerol. *Biochim. Biophys. Acta* **1124**:151–158.

De Vrije, T., De Swart, R. L., Dowhan, W., Tommassen, J., and De Kruijff, B. 1988, Phosphatidylglycerol is involved in protein translocation across *Escherichia coli* inner membranes, *Nature* **334**:173–175.

De Wolf, F. A., Demel, R. A., Bets, D., Van Katz, C., and De Kruijff, B., 1991, Characterization of the interaction of doxorubicin with (poly)phosphoinositides in model systems: Evidence for specific interaction with phosphatidyl inositol monophosphate and -diphosphate, *FEBS Let.* **288**:237–240.

Ducharme, D., Salesse, C., and Leblanc, R. M., 1985, Ellipsometric studies of rod outer segment phospholipids at the nitrogen–water interface, *Thin Solid Films* **132**:83–90.

Duplaa, H., Convert, O., Sautereau, A. M., Tocanne, J. F., and Chassaing, G., 1992, Binding of substance P to monolayers and vesicles made of phosphatidylcholine and/or phosphatidylserine, *Biochim. Biophys. Acta* **1107**:12–22.

Eklund, K. K., Vuorinen, J., Mikkola, J., Virtanen, J. A., and Kinnunen, P.K.J., 1988, Ca^{2+}-induced lateral phase separation in phosphatidic acid–phosphatidylcholine monolayers as revealed by fluorescence microscopy, *Biochemistry* **27**:3433–3437.

Fidelio, G. D., Ariga, T., and Maggio, B., 1991, Molecular parameters of gangliosides in monolayers comparative evaluation of suitable purification procedures, *J. Biochem.* **110**:12–16.

Frenette, M., Knibiehler, M., Baty, D., Géli, V., Pattus, F., Verger, R., and Lazdunski, C., 1989, Interactions of colicin A domains with phospholipid monolayers and liposomes: Relevance to the mechanism of action, *Biochemistry* **28**:2509–2514.

Fukuda, K., Nakahara, H., and Kato, T., 1976, Monolayers and multilayers of anthraquinone derivatives containing long alkyl chains, *J. Colloid Interface Sci.* **54**:431–437.

Gaines, G. L., 1966, *Insoluble Monolayers at Gas–Liquid Interfaces*, Wiley, New York.

Gallay, J., De Kruijff, B., and Demel, R. A., 1984, Sterol–phospholipid interactions in model membranes: Effects of polar group substitutions in the cholesterol side chain at C20 and C22, *Biochim. Biophys. Acta* **769**:96–104.

Gieles, P.M.C., 1987, Thesis, Technical University Eindhoven, The Netherlands.

Gold, J., and Phillips, M. C., 1990, Effect of membrane lipid composition on the kinetics of cholesterol exchange between lipoproteins and different species of red blood cells, *Biochim. Biophys. Acta* **1027**:85–92.

Gorter, E., and Grendel, F., 1925, On bimolecular layers of lipids on the chromocytes of the blood, *J. Exp. Med.* **41**:439–443.

Gosh, D., Williams, M. A., and Tinoco, J., 1973, The influence of lecithin structure on their monolayer behavior and interactions with cholesterol, *Biochim. Biophys. Acta* **291**:351–362.

Graham, D. E., and Phillips, M. C., 1980, Proteins at liquid interfaces dilatational properties, *J. Colloid Interface Sci.* **76**:227–239.

Grainger, D. W., Reichert, A., Ringsdorf, H., and Salesse, C., 1990, Hydrolytic action of phospha-
  lipase A$_2$ in monolayers in the phase transition region: Direct observation of enzyme domain
  formation using fluorescence microscopy, Biochim. Biophys. Acta **1023**:365-379.
Grönberg, L., Ruan, Z., Bittman, R., and Slotte, J. P., 1991, Interaction of cholesterol with
  synthetic sphingomyelin derivatives in mixed monolayers, Biochemistry **30**:10746-10754.
Gruen, D.W.R., and Wolfe, J., 1982, Lateral tensions and pressures in membranes and lipid monolayers, Biochim. Biophys. Acta **688**:572-580.
Hawgood, S., Benson, B. J., Schilling, J., Damm, D., Clements, J. A., and White, R. T., 1987,
  Nucleotide and amino acid sequences of pulmonary surfactant protein SP18 and evidence for
  cooperation between SP18 and SP28-36 in surfactant lipid adsorption, Proc. Natl. Acad. Sci.
  USA **84**:66-70.
Heckl, W. M., Lösche, M., Scheer, H., and Möhwald, H., 1985, Protein–lipid interactions in
  phospholipid monolayers containing the bacterial antenna protein B800–850, Biochim. Biophys. Acta **810**:73-83.
Heckl, W. M., Zaba, B. N., and Möhwald, H., 1987, Interactions of cytochromes b5 and c with
  phospholipid monolayers, Biochim. Biophys. Acta **903**:166-176.
Helm, C. A., Tippmann-Krayer, P., Möhwald, H., Als-Nielsen, H., and Kjaer, K., 1991, Phases of
  phosphatidylethanolamine monolayers studied by synchotron X-ray scattering, Biophys. J.
  60:1457-1476.
Heyn, S. P., Egger, M., and Gaub, H. E., 1990, Lipid and lipid–protein monolayers spread from a
  vesicle suspension: A microfluorescence film balance study, J. Phys. Chem. **94**:5073-5078.
Heyn, S. P., Tillmann, R. W., Egger, M., and Gaub, H. E., 1991, A miniaturized microfluorescence
  film balance for protein-containing lipid monolayers spread from a vesicle suspension, J.
  Biochem. Biophys. Methods **22**:145-158.
Hirasawa, K., Irvine, R. F., and Dawson, R.M.C., 1981, The hydrolysis of phosphatidylinositol
  monolayers at the air–water interface by the calcium ion-dependent phosphatidylinositol phos-
  phodiesterase of pig brain, Biochem. J. **193**:607-614.
Horowitz, A. D., Elledge, B., Whitsett, J. A., and Baatz, J. E., 1992, Effect of lung surfactant
  proteolipid SP-C on the organization of model membrane lipids: A fluorescence study, Biochim.
  Biophys. Acta **1107**:44-54.
Hoyt, D. W., Cyr, D. M., Gierasch, L. M., and Douglas, M. G., 1991, Interaction of peptides
  corresponding to mitochondrial presequences with membranes, J. Biol. Chem. **266**:21693-
  21699.
Jackson, R. L., Pattus, F., and Demel, R. A., 1979, Interaction of plasma apolipoproteins with lipid
  monolayers, Biochim. Biophys. Acta **556**:369-387.
Jackson, R. L., Ponce, E., McLean, L. R., and Demel, R. A., 1986, Comparison of the tri-
  cylglycerol hydrolase activity of human post-heparin plasma lipoprotein lipase and hepatic-
  triacylglycerol lipase. A monolayer study, Biochemistry **25**:1166-1170.
Jentoft, N., and Dearborn, D. G., 1979, Labeling of proteins by reductive methylation using sodium
  cyanoborohydride, J. Biol. Chem. **254**:4359-4365.
Johnson, S. J., Bayerl, T. M., Weihan, W., Noack, H., Penfold, J., Thomas, R. K., Kanellas, D.,
  Rennie, A. R., and Sackmann, E., 1991, Coupling of spectrin and polylysine to phospholipid
  monolayers studied by specular reflection of neurons, Biophys. J. **60**:1017-1025.
Jones, J. D., McKnight, C. J., and Gierasch, L. M., 1990, Biophysical; studies of signal peptides:
  Implications for signal sequence functions and involvement of lipid in protein export, J. Bio-
  energ. Biomembr. **22**:213-232.
Joos, P., and Demel, R. A., 1969, The interaction energetics of cholesterol and lecithin in spread
  mixed monolayers at the air–water interface, Biochim. Biophys. Acta **183**:447-457.
Kalb, E., Frey, S., and Tamm, L. K., 1992, Formation of supported planar bilayers by fusion of
  vesicles to supported phospholipid monolayers, Biochim. Biophys. Acta **1103**:307-316.

Keller, R.C.A., Killian, J. A.. and De Kruijff, B., 1992, Anionic phospholipids are essential for α-helix formation of the signal peptide of prePhoE upon interaction with phospholipid vesicles, *Biochemistry* **31**:1672–1677.

Knobler, C. H., 1990, Seeing phenomena in flatland. Studies of monolayers by fluorescence microscopy, *Science* **249**:870–874.

Kozarac, Z., Dhathathreyan, A., Möbius, D., 1988, Adsorption of cytochrome c to phospholipid monolayers studied by reflection spectroscopy, *FEBS Lett.* **229**:372–376.

Krebs, K. E., Ibdah, J. A., and Phillips, M. C., 1988, A comparison of the surface activities of human apolipoproteins A-I and A-II at the air–water interface, *Biochim. Biophys. Acta* **959**:229–237.

Kuroki, Y., and Akino, T., 1991, Pulmonary surfactant protein A specifically binds dipalmitoylphosphatidylcholine, *J. Biol. Chem.* **266**:3068–3073.

Laboda, H. M., Glick, J. M.. and Philips, M. C., 1986, Hydrolysis of lipid monolayers and the substrate specificity of hepatic lipase, *Biochim. Biophys. Acta* **876**:233–242.

Laboda, H. M., Glick, J. M., and Phillips, M. C., 1988, Influence of the structure of the lipid–water interface on the activity of hepatic lipase, *Biochemistry* **27**:2313–2319.

Lill, R., Dowhan, W., and Wickner, W., 1990, The ATPase activity of SecA is regulated by acidic phospholipids SecY and the leader and mature domains of precursor proteins, *Cell* **60**:271–280.

Lundberg, B., Svens, E., and Ekman, S., 1978, The hydration of phospholipids and phospholipid—cholesterol complexes, *Chem. Phys. Lipids* **22**:285–292.

McConnell, H. M., Tamm, L. K., and Weis, R. M., 1984, Periodic structures in lipid monolayer phase transitions, *Proc. Natl. Acad. Sci. USA* **81**:3249–3253.

MacDonald, R. C., and Simon, S. A., 1987, Lipid monolayer states and their relationship to bilayers, *Proc. Natl. Acad. Sci. USA* **84**:4089–4093.

McKnight, C. J., Briggs, M. S., and Gierasch, L. M., 1989, Functional and non-functional LamB signal sequences can be distinguished by their biophysical properties, *J. Biol. Chem.* **264**:17293–17297.

McLean, L. R., Demel, R. A., Socorro, L., Shinomiya, M., and Jackson, R. L., 1986, Mechanism of action of lipoprotein lipase, *Methods Enzymol.* **129**:738–763.

Malcolm, B. R., 1968, Molecular structure and deuterium exchange in monolayers of synthetic polypeptides, *Proc. R. Soc. A London Ser.* **305**:363–385.

Mattai, J., Hauser, H., Demel, R. A., and Shipley, G. G., 1989, Interactions of metal ions with phosphatidylserine bilayer membranes: Effect of hydrocarbon chain unsaturation, *Biochemistry* **28**:2322–2330.

Mayer, L. D., Nelsestuen, G. L., and Brockman, H. L., 1983, Prothrombin association with phospholipid monolayers, *Biochemistry* **22**:316–324.

Meller, P., 1985, Thesis, University of Munich, Germany.

Moreau, H., Pieroni, G., Jolivet-Reynaud, C., Alouf, J. E., and Verger, R., 1988, A new kinetic approach for studying phospholipase C (*Clostridium perfingens* α toxin) activity on phospholipid monolayers, *Biochemistry* **27**:2319–2323.

Mozaffary, H., 1991, On the sign and origin of the surface potential of phospholipid monolayers, *Chem. Phys. Lipids* **59**:39–47.

Nag, K., Boland, C., Rich, N., and Keough, K.M.W., 1991, Epifluorescence microscopic observation of monolayers of dipalmitoylphosphatidylcholine: Dependence of domain size on compression rates, *Biochim. Biophys. Acta* **1068**:157–160.

Nicolay, K., Sauterau, A. M., Tocanne, J. F., Brasseur, R., Huart, R., Ruysschaert, J. M., and De Kruijff, B., 1988, A comparative model membrane study on structural effects of membrane-active positively charged anti-tumor drugs, *Biochim. Biophys. Acta* **940**:197–208.

Oosterlaken-Dijksterhuis, M. A., Haagsman, H. P., Van Golde, L.M.G., and Demel, R. A., 1991,

Interaction of lipid vesicles with monomolecular layers containing lung surfactant proteins SP-B or SP-C, *Biochemistry* **30**:8276–8281.

Oosterlaken-Dijksterhuis, M. A., Haagsman, H. p., Van Golde, L.M.G., and Demel, R. A., 1991b, Characterization of lipid insertion into monomolecular layers mediated by lung surfactant proteins SP-B and SP-C, *Biochemistry* **30**:10965–10971.

Pattus, F., Slotboom, A. J., and De Haas, G. H., 1979, Regulation of phospholipase A$_2$ activity by the lipid–water interface: A monolayer approach, *Biochemistry* **18**:2691–2697.

Pattus, F., Rothen, C., Streit, M., and Zahler, P., 1981, Further studies on the spreading of biomembranes at the air–water interface. Structure, composition, and enzyme activities of human erythrocyte and sarcoplasmic reticulum membrane films, *Biochim. Biophys. Acta* **647**:29–39.

Peeters, R. A., Veerkamp, J. H., and Demel, R. A., 1989, Are fatty acid-binding proteins involved in fatty acid transfer? *Biochim. Biophys. Acta* **1002**:8–13.

Peters, R., and Beck, K., 1983, Translational diffusion in phospholipid monolayers measured by fluorescence microphotolysis, *Proc. Natl. Acad. Sci. USA* **80**:7183–7187.

Pethica, B. A., 1955, The thermodynamics of monolayer penetration at constant area, *Trans. Faraday Soc.* **51**:1402–1407.

Phillips, M. C., 1970, Molecular interactions in mixed lecithin systems, *Biochim. Biophys. Acta* **196**:35–44.

Phillips, M. C., 1972, The physical state of phospholipids and cholesterol in monolayers, bilayers, and membranes, in: *Progress in Surface and Membrane Science* (J. F. Danielli, M. D. Rosenberg, and D. A. Cadenhead, eds.), Vol. 5, pp. 139–221, Academic Press, New York.

Phillips, M. C, and Chapman, D., 1968, Monolayer characteristics of saturated 1,2-diacylphosphatidylcholines and phosphatidylethanolamines at the air–water interface, *Biochim. Biophys. Acta* **163**:301–313.

Phillips, M. C., and Krebs, K. E., 1986, Studies of apolipoproteins at the air–water interface, *Methods Ezymol.* **128**:387–403.

Pilon, M., Jordi, W., De Kruijff, B., and Demel, R. A., 1987, Interactions of mitochondrial precursor protein apocytochrome c with phosphatidylserine in model membranes, *Biochim. Biophys. Acta* **902**:207–216.

Pisarchick, M., and Thompson, N. L., 1990, Binding of a monoclonal antibody and its Fab fragment to supported phospholipid monolayers measured by total internal reflection fluorescence microscopy, *Biophys. J.* **58**:1235–1249.

Popot, J. L., Demel, R. A., Sobel, A., Van Deenen, L.L.M., and Changeux, J. P. 1978, Interaction of the acetylcholine (nicotinic) receptor protein from *Torpedo marmorata* electric organ with monolayers of pure lipids, *Eur. J. Biochem.* **85**:27–42.

Possmayer, F., 1988, A proposed nomenclature for pulmonary surfactant associated proteins, *Annu. Rev. Respir. Dis.* **138**:990–998.

Quinn, P. J., and Dawson, R.M.C., 1969, Interactions of cytochrome c and [^{14}C] carboxymethylated cytochrome c with monolayers of phosphatidylcholine, phosphatidic acid and cardiolipin, *Biochem. J.* **115**:65–75.

Quinn, P. J., and Dawson, R.M.C., 1970, An analysis of the interaction of protein with lipid monolayers at the air–water interface, *Biochem. J.* **116**:617–680.

Reinhardt-Schlegel, H., Kawamura, Y., Furuno, T., and Sasabe, H., 1991, Microstructure of phospholipid monolayers studied by dark field electron and fluorescence microscopy, *J. Colloid Interface Sci.* **147**:295–306.

Ries, H. E., Matsumoto, M., Uyeda, N., and Suito, E., 1975, Monomolecular layers viewed by electron microscopy, *Adv. Chem. Ser.* **144**:286–293.

Rojo, M., Hovius, R. E., Demel, R. A., Nicolay, K., and Walliman, T., 1991, Mitochondrial creatine kinase mediates contact formation between mitochondial membranes, *J. Biol. Chem.* **266**:20290–20295.

Rouser, G., 1983, Membrane composition, structure and function, in *Membrane Fluidity in Biology* (R. C. Aloia, ed.), Vol. 1, Academic Press, New York.

Salesse, C. S., Ducharme, D., LeBlanc, R. M., and Boucher, F., 1990, Estimation of disk membrane lateral pressure and molecular area of rhodopsin by the measurement of its orientation at the nitrogen–water interface from an ellipsometric study, *Biochemistry* **29**:4567–4575.

Salesse, C. S., Ducharme, D., and LeBlanc, R. M., 1987, Direct evidence for the formation of a monolayer from a bilayer. An ellipsometric study at the nitrogen–water interface, *Biophys. J.* **52**:351–352.

Sasaki, T., and Demel, R. A., 1985, Net mass transfer of galactosyl ceramide stimulated by glycolipid transfer protein from pig brain, *Biochemistry* **24**:1079–1083.

Schiavo, G., Demel, R. A., and Montecucco, C., 1991, On the role of polysialoglycosphingolipids as tetanus toxin receptors. A study with lipid monolayers, *Eur. J. Biochem.* **199**:705–711.

Schürholz, T., and Schindler, H., 1991, Lipid–protein surface films generated from membrane vesicles, *Eur. Biophys. J.* **20**:71–78.

Seelig, A., 1987, Local anesthetics and pressure, a comparison of dibucarine binding to lipid monolayers and bilayers, *Biochim. Biophys, Acta* **899**:196–204.

Seelig, A., and MacDonald, P. M., 1989, Binding of a neuropeptide, substance P, to neutral and negatively charged lipids, *Biochemistry* **28**:2490–2496.

Shiffer, K. A., Goerke, J., Düzgünes, N., Fedor, J., and Shohet, S. B., 1988, Interaction of erythrocyte protein 4.1 with phospholipids. A monolayer and liposome study, *Biochim. Biophys. Acta* **937**:269–280.

Slotte, J. P., 1992, Enzyme catalyzed oxidation of cholesterol in mixed phospholipid monolayers reveals the stoichiometry at which free cholesterol clusters disappear, *Biochemistry* **31**:5472–5477.

Standish, M. M., and Pethica, B. A., 1967, Interactions in phospholipid–cholesterol mixed monolayers at the air–water interface, *Biochim. Biophys. Acta* **144**:659–665.

Subramaniam, S., Seul, M., and McConnell, H. M., 1986, Lateral diffusion of specific antibodies bound to lipid monolayers on alkylated substrates, *Proc. Natl. Acad. Sci. USA* **83**:1169–1173.

Sugi, M., 1985, Langmuir–Blodgett films. A course towards molecular electronics, *J. Mol. Electron.* **1**:3–17.

Tamm, L. K., 1986, Incorporation of a synthetic mitochondrial signal peptide into charged and uncharged phospholipid monolayers, *Biochemistry* **25**:7470–7476.

Tamm, L. K., Tomich, J. M., and Saier, M. H., 1989, Membrane incorporation and induction of secondary structure of synthetic peptides corresponding to the N-terminal signal sequences, of the glucitol and mannitol permeases of *Escherichia coli*, *J. Biol. Chem.* **264**:2587–2592.

Taylor, D. M., De Olivera, O. N., and Morgan, H., 1990, Models for interpreting surface potential measurements and their application to phospholipid monolayers, *J. Colloid Interface Sci.* **139**:508–518.

Teerlink, T., De Kruijff, B., and Demel, R. A., 1980, The action of pimaricin, etruscomycin and amphotericin B on liposomes with varying sterol content, *Biochim. Biophys. Acta* **599**:484–492.

Theunissen, J.J.H., Jackson, R. L., Kempen, H.J.M., and Demel, R. A., 1986, Membrane properties of oxysterols. Interfacial orientation influence on membrane permeability and redistribution between membranes, *Biochim. Biophys. Acta* **860**:66–74.

Thuren, T., Eklund, K. K., Virtanen, J. A., and Kinnunen, P.K.J., 1990, Hydrolysis of supported pyrene–phospholipid monolayers by phospholipase A_2, *Chem. Phys. Lipids* **55**:55–60.

Thuren, T., Wilcox, R. W., Sisson, P., and Waite, M., 1991, Hepatic lipase hydrolysis of lipid monolayers, *J. Biol. Chem.* **266**:4853–4861.

Tournois, H., Gieles, P., Demel, R. A., De Gier, J., and De Kruijff, B., 1989, Interfacial properties of gramicidin and gramicidin–lipid mixtures measured with static and dynamic monolayer techniques, *Biophys. J.* **55**:557–569.

Toyoshima, C., and Unwin, N., 1988, Ion channel of acetylcholine receptor reconstructed from images of postsynaptic membranes, *Nature* **336**:247–250.

Träuble, H., 1971, Phasenumwandlungen in lipiden. Mögliche schaltprozesse in biologischen membranen, *Naturwissenschaften* **58**:277–281.

Trurnit, H. H., 1960, A theory and method for spreading of protein monolayers, *J. Colloid Sci.* **14**:1–13.

Uzgiris, E. E., 1987, Self-organization of IgE immunoglobulins on phospholipid films, *Biochem. J.* **242**:293–296.

Van Amerongen, A., Demel, R. A., Westerman, J., and Wirtz, K.W.A., 1989, Transfer of cholesterol and oxysterol derivatives by the non-specific lipid transfer protein (sterol carrier protein 2). A study on its mode of action, *Biochim. Biophys. Acta* **1004**:36–43.

Van Deenen, L.L.M., and De Gier, J., 1964, Chemical composition and metabolism of lipids in red cells of various animal species, in: *The Red Blood Cell* (C. Bishop and D. M. Surgenor, eds.), pp. 243–302, Academic Press, New York.

Van den Tempel, M., and Lucassen-Reynders, E. H., 1983, Relaxation processes at fluid interfaces, *Adv. Colloid Interface Sci.* **18**:281–286.

Van Golde, L.M.G., Batenburg, J. J., and Robertson, B., 1988, The pulmonary surfactant system: Biochemical aspects and functional significance, *Physiol. Rev.* **68**:374–455.

Van Liempd, J.P.J.G., Boonman, A.A.H., Demel, R. A., Gieles, P.M.C., and Gorell, T.C.M., 1987, Non-selective squeeze-out of dioleoylphosphatidylcholine and dioleoylphosphatidyl glycerol from binary mixed monolayers with dipalmitoylphosphatidylcholine, *Biochim. Biophys. Acta* **897**:495–501.

Van't Hof, R., Van Klompenburg, W., Pilon, M., Kozubek, A., De Korte-Kool, G., Demel, R. A., Weisbeek, P. J., and De Kruijff, B., 1993, The transit sequence mediates the specific interaction of the precursor of ferredoxin with chloroplast envelope membrane lipids, *J. Biol. Chem.* **268**:4037–4042.

Verger, R., and De Haas, G. H., 1976, Interfacial enzyme kinetics of lipolysis, *Annu. Rev. Biophys. Bioenerg.* **5**:77–117.

Weis, R. M., 1991, Fluorescence microscopy of phospholipid monolayer phase transitions, *Chem. Phys. Lipids* **57**:227–239.

Weis, R. M., and McConnell, H. M., 1985, Cholesterol stabilized the crystal–liquid interface in phospholipid monolayers, *Biophys. J.* **47**:44a.

Wirtz, K.W.A., 1991, Phospholipid transfer proteins, *Annu. Rev. Biochem.* **60**:73–99.

Yedgar, S., Cohen, R., Gatt, S., and Barenholz, Y., 1982, Hydrolysis of monomolecular layers of synthetic sphingomyelins by sphingomyelinase of *Staphylococcus aureus*, *Biochem. J.* **201**:597–603.

Zwaal, R.F.A., Demel, R. A., Roelofsen, B. and Van Deenen, L.L.M., 1976, The lipid bilayer concept of cell membranes, *Trends Biochem. Sci.* **1**:112–114.

Chapter 4

Differential Scanning and Dynamic Calorimetric Studies of Cooperative Phase Transitions in Phospholipid Bilayer Membranes

Qiang Ye and Rodney L. Biltonen

1. INTRODUCTION

In this chapter, we will describe the application of equilibrium and dynamic calorimetric techniques to investigate the thermodynamic and dynamic properties of lipid bilayers. Although lipids undergo several types of temperature- and pressure-dependent phase transitions, we will emphasize the gel–liquid-crystalline transition because it has been the most extensively investigated with regard to the effects of temperature and pressure on its equilibrium poise. In addition, the gel–liquid-crystalline transition is also the lipid transition whose relaxation kinetics have been mainly investigated using a variety of techniques.

We will begin our discussion with a statement regarding the potential biological significance of the existence of the gel–liquid-crystalline transition and the usefulness in understanding its physical nature. We will then review the

Qiang Ye and Rodney L. Biltonen   Departments of Biochemistry and Pharmacology, University of Virginia Health Sciences Center, Charlottesville, Virginia 22908.

Subcellular Biochemistry, Volume 23: Physicochemical Methods in the Study of Biomembranes, edited by Herwig J. Hilderson and Gregory B. Ralston. Plenum Press, New York, 1994.

technique of differential scanning calorimetry (DSC) and discuss results obtained with single and binary lipid systems. A special section regarding the effects of pressure on the lipid gel–liquid-crystalline transition will then be presented. Finally, we will describe the technique of volume perturbation calorimetry and its application to obtain dynamic information about the lipid phase transition. In these discussions, most of the data presented will be from this laboratory. This is in part related to the fact that such data are most accessible to us, but it is also related to the facts that this laboratory possesses one of the few differential scanning calorimeters that can operate at high pressures (Mountcastle et al., 1978) and that the volume-perturbation calorimeter is an instrument unique to this laboratory (van Osdol et al., 1989). Nevertheless, it is hoped that we will present a broad overview of the usefulness of the application of dynamic calorimetry to the study of lipid phase transitions.

2. THE GEL–LIQUID-CRYSTALLINE TRANSITION IN BIOLOGICAL MEMBRANES

The first experimental evidence that the lipid matrix of a biological membrane can undergo a gel–liquid-crystalline transition similar to that reported for aqueous dispersions of pure lipid bilayers was provided by Steim et al. (1969). They showed using DSC that bilayers formed from lipids extracted from the mycoplasma *Acholeplasma laidlawii* exhibited a phaselike transition with two heat capacity maxima in the temperature range of 30–70°C. The first of these maxima was totally reversible, associated with the lipid, and occurred at a temperature a few degrees below the growth temperature of the organism. The implied biological relevance of this lipid transition has prompted decades of active investigation into its physical nature, its relationship to associated membrane protein function, and its biological significance in general. The subsequent understanding of the cooperative nature of lipid phase transitions and their influence on protein binding, aggregation, and function had a major impact on changing the traditional view that the lipid matrix played only a passive role in membrane functions. It is now generally accepted that the physical and functional properties of biological membranes are a reflection of both the local and global organization of the lipids and proteins within the membrane (Singer and Nicolson, 1972; Luzzati and Tardieu, 1974; Melchior and Steim, 1976; Nagle, 1980; Cevc and Marsh, 1987; McElhaney, 1989; Biltonen, 1990; Thompson et al., 1992).

Numerous examples demonstrating that protein functions can be modulated by the gel-to-crystalline phase transition of model membranes exist, but the physiological relevance of these correlations is still debated. These examples include the passive diffusion of sodium which exhibits a maximum near the transition temperature (Papahadjopoulos et al., 1973; Mouritsen and Jorgensen, 1991), the observation by Wickner (1975) that incorporation of a 5-kDa viral protein into lipid vesicles occurred most readily in the transition range, and the observation that phospholipase A_2-catalyzed hydrolysis was most rapid in the transition range of the lipid substrate (Op den Kamp et al., 1975; Lichtenberg et al., 1981). However, it could be argued that these correlations are not physiologically relevant in mammalian systems, particularly because the phase transition temperature of many lipids extracted from the cells of these systems have a gel-to-liquid-crystalline phase transition at temperatures well below physiological ranges (Nelson, 1967; Keenan and Moore, 1970; Kleinig, 1970; van Deenen and de Gier, 1974; Thompson and Huang, 1986). One exception is the lipids of the red blood cell which, after removal of cholesterol, exhibit a broad transition centered at approximately 33°C (Jackson et al., 1973). But in this case it might be argued that the transition is not significant because in the presence of the cholesterol the transition cannot be observed, as is the case with the lipid matrix of many cellular systems. However, the absence of a sharp transition does not preclude its existence. This is particularly true for complex lipid mixtures where the degree of cooperativity of a phaselike transition will be greatly reduced. Even in the case of binary lipid mixtures which are highly immiscible, the transition in multilamellar liposomes will occur over a range of 30 to 40°C, compared to a few tenths of a degree for a single-component system (Mabrey and Sturtevant, 1976; Biltonen, 1990). Furthermore, as the system becomes less apparently cooperative, fluctuations associated with the phase transition could extend many degrees above the transition midpoint.

Temperature and pressure are complementary thermodynamic variables. For lipid phase transitions, an increase in temperature results in a higher degree of melting whereas an increase in pressure induces formation of the gel phase (Trudell et al., 1975; Mountcastle et al., 1978). Thus, temperature and pressure have antagonistic effects. An important observation indicating the significance of the gel–liquid transition of lipids in biological functions is that microorganisms and fish can adapt to survival at high pressures by altering their lipid composition in a manner as to lower the melting temperature of the membrane lipid (Melchior and Steim, 1976; Delong and Yaganos, 1985). Likewise, the function of many enzymes as well as genetic transcription are sensitive to pressure changes (Somero, 1992). Finally, it has been suggested that plant sensitivity to dehydration is the result of an increase in the melting temperature of the lipid such that it transforms from primarily a liquid state to a gel state (Crowe et al., 1992). It is thought that the sugar trehalose, which allows the plants to survive, stabilizes the liquid state of the lipid by replacing the water at the polar surface of the membrane (Crowe et al., 1992).

One characteristic of the existence of the transition which could be important in determining the coupling of a protein function to the lipid structure is the fluctuation behavior of the lipid matrix (Freire and Biltonen, 1978; Lichtenberg *et al.*, 1981; Ruggiero and Hudson, 1989; Mouritsen and Jorgensen, 1991). These fluctuations can be either dynamic or static, but both are related to the size of clusters or domains of like lipids. These fluctuations produce a dynamic heterogeneity in the system that could greatly influence membrane functions. For example, Mouritsen and Jorgensen (1991) have suggested that the rate of passive sodium diffusion is quantitatively related to the number and size of gel/liquid clusters implying a correlation between the boundary regions in the lipid matrix and diffusion. A similar correlation can be made between any function which exhibits a maximum in a phase transition region.

Another type of phase transition which likely plays an important role in membrane function is compositional lateral phase separation. Such lateral phase separation can be associated exclusively with lipids or could include proteins. The potential of lateral phase separation and domain structure has been discussed in a recent edition of *Comments in Molecular and Cellular Biophysics* (Vol. 8, 1992). A rapid increase in the activity of phospholipase A_2 toward vesicular substrates is associated with lateral phase separation of reaction products from the phospholipid (Burack *et al.*, 1993). Although it is generally though that fluid state lipids are homogeneously dispersed in the plane of the membrane, recent evidence on Ca^{2+} binding to phosphatidylcholine/phosphatidylserine systems indicated segregation of the lipid components in the fluid state (J. Huang *et al.*, 1993). Such information can, in principle, be obtained from experimentally determined phase diagrams.

3. EQUILIBRIUM PROPERTIES OF MEMBRANE PHASE TRANSITIONS

3.1. Differential Scanning Calorimetry

A differential scanning calorimeter measures the heat capacity of a sample material as a continuous function of temperature. The uniqueness of DSC lies in its capability of determining the energetics of a phase transition process directly from a measurement of the heat capacity of the system:

$$\Delta H = \int_{T_0}^{T} C_p(T) dT \tag{1}$$

$$\Delta S = \int_{T_0}^{T} [C_p(T)/T] dT \tag{2}$$

$$\Delta G = \Delta H - T\Delta S \tag{3}$$

where $C_p(T)$, ΔH, ΔS, and ΔG represent the excess heat capacity, enthalpy change, entropy change, and Gibbs free energy change associated with the transition, respectively; T represents any given temperature and T_0 a reference at which all lipid molecules exist in a single well-defined state.

All scanning calorimeters currently in use employ a differential method, i.e., the sample heat capacity is measured relative to the heat capacity of some reference material such as a water or buffer solution. Two basic designs for DSC exist and both take advantage of the fact that two materials of equal amount at the same initial temperature will have the same final temperature after absorbing equal amounts of heat, if and only if their heat capacities are identical. Therefore, when a sample and reference material, initially equilibrated at the same temperature, are both heated with the same power input, any difference in heat capacity will produce a difference in temperature between sample and reference. This temperature difference can be measured with a thermopile which generates an electrical voltage proportional to the temperature difference.

One type of DSC design, called "power compensation" or "thermal null" design, employs an active feedback circuit to supply power to an auxiliary sample heater as necessary to eliminate any temperature difference between the sample and reference compartment. The amount of power thus supplied is a direct measure of the sample heat capacity. Examples of this type of design are the Privalov and Khechinashvili (1974) calorimeter and the one manufactured by Microcal Inc. The other type of DSC design is of the "heat conduction" type. It employs no active feedback circuit but simply records the thermopile voltage generated by the temperature difference between the sample and reference cells on temperature scanning. An example of this type of design is the heat conduction DSC constructed in our laboratory (Suurkuusk et al., 1976). A commercial version of this type of DSC is manufactured by Hart Scientific. The sample heat capacity is calculated according to the following differential equation describing the thermal properties of the calorimeter system:

$$C_p = \epsilon/\alpha(V + \tau \cdot dV/dt - A \cdot V) \tag{4}$$

where ϵ is a calibration constant equal to the ratio of the thermopile thermal conductance to the Seebeck coefficient, α the temperature scanning rate or the time derivative of temperature (dT/dt), V the measured thermopile voltage, τ the cell thermal response time which is equal to the ratio of the cell heat capacity to the thermopile thermal conductance, and A a small, higher-order correction term related to the temperature dependence of the Seebeck coefficient.

The major difference between the head conduction design and the power compensation design is that the former has a longer thermal response time, usually ~ 40–160 sec, whereas the latter has a shorter thermal response time, usually ~ 8–30 sec. A long thermal response time does not affect the measurement of total transition enthalpy but does cause disproportionality between the direct voltage signal and the true heat capacity. In order to obtain the exact shape of the transition heat capacity function, the DSC must be run at a very low scan rate or the thermal response time carefully corrected (Biltonen, 1990). This point has unfortunately been largely ignored in the literature and can cause confusion in the interpretation of DSC data.

3.1.1. Phase Transition of One-Component Lipid Membranes

During the last two decades, DSC has been a popular and extensively used technique in studying the phase transition phenomenon of both model and biological membranes. Various kinds of phase transitions have been observed in lipid bilayers, including the subtransition (solid crystal–tilted gel, L_c–L_β), pretransition (tilted gel–rippled gel, L_β–$P_{\beta'}$), main transition (gel–liquid crystalline, L_β–L_α or $P_{\beta'}$–L_α), and lamellar to hexagonal transition (lamellar-hexagonal, L_α–H_{II}). The phase transition temperature is usually defined as the temperature of 50% completion of the transition or, approximately, the temperature of maximum heat capacity. The most intensive investigations have been focused on the main transition. The energetic and other important thermodynamic parameters such as the transition temperature associated with the transition have been determined using DSC for many lipid species found in biomembranes. An excellent reference for this information exists (Marsh, 1990).

The main transition is dominated by a highly cooperative reversible melting and freezing of the hydrocarbon chains of phospholipid molecules and is characterized by a sharp excess heat capacity curve. An example of a lipid phase transition monitored by DSC is shown in Figure 1 for multilamellar vesicles (MLVs) made of dipalmitoylphosphatidylcholine (DPPC). The large heat capacity maximum, $C_{p,max}(T)$ of about 100 kcal/mol · °C at the transition temperature, T_m, of 41.3°C and the narrow transition width at half-height, $\Delta T_{1/2}$, of about 0.076°C show that the transition is highly cooperative. The transition enthalpy, ΔH, of 8.7 kcal/mol is given by the area underneath the excess heat capacity curve. Albon and Sturtevant (1978) have suggested that the transition is first-order and the finite transition width is due to minor contamination of the lipid sample. However, the facts that $C_p(T)$ exhibits no discontinuity and is asymmetric in shape are also consistent with the existence of large thermodynamic fluctuations both below and above the transition temperature. Together with a careful examination of the properties of the transition heat capacity of DPPC small unilamellar vesicles (SUVs), Biltonen (1990) suggested that the main transition of DPPC MLVs is more properly described as a weakly first-order transition or a

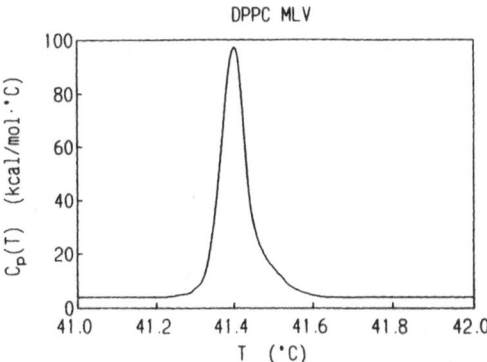

FIGURE 1. The excess heat capacity $C_p(T)$, of DPPC MLVs as a function of temperature obtained with a heat conduction calorimeter (Suurkuusk *et al.*, 1976) at a scan rate of 0.1°C/hr.

continuous order transition. A similar conclusion has been reached by Corvera *et al.* (1993) from Monte Carlo simulation studies.

The ΔH and T_m of the main transition are influenced greatly by lipid acyl chain length. In general, the longer the acyl chain, the greater are the transition temperature and enthalpy. An example of this general trend is shown in Table I for saturated lipids with identical diacyl chains. The extent of unsaturation is also an important parameter in determining the phase behavior of lipids. The higher the degree of unsaturation in the acyl chain, the lower is the main transition temperature (Ladbrooke and Chapman, 1969; Coolbear *et al.*, 1983; Cevc and Marsh, 1987). Moreover, the main transition temperature is also very sensitive to the position of the unsaturated carbon bond in the acyl chain, with the lowest T_m observed when the unsaturated carbon bond is located in the middle of the lipid acyl chain (Silvius, 1982). In lipids with mixed acyl chains, the intramolecular chain length difference (ΔL) is also an important parameter in determining the bilayer phase behavior. An initial increase in ΔL results in a decreased T_m up to a

Table I
The Main Transition Temperature of DC_nPC, DC_nPE, and DC_nPG[a,b]

Acyl chain/head group (DC_n-):	12	14	16	18
PC	−1.8°C	23.9°C	41.4°C	54.9°C
PE	30.2	49.3	64.0	73.9
PG (pH 7.0)	0	23.7	41.5	54.5
(pH 1.1)		42	57	

[a]PC, phosphatidylcholine; PE, phosphatidylethanolamine; PG, phosphatidylglycerol; T_m, main transition temperature; n, number of carbon units in the acyl chain.
[b]Data from Mabrey and Sturtevant (1976) for PC, from Cevc and Marsh (1985) for PE, from Findlay and Barton (1978) for PG at pH 7.0 and from Cevc *et al.* (1980) for PG at pH 1.1.

minimum T_m and a further increase in ΔL will cause T_m to again increase, presumably because of lipid chain–chain interdigitation in the bilayer (Mason *et al.*, 1981; Lin *et al.*, 1991). From their extensive studies of a variety of mixed chain phosphatidylcholines, Huang and his co-workers (Huang, 1991) have deduced, empirically, a linear relationship between the main transition temperature and the ratio of the difference of chain length over the length of the shorter chain. This linear relationship is also found to hold for mixed chain phosphatidylethanolamines. For more detailed information, the reader is referred to the recent series of papers published by Huang and co-workers (Wang *et al.*, 1990; Lin *et al.*, 1991; Huang, 1991; Slater *et al.*, 1992; C. Huang *et al.*, 1993).

Lipid head groups also play an important role in determining the thermotropic properties of lipid bilayers. The influence of lipid head groups on the phase behavior of bilayer membranes can be seen by comparing the transition temperatures of PC and phosphatidylthanolamine (PE) (Table I). Although the head group and the diacylglycerol part of PC and PE have been shown to adopt a similar conformation in the bilayer membrane (Hauser *et al.*, 1981; Seelig *et al.*, 1987), considerable difference in the main transition temperature exists between PC and PE. The differences in various intermolecular interactions such as the head group steric effect and hydration of the head groups seem mostly important for the lipid packing properties and the bulk phase behavior. The relatively higher T_m of $DC_n PE$ than that of $DC_n PC$ has been suggested to result from a strong, direct hydrogen bonding between phosphate and ammonium groups of adjacent PE head groups (Hauser *et al.*, 1981). For charged lipids, the chemical nature of the head group is reflected in the sensitivity of the transition temperature to pH, ionic strength, and presence of multivalent cations (Cevc, 1987; Marsh, 1990, and references therein). An example of charge effects of the lipid phase transition is shown by the strong pH dependence of the T_m of phosphatidylglycerols (PG) (Table I).

The gel–liquid-crystalline transitions of small and large unilamellar vesicles with diameters of less than 300 Å (SUVs) or larger than 700 Å (LUVs) are also highly cooperative although their transition heat capacity functions are less sharp than those of MLVs. The main transition temperature of unilamellar vesicles has been found to be strongly dependent on the diameter of the vesicle, with a T_m of $\sim 37°C$ for DPPC SUVs having diameters of ~ 200 Å and a T_m of $\sim 41°C$ for DPPC LUVs having a diameter of ~ 1000 Å. This vesicle size dependence of the T_m is most profound for vesicles smaller than 400 Å and levels off as the vesicle size further increases (Lichtenberg *et al.*, 1981). It should be mentioned that at low temperatures ($<T_m$), DPPC SUVs with diameters of < 300 Å are highly unstable and have a strong tendency to aggregate and fuse into the more stable LUVs with diameters of > 700 Å. On the other hand, DPPC SUVs are quite stable when kept at temperatures above T_m (Suurkuusk *et al.*, 1976; Wong and Thompson, 1982).

3.1.2. Phase Transitions of Multicomponent Lipid Membranes

Cellular membranes of living organisms are a mixture of many lipid species having different chain length and various extents of unsaturation in the acyl chain. Binary lipid systems are the simplest multicomponent systems with which to model biological membranes. The detailed molecular structure and thermo-dynamics of binary mixture membranes provide fundamental information about the molecular interactions between unlike lipid pairs. This information is crucial for the understanding of various types of unlike nearest-neighbor interaction terms in a more realistic many-component membrane. Unfortunately, the phase transition properties of binary lipid mixtures are complicated by the presence of the second lipid component. Thus, it is more difficult to draw general conclusions for binary mixtures than single-component systems although some effort has been made to predict the miscibility of lipid pairs (Curatolo et al., 1985).

A popular and convenient way to describe the physical properties of a binary lipid system is a phase diagram. Many binary lipid systems have been examined using DSC and the results interpreted within the context of regular solution theory (Mabrey and Sturtevant, 1976; Lee, 1978; Ipsen and Mouritsen, 1988). Figure 2 shows some examples of heat capacity curve obtained by DSC for binary lipid mixtures of dimyristoylphosphatidylcholine/distearoylphospha-tidylcholine (DMPC–DSPC), DMPC–DPPC, and DPPC–DSPC. In the DMPC–DSPC system, a progressively broader transition is observed as the molar ratio of DSPC is increased until two maxima in the heat capacity function become obvious for the 1:1 DMPC–DSPC mixture; the entire transition of the 1:1 DMPC–DSPC mixture spans nearly 20°C. The appearance of two maxima in the heat capacity function is a clear indication of compositional phase separation in the bilayer during the transition process. On the other hand, there is only one heat capacity maximum observed for 1:1 DMPC–DPPC and DPPC–DSPC mixtures. In fact, the heat capacity function of DMPC–DPPC and DPPC–DSPC mixtures exhibit only one maximum at all molar ratios (Mabrey and Sturtevant, 1976; Ye, 1992), indicating good mixing of unlike lipids in these two systems.

From the DSC data at each molar ratio, the starting and completion tem-peratures of the transition can be identified and the phase diagrams for these binary systems then constructed as described by Mabrey and Sturtevant (1976). The phase diagrams shown in Figure 3 reveal many basic properties of the gel–liquid-crystalline transition of the three binary lipid systems. For instance, the nonhorizontal solidus and liquidus lines in the phase diagrams of DMPC–DPPC and DPPC–DSPC systems indicate relatively good mixing of the two lipid spe-cies in the two binary systems in both gel and liquid-crystalline states at all molar ratios. The DMPC–DPPC and DPPC–DSPC systems have been well char-acterized as nearly ideal-mixing systems (Shimshick and McConnell, 1973;

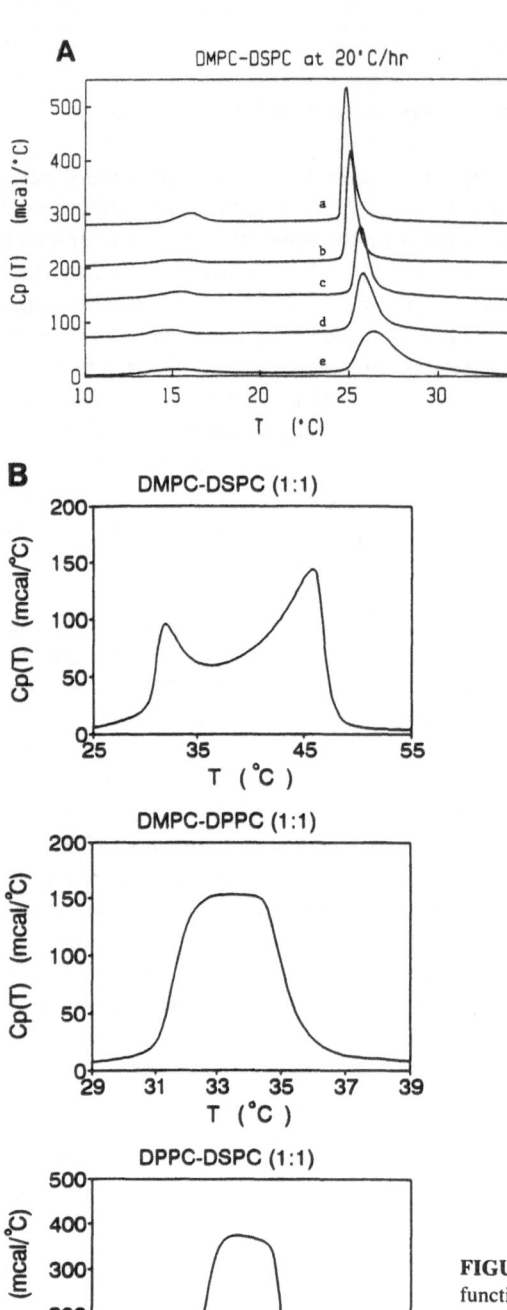

FIGURE 2. (A) The excess heat capacity as a function of temperature obtained by DSC at a scan rate of 20°C/hr for MLVs of DMPC–DSPC containing 0 (a), 1.3 (b), 6 (c), 9 (d), and 12 (e) mol% DSPC. (B) The excess heat capacity as a function of temperature obtained by DSC for MLVs of 1:1 DMPC–DSPC, DMPC–DPPC, and DPPC–DSPC. (From Ye, 1992)

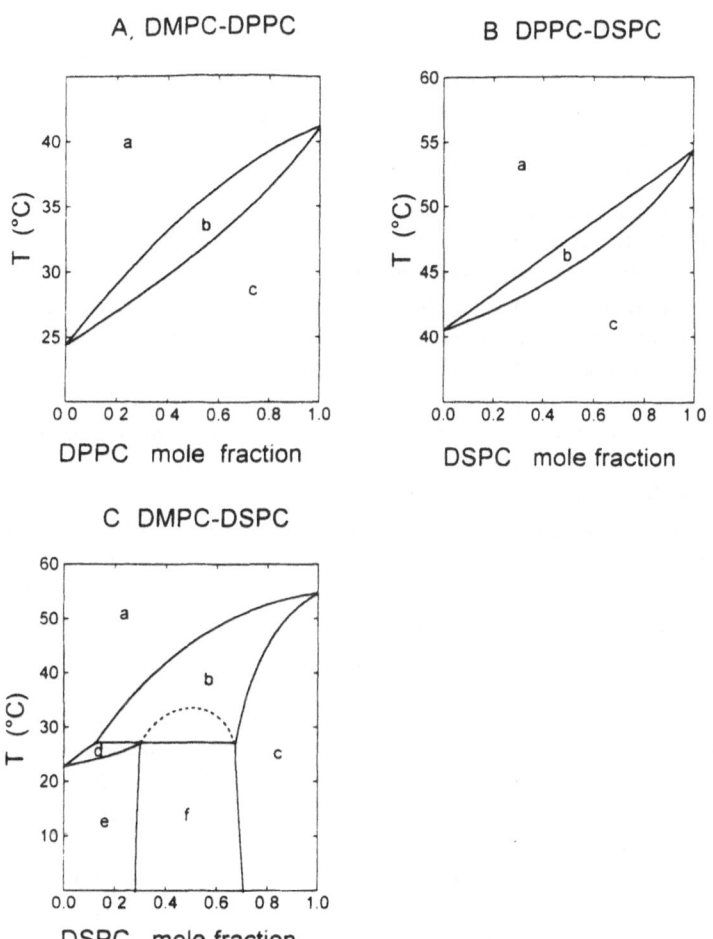

FIGURE 3. Phase diagrams for DMPC–DPPC, DPPC–DSPC, and DMPC–DSPC mixtures. (A) DMPC–DPPC, adapted from Mabrey and Sturtevant (1976). (a) Liquid-crystalline phase; (b) coexistence region of gel and liquid-crystalline phases; (c) gel phase. (B) DPPC–DSPC, adapted from Shimshick and McConnell (1973). (a) Liquid-crystalline phase; (b) coexistence region of gel and liquid-crystalline phases; (c) gel phase. (C) DMPC–DSPC, adapted from Knoll *et al.* (1981). (a) Liquid-crystalline phase; (b) coexistence region of gel 1 and liquid-crystalline phases; the maximum of the dotted line indicates a latent critical demixing point hidden within the coexistence region; (c) gel 1 phase; (d) coexistence region of liquid-crystalline and gel 2 phases; (e) gel 2 phase; (f) coexistence region of gel 1 and gel 2 phases.

Mabrey and Sturtevant, 1976). On the other hand, the existence of a horizontal solidus line within the molar ratio region of $\sim$ 30 to 65 mol% DSPC in the phase diagram of the DMPC–DSPC system suggests an immiscibility of the two lipid species in the gel state. The DMPC–DSPC system has been characterized as a peritectic system with nearly ideal mixing in the liquid-crystalline state but highly nonideal mixing in the gel state (Shimshick and McConnell, 1973).

Lateral gel–liquid-crystalline phase separation has been suggested to occur in these three binary lipid systems within the phase transition region where gel and liquid-crystalline molecules coexist (Shimshick and McConnell, 1973; Knoll et al., 1981). It has been concluded that in the phase transition region, the two lipid species in the DMPC–DPPC and DMPC–DSPC mixtures do not distribute randomly in the bilayer, indicating a separation of the lipids into compositionally different domains. The neutron scattering data of Knoll et al. (1981) also suggested the existence of liquid-crystalline state immiscibility in DMPC–DSPC. This finding has been interpreted to be the result of long-range concentration fluctuations related to the existence of a latent gel state critical demixing point hidden within the gel–liquid-crystalline coexistence region, as shown by the dotted line in Figure 3C (Knoll et al., 1983).

Lipid miscibility in a bilayer membrane is determined by the magnitude of unlike nearest-neighbor interactions. The lipid chain length difference, types and charges of the head groups, and the extent of chain unsaturation in a binary system all play important roles in determining the interaction energies between lipids. An increase in chain length difference between lipid pairs with identical head groups usually results in a decreased miscibility, as shown by the examples of DMPC–DPPC and DMPC–DSPC mixtures above. In addition, various complicated mixing behavior has been reported for binary lipid systems containing mixed chain lipids (Curatolo et al., 1985; Shaukat et al., 1989; Slater et al., 1992). For instance, when lipids with nonidentical chain lengths are mixed with those with identical chains, an interesting eutectic behavior has been observed in the lipid main transition (Lin et al., 1991). This eutectic behavior is interpreted to originate from gel-state chain–chain interdigitation caused by the large intramolecular chain length difference.

The head group difference between DC_nPC and DC_nPG (at neutral pH) does not seem to prevent these two species from mixing well with each other. In fact, these two lipids mix with each other so well that the main transition heat capacity profiles of DMPC–DMPG mixtures appear very similar to those of pure DMPC or DMPG (Sixl and Watts, 1983). On the other hand, the head group difference between DC_nPC and DC_nPE results in much less ideal mixing of the lipid pair (Shimshick and McConnell, 1973; Marsh, 1990, and references therein).

In an attempt to reach some general conclusions regarding phase miscibility of binary PC mixtures, Curatolo et al. (1985) studied 21 pairs of binary PC mixtures. They concluded that binary mixtures of PCs exhibit gel-state mis-

cibility if the difference in the gel–liquid-crystalline transition temperature between the two lipid components is less than 33°C. However, Shaukat *et al.*
(1989) deduced from their DSC experiments that DMPC is not completely mis-
cible in the gel state with C(18):C(11:1)PC, C(18):C(10)PC or C(18):C(11)PC,
in 1:1 mixtures even though the differences in transition temperatures are only
10.6, 5.3, and 2.5°C for these three lipid pairs, respectively. Therefore, the
above conclusion reached by Curatolo *et al.* (1985) is not quite valid.

Studies of binary mixtures of phospholipids have shown that lateral phase
separation or formation of compositionally different domains is a common phenomenon. In many cases, the distribution of various lipids in the bilayer plane is
not physicochemically ideal (Shimshick and McConnell, 1973; Knoll *et al.*,
1981; Curatolo *et al.*, 1985). The extent of nonideality of lipid–lipid interactions
in binary lipid systems is dependent on both the physical state of the lipid bilayer
and the molar ratio of the compositional lipids (Von Dreele, 1978). These investigations indicate that the phase behavior of binary mixture systems are very
dependent on the particular lipid components involved and generalizations are
difficult to make.

3.2. DSC Studies at High Pressure

Investigations of the effects of pressure on the phase transition properties of
lipid bilayers have been mostly limited to single-component systems. The pressure dependence of the gel–liquid-crystalline transition temperature of single-
component lipid bilayers is described by the Clausius–Clapeyron equation:

$$dT_m/dP = \Delta V/\Delta S = T_m\Delta V/\Delta H \tag{5}$$

where T_m, P, ΔV, ΔS, and ΔH are the transition temperature, pressure, molar
volume, molar entropy, and molar enthalpy changes of the transition, respec-
tively. If $\Delta V \neq 0$, the transition can be induced isothermally by a change in
pressure. Application of the Clausius–Clapeyron equation allows the estimation
of the molar volume change from the pressure dependence of the transition
temperature. Conversely, the molar volume change, measured by other methods
such as dilatometry (Nagle and Wilkinson, 1978), can be used to predict the
variation of the phase transition temperature with pressure.

The pressure dependence of the gel–liquid-crystalline transition temperature has been determined for many one-component lipid systems in which the
very sharp, weakly first-order transition nature allows an accurate determination
of the transition temperature. A linear relationship between T_m and P has been
found for saturated phospholipids. The slope dT_m/dP increases as the acyl chains
become longer but is usually within the range of 0.015–0.029°C/atm regardless
of the type of polar head groups (Marsh, 1990, and references therein). A careful

DSC study by Mountcastle *et al.* (1978) showed that neither the enthalpy change nor the shape of the excess heat capacity of the transition is altered by pressure up to 136 atm in DPPC multilamellar vesicles. The fact that no change in the shape of the heat capacity function by pressure was observed indicates that moderate pressure does not alter the cooperativity of the transition. Therefore, the magnitude of the unfavorable Gibbs free energy between unlike nearest neighbors, which determines the degree of cooperativity of the bilayer, is apparently unchanged by pressure up to 136 atm. Furthermore, the molar volume change associated with the gel–liquid-crystalline transition of DPPC multilamellar vesicles has been found to be independent of pressure up to 300 atm (Macdonald, 1978).

Because of a generally much broader phase transition and more complicated phase behavior of binary lipid systems than single-component systems, it has been a challenge in experimental design and methodology to study the effect of pressure on the phase transition properties of binary lipid systems. Using the TEMPO ESR spectrum as a measure of the degree of melting of the bilayer membrane, Trudell *et al.* (1975) found that both the solidus and liquidus lines on the phase diagram of the DMPC–DPPC system are equivalently shifted to higher temperatures by pressure, but the shape of the phase diagram is nearly unaltered by moderate pressures up to 137 atm. However, the possible alteration of the lipid–lipid interactions between unlike nearest neighbors by pressure in binary lipid systems, as may be reflected by the change in the shape of the phase transition heat capacity function (or other intrinsic parameters which describe the phase transition process), has not been critically examined.

The development of high-pressure DSC (Mountcastle *et al.*, 1978) has made it possible to study the effects of pressure on the phase transition properties of binary lipid mixtures. The effects of hydrostatic pressures up to 200 atm on the gel–liquid-crystalline transition temperature and the shape of the transition heat capacity function of the DMPC–DSPC, DMPC–DPPC, and DPPC–DSPC systems have been recently examined using high-pressure DSC (Ye *et al.*, 1993). The upper panels of Figure 4A–C show some typical results of the excess heat capacity, $C_p(T)$ as a function of temperature for MLVs of 1:1 DMPC–DPPC, DPPC–DSPC, and DMPC–DSPC at various pressures. It can be seen that a higher pressure shifts the overall phase transition toward higher temperature, but the shape of the excess heat capacity function remains unchanged within experimental error. A linear relationship between the nominal transition temperature, T_m. corresponding to the temperature of a heat capacity maximum and the applied pressure, P, is obtained. Note that two nominal transition temperatures, T_{m1} and T_{m2}, are defined for 1:1 DMPC–DSPC, corresponding to the heat capacity peaks at lower and higher temperatures, respectively. For 1:1 DMPC–DPPC, $dT_m/dP = 0.0228\pm0.0003°C/atm$, and for 1:1 DPPC–DSPC, $dT_m/dP = 0.0246\pm0.0004°C/atm$ (middle panels of Figure 4A and B). For 1:1 DMPC–

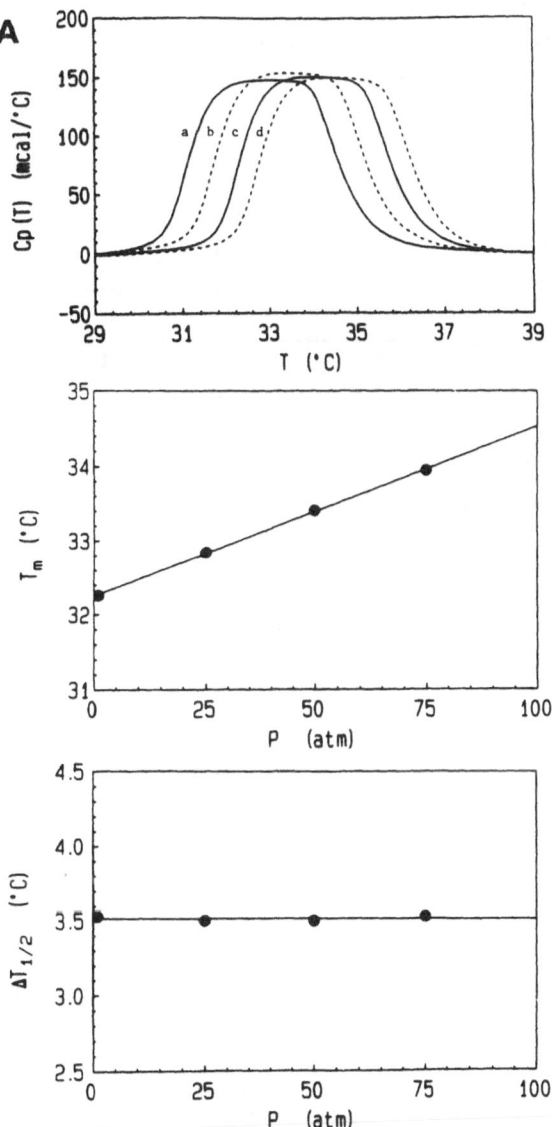

FIGURE 4. (A) The excess heat capacity as a function of temperature obtained by high-pressure DSC at a scan rate of 20°C/hr for 1:1 DMPC–DPPC MLVs at air pressure of 1 atm (a) or helium pressures of 25 (b), 50 (c), and 75 (d) atm. Note that the shape of the heat capacity function remains unchanged as pressure varies. (Middle) The nominal transition temperature T_m and (lower) the transition half-width $\Delta T_{1/2}$ as a function of pressure for 1:1 DMPC–DPPC MLVs. Solid lines with slopes of $dT_m/dP = 0.0228 \pm 0.0003$°C/atm for T_m at $d(\Delta T_{1/2})/dP \approx 0$ for $\Delta T_{1/2}$ represent the least-squares regression of the data. T_m is defined as the temperature corresponding to the maximum excess heat capacity value, and $\Delta T_{1/2}$ the temperature span at the half height of the maximum excess heat capacity. (From Ye, 1992)

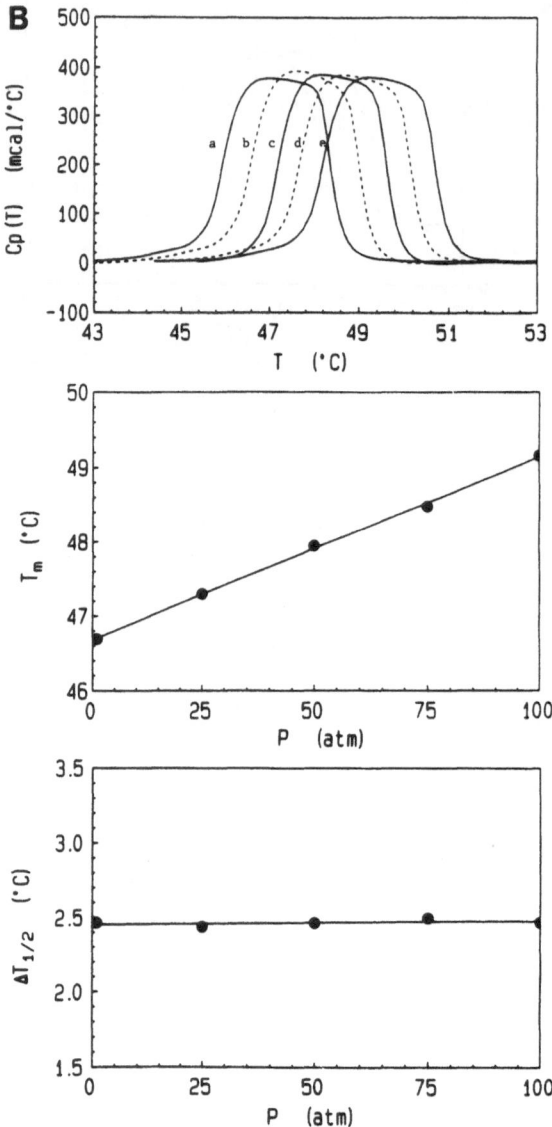

FIGURE 4. (B) (Upper) The excess heat capacity as a function of temperature obtained by high-pressure DSC at a scan rate of 10°C/hr for 1:1 DPPC–DSPC MLVs at air pressure of 1 atm (a) or helium pressures of 25 (b), 50 (c), 75 (d), and 100 (e) atm. Note that the shape of the heat capacity function remains unchanged as pressure varies. (Middle) The nominal transition temperature T_m and (lower) the transition half-width $\Delta T_{1/2}$ as a function of pressure for 1:1 DPPC–DSPC MLVs. Solid lines with slopes of $dT_m/dP = 0.0246 \pm 0.0004$°C/atm for T_m and $d(\Delta T_{1/2})/dP \approx 0$ for $\Delta T_{1/2}$ represent the least-squares regression of the data. T_m and $\Delta T_{1/2}$ are defined in panel A. (From Ye, 1992)

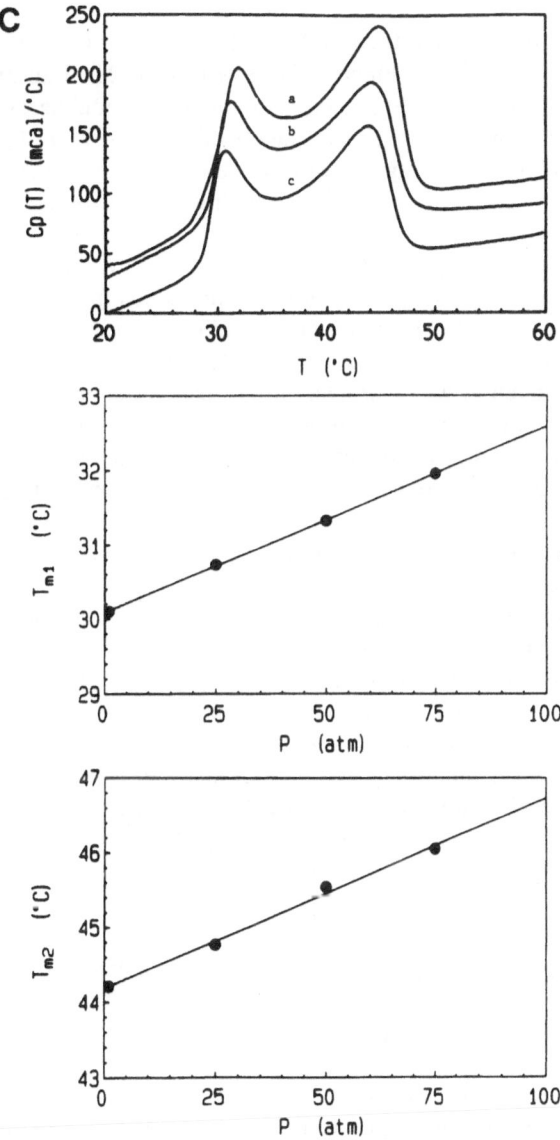

FIGURE 4. (C) (Upper) The excess heat capacity as a function of temperature obtained by high-pressure DSC at a scan rate of 10°C/hr for 1:1 DMPC–DSPC MLVs at air pressure of 1 atm (a) or helium pressures of 20 (b) and 45 (c) atm. Note that the transition is more than 20°C broad and the heat capacity baseline is not a linear function of temperature; the shape of the heat capacity function remains essentially unchanged as pressure varies. (Middle) Temperatures T_{m1} and (lower) T_{m2} corresponding to the two maximum values in the excess heat capacity function of 1:1 DMPC–DSPC MLVs as a function of pressure P. Solid lines with slopes of 0.0250 ± 0.0002°C/atm for T_{m1} and 0.0255 ± 0.0011°C/atm for T_{m2} represent the linear least-squares regression of the data. (From Ye, 1992)

DSPC, $dT_{m1}/dP = 0.0250 \pm 0.0002$ and $dT_{m2}/dP = 0.0255 \pm 0.0011°C/atm$ (middle and lower panels of Figure 4C), respectively. The transition half-width, $\Delta T_{1/2}$, defined as the temperature span at the half-maximal heat capacity height, appears unaltered by pressure (lower panels of Figure 4A and B). No change in the shape of the heat capacity function for diacyl PCs has been found for pressures up to 200 atm. The transition enthalpy change is also independent of pressure.

The degree of cooperativity and the shape of the excess heat capacity function are determined by the magnitude of the Gibbs free energy of interaction between unlike nearest neighbors. These unlike nearest-neighbor pairs can be formed by two lipid molecules of either different type or different state. Since there are six different ways to form an unlike nearest-neighbor pair in the triangular lattice of a binary system, any alteration of these unlike nearest-neighbor interactions by pressure will be reflected in the shape change of the excess heat capacity function. An extreme example is a situation where the pressure would cause a large change in lipid—lipid interactions leading to complete compositional phase separation of the two unlike lipid molecules resulting in two non-overlapping peaks in the heat capacity function.

The change of the transition enthalpy with pressure is determined by $P \cdot \Delta V$. For the gel–liquid-crystalline transition of DMPC, DPPC, and DSPC, ΔV is 18.3, 27.2, and 35.6 ml/mol, respectively (Nagle and Wilkinson, 1978; Wilkinson and Nagle, 1979). If $P = 200$ atm and $\Delta V = 18.3–35.6$ ml/mol, $P \cdot \Delta V = 90–170$ cal/mol, which is much smaller than the transition enthalpy of 5400, 8700, and 10600 cal/mol of DMPC, DPPC, and DSPC (Mabrey and Sturtevant, 1976). These results show that a pressure change of 200 atm alters the transition enthalpy by $\sim 1\%$ for both one- and two-component lipid systems. Therefore, it is not surprising that no significant change of the transition enthalpy by pressure up to 200 atm was observed in the pressure-DSC experiment of lipid bilayers.

4. DYNAMIC PROPERTIES OF MEMBRANE PHASE TRANSITION

4.1. Volume-Perturbation Calorimeter

Most studies of the lipid phase transition kinetics have used large perturbations to alter the equilibrium poise of the lipid system so that the system has been forced to respond beyond its linear response range (Yager and Peticolas, 1982). The overall kinetic relaxation rate thus measured becomes strongly dependent on the magnitude of the external perturbation and therefore may not be the best representation of the intrinsic relaxation rate of the system (Ye *et al.*, 1991). Techniques involving the use of optical probes, on the other hand, suffer from the artifacts induced by the contamination of the system by external probes. A volume-perturbation calorimeter (VPC), based on the original design of Clegg and Maxfield (1976) as modified by Halvorson (1979), does not have these limitations (Johnson *et al.*, 1983). The detailed description of its design, calibration, and operation and the method of data analysis can be found in van Osdol *et al.* (1989). Briefly, the instrument uses the voltage-dependent extension of a stack of piezoelectric crystals to generate a small, adiabatic perturbation in the volume of an aqueous sample. The pressure and temperature responses following a volume perturbation are recorded in the time domain and then converted into and analyzed in the frequency domain using Fourier series technique. Two transfer functions characterizing the response of the system are formed: the ratio of the temperature to pressure changes (dT/dP) and the ratio of the pressure to volume changes (dP/dV). These transfer functions characterize the "dynamic" or "frequency-dependent" heat capacity, $C_p(\omega)$, and bulk modulus (inverse of the isothermal compressibility), $B_T(\omega)$, respectively. It should be noted that $C_p(\omega)$ is frequency dependent and therefore corresponds to the dynamic heat capacity of the sample at a given temperature. The term *dynamic heat capacity* is introduced to distinguish it from the heat capacity measured by DSC experiments which is the equilibrium heat capacity.

The frequency-dependent heat capacity transfer function is calculated from the amplitude ratio and the phase angle shift of the corresponding harmonics of the series for temperature and pressure, i.e., $C_p(\omega) = A(\omega)e^{j\theta}$, with $A(\omega) = dT/dP$, $\theta(\omega) = \theta_T\theta_P$ and $j = \sqrt{-1}$. The same procedure using the Fourier series for pressure and voltage yields the transfer function $B_T(\omega)$ assuming that the volume change is proportional to the voltage change. The relaxation times for the system under stationary perturbation can be obtained from the nonlinear least-squares analysis of either $A(\omega)$ or $\theta(\omega)$. Assuming a multiexponential relaxation process, the normalized amplitude spectra for each transfer function are fit to

$$A(\omega) = \{(\Sigma\alpha_i\omega\tau_i/[1+(\omega\tau_i)^2])^2+(\Sigma\alpha_i/[1+(\omega\tau_i)^2])^2\}^{1/2}] \qquad (6)$$

and the phase shift data are fit to

$$\theta(\omega) = \tan^{-1}\{(\Sigma\alpha_i\omega\tau_i/[1+(\omega\tau_i)^2])/(\Sigma\alpha_i/[1+(\omega\tau_i)^2])\}] \qquad (7)$$

using a nonlinear least-squares analysis algorithm (Johnson and Fraiser, 1985), where α_i and τ_i are the amplitude and relaxation time of the assumed ith independent relaxation process, respectively, and ω is the perturbation frequency.

In the following section, the results of phase transition kinetics of one- and two-component PC bilayers obtained with the VPC will be discussed.

4.1.1. Phase Transition Kinetics of One-Component Lipid Membranes

The kinetic results of single-component lipid bilayers reported by different research groups using various techniques, and the application of VPC for the kinetic experiments on lipid phase transitions have been reviewed (Caffrey, 1989;

van Osdol *et al.*, 1989). A wide range of relaxation times, from nanoseconds to tens of seconds, have been documented. The studies of the relaxation dynamics of the main phase transition of five homologous PC (DMPC, $DC_{15}PC$, DPPC, $DC_{17}PC$, and DSPC) MLVs using the VPC have been reported previously (van Osdol *et al.*, 1991; Ye *et al.*, 1991) and will be briefly summarized below.

Figure 5 shows a typical result for the relaxation amplitude of the transfer function, $A(\omega)$, versus the scaled temperature, T, for DPPC MLVs at four perturbation frequencies (0.01, 0.1, 1, and 10 Hz). The ordinate is normalized by the response due to water to correct for the instrumental effects which are not relevant to the relaxation dynamics of the lipid (van Osdol *et al.*, 1991). At a perturbation frequency of 0.01 Hz, the shape of the relaxation amplitude as a function of the scaled temperature is very similar to the corresponding excess heat capacity curve obtained by DSC (see Figure 1). The scaled temperature makes a correction for the average pressure effect on the transition temperature so that the equilibrium transition profile is reduced to a common temperature scale at all pressures (Mountcastle *et al.*, 1978; van Osdol *et al.*, 1991). Data obtained for the other lipids are similar.

The frequency spectrum of the relaxation amplitude of DPPC MLVs at T_m is

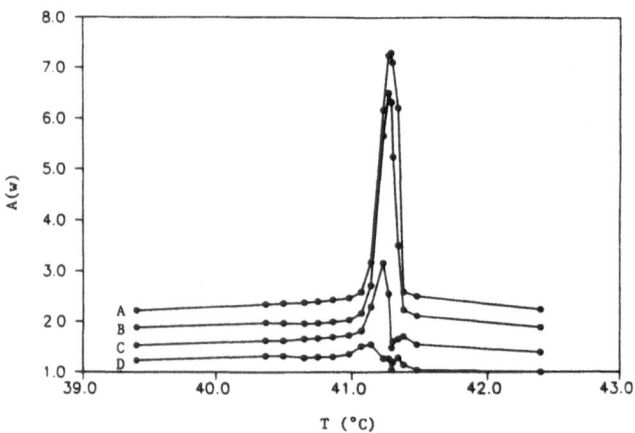

FIGURE 5. The relaxation amplitude, A (ω), of DPPC MLVs, as a function of the scaled temperature, T, at four perturbation frequencies: (A) 0.01 Hz, (B) 0.1 Hz, (C) 1 Hz, (D) 10 Hz. Curves A–C have been displaced on the ordinate for a cleaner presentation. The ordinate is in units of the response due to water (0.003°C/atm at 41°C) to account for the instrumental response. Error bars have been omitted for clarity. The scaled temperature is defined as $T = T_0 - (P - 1) \cdot (dT_m/dP)$, where T_0 is the actual mean temperature of the experiment, P the mean pressure, and $dT_m/dP = 0.024°C/atm$, the pressure dependence of the phase transition temperature (Mountcastle *et al.*, 1978). This scaling procedure reduces the equilibrium transition profile to a common temperature scale for all pressures. Adapted from van Osdol *et al.* (1991).

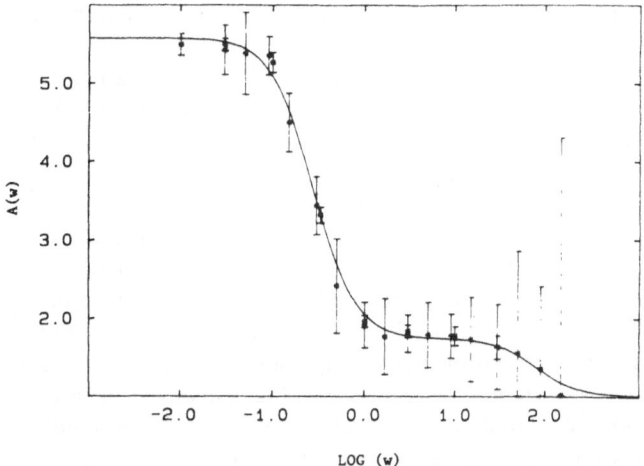

FIGURE 6. The relaxation amplitude of DPPC MLVs at T_m, as a function of $\log_{10}$ of the perturbation frequency. The ordinate is in units of the water response. The solid line represents the best fit of the data to two exponential decay processes, the faster one accounting for the relaxation baseline observed at higher frequencies and not pertinent to the relaxation of the lipid phase transition itself. The standard errors in the data are indicated by the vertical bars.

plotted in Figure 6. The relaxation time is estimated by a nonlinear least-squares fitting assuming simple exponential decays for the transfer function (Johnson and Frasier, 1985; van Osdol *et al.*, 1991). The relaxation time can be obtained by fitting either the amplitude or the phase angle spectra since, in principle, both are the result of the same physical phenomena (van Osdol *et al.*, 1991). Figure 7 shows the relaxation time as a function of the fractional degree of melting for the five PCs studied. These results show that the maximum relaxation time appears

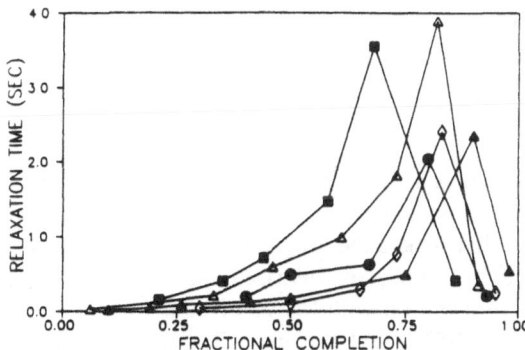

FIGURE 7. The relaxation time as a function of the fractional degree of completion of the transition for MLVs of DMPC (●), $DC_{15}PC$ (■), DPPC (△), $DC_{17}PC$ (◇), and DSPC (▲). Error bars have been omitted for clarity. Adapted from van Osdol *et al.* (1991).

to be independent of the lipid chain length and has a common value of 2–4 sec for all five lipids.

The relaxation kinetics of the lipid phase transition have also been analyzed in terms of the classical Kolmogorov–Avrami kinetic theory (Yang and Nagle, 1988; Ye *et al.*, 1991), which is described in the time domain as: $f = 1 - \exp[(t/\tau)^n]$, where f, t, and τ are the fractional completion of the transient change of the transition, time, and the characteristic time constant, respectively. It was found for MLVs of DMPC and DPPC that the relaxation process of the gel–liquid-crystalline transition is better fit with an effective dimensionality of $n = 2$ rather than $n = 1$ whereas for DSPC MLVs, an effective dimensionality of ~ 1.5 was estimated to best fit the data (Ye *et al.*, 1991). These results indicate that the gel–liquid-crystalline transition of some lipid bilayers follows the classical Kolmogorov–Avrami kinetic model with an effective dimensionality > 1 and the assumption of simple exponential decay ($n = 1$) commonly used in data analysis may not always be valid for lipid transitions. The authors suggest that the fractional dimensionality of ~ 2 for DMPC and DPPC MLVs implies that the growing domains are very compact and nearly circular in shape in the bilayer plane. The compactness of domains indicates a highly unfavorable Gibbs free energy of interaction between unlike nearest neighbors and therefore a high degree of cooperativity in the transition process. The apparently lower fractional dimensionality of DSPC MLVs in contrast to those of DMPC and DPPC MLVs indicates less compact domains and therefore a lower cooperativity for the transition. This conclusion is consistent with the temperature-dependent heat capacity results of Mabrey and Sturtevant (1976) who found that the transition cooperativity of saturated diacyl PCs decreases as the acyl chains become longer. Application of the dimensionality analysis to the transition kinetics of the local anesthetic dibucaine–DPPC system (van Osdol *et al.*, 1992) shows that the fractional dimensionality of the relaxation process decreases monotonically from ~ 2 to ~ 1 as the nominal anesthetic/lipid mole ratio increases from 0 to 0.027. These results are consistent with the hypothesis that incorporation of dibucaine into DPPC bilayers reduces the average cluster size and causes the fluctuating lipid clusters to become more ramified.

4.1.2. Phase Transition Kinetics of Two-Component Lipid Membranes

Kinetic experiments with binary lipid mixtures are much more complex than single-component systems and a greater challenge to methodology, experimental design, and interpretation. It is the unique feature of the VPC to monitor the kinetic energetics of the transition that makes the kinetic experiments of binary lipid systems possible (van Osdol *et al.*, 1989; Ye, 1992). In this section, we will discuss the results of kinetic experimentation with MLVs of binary mixtures of DMPC–DSPC, DMPC–DPPC, and DPPC–DSPC using VPC. Detailed results of the entire series of study can be found in Biltonen and Ye (1993).

Figure 8 shows the frequency-dependent relaxation amplitude spectra for pure DMPC and a 94:6 DMPC–DSPC mixture, at their respective T_m. It can be seen that the frequency spectra of 94:6 DMPC–DSPC are shifted to the high-frequency side relative to those of pure DMPC, indicating a faster average relaxation rate in the lipid mixture than in the pure lipid. Since the relaxation process of binary lipid mixtures cannot be well fit with a single exponential

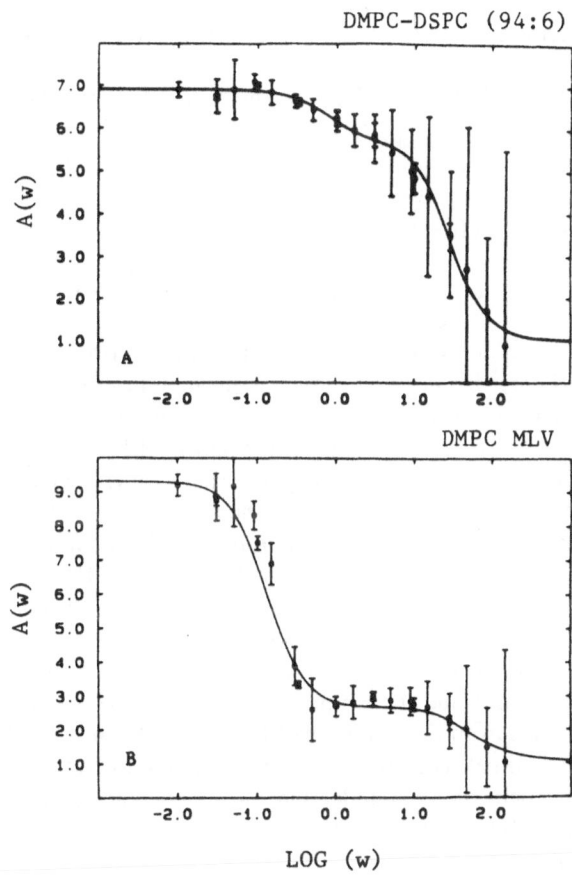

FIGURE 8. The relaxation amplitude as a function of $\log_{10}$ of the perturbation frequency for (A) 94:6 DMPC–DSPC and (B) pure DMPC at their respective T_m. The ordinate is in units of the response due to water at the respective temperature. The solid line in each graph represents the best fit of the data to a sum of multiple (two for DMPC and three for DMPC–DSPC) exponential decay processes, the fastest relaxation process containing the baseline Joule–Thompson effect (Ye *et al.*, 1991). The standard errors in the data are indicated by the vertical bars. Note that the frequency spectrum of the DMPC–DSPC mixture is significantly shifted toward the high-frequency region relative to that of DMPC, corresponding to a relaxation time almost three orders of magnitude shorter in the DMPC–DSPC mixture than in pure DMPC. (From Biltonen and Ye, 1993)

decay of any dimensionality (Ye, 1992), all data presented for the binary lipid systems have been analyzed as a sum of several independent, one-dimensional exponential decays (solid curves in Figure 8). A moment analysis procedure, in analogy to that of Sturgill and Biltonen (1976), has been developed to generate characteristic parameters for complex relaxation processes (Ye and Biltonen, in preparation). Briefly, the first, second, and third moments of the derivative relaxation spectrum with regard to the $\log_{10}$ of the perturbation frequency are calculated and used to describe a multiexponential decay process. These moments are defined in terms of the relaxation amplitude spectra [Equation (6)] as follows:

$$\text{first moment: } u_1 = \int y \cdot A'(\omega) \cdot dy / \int A'(\omega) \cdot dy] \tag{8}$$

$$i^{\text{th}} \text{ moment } (i > 1): u_i = \int [y - u_1]^i \cdot A'(\omega) \cdot dy / \int A'(\omega) \cdot dy \tag{9}$$

where ω = perturbation frequency, $y = \log_{10}(\omega)$, and $A'(\omega) = dA(\omega)/d \log_{10}(\omega)$.

$$\text{mean relaxation time: } \tau^* = 1/(\pi \cdot 10^{u_1})] \tag{10}$$

With the above definitions, the first moment describes a mean relaxation frequency which can be used to define a mean relaxation time; the second moment characterizes the distribution of the relaxation times or the fractional dimensionality of the relaxation process (Ye et al., 1991; Ye and Biltonen, in preparation); and the third moment characterizes the asymmetry of the distribution of the relaxation times. An example of the application of moment analysis to the study of phase transition kinetics in lipid bilayers is given in Figure 9. For all three lipid samples, pure DMPC, 94:6 and 6:94 DMPC–DSPC, the relaxation amplitude profile of 0.01 Hz perturbation frequency (upper panels of Figure 9) is very similar to the shape of the corresponding heat capacity curve of the transition (Ye, 1992). The mean relaxation time (lower panels of Figure 9) reaches a pronounced maximum value at a temperature near T_m and declines quickly as the temperature deviates from T_m. Addition of 6 mol% DSPC into DMPC results in a decrease of the maximal mean relaxation time relative to that of pure DMPC by more than two orders of magnitude. On the other hand, addition of 6 mol% DMPC and DSPC had an effect on the mean relaxation rate at least one order of magnitude smaller.

Similar kinetic experiments and data analysis have been conducted over the whole range of DMPC/DSPC mole ratios. Table II presents a summary of the results obtained from these studies. It can be seen that the mean relaxation times of binary mixtures are much shorter than those of pure lipids. The mean relaxation time at the T_m, τ_m^*, decreases sharply at the initial doping of DMPC with DSPC, whereas τ_m^* decreases less sharply at the initial doping of DSPC with DMPC. τ_m^* levels off in the compositional region between 6 mol% < DSPC < 50

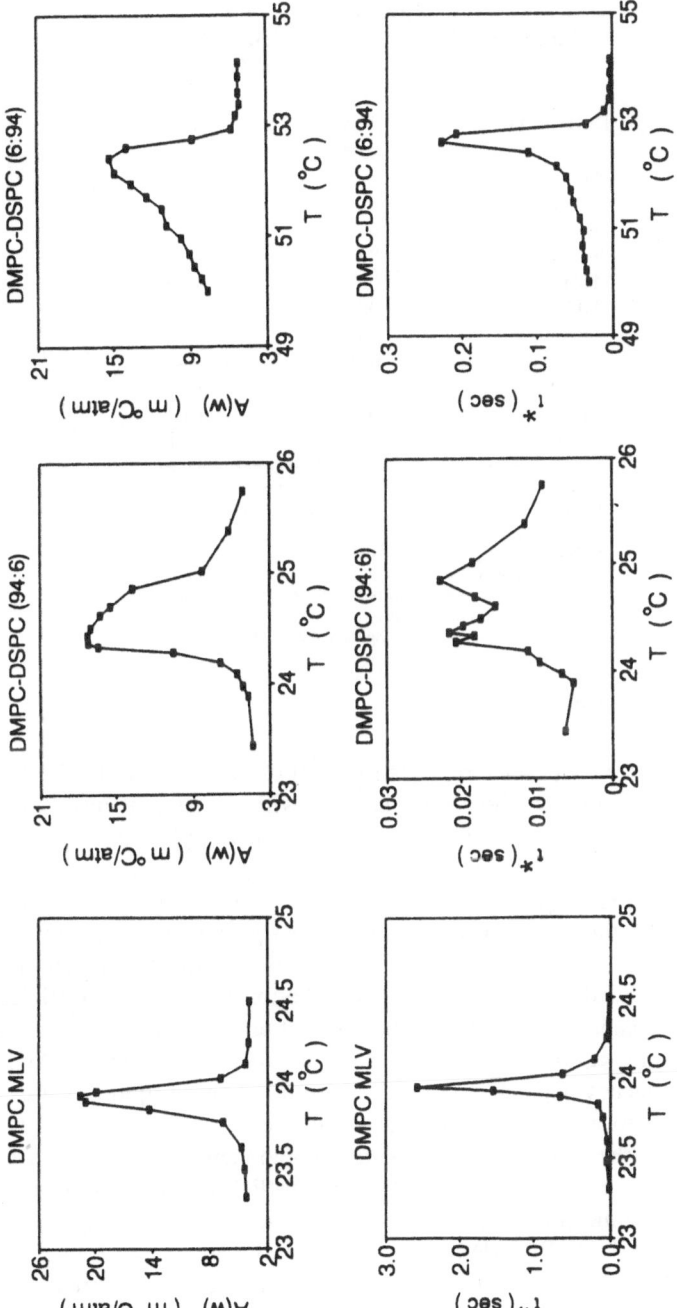

FIGURE 9. (Upper row) The relaxation amplitude at a perturbation frequency of 0.01 Hz and (lower row) the mean relaxation time as a function of the scaled temperature for DMPC, 94:6 DMPC-DSPC, and 6:94 DMPC-DSPC. Note that graphs are plotted on different scales. Error bars are omitted for clarity. The scaled temperature is defined in the legend of Figure 5.

Table II
Summary of Kinetic Relaxation Data
for DMPC–DSPC Mixtures at the T_m

DMPC–DSPC	τ^* (sec)	$\Delta T_{1/2}{}^a$
100:0	2.72[b]	0.15
98.7:1.3	0.17	0.25
94:6	0.022	0.55
91:9	0.025	0.82
88:12	0.023	0.86
80:20	0.021	1.55
50:50	0.045[c]	—
	0.091[d]	—
20:80	0.14	3.15
6:94	0.45	1.65
0:100	2.43	0.41

[a] $\Delta T_{1/2}$ is the temperature span corresponding to the half-height of the relaxation amplitude curve at 0.01 Hz perturbation frequency.
[b] Calculated using the dominant slow relaxation component only. τ^* would be reduced to 1.16 if the minor, fast relaxation component, which is not well understood, is also used in the moment analysis.
[c] Corresponding to the low-temperature peak in the heat capacity function.
[d] Corresponding to the high-temperature peak in the heat capacity function.

mol%. Furthermore, our collective data for DMPC–DSPC, DMPC–DPPC, and DPPC–DSPC systems have shown that mixtures with a small amount of lipid with higher melting temperature, T_m, in a dominant amount of lipid with lower T_m have faster mean relaxation rates than those with a small amount of lipid with lower T_m in a dominant amount of lipid with higher T_m (Ye and Biltonen, in preparation).

In Figure 10, the relaxation amplitudes (upper panels) of 1:1 DMPC–DSPC and DMPC–DPPC mixtures at 0.01 Hz perturbation frequency are shown as a function of the scaled temperature. At a perturbation frequency of 0.01 Hz, the shape of the relaxation amplitude as a function of temperature is very similar to the corresponding equilibrium heat capacity curve as obtained by DSC (Ye, 1992). The frequency-dependent relaxation spectra of these lipid mixtures were least-squares fit to a sum of three exponential decays. The middle panels of Figure 10 show the longest relaxation time components as a function of the scaled temperature for 1:1 DMPC–DSPC and DMPC–DPPC. A similar profile is obtained for the mean relaxation time (lower panels of Figure 10). The kinetic characteristics of 1:1 DPPC–DSPC are very similar to those of 1:1 DMPC–DPPC (Ye, 1992). The most interesting finding here is the occurrence of two relaxation time maxima in the phase transition region of all three 1:1 PC mixtures

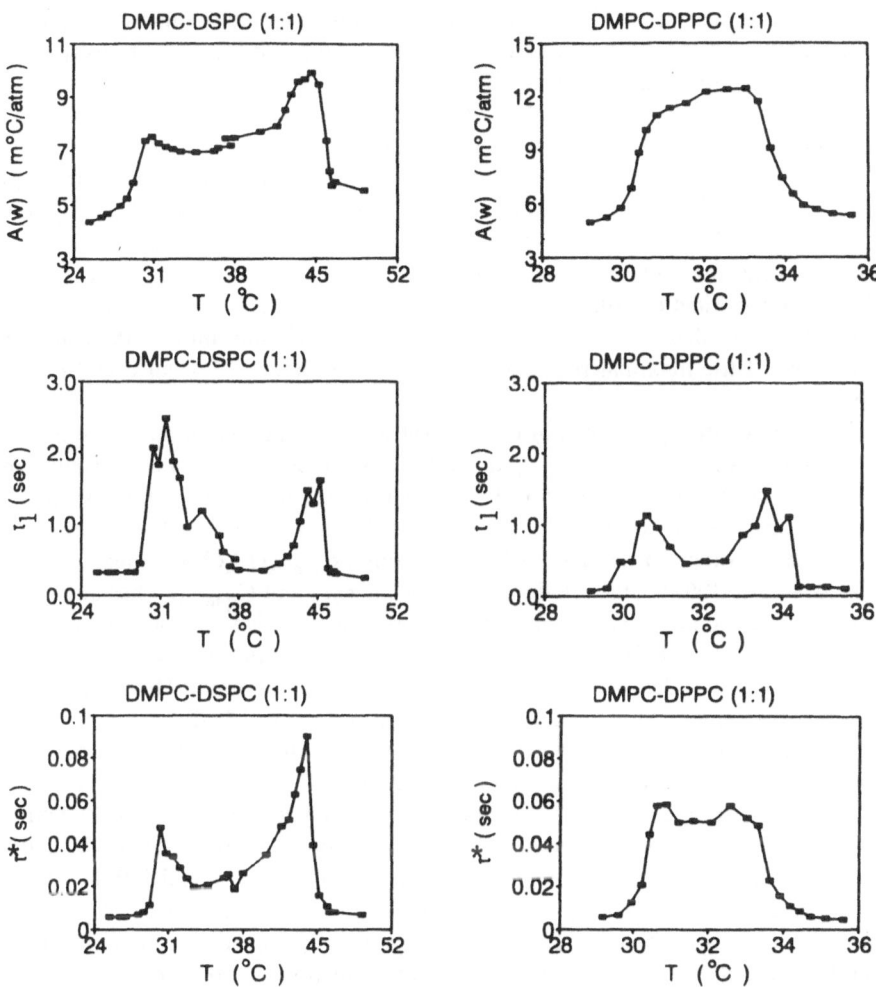

FIGURE 10. (Upper row) The relaxation amplitude at a perturbation frequency of 0.01 Hz, (middle row) the longest relaxation time component obtained from the least-squares fit of the relaxation spectrum by a sum of three exponential decays, and (lower row) the mean relaxation time, as a function of the scaled temperature for 1:1 DMPC–DSPC and 1:1 DMPC–DPPC. Error bars are omitted for clarity. The scaled temperature is defined in the legend of Figure 5. (From Biltonen and Ye, 1993)

studied, even though the corresponding heat capacity function may have only one maximum. This is true whether the relaxation process is characterized by the slowest step or by the mean relaxation time. Furthermore, the presence of bimaximal relaxation times is also observed if the second slowest relaxation step is examined (Ye, 1992). These two relaxation time maxima correspond to temperatures where the two maxima of the equilibrium heat capacity function occur for 1:1 DMPC–DSPC, or the onset and completion edges of the phase transition for 1:1 DMPC–DPPC and DPPC–DSPC.

From these results, several major points can be made. First, the long relaxation time on the order of seconds for MLVs of pure PCs (van Osdol et al., 1991) is unlikely the result of bilayer–bilayer interactions since almost three orders of magnitude change in the relaxation time has been observed with addition of only 6 mol% DSPC into DMPC. This conclusion supports the earlier suggestion that the rate-limiting step is not solely determined by bilayer–bilayer interactions based on the kinetic results of a dibucaine–DPPC mixture system (van Osdol et al., 1991). In that case, addition of 1–5 mol% dibucaine into DPPC causes a decrease of the maximal relaxation time by a factor of ~ 2 (van Osdol et al., 1992).

Second, the appearance of two relaxation time maxima in the phase transition region of all 1:1 binary lipid mixtures studied suggests the existence of dynamic lateral phase separation. Recall that in the one-component PC systems, the maximal relaxation time, τ_{max}, has been found to correspond approximately to the temperature of the maximal heat capacity, $C_{p,max}(T)$ of the transition (van Osdol et al., 1991). This relationship between τ_{max} and $C_{p,max}(T)$ has also been observed in some binary lipid systems with one of the two lipids as a minor component (Ye, 1992). If the two maxima in the transition heat capacity of 1:1 DMPC–DSPC reflect the existence of two laterally separated DMPC-rich and DSPC-rich phases due to nonideal mixing of DMPC and DSPC molecules, the kinetic behavior of the DMPC-rich and DSPC-rich phases would likely be similar to that of a binary lipid mixture with one of the two lipids as a minor component. In this case, it would not be surprising that the 1:1 DMPC–DSPC mixture shows two relaxation time maxima corresponding to the two peaks in the equilibrium heat capacity function. For 1:1 DMPC–DPPC and DPPC–DSPC mixtures, similar arguments could also be applied except that the two peaks in the transition heat capacity are presumably not well separated as in 1:1 DMPC–DSPC so that only a plateau in $C_p(T)$ of the transition is observed.

Third, if the above argument of dynamic lateral phase separation is not applicable to the nearly ideal-mixing systems of DMPC–DPPC and DPPC–DSPC (Shimshick and McConnell, 1973; Mabrey and Sturtevant, 1976; Knoll et al., 1981), an alternative interpretation for the appearance of bimaximal relaxation times in 1:1 DMPC–DPPC and DPPC–DSPC mixtures would be that molecular diffusion plays a significant role in determining the phase transition kinetics of 1:1 binary lipid mixtures. This interpretation is also consistent with the kinetic data of 1:1 DMPC–DSPC. The question whether dynamic phase separation, molecular diffusion, or both play major roles in the relaxation behavior of binary lipid systems will only be answered by further study.

4.2. AC Calorimetry and Multifrequency Calorimetry

The technique of measuring specific heat capacity by monitoring amplitudes of periodic heat pulses propagated across the bulk sample material was first discussed by Sullivan and Seidel (1968) and was subsequently called "ac calorimetry." Since then a number of articles have been published using ac calorimetry for studies of phase transition phenomena in various materials including biological substances (Tanasijczuk and Oja, 1978; Black and Dixon, 1981). The design and operation of ac calorimetry have been discussed in several publications (Sullivan and Seidel, 1968; Lewis, 1970; Schartz, 1971). Briefly, this technique consists of measuring the thermal response of a sample of an oscillating heat signal in the form of temperature waves propagating through the sample. Heat pulses are supplied from one side of a thin sample chamber and detected on the opposite side of the sample with a temperature sensor. The temperature response on the measuring side are attenuated in amplitude and shifted in phase with respect to the temperature oscillations on the excitation side. The magnitudes of amplitude attenuation and phase shift depend strongly on the thickness of the sample chamber, the heat capacity and heat conduction of the sample, and the heat pulse frequency. The sample heat capacity is calculated from the measured temperature signal according to a one-dimensional heat transfer equation.

Some examples of studying lipid phase transitions using ac calorimetry are those by Black and Dixon (1981) on DMPC MLVs and Hatta et al. (1983, 1984) on DPPC MLVs. At 0.4 Hz oscillating frequency and in the heating mode, Black and Dixon (1981) found that DMPC MLVs exhibited a T_m 23.9°C with $\Delta T_{1/2} = 0.15$–0.2°C and $\Delta H = 4.8$ kcal/mol, which are in good agreement with the DSC results of Mabrey and Sturtevant (1976) and Lentz et al. (1978). However, the cooling scan showed significantly lower values: $T_m = 23.4$–23.8°C with $\Delta T_{1/2} = 0.25$–0.40°C and $\Delta H - 1.9$ kcal/mol. These authors have interpreted this apparent hysteresis in the main transition in terms of a nucleation and subsequent annealing mechanism which prescribes the formation of small ordered domains in the bilayer on freezing the lipid acyl chains (Black and Dixon, 1981). This type of apparent hysteresis in the lipid main transition has also been reported to occur in DPPC MLVs (Tenchov et al., 1989). Based on the analysis of the 0.6-Hz ac calorimetric data for DPPC MLVs, Hatta et al. (1983, 1984) have proposed that the main transition of DPPC MLVs is of a weak first-order nature. This proposal is supported by the DSC results obtained with DPPC MLVs at a very slow scan rate, 0.1°C/hr (Biltonen, 1990) and the results from recent Monte Carlo simulations (Corvera et al., 1993).

Since the conventional ac calorimetry has been restricted to operate at a single or very narrow range of perturbation frequency, Freire and co-workers designed a modified version of an ac calorimeter which operates over a broader frequency range of the perturbation (Mayorga *et al.*, 1988). This modified instrument is called a "multifrequency calorimeter" (MFC) and is capable of directly accessing the frequency spectrum of the enthalpy fluctuations. From this frequency spectrum, the amplitude and time scale of the enthalpy fluctuations associated with phase transitions of the sample can be obtained, as in the case of VPC. The difference between MFC and VPD is that the former measures the autocurrelation function of the enthalpy fluctuations whereas the latter measures the cross-correlation function of the enthalpy fluctuation and the density fluctuation. It should be noted that both the enthalpy fluctuations, which are represented by the heat capacity, and the density fluctuations, which are represented by the compressibility, originate primarily from conformational isomerization of the hydrocarbon chains. In fact, the relaxation times obtained for phase transitions of unilamellar phospholipid bilayers using VPC and MFC have been found to be in good agreement with each other (van Osdol *et al.*, 1991).

5. CALORIMETRIC STUDIES OF LIPID–PROTEIN INTERACTIONS

5.1. Isothermal Titration Calorimetry

In the previous sections, we described equilibrium thermodynamic experimentation using DSC and dynamic thermodynamic experimentation using VPC. Both techniques depend on temperature variations as monitors of physicochemical reactions and usually work under well-established premixing and pre-equilibrium conditions. Another category of experimental calorimetry concerns the heat production associated with a chemical reaction or physical interaction. This technique is often called *isothermal titration calorimetry* (ITC) since the chemical reaction is usually initiated under isothermal conditions by titrating one reactant or solution into another reactant or solution. One classic type of titration calorimeter works under a (pseudo) adiabatic condition. The heat generated is monitored by the temperature change of the system. The total heat generated is calculated from the product of the measured temperature change and the known heat capacity of the system. The commercially available Tronac 450 Titration calorimeter is an example of this type of adiabatic titration calorimeter.

Another popular type of titration calorimeter works under isothermal or psuedoisothermal conditions. The heat production associated with a reaction is determined by measurement of the temperature difference between the reaction cell and the heat sink. There are two different techniques by which this temperature signal can be used to determine the total heat effect. One involves an active use of a Peltier heating–cooling device controlled by a feedback circuit. As the temperature is changed by heat release or absorption of the reaction, a current is applied to the Peltier device to maintain the reaction system at constant temperature. The total electric power input to the device during the reaction course is equal to the reaction heat effect. An example of this type of calorimeter is the Tronac 550 Titration calorimeter. The more recent Microcal MC-2 high-sensitivity titration calorimeter is similar in principle, but distinct in design. With this calorimeter, both the sample and the reference cells are heated at a very slow, but constant rate. On addition of titrant into the sample cell, heat will be generated or absorbed by the reaction. This heat of reaction is compensated by electrical heating of the appropriate cell to maintain a zero temperature difference between cells. The necessary power to do this is the measure of the heat of the reaction. Thus, the system is not strictly isothermal, but the high sensitivity of the instrument keeps the temperature change in the system very small.

Another ITC approach uses a Peltier device passively. During the reaction course, the device generates a voltage proportional to the temperature difference between the reaction cell and the sink, which according to Newton's law of heat transfer is directly proportional to the rate of heat flow between the cell and the sink. The integral of the rate of heat flow over the time of the experiment yields the heat of the reaction. The sensitivity of the calorimeter is usually improved by adopting a twin-cell or differential design, in which the two temperature sensors of the reaction cell and reference cell are connected in electrical opposition so that the effects of thermal fluctuations in the heat sink on the two cells tend to cancel each other out. The voltage then measured is proportional to the difference in temperature between the reaction and reference cell. In this design, the voltage (proportional to the rate of heat flow) is monitored as a function of time and the integration of the voltage—time curve provides an estimate of the total heat production. With appropriate correction for the finite dynamic response of the calorimeter, the time-resolved or kinetic energetics of the reaction can also be obtained. The calorimeter in this case is not strictly under isothermal conditions since the temperature of the sample changes during the reaction process although being identical at the beginning and end of the experiment. Examples of this type of calorimeter design are the older LKB batch and flow microcalorimeters and the more recent Termometric Thermal Activity Monitors, which are also referred to as heat-leak calorimeters. The readers are referred to Suurkuusk and Wadso (1982), Nordmark et al. (1984), Wiseman et al. (1989), Backman and Wadso (1991), and Bastos et al. (1991) for further details of the design and application of commonly used titration calorimeters. Older reviews of reaction calorimetry for biological systems include Langerman and Biltonen (1979), Biltonen and Langerman (1979), and Beezer (1980).

5.2. Calorimetry of Lipid–Protein Interactions

Calorimetric investigations of lipid–protein or lipid–peptide interactions have been mainly performed with two types of calorimeter, one being a differential scanning calorimeter (Papahadjopoulos *et al.*, 1975; Silvius, 1982; Freire *et al.*, 1983; McElhaney, 1986) and the other being a titration calorimeter (Myers *et al.*, 1987; Beschiaschvili and Seelig, 1992; Heimberg and Biltonen, 1994). DSC is used to investigate the influence of protein on the membrane phase transition properties from which the nature of lipid–protein interactions can be assessed. To date, lipid–protein interactions have mostly been examined using simple membrane systems containing only one lipid species and protein due to the general complexity of multicomponent lipid systems. A general finding is that on incorporation of protein, the membrane phase transition is broadened relative to that of the pure lipid membrane, with a concomitant change of the phase transition enthalpy and a shift of the membrane transition temperature relative to that of the pure lipid membrane. In many cases, multiple peaks in the transition heat capacity function are observed, with one of the peaks corresponding to the heat capacity peak of the pure lipid system. A popular interpretation of the appearance of multiple heat capacity peaks in the lipid–protein system in contrast to a single sharp peak in the one-component lipid system is that the binding of the protein to the lipid bilayer has induced conformational or chemical phase separations in the bilayer membrane, with some phases lipid-rich and others protein-rich. The method of decomposing the broad transition heat capacity peak into several subpeaks has been used to analyze in a more quantitative way the details of putative multiple domain populations coexisting in the bilayer membrane (Papahadjopoulos *et al.*, 1975; Silvius, 1982; Rigell *et al.*, 1985; Zhang *et al.*, 1992).

A systematic DSC study by Papahadjopoulos *et al.* (1975) on the phase behavior of phospholipid membranes containing a variety of proteins suggested the lipid–protein interactions could be divided into three general groups. Group 1 proteins, including ribonuclease and the peptide polylysine, cause an increase in the transition enthalpy of the bilayer membrane with or without significant changes in the transition temperature. In addition, these proteins show minimal effects on vesicle permeability and area expansion of monolayers. The lipid–protein interaction in these cases is represented by a simple surface binding of the protein to the lipid bilayer without protein penetration into the acyl chain region. Group 2 proteins, including cytochrome *c* and basic myelin protein (A1 protein), induce a drastic decrease in both the transition temperature and enthalpy, and also a large increase in the vesicle permeability and area expansion of closely packed monolayers at the air–water interface. The lipid–protein interaction involving Group 2 proteins is represented by surface binding of the protein followed by partial penetration. Group 3 proteins, including proteolipid apoprotein and gramicidin A, have no appreciable effects on the membrane transition temperature but cause a linear decrease of the transition enthalpy, proportional to the mole fraction of protein. The protein in this case is thought to be embedded within the bilayer, interacting locally with the surrounding lipid molecules, while the rest of the bilayer is largely unperturbed. These conclusions have been corroborated by studies using similar or different techniques (Boggs *et al.*, 1981a,b; McElhaney, 1986).

On the other hand, ITC is usually used to quantify directly the energetics of the lipid–protein binding process. One classical application of titration calorimetry is to determine the equilibrium binding constant of two reactants. With either one of the reactants as a titrant, an equilibrium binding curve is generated by measuring the heat of mixing of the two reactants as a function of the titrant concentration. Both the enthalpy change and the binding constant can be estimated from such data. In cases where the protein binds to the lipid membrane and induces subsequent cooperative conformational phase separation in the membrane (Bazzi and Nelsestuem, 1991), the heat production or absorption associated with the membrane phase separation will be simultaneously measured. The binding curve so generated is more complex and contains information on both the reversible binding of the protein to the bilayer membrane and the energetics of the membrane conformationl change. An example of protein binding being coupled strongly to thermotropic structural changes of the lipid bilayer is the study of cytochrome *c* binding to DMPG (Heimberg and Biltonen, 1994).

6. CONCLUDING REMARKS

In this chapter, we have briefly reviewed various calorimetric techniques for static and dynamic studies of cooperative phase transitions in phospholipid bilayer membranes. The method of using the energetics as a monitor of the phase transition is a well-known and powerful technique because structural reorganization during the phase transition is generally associated with large energy changes. DSC is a popular technique to monitor the equilibrium thermodynamic changes associated with phase transition in lipid systems. VPC provides a unique energetic method to study the phase transition kinetics of lipids. ITC is a useful tool for monitoring the energetics of lipid–protein interactions and the coupling between protein binding and the lipid structural transitions.

The concepts of lateral phase separations have been widely established experimentally in various lipid mixture bilayers but have only so far been mostly derived in principle for biomembranes based on physical and chemical studies of simple lipids and lipid mixture bilayer models (Gennis, 1989). Because of the highly dynamic properties of biological membranes, the rate and extent of molecular motion, particularly in a cooperative fashion in the bilayer system, may be essentially important for biological functions (Biltonen, 1990; Thompson *et al.*, 1992). It has been suggested that well-organized functional units are formed in biomembranes by the heterogeneous distribution of lipids and proteins (Sack-

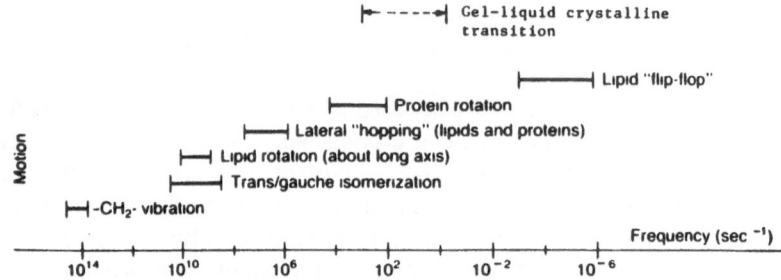

FIGURE 11. The characteristic frequencies of molecular motions of proteins and lipids in bio-membranes, adapted from Gennis (1989). The dashed bar represents the frequency scale of the gel–liquid-crystalline transition of lipid bilayers. Note that boundaries are very approximate.

mann, 1984; Gennis, 1989). The dynamics of membrane components have drawn considerable attention during the past two decades since the proposal of the fluid mosaic model which emphasized the importance of the molecular mobility of membrane components by conceptualizing the membrane as a sea of lipids in which embedded proteins are freely floating. Many sophisticated techniques of optical and magnetic resonance spectroscopy have been developed and applied to investigate the qualitative and quantitative aspects of the membrane dynamics. A wide range of motions in the membrane from 10^{-14} sec for molecular vibrations to as long as many days for transbilayer flip-flop have been characterized, as shown in Figure 11. It is interesting to note that the frequency range marked by the dashed line characterizing the cooperative gel–liquid-crystalline transition as measured by VPC happens to fill in the gap in the original frequency assignment plot drawn by Gennis (1989). This graph indicates that the complex membrane function in biological systems requires a wide spectrum of molecular motions. A comprehensive study of many physical and chemical aspects of the membrane system will need to be carefully carried out in order to answer specific questions regarding biomembrane function. This is certainly true for the studies on the phase transition kinetics of multicomponent phospholipid systems. Many experimental and theoretical investigations on density fluctuations and kinetic phase separations in more complex membrane system consisting of a greater diversity of natural lipid species are necessary to construct a more realistic model of a dynamic bilayer membrane.

ACKNOWLEDGMENTS. Supported by grants from NIH (GM37658) and NSF (PCM8300056 and DMB9005374).

7. REFERENCES

Albon, N., and Sturtevant, J. M., 1978, Nature of the gel to liquid crystal transition of synthetic phosphatidylcholines, *Proc. Natl. Acad. Sci. U.S.A.* **75:**2258–2260.

Backman, P., and Wadso, I., 1991, Cell growth experiments using a microcalorimetric vessel equipped with oxygen and pH electrodes, *J. Biochem. Biophys. Methods* **23:**283–293.

Bastos, M., Hagg, S., Lonnbro, P., and Wadso, I., 1991, Fast titration experiments using heat conduction microcalorimeters, *J. Biochem. Biophys. Methods* **23:**255–258.

Bazzi, M. D., and Nelsestuem, G. L., 1991, Extensive segregation of acidic phospholipids in membranes induced by protein kinase C and related proteins, *Biochemistry* **30:**7961–7969.

Beezer, A. E., ed., 1980, *Biological Microcalorimetry,* Academic Press, New York.

Beschiaschvili, G., and Seelig, J., 1992, Peptide binding to lipid bilayers. Nonclassical hydrophobic effect and membrane-induced pK shifts, *Biochemistry* **31:**10044–10053.

Biltonen, R. L., 1990, A statistical-thermodynamic view of cooperative structural changes in phospholipid bilayer membranes: Their potential role in biological function, *J. Chem. Thermodyn.* **22:**1–19.

Biltonen, R. L., and Langerman, N., 1979, Microcalorimetry for biological chemistry: Experimental design, data analysis, and interpretation, *Methods Enzymol.* **61:**287–318.

Biltonen, R. L., and Ye, Q., 1993, Kinetics of the gel-to-liquid phase transition of binary lipid bilayers using volume perturbation calorimetry, *Prog. Colloid. Polymer Sci.* **93:**112–117.

Black, S. G., and Dixon, G. S., 1981, AC calorimetry of odimyristoylphosphatidylcholine multilayers: Hysteresis and annealing near the gel to liquid-crystal transition, *Biochemistry* **20:**6740–6744.

Boggs, J. M., Stamp, D., and Moscarello, M. A., 1981a, Interaction of myelin basic protein with dipalmitoylphosphatidylglycerol: Dependence on the lipid phase and investigation of a metastable state, *Biochemistry* **20:**6066–6072.

Boggs, J. M., Wood, D. D., and Moscarello, M. A., 1981b, Hydrophobic and electrostatic interactions of myelin basic proteins with lipid. Participation of N-terminal and C-terminal portions, *Biochemistry* **20:**1065–1073.

Burack, W. R., Yuan, Q., and Biltonen, R. L., 1993, Role of lateral phase separation in the modulation of phospholipase A2 activity, *Biochemistry* **32:**583–589.

Caffrey, M., 1989, The study of lipid phase transition kinetics by time-resolved x-ray diffraction, *Annu. Rev. Biophys. Biophys. Chem.* **18:**159–186.

Cevc, G., 1987, How membrane chain melting properties are regulated by the polar surface of the lipid bilayer, *Biochemistry* **26:**6305–6310.

Cevc, G., and Marsh, D., 1985, Hydration of noncharged lipid bilayer membranes. Theory and experiments with phosphatidylethanolamines, *Biophys. J.* **47:**31–31.

Cevc, G., and Marsh, D., 1987, *Phospholipid Bilayers: Physical Principles and Models,* Wiley, New York.

Cevc, G., Watts, A., and Marsh, D., 1980, Non-electrostatic contribution to the titration of the ordered-fluid phase transition of phosphatidylglycerol bilayers, *FEBS Lett.* **120:** 267–270.

Clegg, R. M., and Maxfield, B. W., 1976, Chemical kinetics studies by a new small-pressure perturbation method, *Rev. Sci. Instrum.* **47:**1383–1393.

Coolbear, K. P., Berde, C. B., and Keough, K. M., 1983, Gel to liquid-crystalline phase transitions of aqueous dispersions of polyunsaturated mixed-acid phophatidylcholines, *Biochemistry* **22:**1466–1473.

Corvera, E., Laradji, M., and Zuckermann, M. J., 1993, Application of finite size scaling to the Pink model for lipid bilayers, *Phy. Rev. E.,* **47:**696–703.

Crowe, J. H., Hoekstra, F. A., and Crowe, L. M., 1992, Anhydrobiosis, *Annu. Rev. Physiol.* **54:**579–599.

Curatolo, W. Sears, B., and Neuringer, L. J., 1985, A calorimetry and deuterium NMR study of mixed model membranes of 1-palmitoyl-2-oleoylphosphatidylcholine and saturated phosphatidylcholines, *Biochim. Biophys. Acta* **817**:261-270.

Delong, E. F., and Yaganos, A. A., 1985, Adaptation of the membrane lipids of a deep-sea bacterium to changes in hydrostatic pressure, *Science* **228**:1101-1102.

Findlay, E. J., and Barton, P. G., 1978, Phase behavior of synthetic phosphatidylglycerols and binary mixtures with phosphatidylcholines in the presence and absence of calcium ions, *Biochemistry* **17**:2400-2408.

Freire, E., and Biltonen, R. L., 1978, Estimation of molecular averages and equilibrium fluctuations in lipid bilayer systems from the excess heat capacity function, *Biochim. Biophys. Acta* **514**: 54-68.

Freire, E., Markello, T., Rigell, C., and Holloway, P. W., 1983, Calorimetric and fluorescence characterization of interactions between cytochrome b5 and phosphatidylcholine bilayers, *Biochemistry* **22**:1675-1680.

Gennis, R. B., 1989, *Biomembranes: Molecular Structure and Function*, Springer-Verlag, Berlin.

Halvorson, H. R., 1979, Relaxation kinetics of glutamate dehydrogenase self-association by pressure perturbation, *Biochemistry* **18**:2480-2487.

Hatta, I., Suzuki, K., and Imaizumi, S., 1983, Pseudo-critical heat capacity of single lipid bilayers, *J. Phys. Soc. Jpn.* **52**:2790-2797.

Hatta, I., Imaizumi, S., and Akutsu, Y., 1984, Evidence for weak first-order nature of lipid bilayer phase transition from the analysis of pseudo-critical specific heat, *J. Phys. Soc. Jpn.* **53**:882-888.

Hauser, H., Pascher, I., Person, R. H., and Sundell, S., 1981, Preferred conformation and molecular packing of phosphatidylethanolamine and phosphatidylcholine, *Biochim. Biophys. Acta* **650**: 21-51.

Heimberg, T., and Biltonen, R. L., 1994, The thermotropic behavior of dimryistoylphosphatidylglycerol and its interaction with cytochrome C, *Biochemistry* **33**:9477-9488.

Huang, C., 1991, Empirical estimation of the gel to liquid crystalline phase transition temperature for fully hydrated saturated phosphatidylcholines, *Biochemistry* **30**:26-30.

Huang, C., Li, S., Wang, Z., and Lin, H., 1993, Dependence of the bilayer phase transition temperatures on the structural parameters of phosphatidylcholines, *Lipids* **28**:365-370.

Huang, J., Swanson, J. E., Dibble, A.R.G., Hinderliter, A. K., and Feigenson, G., 1993, Nonideal mixing of phosphatidylserine and phosphatidylcholine in the fluid lamellar phase, *Biophys. J.* **64**:413-425.

Ipsen, J. H., and Mouritsen, O. G., 1988, Modeling the phase equilibria in two-component membranes of phospholipids with different acyl chain lengths, *Biochim. Biophys. Acta* **944**:121.

Jackson, W. M., Kostyla, J., Nordin, J. H., and Brandts, J. F., 1973, Calorimetric study of protein transitions in human erythrocyte ghosts, *Biochemistry* **12**:3662-3667.

Johnson, M. L., and Frasier, S. G., 1985, Nonlinear least-square analysis, *Methods Enzymol.* **117**:301-342.

Johnson, M. L., Winter, T. C., and Biltonen, R. L., 1983, The measurement of the kinetics of lipid phase transitions: A volume-perturbation kinetic calorimeter, *Anal. Biochem.* **128**:1-6.

Keenan, T. W., and Moore, D. J., 1970, Phospholipid class and fatty acid composition of Golgi apparatus isolated from rat liver and comparison with other cell fractions, *Biochemistry* **9**: 19-25.

Kleinig, H., 1970, Nuclear membranes from mammalian liver, *J. Cell Biol.* **46**:396-402.

Knoll, W., Ibel, K., and Sackmann, E., 1981, Small-angle neutron scattering study of lipid phase diagrams by the contrast variation method, *Biochemistry* **20**:6379-6383.

Knoll, W., Schmidt, G., and Sackmann, E., 1983, Critical demixing in fluid bilayers of phospholipid mixtures. A neutron diffraction study, *J. Chem. Phys.* **79**:3439-3442.

Ladbrooke, B. D., and Chapman, D., 1969, Thermal analysis of lipids, proteins and biological membranes. A review and summary of some recent studies, *Chem. Phys. Lipids* **3**:304–356.

Langerman, N., and Biltonen, R. L., 1979, Microcalorimeters for biological chemistry: Applications, instrumentation and experimental design, *Methods Enzymol.* **61**:261–286.

Lee, A. G., 1978, Calculation of phase diagrams for non-ideal mixtures of lipids, and a possible nonrandom distribution of lipids in lipid mixtures in the liquid crystalline phase, *Biochim. Biophys. Acta* **507**:433–444.

Lentz, B. R., Freire, E., and Biltonen, R. L., 1978, Fluorescence and calorimetric studies of phase transitions in phosphatidylcholine multilayers: kinetics of the pretransition, *Biochemistry* **17**:4475–4480.

Lewis, E. A., 1970, Heat capacity of gadolinium near the Curie point, *Phys. Rev. B* **1**,4368–4377.

Lichtenberg, D., Freire, E., Schmidt, C. F., Barenholz, Y., Felgner, P. L., and Thompson, T. E., 1981, Effect of surface curvature on stability, thermodynamic behavior, and osmotic activity of dipalmitoylphosphatidylcholine single lamellar vesicles, *Biochemistry* **20**:3462–3467.

Lin, H. N., Wang, Z. Q., Huang, C. H., 1990, The influence of acyl chain-length asymmetry on the phase transition parameters of phosphatidylcholine dispersions, *Biochim. Biophys. Acta* **1067**:17–28.

Luzzati, V., and Tardieu, A., 1974, Lipid phases: Structure and structural transitions, *Annu. Rev. Phys. Chem.* **25**:79–94.

Mabrey, S., and Sturtevant, J. M., 1976, Investigation of phase transition of lipids and lipid mixtures by high-sensitivity differential scanning calorimetry, *Proc. Natl. Acad. Sci. USA* **73**:3862–3866.

Macdonald, A. G., 1978, A dilatometric investigation of the effect of general anesthetics, alcohols and hydrostatic pressure on the phase transition in smectic mesophases of dipalmitoyl phosphatidylcholine, *Biochim. Biophys. Acta* **507**:26–37.

McElhaney, R. N., 1986, Differential scanning calorimetric studies of lipid–protein interactions in model membrane systems, *Biochim. Biophys. Acta* **864**:361–421.

McElhaney, R. N., 1989, The influence of membrane lipid composition and physical properties of membrane structure and function in *Acholeplasma Ladilawii, Microbiology* **17**:1–32.

Marsh, D., 1990, *Handbook of Lipid Bilayers*, CRC Press, Boca Raton, Fla.

Mason, J. T., Huang, C.-H., and Biltonen, R. L., 1981, Calorimetric investigations of saturated mixed-chain phosphatidylcholine bilayer dispersions, *Biochemistry* **20**:6086–6092.

Mayorga, O. L., Van Osdol, W. W., Lacomba, J. L., and Freire, E., 1988, Frequency spectrum of enthalpy fluctuations associated with macromolecular transitions, *Proc. Natl. Acad. Sci USA* **85**:9514–9518.

Melchior, D. L., and Steim, J. M., 1976, Thermotropic transitions in biomembranes, *Annu. Rev. Biophys. Bioeng.* **5**:205–238.

Mountcastle, D. B., Biltonen, R. L., and Halsey, M. J., 1978, Effects of anesthetics and pressure on the thermotropic behavior of multilamellar dipalmitoyl phosphatidylcholine liposomes, *Proc. Natl. Acad. Sci. USA* **75**:4906–4910.

Mouritsen, O. G., and Jorgensen, K., 1991, Dynamic lipid-bilayer heterogeneity: A mesoscopic vehicle for membrane function? *BioEssays* **14**:129–136.

Myers, M., Mayorga, O., Emtage J., and Freire, E., 1987, Thermodynamic characterization of interaction between ornithine transcarbamylase leader peptide and phospholipid bilayer membranes, *Biochemistry* **26**:4309–4315.

Nagle, J. F., 1980, Theory of the main lipid bilayer phase transition, *Annu. Rev. Phys. Chem.* **31**:157–195.

Nagle, J. F., and Wilkinson, D. A., 1978, Lecithin bilayers: Density measurements and molecular interactions, *Biophys. J.* **23**:159–175.

Nelson, G. J., 1967, Lipid composition of erythrocytes in various mammalian species, *Biochim. Biophys. Acta* **144**:221–232.

Nordmark, M. G., Laynez, J., Schon, A., Suurkuusk, J., and Wadso, I., 1984, Design and testing of a new microcalorimetric vessel for use with living cellular systems and in titration experiments, *J. Biochem. Biophys. Methods* **10**:187–202.

Op den Kamp, J. A., Kauerz, M. T., and van Deenen, L. L., 1975, Action of pancreatic phospholipase A2 on phosphatidylcholine bilayers in different physical states, *Biochim. Biophys. Acta* **406**:169–177.

Papahadjopoulous, D., Jacobsen, K., Nir, S. and Isac, T. 1973, Phase transitions in phospholipid vessicles, fluoresence polarization and permeability measurements concerning the effects of temperature and cholesterol. *Biochim. Biophys. Acta,* **311**:330–348.

Papahadjopoulos, D., Moscarello, M., Eylar, E. H., and Isac, T., 1975, Effects of proteins on thermotropic phase transitions of phospholipid membranes, *Biochim. Biophys. Acta* **401**:317–335.

Privalov, P. L., and Khechinashvili, N. N., 1974, A thermodynamic approach to the problem of stabilization of globular protein structure: A calorimetric study, *J. Mol. Biol.* **86**:665–684.

Rigell, C. W., de Saussure, C., and Freire, E., 1985, Protein and lipid structural transitions in cytochrome *c* oxidase–dimyristoylphosphatidylcholine reconstitutions, *Biochemistry* **24**:5638–5646.

Ruggiero, A., and Hudson, B., 1989, Critical density fluctuations in lipid bilayers detected by fluorescence lifetime heterogeneity, *Biophys. J.* **55**:1111–1124.

Sackmann, E., 1984, Physical basis of trigger processes and membrane structures, *Biol. Membr.* **5**:105–143.

Schartz, P., 1971, Order–disorder transition in NH_4Cl. III. Specific heat, *Phys. Rev. B.* **4**:920–928.

Seelig, J., Macdonald, P. M., and Scherer, P. G., 1987, Phospholipid head groups as sensors of electric charge in membranes, *Biochemistry* **26**:7535–7541.

Shaukat, A., Lin, H., Bittman, R., and Huang, C.-H., 1989, Binary mixtures of saturated and unsaturated mixed-chain phosphatidylcholines. A differential scanning calorimetry study, *Biochemistry* **28**:522–528.

Shimshick, E. J., and McConnell, H. M., 1973, Lateral phase separation in phospholipid membranes, *Biochemistry* **12**:2351–2360.

Silvius, J. R., 1982, Thermotropic phase transitions of pure lipids in membranes and their modification by membrane proteins, in: *Lipid–Protein Interactions,* Vol. 2 (P.C. Jost and O. H. Griffith, eds.), pp. 235–281, Wiley, New York.

Singer, S. J., and Nicolson, G. L., 1972, The fluid mosaic model of the structure of cell membranes, *Science* **175**:720–731.

Sixl, F., and Watts, A., 1983, Headgroup interactions in mixed phospholipid bilayers, *Proc. Natl. Acad. Sci. USA* **80**:1613–1615.

Slater, J. L., Huang, C. H., and Levin, I. W., 1992, Interdigitated bilayer packing motifs: Raman spectroscopic studies of the eutectic phase behavior of the 1-stearoyl-2-caprylphosphatidylcholine/dimyristoylphosphatidyl choline binary mixture, *Biochim. Biophys. Acta* **1106**:242–250.

Somero, G. N., 1992, Adaptations to high hydrostatic pressure, *Annu. Rev. Physiol.* **54**:557–577.

Steim, J. M., Tourtellotte, M. E., Reinert, J. C., McElhaney, R. N., and Rader, R. L., 1969, Calorimetric evidence for the liquid-crystalline state of lipids in a biomembrane, *Proc. Natl. Acad. Sci. USA* **63**:104–109.

Sturgill, T., and Biltonen, R. L., 1976, Analysis of derivative binding isotherms. Theoretical considerations, *Biopolymers* **15**:337–354.

Sullivan, P. F., and Seidel, G., 1968, Steady-state, ac-temperature calorimetry, *Phys. Rev.* **173**:679–685.

Suurkuusk, J., and Wadso, I., 1982, A multichannel microcalorimetry system, *Chem. Scr.* **20**:155–163.

Suurkuusk, J., Lentz, B., Barenholz, Y., Biltonen, R. L., and Thompson, T. E., 1976, A calorimetric and fluorescent probe study of the gel–liquid crystalline phase transition in small single-lamellar dipalmitoylphosphatidylcholine vesicles, *Biochemistry* **15**:1393–1401.

Tanasijczuk, O. S., and Oja, T., 1978, High resolution calorimeter for the investigation of melting in organic and biological materials, *Rev. Sci. Instrum.* **49**:1545–1548.

Tenchov, B. G., Yao, H., and Hatta, I., 1989, Time-resolved x-ray diffraction and calorimetric studies at low scan rates, *Biophys. J.* **56**:757–768.

Thompson, T. E., and Huang, C. H., 1986, Composition and dynamics of lipids in biomembranes, in: *Physiology of Membrane Disorder,* 2nd ed. (T. E. Andreoli, J. F. Hoffman, D. D. Franestil, and S. G. Schultz, eds.), pp. 25–44, Plenum Press, New York.

Thompson, T. E., Sankaram, M. B., and Biltonen, R. L., 1992, Biological membrane domains: Functional significance, *Comments Mol. Cell. Biophys.* **8**:1–15.

Trudell, J. R., Payan, D. G., Chin, J. H., and Cohen, E. N., 1975, The antagonistic effect of an inhalation anesthetic and high pressure on the phase diagram of mixed dipalmitoyl-dimyristoylphosphatidylcholine bilayers, *Proc. Natl. Acad. Sci. USA* **72**:210–213.

van Deenen, L.L.M., and de Gier, J., 1974, Lipids of the red cell membrane, in: *The Red Blood Cell,* Vol. I (D. N. Surgenor, ed.), pp. 147–211, Academic Press, New York.

van Osdol, W. W., Biltonen, R. L., and Johnson, M. L., 1989, Measuring the kinetics of membrane phase transitions, *J. Biochem. Biophys. Methods* **20**:1–46.

van Osdol, W. W., Johnson, M. L., Ye, Q., and Biltonen, R. L., 1991, Relaxation dynamics of the gel to liquid crystalline transition of phosphatidylcholine bilayers. Effects of chainlength and vesicle size, *Biophys. J.* **59**:775–785.

van Osdol, W. W., Ye, Q., Johnson, M. L., and Biltonen, R. L., 1992, Effects of the anesthetic dibucaine on the kinetics of the gel–liquid crystalline transition of dipalmitoylphosphatidylcholine multilamellar vesicles, *Biophys. J.* **63**:1011–1017.

Von Dreele, P. H., 1978, Estimation of lateral species separation from phase transitions in nonideal two-dimensional lipid mixtures, *Biochemistry* **17**:3939–3943.

Wang, Z. Q., Lin, H. N., and Huang, C. H., 1990, Differential scanning calorimetric study of a homologous series of fully hydrated saturated mixed-chain $C(X):C(X + 6)$ phosphatidylcholines, *Biochemistry* **29**:7072–7076.

Wickner, W., 1975, Asymmetric orientation of phage M13 coat protein in Escherichia coli cytoplasmic membranes and in synthetic lipid vesicles, *Proc. Natl. Acad. Sci. USA* **73**:1159–1163.

Wilkinson, D. A., and Nagle, J. F., 1979, Dilatometric study of binary mixtures of phosphatidylcholines, *Biochemistry* **18**:4244–4249.

Wiseman, T., Williston, S., Brandts, J. F., and Lin, L. N., 1989, Rapid measurement of binding constants and heats of binding using a new titration calorimeter, *Anal. Biochem.* **179**:131–137.

Wong, M., and Thompson, T. E., 1982, Aggregation of dipalmitoylphosphatidylcholine vesicles, *Biochemistry* **21**:4133–4139.

Yager, P., and Peticolas, W. L., 1982, The kinetics of the main phase transition of aqueous dispersions of phospholipids induced by pressure jump and monitored by Raman spectroscopy, *Biochim. Biophys. Acta* **688**:775–785.

Yang, C. P., and Nagle, J. F., 1988. Phase transformation in lipids follow classical kinetics with small diimensionalities, *Phys. Rev. A.* **37**:3993–4000.

Ye, Q., 1992, Phase transition kinetics of multicomponent lipid membranes, Ph.D. dissertation, University of Virginia, Charlottesville.

Ye, Q., van Osdol, W. W., and Biltonen, R. L., 1991, Gel–liquid crystalline transition of some

multilamellar lipid bilayers follows classical kinetics with a fractional dimensionality of approximately two, *Biophys. J.* **60**:1002–1007.

Ye, Q., Thompson, K. K., and Biltonen, R. L., 1993, Effects of moderate pressure on the gel–liquid
crystalline transition of binary phosphataidylcholine bilayers, *Biophys. J.* **64**:A71.

Zhang, Y. P., Lewis, R. N., Hodges, R. S., and McElhaney, R. N., 1992, Interaction of a peptide
model of a hydrophobic transmembrane alpha-helical segment of a membrane protein with
phosphatidylcholine bilayers: Differential scanning calorimetric and FTIR spectroscopic studies,
Biochemistry **31**:11579–11588.

Chapter 5

Ektacytometry of Red Cells

Robert M. Johnson

The ektacytometer measures the deformation of a population of red cells or red cell membranes in response to shear. The basic configuration developed by Bessis and Mohandas (1974, 1975a) comprises a Couette viscometer, with two concentric cylinders constructed of an optically clear material (Figure 1). The cells are suspended in a viscous medium and introduced into the gap between the cylinders (typically 0.5 mm). Cell elongation under shear after cylinder rotation begins is monitored by the diffraction of a laser beam directed normal to the cylinder axis. Some distinctive features of the ektacytometer are (1) cell populations are observed rather than individual cells; (2) cell deformability can be monitored continuously rather than point by point; (3) the viscometer of the ektacytometer is constructed as a flow-though chamber, so that the cell response to varying conditions can be easily determined; (4) flow in the viscometer is laminar, so that the forces acting on the red cells are readily calculated; (5) the cells are sufficiently dilute to make both aggregation and cell–cell collisions negligible factors.

Both intact erythrocytes and ghosts can be studied in the ektacytometer. While the primary focus of ektacytometric work has often been whole cell deformability, proper selection of methods can produce much useful information

Robert M. Johnson   Department of Biochemistry, Wayne State University, Detroit, Michigan 48201.

Subcellular Biochemistry, Volume 23: Physicochemical Methods in the Study of Biomembranes, edited by Herwig J. Hilderson and Gregory B. Ralston. Plenum Press, New York, 1994.

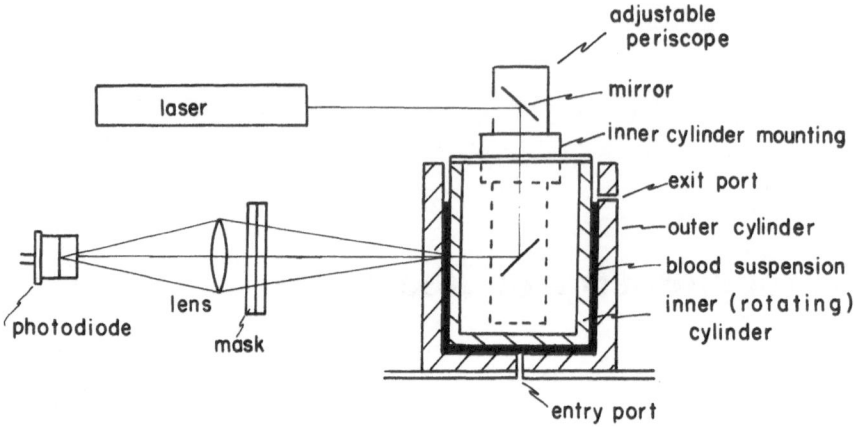

FIGURE 1. Schematic of the ektacytometer. The configuration of the Technicon instrument is shown. The two cylinders are made of polished lucite, and the inner cylinder is mounted on a bearing on the central shaft which also contains a periscope to direct the 1-mW helium–neon laser beam through the blood suspension.

about the membrane itself. This review will emphasize our knowledge about the red cell membrane that has been gained by ektacytometric techniques.

1. HISTORY

Ektacytometry evolved from a collaboration (Bessis and Mohandas, 1974, 1975a, b) between Marcel Bessis and Narla Mohandas at the Institut de Pathologie Cellulaire, Paris. It was a combination of rheological techniques with the earlier observations of Tzanck and Bessis (1947) that the axes of elliptical red cells can be measured by the diffraction methods pioneered by Pijper. In the initial years of its use, it was applied to shape-altered and dehydrated red cells and various hemolytic anemias (Bessis and Mohandas, 1975b) and sickle cells (Bessis and Mohandas, 1976). Important technical advances were developed in collaboration with scientists from Technicon Instruments, including the automated image analyzer (Groner *et al.*, 1980) and provisions for constructing a flow gradient within the measuring chamber. Narla Mohandas moved to the University of California at San Francisco in 1976, where he established highly productive collaborations with M. Clark and S. Shohet. They established the utility of the osmotic gradient scan and developed the membrane fragmentation technique to detect defects in membrane protein interactions. These methods have since received extensive application in many areas of hematology. These developments were paralleled at the Institut de Pathologie Cellulaire, where

Bessis's group developed the oxygen scan methodology. The ektacytometer has been used in more than 200 published investigations and has become one of the major methods of assessing the deformability of the erythrocyte and the rigidity and mechanical stability of the red cell membrane. Some earlier reviews include Mohandas *et al.* (1979), Evans and Mohandas (1986), Mohandas (1988), Johnson (1989), and Scheven (1989).

2. INSTRUMENTATION

Ektacytometers have been made in a number of different laboratories around the world (Stozicky *et al.*, 1979; Gao, 1989), following the original designs implemented by Bessis's group at the Institut de Pathologie Cellulaire (Bessis and Mohandas, 1974, 1975a; Cavadini, 1975). The essential elements of the instrument are a Couette viscometer of transparent materials and a means of passing a laser beam through the layer of sheared erythrocytes in the cylinder gap (Figure 1). In the past, ektacytometers were available commercially from Technicon (511 Benedict Ave., Tarrytown, NY 10591), and many of the published studies have used this instrument. Unfortunately, this source is no longer available, but it is likely that other commercial versions will become available in the next few years. In addition to the basic configuration, the Technicon instrument includes two pumps to generate buffer gradients, a microprocessor that controls the viscometer motor and gradient pumps, an image analyzer, and a keyboard with a display that will show the ektacytometric index or calibration information on demand. The concentric cylinder viscometer diameter is 50.7 mm and the cylinder gap is 0.5 ± 0.02 mm. The available rotation speeds are 0–256 rpm (inner cylinder), settable to 1 rpm. The instrument also has a relaxation mode, in which the decay time of the cell deformation is determined. The cylinder is started and stopped 25 times in succession at 1 sec intervals, and the resulting decay curves of the index are averaged. In principle, the dynamic rigidity of the erythrocyte can then be determined, but this capability has not had many applications.

2.1. Principle

Cell elongation is dependent on the external factors of shear rate and medium viscosity. It can be shown that the shear rate is nearly constant across the gap of a Couette viscometer (see, for example, Stolitz *et al.*, 1981; Scheven, 1989). Therefore, all of the cells in the gap experience the same shear rate, which is (in $\sec^{-1}$):

$$r = R\, \Omega/\epsilon \qquad\qquad (1)$$

where R = cylinder radius (cm), ϵ = cylinder gap (cm), Ω = angular velocity (radians) = $(2\pi/60)$ × rotations per min.

$$\text{shear stress (dynes/cm}^2) = r \times \eta \tag{2}$$

where η = viscosity of the suspending medium (poise).

An additional requirement for elongation is that the external viscosity (η) must exceed the internal viscosity (η_i). If $\eta < \eta_i$, the cells will act as solid particles and tumble in the shear flow. If $\eta > \eta_i$, the cells will elongate. As pointed out by Fischer *et al.* (1978), in this respect the erythrocyte resembles a suspended liquid droplet. Concomitant with elongation, the membrane of the cell rotates around the periphery of the cell, in the motion called "tank-treading." To ensure that elongation takes place, the viscosity of the medium is increased by the addition of dextran, polyvinylpyrrolidone (PVP), or arabinogalactan (Stractan, Larex-O).

2.2. Quantitation of Red Cell Deformation

At the rotation speeds and viscosities used in ektacytometry, the flow between the cylinders is laminar. The cells elongate and orient themselves with their long axis normal to the rotational axis. Because of its high hemoglobin concentration, the interior of the red cell has a refractive index greater than that of the medium, and the erythrocyte will scatter light. The basic information about cell shape is derived from the diffraction pattern, which is circular when the cells are at rest, but becomes elliptical as they are stressed and elongate. The diffracted light is a sum of the scattered light from each particle in the suspension, and in principle contains detailed information about the heterogeneity of the population. In practice, it has been a valuable simplification to ignore the heterogeneity and to characterize the population by a single number, the deformability index (DI). The assumption of homogeneity is, in any case, reasonably accurate for normal blood samples, in which only the immature reticulocytes and the senescent dense cells have unusual deformability properties. These comprise no more than 2–3% of the total erythrocytes.

2.3. Automated Image Analysis

When a uniform population of red cells is sheared in the Couette viscometer, they become ellipsoids with an ellipticity ratio $(L - W)/(L + W)$, where L = the length of the major axis and W = the length of the minor axis. Diffraction theory predicts that the ellipticity of the scattered light pattern will have the same value as the ellipticity of the scattering particles, but rotated 90° (Figure 2). In the

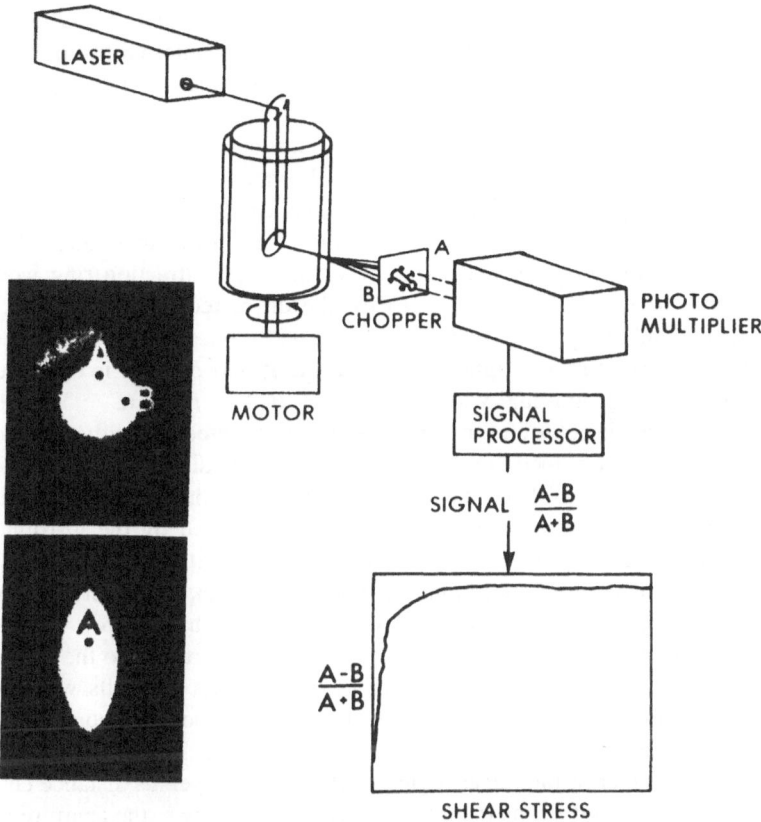

FIGURE 2. Ektacytometer with image analyzer. Inset photographs indicate the points where scattered light intensity is measured (Mohandas *et al.*, 1980).

earliest applications, the diffracted light was projected on a screen and photographed.

Experimental applications of ektacytometry are greatly facilitated by the automated image analyzer (Bessis *et al.*, 1980; Groner *et al.*, 1980; Mohandas *et al.*, 1980) which allows continuous monitoring of the diffraction pattern. The principle is shown in Figure 2. Photodiodes are placed behind a mask with two holes. The hole designated A is on the major axis of the ellipse produced by stressed cells, and the second, designated B, is on the minor axis. Both are equidistant from the center of the diffraction pattern. The light intensities are passed to the microprocessor which calculates the index $(A - B)/(A + B)$. An output voltage of 0 to 10 V proportional to the index is produced by the Technicon apparatus, which can be sent to an x–y recorder. For unstressed cells, the pattern is circular and the light intensity in each hole is equal. Therefore, $(A - B/A + B) = 0$. As the cells are stressed, the pattern becomes ellipsoidal, and the light intensity in A increases as the intensity in B declines, and $(A - B/A + B)$ becomes positive.

It can be shown that this image analyzer produces a quantitatively accurate measure of the ellipticity for normal red cells. Mohandas and co-workers (Groner *et al.*, 1980) first photographed elongated red cells at defined shears in the rheoscope and measured their ellipticity ratio. Cells were then sheared in the ektacytometer, the length L and width W of the first diffraction ring were obtained, and the ratio $(L - W/L + W)$ for the diffracted light calculated. This value was found to equal the cell ellipticity directly observed in the rheoscope. They then showed that the light intensity at the A and B holes is equivalent to measuring ellipse length and width, i.e., $(A - B/A + B) = (L - W/L + W)$, provided that the detector and the viscometer are oriented correctly. The correct positioning was determined empirically, but a theoretical development based on the light scattering of prolate ellipsoids predicted a similar critical placement of the detector (Groner *et al.*, 1980). The index calculated by the instrument is therefore a quantitative measure of the red cell ellipticity. The index has been given various names: DI (deformation index) or EI (ektacytometric index, or elongation index). For consistency, DI will be used in this review.

The size of the diffraction image is inversely proportional to the size of the cell. As a result, red cells that are smaller than human red cells will give an erroneously large value of the DI. This can be understood and corrected if the light intensity across the image is considered (Figure 3). The analyzer holes are placed on the center of the linear region of the intensity versus distance curve of human red cells. In effect, as the image widens and narrows, the aperture moves up and down the intensity curve. The placement of the holes has been chosen so that for the maximal observed widening or narrowing of the image of normal human erythrocytes, the holes remain on the linear portion of the intensity curve. For smaller cells with a larger diffraction image, this condition can be obtained by moving the image analyzer closer to the viscometer, so that it subtends a smaller solid angle of the diffracted light.

The ratio $(A - B/A + B)$ of the image analyzer equals $(L - W/L + W)$ of the diffracted light only when the diffraction image is elliptical. For heterogeneous cell suspensions, notably sickle cells, the diffracted light pattern will be more complex.

2.4. Other Detectors

A video camera interfaced with a microcomputer can also be used for analysis of the diffracted light pattern. Wolf *et al.* (1992) have constructed an

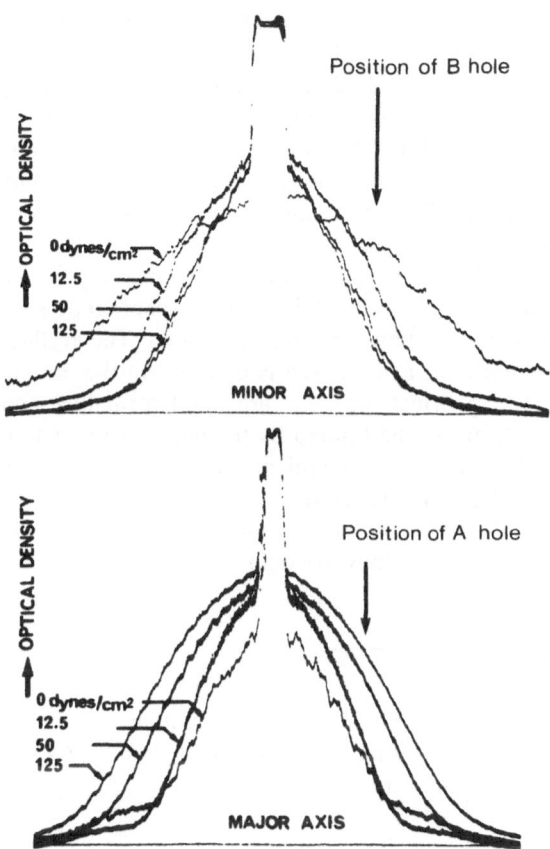

FIGURE 3. Optical density scans of the major and minor axis of the diffraction patterns of normal human red cells under different applied stress values. The mask apertures are placed at the midpoint of the slope of the curve for unstressed cells. The x-axes are not to scale. (From Bessis and Mohandas, 1975a.)

apparatus in which the light intensity at each pixel is recorded and an associated computer program identifies the contours of equal brightness to calculate L and W. The response time of this detector is slow (4 sec) because of the computing time required, but it should be possible to improve this in the future. LORCA, a commercial instrument with video detection built around the Contraves LS-30 viscosimeter, is available from RR Mechatronics (Hoorn, the Netherlands). Although it has not been used in many published investigations, this instrument appears to produce results similar to those of the standard ektacytometer (Hardeman *et al.*, 1987, 1988).

3. FACTORS DETERMINING ERYTHROCYTE DEFORMABILITY

For an individual red cell, deformability is determined by three factors: internal viscosity, geometry, and membrane rigidity (Chien *et al.*, 1970; Mohandas *et al.*, 1979; Chien, 1987). Internal viscosity is related to hemoglobin concentration, usually given as the mean corpuscular hemoglobin concentration (MCHC). The relationship is non linear, with viscosity increasing rapidly at higher hemoglobin concentrations (Chien *et al.*, 1970). The geometry of the cell can be expressed as the surface-to-volume ratio, since the availability of excess membrane is a prerequisite for cell elongation. A sphere represents the maximum volume that can be bounded by a given surface area. The erythrocyte membrane is essentially inextensible, and volume is constant in the absence of net cation movement. Hence, spherical red cells are nondeformable, and the attainable elongation under shear will be limited by the degree to which the cell approximates a sphere. This factor is most evident in the hypotonic region of the osmotic scan (see below). The rigidity of the membrane is not usually a significant influence on the deformability of the whole cell. In unusual circumstances, however, large increases in membrane rigidity can affect the global deformability.

4. MEMBRANE PROPERTIES

Using the full range of ektacytometric techniques, a significant amount of information about the red cell membrane can be obtained. As will be shown, the ektacytometer can provide estimates of:

1. *Membrane deformability,* in the ramp assay
2. *Mechanical strength,* assayed as the fragmentation time under shear
3. *Surface-to-volume ratio,* from the position of O_{min} in osmotic scan ektacytometry
4. *Surface area,* which can be estimated from the maximum value of the osmoscan, if membrane deformability is normal.

Surface area and surface-to-volume ratio can be obtained by other methods, although the ektacytometer does have advantages of speed and convenience. Membrane strength and rigidity are not, however, readily available by other methods. It is of interest to consider how these ektacytometric parameters are related to fundamental membrane properties.

The intrinsic properties of the membrane as a deformable material, independent of any particular geometry, can be obtained by micropipet techniques (Evans and Skalak, 1980; Berk *et al.*, 1988; Evans, 1988), in which small sections of membrane are aspirated into narrow-bore micropipets (1 μm). Although the amount of labor required is considerable, the precision of the values obtained is unequaled by other methods, and more importantly, the intrinsic material properties of the membrane are measured. The information obtained includes the shear extensional modulus (μ), surface viscosity (η_e), area compressibility, and the bending modulus. By stretching an erythrocyte between two pipets, a measure of the yield strength can be obtained. It is also possible to draw an erythrocyte into a micropipet until the portion of the cell outside the pipet assumes a spherical shape. The geometry is then very simple (a sphere attached to a cylinder), and the volume and surface of the cell can be calculated. Finally, it is possible to observe the irreversible plastic deformation that ensues when deformation is maintained for periods of 10 min or more, although this does not occur in the tank-treading erythrocyte under shear in the ektacytometer.

For short periods of deformation, which are the times of interest for ektacytometry, the membrane acts as an elastic solid. Then the relevant properties are μ, the elastic shear modulus for surface deformation without area change, and η_e. the coefficient of surface viscosity (Mohandas, 1988). Changes in the bending modulus or area compressibility modulus can in general be neglected, the bending modulus because it is so small and the area coefficient because it is so large (the membrane is essentially incompressible). Rigidity as measured in the ektacytometer is related to intrinsic material properties in a complex and not completely understood way, and it is also influenced by geometrical factors (Mohandas, 1988). Unpublished observations by Evans and Mohandas (cited in Heath et al., 1982) suggest that μ is an important influence on ektacytometric deformability. It remains true, however, that systematic studies of the effect of alterations in membrane material properties like μ and η_e on shear deformability have yet to be done. The membrane rigidity is probably best regarded as an indicator of structural changes within the membrane that affect the red cell's mechanical response.

5. EKTACYTOMETRIC ASSAYS

While the output of the ektacytometer is always the DI, the cells in the Couette can be subjected to various perturbations. Shear stress can be varied, and the conditions in the viscometer can be varied in many ways. Osmolality, oxygen pressure, buffer composition, and pH are some of the variables that have been used. In addition, kinetic information can be determined, such as the fragmentation time of sheared membranes or relaxation times. These will be considered in turn. Many of the technical details are available elsewhere (Clark et al., 1983; Mentzer et al., 1984; Mohandas, 1988; Johnson, 1989).

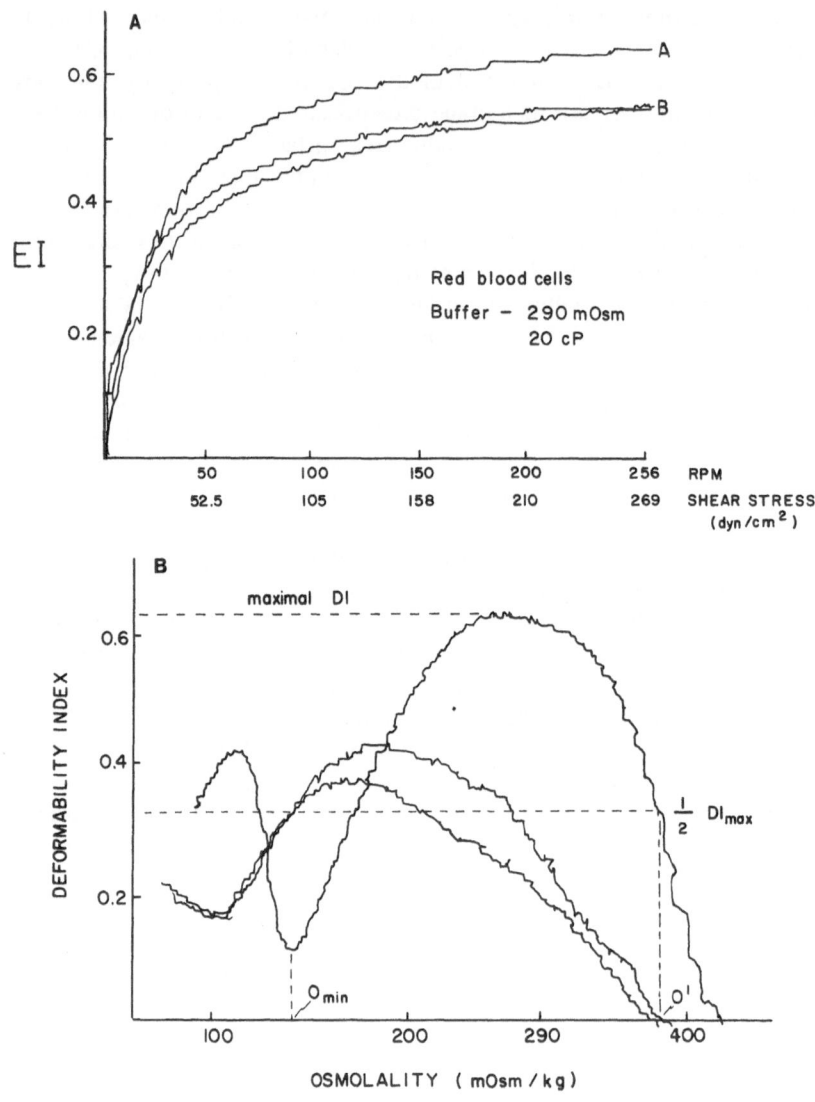

FIGURE 4.

5.1. DI as a Function of Applied Shear (Ramp Assay)

In this experiment, the response of erythrocytes or ghosts to variable shear is determined. For intact cells, a suspension of red cells in isotonic dextran or PVP buffer with a viscosity of about 20 cP (about 120 μl whole blood in 10 ml) is

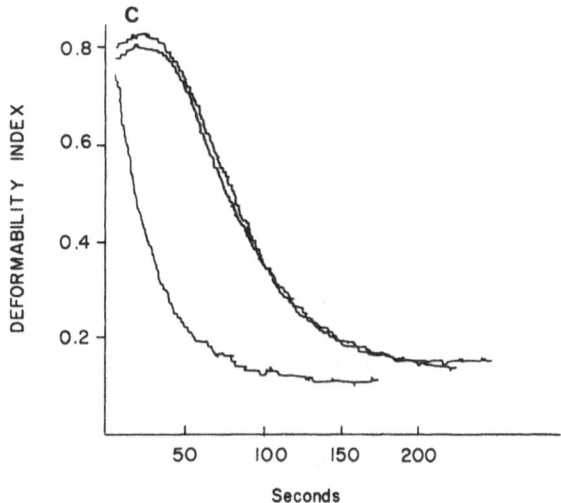

FIGURE 4. (A) DI as a function of applied stress for (A) normal and (B) two examples of slightly nondeformable pathological red cells. (B) Osmoscans for a normal and two dehydrated cell samples. The parameters O' ($= O_{hyper}$) and O_{min} are indicated for the normal curve. (Modified from Clark *et al.*, 1983). (C) Ghost fragmentation. Two curves for normal membranes are shown, together with the fragmentation curve of membranes with a defect in the self-association site of spectrin. These red cells are fragile and are associated with hemolytic anemia.

loaded into the stationary viscometer. The DI as a function of viscometer rotation speed is recorded as the rotation speed is increased. This "ramp-up" to 256 rpm at a rate of 1 rpm/sec is programmed into the microprocessor of the Technicon ektacytometer. The shear rate for a given rpm can be readily calculated and entered on the *x*-axis using Equation (2) (Figure 4A). The plateau of the curve gives the average deformability of the cell population. However, the most useful information to be derived from the ramp is an estimate of membrane deformability. The initial rise of the index for erythrocytes is strongly influenced by the rigidity of the membrane (Mohandas *et al.*, 1980). It was later shown that the ramp-up assay can also be done with resealed ghosts (Heath *et al.*, 1982), which eliminate the confounding effect of shape and the cytoplasmic content of the erythrocyte on deformability. It is usual to plot the index versus the logarithm of applied shear, which produces a straight line over much of the initial part of the rise. A shift of the line to higher stresses indicates that the membranes being tested deform less than normal at a given shear force. However, membrane loss can influence the deformability index of ghosts (Chasis *et al.*, 1988a). This is indicated when the plateau value for a ghost sample is lower than normal. In this case, the initial slope of the index versus stress curve must be interpreted with caution.

It is also possible to do a "ramp-down" from 256 rpm to zero at 1 rpm/sec. This can be used to detect hysteresis in cell deformability.

5.2. Osmotic Gradient Ektacytometry

In this procedure, cells are mixed with a buffer–PVP stream with an increasing linear gradient of osmolality before entering the viscometer. In the Technicon apparatus, the gradient is mixed and introduced into the viscometer by a multi-channel peristaltic pump. Typically the gradient maker vessels contain 20 ml of low-tonicity (38 mosm) solution and 8 ml of high-tonicity (775 mosm) PVP. A suspension of red cells in isotonic buffer (100 μl of blood in 3 ml isotonic PVP) is added through a thin tube added to the mixed stream at a T-joint near the chamber entry port. An entire scan from low to high osmolality requires about 10 min. As there is no convenient way to monitor osmolality continuously, the desired solution osmolality is obtained by the addition of electrolytes and con-ductivity is measured instead. Two platinum wire electrodes are built into the wall of the viscometer, and the output of the instrument's conductivity circuit appears as a voltage at the x-axis output. Earlier problems (Johnson, 1989) with nonlinearity of response have been solved, and the x-axis is linear in osmolality. The output voltage is calibrated by manually filling the viscometer with PVP–NaCl solutions whose osmolalities have been independently determined.

The resulting plot of DI versus osmolality (Figure 4B) is often called an osmoscan. The maximum DI is reached around 290 mosm, the physiological value, showing that normal erythrocytes are optimized for maximal flexibility. The significant parameters derived from the osmoscan are: O_{min}, O' [also called O_{hyper} (Mohandas, 1988], and the maximal value of the index, DI_{max}. Deviations of the curves from the normal shape are indicative of cell heterogeneity and can be diagnostic for various hemolytic anemias.

The three elements determining the deformability of red cells—internal viscosity, surface-to-volume ratio, and membrane rigidity—affect different parts of the osmoscan and can therefore be evaluated independently of each other. A classic analysis of these curves is found in Clark *et al.*, (1983).

Two features of red cell physiology are relevant in understanding these curves. First, biological membranes cannot change their area to any significant degree. Second, they are highly permeable to water, but impermeant to the internal solutes of the red cell, at least over the time span of the ektacytometric experiment. Hence, any change in the external osmolality induces a compensa-tory loss or gain of water to equilibrate the internal and external osmotic pres-sure, which typically requires less than 5 sec.

O' or O_{hyper} is an estimate of internal viscosity and, indirectly, MCHC. Internal viscosity is determined by the concentration of hemoglobin in the cell, which ranges from about 29 to 37% in a population of normal erythrocytes.

Therefore, internal viscosity varies from 5 to 17 cP (Chien *et al.*, 1970). The average MCHC is about 35%, creating an average internal viscosity of 11 cP. As medium osmolality increases, water exits the erythrocytes following the imposed osmotic gradient, increasing hemoglobin concentration and internal viscosity, making the cell less and less deformable. At some point, deformability drops to zero, and the osmoscan crosses the *x*-axis. If a cell is dehydrated to begin with, i.e., has a high MCHC, less additional dehydration will be needed to reach a state of zero deformability in the ektacytometer. This will be manifested in a left shift of the high-osmolality region of the curve. Similarly, overhydrated (low MCHC) cells will require more water loss to concentrate their hemoglobin to the point of nondeformability, right-shifting the high-osmolality region. In this way, the osmoscan can be used to estimate MCHC. Clark *et al.* (1983) found that the osmolality at which DI was midway between DI_{max} and zero was more reproducible than the actual crossover point, and this is defined as O' or O_{hyper} (Figure 4B). A linear relationship was found between O_{hyper} and $1/MCHC$ (Figure 5).

O_{min} is determined by the surface-to-volume ratio. As osmolality is reduced, cells take up water and swell to a volume dictated by the external osmolality and their original content of solutes. As osmolality is reduced further, the cells eventually attain a spherical shape. At this osmolality (O_{min}), DI is zero. This is so because both surface and volume are invariant: the surface because of the fundamental nature of biological membranes, and the volume because a sphere is the maximum volume that can be bounded by a given surface. The cell volume at this point is termed the critical hemolytic volume (CHV). Erythrocytes are nondeformable at their CHV, since any shape other than a sphere would require an increase in the invariant membrane surface area. Any further decline in external osmolality lyses the cells. If cells are already swollen, relative to their surface area, in isotonic conditions, a relatively smaller degree of additional swelling will be required to attain the CHV, resulting in a right shift of the low-osmolality branch of the curve. Similarly, cells with a relative excess of surface will require more swelling to achieve the CHV, resulting in a left shift of O_{min}. The position of O_{min} is directly related to membrane surface area, quantitated by lipid per cell, in cells with equal volume (Figure 5). The surface-to-volume ratio can be determined by the osmotic fragility test. In addition, O_{min} and the midpoint of the osmotic fragility curve are well-correlated, which is experimental confirmation that O_{min} measures the surface-to-volume ratio.

Finally, DI_{max}, the maximum value of the index, is influenced by membrane deformability and the surface area of the cell. If it can be established that membrane deformability is normal (using a ramp assay, for example), then DI_{max} is directly proportional to surface area (Chasis *et al.*, 1988a).

These correlations are shown in Figure 5. The results of many experimental manipulations affecting red cell deformability can be fitted to an explanatory model that includes these factors. For example, changes in internal viscosity

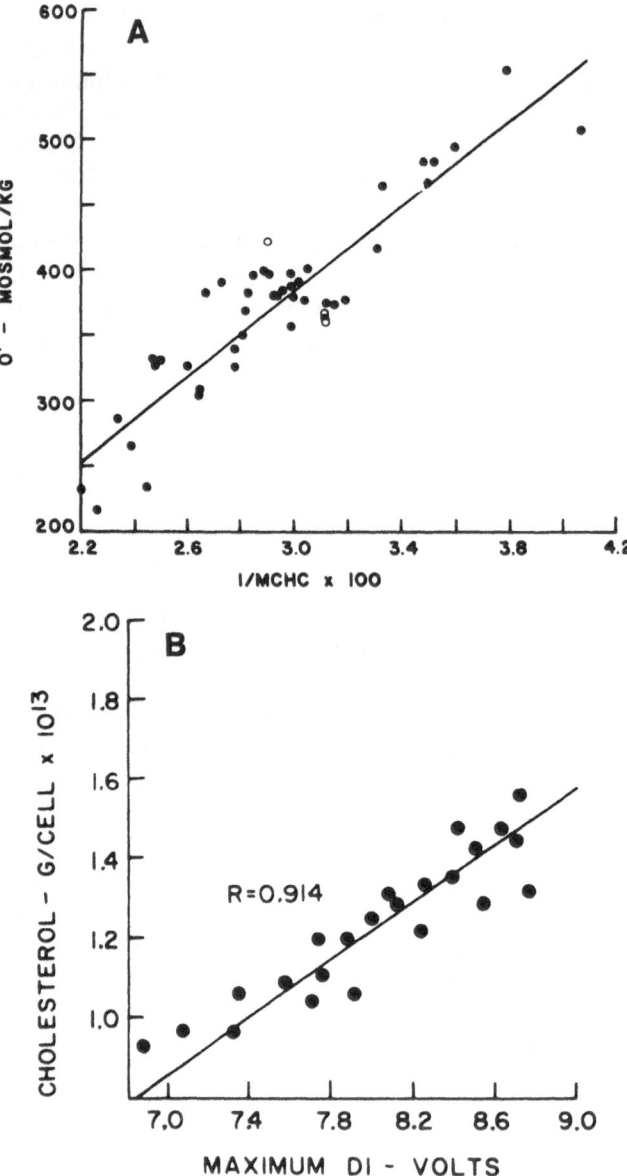

FIGURE 5. (A) Graph of the osmolality (O') at which DI reached $DI_{max}/2$ in hypertonic media versus 1/MCHC. The open symbols represent lysophospholipid-treated cells that have lost varying amount of membrane with a significant volume change. The filled symbols are red cells with varying degrees of hydration. (B) Relationship between membrane cholesterol and DI_{max}. (In this graph, the DI is given as a voltage output.) (Both from Clark *et al.*, 1983.)

brought about by hydration (Clark *et al.*, 1983) or dehydration (Clark *et al.*, 1978, 1981, 1982) shift O_{hyper}. Reduction of surface area by vesiculation (Clark *et al.*, 1983) shifts O_{min} upward. It is possible to predict the osmoscan for a population of red cells, if a single value for surface area and for isotonic volume and a Gaussian distribution in mean intracellular hemoglobin concentration (MCHC) are assumed (Clark, 1989a). The rheological response of red cells to pH (Johnson, 1985) is also rationalized by this model.

5.3. Membrane Fragmentation

Because the diffracted light from the ektacytometer is measured continuously, it is possible to determine the time required to fragment cells or ghosts under high shear. This quantitative measurement of mechanical stability, which is difficult to obtain by other methods, is one of the most valuable features of the ektacytometer.

High shears (1300–1500 dyn/cm^2) are required to fragment erythrocytes. Mohandas and co-workers (Tchernia *et al.*, 1981) found that erythrocytes fragment at different rates, depending on their MCHC. It appears that internal viscosity, which is essentially determined by the cytoplasmic hemoglobin concentration, influences the extent of cell elongation. At a given shear stress, cells with differing MCHC will elongate to different degrees and therefore their membranes will experience different degrees of strain. In this way, MCHC obscures the effect of intrinsic membrane stability on the fragmentation time. The effect of MCHC can be eliminated by using resealed ghosts (Mohandas *et al.*, 1982). Of the three factors that determine erythrocyte flexibility, two are reduced to negligible levels by the process of ghosting. Because ghosts have low hemoglobin content, their internal viscosity is very low. For resealed erythrocyte ghosts, the surface-to-volume ratio is also high, eliminating the effect of this factor on deformability. The absolute value of the DI at the beginning of the fragmentation curve is partially dependent on membrane surface area, just as is the plateau value of the ramp assay (Mohandas, 1988) and these values can be shown to be the same (Messmann *et al.*, 1990). Nevertheless, the process of ghosting more nearly equalizes the strain on all membranes at a given applied shear (Heath *et al.*, 1982). In order to generate the refractive index difference required for light scattering, resealed ghosts are placed in a medium of high refractive index. The refractive index depends linearly on the weight percentage of polymer in solution, independent of molecular weight, while viscosity increases exponentially with molecular weight. The choice of polymer is then determined by the requirement for a high refractive index with an appropriate viscosity. Dextran T-40 (average MW = 40,000) can be prepared at 35–40% concentrations to fulfill both requirements. In contrast, PVP (MW 360,000) at a similar concentration by weight would have an unworkably high viscosity. Stractan (Mohandas *et al.*,

1982) can also be used in place of dextran. Details of ghost preparation can be found elsewhere (Heath *et al.*, 1982; Johnson, 1989). The resulting curve (Figure 4C) defines a $t_{1/2}$ for fragmentation, which is a measure of the membrane's mechanical stability.

After Mohandas and co-workers established that ghosts could be successfully resealed and subjected to shear, membrane fragmentation times have become widely used in the detection of genetic membrane defects. A lowered $t_{1/2}$ is a simple and definitive diagnostic test for an unstable membrane skeleton. The fragmentation test can also detect abnormalities in erythrocytes of normal appearance, and has been useful in guiding investigations of the membrane skeleton.

5.4. Oxygen Scans

A number of different laboratories have made modifications to the ektacytometer to control oxygen pressure in the viscometer. All of these modifications have been made for the study of sickle cells, as this is the only case where oxygenation controls cell deformability. The deformability of normal erythrocytes is not affected by deoxygenation (Bessis *et al.*, 1982, 1983).

In one modification (Bessis *et al.*, 1983; Johnson, 1989), deoxygenated cells are mixed with a stream of buffer of varying pO_2. Two bottles of buffer with PVP or dextran in a 37°C bath are gassed with either N_2 or air. The gradient is made by two peristaltic pumps driven by stepping motors which gradually increase the proportion of air-gassed buffer in the stream of buffer entering the viscometer. The blood sample, appropriately diluted and deoxygenated in a tonometer at 37°C, is added to the mixed buffer stream by a third peristaltic pump. The cell suspension passes through a heating coil at 37°C whose length is adjusted to give a transit time of 1 min, to permit complete mixing of the cells and the O_2 gradient. The cells then enter the viscometer where O_2 concentration is measured with a Clark electrode installed in the outer cylinder. The outputs of the electrode and the image analyzer are fed into an x–y recorder to give a continuous record of deformability as a function of O_2. The entire apparatus is maintained at 37°C. The concentration of cells in the viscometer is approximately 25×10^6 ml^{-1}.

Another experimental system was described by Sorette *et al.* (1987) in which sickle cells, mixed with PVP or dextran buffer, are deoxygenated in a gas–porous fiber gas exchange system before entering the ektacytometer. Since oxygen tension was also measured outside the ektacytometer, the procedure required careful calibration to estimate the pO_2 within the viscometer. It should be possible to determine the actual pO_2 at the time of measurement with an installed Clark electrode, however. This method appears to be more convenient than the original deoxygenation system described above.

6. USE OF THE EKTACYTOMETER FOR DETECTION AND ANALYSIS OF MEMBRANE DEFECTS

6.1 Hereditary Elliptocytosis

When Mohandas *et al.* (1982) showed that a diffraction pattern could be obtained from resealed pink ghosts, the way was open to the analysis of membrane stability. Continuous shear stresses of 600 dyn/cm^2 were applied to the ghosts in Dextran T-40. There was an immediate rise in the deformability index as the ghosts deformed, followed by a slower decline as they fragmented. The $t_{1/2}$ for ghosts from individuals with dominant hereditary elliptocytosis (HE) without anemia was shortened, indicating membrane instability. When ghosts were tested from an individual with a condition called hereditary pyropoikilocytosis (HPP), characterized by abnormally shaped erythrocytes and chronic hemolysis, $t_{1/2}$ was found to be markedly shortened (Mentzer *et al.*, 1984). Membranes from nonhereditary elliptocytosis (myelofibrosis) had a normal $t_{1/2}$.

As is well known, the red cell has a network of proteins on the cytoplasmic surface, the membrane skeleton, that supports the lipid bilayer and is essential for the cell's structure. The major components are a number of spectrin tetramers linked to a junctional complex containing short actin filaments and other proteins, including protein 4.1, 4.9, and 2.9 (Bennett, 1989). Electron microscopy of skeletons reveals a regular lattice organization with five to six spectrin molecules attached to each junction. Since the membrane skeleton is required to maintain the integrity of the red cell membrane, it was reasonable to assume that the reduction in $t_{1/2}$ reflected a defect in the membrane skeleton. The molecular basis for membrane instability in both HPP (Liu *et al.*, 1981) and HE (Liu *et al.*, 1982) was clarified by Liu *et al.* who showed that spectrin from these erythrocytes had a self-association defect. Spectrin normally exists in the membrane as a head-to-head association of two $\alpha\beta$-dimers to form a tetramer. The tails of the tetramers have the junctional binding sites. In HE and HPP, the self-association constant is reduced and some of the spectrin is dimeric.

Mentzer *et al.* (1984) measured the amount of dimeric spectrin and the $t_{1/2}$ in membranes from a single kindred in which HE, HPP, and normal erythrocytes were found. The reduction in $t_{1/2}$ correlated well with the percentage of dimer in the membranes. For example, when 25% of the spectrin was in dimeric form, $t_{1/2}$ was reduced from 135 to 65 sec. Normally, about 5% of the extracted spectrin is dimeric. There were no clinical consequences of a 50% reduction in $t_{1/2}$, but chronic hemolysis was found in an HPP patient, whose membrane fragmentation time was reduced to 20 sec (15% of control). It now appears that HE is found in erythrocytes heterozygous for a spectrin self-association defect, while HPP is a

consequence of coinheritance of a self-association defect with an additional
spectrin variant, which may reside in the α chain (Guetarni *et al.*, 1990; Baklouti
et al., 1992) (reviews in Delaunay *et al.*, 1990; Palek and Sahr, 1992).

It was unexpected that the osmotic gradient scan (osmoscan) would be
useful in the clinical diagnosis of inherited red cell defects. The utility of the
osmoscan in diagnosis emerged over a number of years of use in laboratories
interested in erythrocyte pathology. Incidental observations of the distinctive
trapezoidal profile of HE cells (Mohandas *et al.*, 1982; Mentzer *et al.*, 1984) had
been made, but the review of Dhermy *et al.* (1986) on the early work on HE,
including ektacytometry, firmly established the utility of the scan for the diagnosis of HE. The trapezoidal osmoscan (Figure 6) is seen even in individuals
with the so-called "nonelliptocytic" HE, whose red cells have a spectrin self-
association defect without any associated morphological change.

The membrane fragmentation time is used in a number of laboratories as a
screening test for defects in the membrane skeleton. For example, at the Chil-
dren's Hospital of Michigan, congenital hemolytic anemias which are not expli-
cable by an enzyme defect or hemoglobinopathy are referred for ektacytometry.
The osmoscan is often diagnostic for HE and HS, and the existence of a membrane skeletal defect is confirmed by a finding of reduced $t_{1/2}$ for fragmentation.

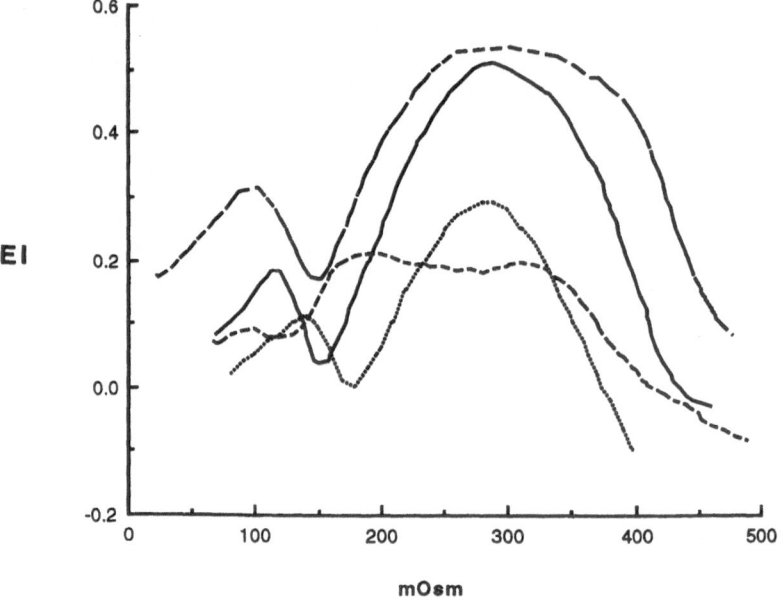

FIGURE 6. Osmoscans in different clinical conditions. ......, HS; ----, HE; ---, glucose-6-
phosphate dehydrogenase deficiency; ———, control.

6.2   Hereditary Spherocytosis

The common genetically dominant hereditary spherocytosis (HS) is characterized by a moderate deficiency of spectrin, although the spectrin itself is normal. In recessive HS, mutations in spectrin can often be identified, the anemia is more severe, and the spectrin deficiency is greater (Agre et al., 1986). Spherocytosis is attributed to loss of surface area by vesiculation, volume remaining approximately constant. Clark et al. (1983) first reported the characteristic osmoscan of HS (Figure 6). The left shift of O_{hyper} indicates the presence of high-MCHC cells, and the right shift of O_{min} is a consequence of a reduced surface-to-volume ratio, because of membrane loss. It will be noted that the DI_{max} is also reduced, indicating that either surface area or membrane deformability is reduced. Subsequently, Mohandas and co-workers established that membrane deformability was normal, and that the reduction in DI_{max} could therefore be used as a measure of surface area (Chasis et al., 1988b). Taking advantage of a large group of patients gathered by P. Agre (Agre et al., 1986) in which individuals with varying amounts of erythrocyte spectrin are found, it was possible to show that membrane stability was reduced in proportion to the spectrin deficiency (Chasis et al., 1988b). This again demonstrates the importance of the spectrin–actin network for the mechanical integrity of the erythrocyte membrane. Like the qualitative defects in the spectrin network introduced by genetic defects in dimer–dimer association (HE), the quantitative defect in HS compromises the ability of the red cell to resist mechanical shear.

It is of interest that Waugh and Agre (1988) studied the intrinsic material properties of HS membranes, and found a reduction in both the extensional shear modulus and the surface viscosity that paralleled the spectrin deficiency.

6.3.   Protein 4.1 Deficiency

The first application of ektacytometry to the analysis of membrane stability was the work of Tchernia et al. (1981) on protein 4.1-deficient erythrocytes. (Féo et al., 1980). Red cell fragmentation could be detected by the disappearance of the elliptical diffraction pattern produced by intact erythrocytes and its replacement with a circular pattern (DI = O) generated by nondeformable cell fragments. They minimized the confounding effects of internal viscosity by selecting cells of equal MCHC with density gradient centrifugation. Heterozygote cells with half the normal amount of protein 4.1 fragmented more quickly than normal erythrocytes, and homozygous erythrocytes with no detectable 4.1 fragmented in a very short period. Later, ghosts of protein 4.1-deficient cells were also shown to be unstable (Takakuwa et al., 1986).

The $t_{1/2}$ has also been used to monitor reconstitution of the membrane skeleton in 4.1-deficient cells (Takakuwa et al., 1986). Normal 4.1 was allowed to bind to ghost membranes prepared from 4.1-deficient erythrocytes. Ektacytometry then demonstrated the restoration of normal membrane stability. Recently, the same system has been used with recombinant protein 4.1, to demonstrate that a distinct 22-amino-acid domain is both necessary and sufficient for recovery of membrane stability (Discher et al., 1992).

Conboy et al. (1991) have recently shown that membrane instability can be caused by either qualitative or quantitative defects in erythrocyte 4.1, as was shown earlier for spectrin.

6.4 Sickle-Cell Anemia

Perhaps the earliest use of the ektacytometer in a clinical area was the observation by Bessis and Mohandas that sickle cells have a lowered DI (Bessis and Mohandas, 1976), as a consequence of their reduced deformability. Most sickle-cell anemia blood samples produce a distinctive diffraction pattern in the ektacytometer (Figure 7). This is a consequence of the presence of a highly rigid and asymmetric population of cells (Bessis and Mohandas, 1977). These cells do not elongate under shear; instead they rotate with their long axes perpendicular to the cylinder axis. Diffracted light from these cells is perpendicular to the light from flexible cells, yielding a cross-shaped diffraction pattern. Consideration of the arrangement of the image analyzer will show that the effect is to lower DI. In principle, the distribution of scattered light contains quantitative information about the nondeformable cell fraction. Newer instruments using video cameras for image analysis may realize this possibility.

Sickle cells also produce abnormal osmoscans, which are, in general, left-shifted and reach their maximal deformation at a lower osmolality as a consequence of dehydration (Gulley et al., 1982; Clark et al., 1983). As noted above, the left shift of O_{min} is an indication of elevated S/V and the left shift of O_{hyper} indicates that MCHC is increased. The osmoscan is not diagnostic for sickle-cell anemia, however, since a similar osmoscan is obtained with HbCC erythrocytes (Clark, 1989b).

6.4.1. Mechanical Properties of Sickle-Cell Membranes

Sickle cells are heterogeneous in many characteristics, as a consequence of variable degrees of damage within the circulation. This heterogeneity is seen in the membrane mechanical properties as well. Membrane deformability ranges from normal to quite rigid in the densest cell fraction (Fortier et al., 1988). When membranes were prepared from sickle anemia blood samples, the fragmentation time was normal, but fractionation on density gradients yielded light cells with longer than normal $t_{1/2}$, and dense cells with short $t_{1/2}$ (Messmann et al., 1990). The 50% reduction in $t_{1/2}$ in dense cells approximates that seen in heterozygotes

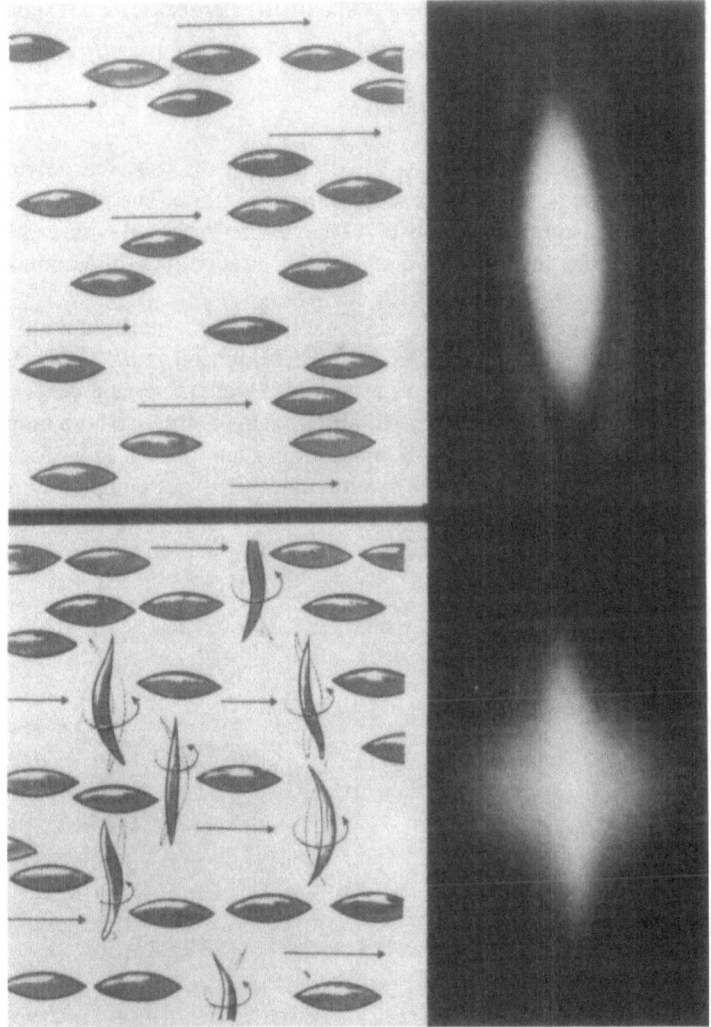

FIGURE 7. The diffraction image of asymmetric rigid erythrocytes. A population of sickle cells containing numerous nondeformable dehydrated cells (irreversibly sickled cells) during shear at 125 dyn/cm². (From Bessis and Mohandas, 1977.)

for spectrin self-association defects (dominant HE). Since heterozygous HE manifests no clinical symptoms, it is unlikely that the membrane instability of dense cells contributes significantly to the hemolysis of sickle-cell anemia. Damaged sickle-cell membranes have increased rigidity and reduced mechanical stability. They therefore resemble β-thalassemia erythrocytes and cells treated with phenylhydrazine (see below), suggesting that oxidative processes are responsible for the membrane damage.

6.4.2.   Antisickling Agents

The major pathogenic property of sickle erythrocytes is that HbS polymerizes at low oxygen pressures, making the cell extremely rigid. The study of this phenomenon was a natural extension of ektacytometry, and Bessis *et al.* (1982, 1983) modified the instrument to allow the observation of continuous pO_2 changes.

6.4.2a.   Oxygen Scan.   As sickle cells are deoxygenated within the viscometer (Féo *et al.*, 1981; Bessis *et al.*, 1982; Johnson *et al.*, 1985; Sorette *et al.*, 1987), DI remains reasonably constant until a characteristic pO_2 is reached, where the index drops dramatically (Figure 8). This is true for both homozygous and heterozygous (AS) cells. This is most simply interpreted as the onset of HbS polymerization (Sorette *et al.*, 1987), although this is not completely consistent with theoretical and NMR determinations of the fraction of polymerized hemoglobin (Noguchi *et al.*, 1980). Shifts in the deoxygenation curve can be used to

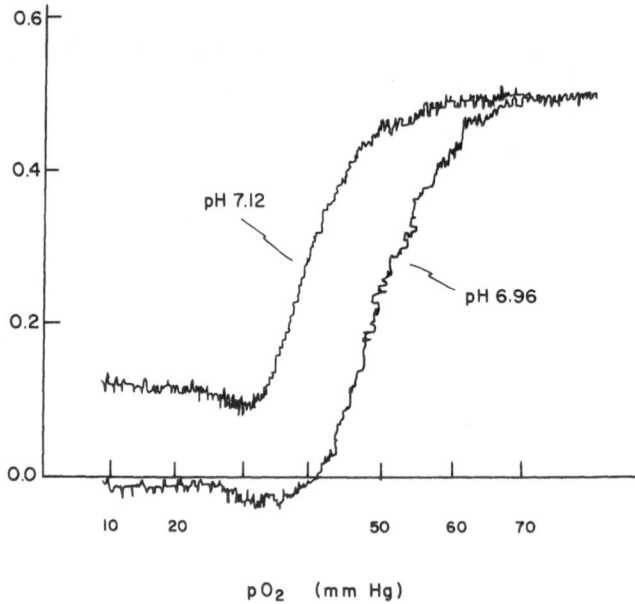

FIGURE 8. EI as a function of oxygen tension for sickle cells at two different pH values.

analyze the effects of antisickling agents on the polymerization process (Féo *et al.*, 1983; Johnson *et al.*, 1985). Antisickling agents can act by inhibiting the formation of deoxy HbS, by increasing the oxygen affinity of HbS, or by inhibiting the polymerization of deoxy HbS. The mechanism can be determined with the oxygen scan: a change in oxygen affinity shifts pEI50, the pO_2 at which 50% of the total observed change in DI occurs, while inhibition of polymerization increases the DI of maximally deoxygenated cells.

In addition to identifying the antisickling mechanism, the ektacytometric assay has the advantage of detecting modifications to membrane by the antisickling agent. Some proposed agents that form covalent bonds can react with components of the membrane to reduce its flexibility. This can look like an antisickling effect, since the rigid membrane does not allow cell deformation to occur even when internal HbS has polymerized. The ektacytometric scan detects this artifact, since the DI of oxygenated cells will be reduced (Johnson *et al.*, 1985). Similarly, other proposed antisickling agents affect cell hydration. These can compromise deformability by increasing the sphericity of the cell (Orringer *et al.*, 1992), making these agents unsuitable for clinical use.

6.4.2b. Osmoscan. A second rheological abnormality is found in sickle cells, in the dehydrated dense fraction observable even in oxygenated samples. While the pathological significance of these cells is uncertain (Ballas *et al.*, 1988; Lande *et al.*, 1988; Ballas, 1991), they are easy to study since elaborate efforts to exclude oxygen from the ektacytometer are not necessary. There have therefore been a number of studies delineating the ability of various compounds to rehydrate this fraction, using oxygen and osmotic scan ektacytometry. An increase in cell water is readily detected by osmotic scan ektacytometry. The cells become more spherocytic, raising O_{min}, and the viscosity of their cytoplasm is lowered, shifting O_{hyper} downward. These effects have been experimentally demonstrated in studies with monensin (Clark *et al.*, 1982). These features of the osmoscan have been used to demonstrate that various proposed compounds either do (Orringer *et al.*, 1992) or do not (Acquaye *et al.*, 1988; Cummings and Ballas, 1990; Abraham *et al.*, 1991) affect cell hydration.

6.4.3. Hydroxyurea

The administration of hydroxyurea increases the HbF levels in the circulation of sickle-cell anemia patients. This is a promising modality for therapy, since the concentration of HbS is lowered and sickling tendency is diminished. Ballas *et al.* (1989) found that the degree of anemia remained the same, but there was significant increase in the mean cell volume (MCV) and a significant decrease in the MCHC. Whole cell deformability improved by twofold, but membrane stability remained within normal limits.

6.4.4. Relationship to Clinical Severity

The presence of a rigid subset of sickle cell lowers the DI. A number of groups have hypothesized that the frequency of rigid cells might be associated with a greater degree of pathology in sickle-cell disease. When this was examined, however, the opposite result was found: a low DI was associated with less severe disease (Ballas *et al.*, 1988; Lande *et al.*, 1988; Ballas, 1991). The explanation for this paradoxical finding remains unknown.

6.5. Thalassemia

The unpaired globin chains of thalassemia are physically and chemically unstable. They tend to denature, aggregate, and precipitate onto the interior surface of the cell membrane. The released heme catalyzes oxidative reactions, further damaging the hemoglobin as well as components of the membrane. When thalassemic membranes were subjected to ektacytometric analysis, the surprising discovery was made that membrane damage was different in α- and β-thalassemia. In both syndromes, there was increased membrane rigidity, but $t_{1/2}$ was shortened in β-thalassemia, while, in contrast, $t_{1/2}$ of α-thalassemic membranes was lengthened (Schrier *et al.*, 1989). Biochemical evidence suggested that the β-globin associated with the membrane in α-thalassemia induced oxidation of spectrin, while α-globin (in β-thalassemia) oxidized protein 4.1. Because of the importance of protein 4.1 to membrane integrity, the oxidative damage in β-thalassemia may explain the shortened $t_{1/2}$ (Advani *et al.*, 1992a). Becker *et al.* (1987) found that a spectrin variant with defective binding to protein 4.1 in a case of recessive HS was associated with a fragile membrane, supporting the idea that this binding interaction is important for membrane stability. Interestingly, this variant appeared to be unusually susceptible to oxidation.

β-thalassemia occurs in mice. In an ingenious application of the transgenic mouse technology, the β-thalassemia has been partially corrected by the introduction of the human β-globin gene into this line (Advani *et al.*, 1992b). Comparison of control, thalassemic, and transgenic mouse erythrocytes showed that membrane-bound α-globin chains, protein 4.1 oxidation, and membrane instability were well-correlated. In every case, the transgenic mouse erythrocyte membranes were intermediate between the control and the uncorrected thalassemic red cells.

Some of these changes can be reproduced experimentally. Ghosts resealed with either α- or β-globin chains had increased membrane rigidity after 20 hr at 37°C (Scott *et al.*, 1992). During this incubation, globin chains became attached to the membrane. Membrane stability was not examined. The contrasting effects of heme oxidation seen in α- and β-thalassemia can be reproduced *in vitro* with normal red cells with the proper choice of oxidizing agents (Schrier and Mohandas, 1992). Treating normal erythrocytes with phenylhydrazine replicated the membrane damage seen in β-thalassemia (an unstable membrane with bound α-globin). Methylhydrazine treatment, in contrast, mimicked α-thalassemia (a more stable membrane with bound β-globin). Both compounds increased the rigidity of the membrane.

6.6. Abnormal Hemoglobins

In addition to sickle-cell anemia, abnormal erythrocyte deformability has been noted in other cells with abnormal hemoglobins (Table I): Hb Bicêtre (Allard et al., 1976), Hb CC (Mohandas et al., 1980; Clark, 1989b), Hb SC (Ballas et al., 1987), Hb J Guantanamo (Wajcman et al., 1988), Hb Athens-Georgia (Mrad et al., 1989), Hb Setif (Noguchi et al., 1991).

Table I
Clinical Conditions for Which Red Cell Ektacytometry Has Been Performed

Condition	References
Hb CC	Mohandas et al. (1980), Ballas et al. (1987), Clark (1989b)
Hb SC	Allard et al. (1976), Mohandas et al. (1980), Ballas et al. (1987), Wajcman et al. (1988), Clark (1989b)
Hb Bicêtre	Allard et al. (1976)
Hb Setif	Noguchi et al. (1991)
Hb SS	Smith et al. (1980), Féo et al. (1981, 1982), Bessis et al. (1982, 1983), Gulley et al. (1982), Johnson et al. (1985), Mohandas et al. (1986), Sorette et al. (1987)
Hb AS	Bessis et al. (1982), Johnson et al. (1985)
Autoimmune hemolytic anemia	Mohandas and De Boisfleury (1977), Ballas et al. (1985)
LCAT deficiency	Jain et al. (1982)
Iron deficiency	Yip et al. (1983)
McLeod syndrome	Ballas et al. (1990)
Acanthocytes	Clark et al. (1989)
Rh null red cells	Ballas et al. (1984)
Diabetes	Bareford et al. (1986), Schwartz et al. (1991)
Arterial disease	Bareford et al. (1985)
Renal failure	Bareford et al. (1985)
Liver failure	Bareford et al. (1985)
Xerocytosis	Clark et al. (1978), Fortier et al. (1988)
Spectrin variants	Many have been studied. See Delaunay et al. (1990) for review
Protein 4.2	Ghanem et al. (1990)

6.7. Band 3 and Ovalocytosis

Hereditary ovalocytosis, characterized by rigid elliptical red cells, is widespread in Southeast Asia. The nondeformability of the cell is a consequence of membrane rigidity, which was established by ektacytometric (Mohandas *et al.*, 1984) and filtration (Saul *et al.*, 1984) methods in 1984. The highly inflexible membrane presents an effective barrier to malarial invasion. Remarkably, the genetic alteration in ovalocytes lies in band 3, not in any of the skeletal proteins. A nine-amino-acid deletion has occurred at the junction between the transmembrane and cytoplasmic domains (Jarolim *et al.*, 1991). It is not understood how this band 3 deletion makes the membrane less flexible, but a number of models have been proposed (Liu *et al.*, 1990; Pinder, 1991; Mohandas *et al.*, 1992; Schofield *et al.*, 1992). Pinder (1991) noted that band 3 is a major attachment site for ankyrin, which in turn is linked to the skeleton at a site on spectrin. The primary sequence of the mutant suggests that it might be unable to flex in response to lateral tension, and Pinder proposed that this restricts the ability of the deformed skeleton to rearrange. Mohandas *et al.* (1992) proposed a different model, which also assumes that ovalocyte band 3 is less flexible, but that the cytoplasmic domain of ovalocyte band 3 extends into and restricts the movement of the spectrin–actin network without a requirement for direct binding to the network via ankyrin.

Pasvol *et al.* (1989) have found that increasing membrane rigidity by liganding monoclonal antibodies to glycophorin protected erythrocytes from invasion by malarial parasites, supporting the idea that membrane rigidity inhibits malarial infection.

6.8. Transfusion

Ektacytometry indicates that banked red cells do not lose any significant degree of deformability at 290 mosm during storage of up to 42 days (Card *et al.*, 1982). Intracellular ATP must decline by more than 80% before any change in deformability is noted. After this point, there is membrane loss, detectable in the osmotic fragility curve and in the deformability index at 190 mosm, but membrane rigidity is unchanged.

6.9. Aging

Characterization of the membrane changes in aged red cells had suffered from the lack of an unambiguous isolation method for old cells. This has been overcome by the development of hypertransfusion methods for mice and biotin tagging for rabbit erythrocytes. Ektacytometric analysis of aged erythrocytes found a progressive loss of surface area and cellular dehydration, but no membrane changes. The parallel loss of volume and surface left the surface-to-volume ratio unchanged. These results were supported by micropipet measurements, which also found normal membrane elasticity (Waugh *et al.*, 1992).

6.10. Other Conditions

The diagnostic usefulness of the osmoscan in HE and HS is not found in other red cell diseases, although osmoscans in many other conditions have been reported (Table I). Although it has not been well-documented, red cells with enzyme deficiencies appear to have an osmotic curve in which the DI is consistently higher than normal, although the landmarks of O_{min} and O_{hyper} are unchanged (Figure 6).

7. TRANSPORT STUDIES

Many compounds affect erythrocyte deformability. The osmoscan can distinguish loss of deformability because of dehydration from that brought about by membrane loss or membrane rigidity. For this reason, it has been possible to study compounds that affect ion transport using the ektacytometer. The rheological consequences of abnormal cation transport (Clark *et al.*, 1978) were determined early in the history of the instrument. Another early result was the resolution of a controversy over the role of calcium in erythrocyte deformability. While calcium was known to compromise the flexibility of red cells, it was widely believed that the action of calcium was to increase membrane rigidity. Support for this idea was drawn from Lorand's observation that, in isolated membranes, spectrin cross-linking by endogenous transglutaminase was activated by calcium. Since the ektacytometer can distinguish the various factors affecting cell rigidity, it was possible for Clark *et al.* (1981) to show that calcium made erythrocytes less deformable by activating the Ca-dependent K channel and dehydrating the cell, and that there was little or no effect on membrane flexibility. Clark and co-workers were also able to determine that the nondeformability of oxygenated sickle cells is related to abnormalities in ion transport (Clark *et al.*, 1983), and to characterize cation transport in ionophore-treated red cells (Clark *et al.*, 1982).

The ektacytometer has been used to study a case of congenital stomatocytosis with hemolytic anemia and cation transport abnormalities (Morlé *et al.*, 1989). The red cells included 15–20% stomatocytes and displayed a marked increase of volume. Erythrocyte Na^+ was high (27 meq/liter) and K^+ low (65 meq/liter) and band 7 was partially absent. The osmoscan of the grossly overhydrated cells provided evidence that the cells were spherocytic, even though their deformability was unaltered at 290 mosm, because of extreme hypochromicity.

8. EXPERIMENTAL ALTERATION OF MEMBRANE MECHANICAL PROPERTIES

8.1. Oxidation

The effects of oxidation on red cell membranes have generated an enormous literature. Components of the membrane become oxidized in various pathological conditions, and the consequences of oxidation by exogenous agents have also been studied. The findings with the *in vivo* oxidized membranes of thalassemia and sickle-cell anemia have been outlined above. There is evidence for membrane oxidation in xerocytosis (Fortier *et al.*, 1988), iron deficiency anemia (Yip *et al.*, 1983), and diabetes (Schwartz *et al.*, 1991) as well.

Snyder *et al.* (1985) have studied the effect of hydrogen peroxide on red cells using many techniques, including ektacytometry. Ektacytometric analysis showed that the observed reduction in cell deformability was a consequence of membrane rigidity (Snyder *et al.*, 1988). Electrophoretic and spectroscopic analysis showed that hydrogen peroxide caused cross-linking between hemoglobin and sulfhydryl groups on membrane skeletal proteins (Snyder *et al.*, 1988), which required the prior oxidation of hemoglobin. The suggestion of a causal relationship was reinforced by the observation of membrane-bound hemoglobin in clinical conditions where the membrane has lost flexibility (thalassemia, sickle-cell anemia, xerocytosis). It is not clear whether hemoglobin binding directly affects membrane rigidity, or alternatively, that binding serves to promote the catalysis of membrane oxidation by heme derivatives. Both alternatives may be true. There is evidence that an increase in MCHC can itself increase rigidity by favoring hemoglobin binding to the membrane (Evans and Mohandas, 1987; Fortier *et al.*, 1988), and many exogenous oxidants can also compromise membrane flexibility. Scott *et al.* (1992) have resealed α- or β-globin chains in ghosts, and found that changes in membrane deformability were not immediate. Incubations of 20 hr at 37°C were required to produce increases in membrane rigidity. In these experiments, it is still not clear whether the rate-limiting step for deformability loss is binding to the membrane, conversion of the heme to a catalytic species, or the actual oxidation of membrane components.

Kuypers *et al.* (1990) have shown that ektacytometry can be used to monitor oxidative stress on red cells in a continuous fashion. Red cells were exposed to oxidizing reagents (hydrogen peroxide, cumene peroxide, and *t*-butyl peroxide) within the viscometer chamber, and a progressive loss of deformability was observed. Sickle cells lost flexibility more rapidly than normal cells in this assay, indicating a greater susceptibility to oxidant damage.

8.2. Glycophorin Ligands

The absence of glycophorins β and γ (also known as glycophorins C and D) is associated with a loss in membrane stability and deformability as determined by the ektacytometer (Reid *et al.*, 1987). Absence of either glycophorin α or δ (glycophorins A and B) had no effect on these properties. Glycophorins C and D are attached to the membrane skeleton through their cytoplasmic domains, which bind to protein 4.1, and the absence of this transverse interaction apparently destabilizes the membrane. However, it must be noted that micropipet work indicates that the intrinsic properties of the membrane are normal in glycophorin-deficient red cells, suggesting that the ektacytometric studies may have been influenced by erythrocyte shape (Nash *et al.*, 1990). This observation may help explain an otherwise puzzling inconsistency in the behavior of the glycophorins. Chasis *et al.* (1988a) have reported that binding of wheat germ agglutinin to the external region of glycophorin A also increases membrane rigidity, but in this case, the glycophorin's interaction with the membrane skeleton increases. Binding of monovalent Fc fragments of antibodies to glycophorin similarly diminished membrane deformability.

Disruption of the band 3–ankyrin interaction was also reported to have a destabilizing effect on the membrane (Low *et al.*, 1991).

8.3. Other Ligands

Ca-calmodulin resealed within ghosts reduces membrane stability (Takakuwa and Mohandas, 1988). This effect is reversible. Calmodulin binds to erythrocyte spectrin (Sobue *et al.*, 1981), but with low affinity. A more likely site for the destabilizing interaction is adducin, a component of the junctional complex of the membrane skeleton. Ca-calmodulin inhibits spectrin binding to the actin–adducin complex (Gardner and Bennett, 1987).

When high concentrations (6–8 mM) of 2,3-diphosphoglycerate (DPG), a compound that dissociates membrane skeletons, is resealed within normal membranes, ghost mechanical stability is reduced to the levels seen in HE or HPP. Addition of physiological levels of 2,3-DPG (2.55 mM) to HE ghosts reduced their stability relative to that of HPP ghosts. This suggests that free 2,3-DPG, present in neonatal RBC as a consequence of diminished binding to HbF, may augment hemolysis in infants with HE (Mentzer *et al.*, 1987).

8.4. Shape and Membrane Flexibility

Shape abnormalities in erythrocytes can occur because of genetic defects in the membrane, or secondarily as a consequence of other pathological conditions. Shape changes can also be induced *in vitro*. These various modes of shape alteration have differing effects on the mechanical properties of the membrane.

1. HE and HS are obvious examples of a genetically induced shape alteration. The membrane properties of these cells have been described. Other genetic changes associated with a nondiscoidal morphology include ovalocytosis and stomatocytosis.

2. As noted above (Section 6.10), a host of pathological conditions can alter the shape and deformability of circulating red cells. Few of these cases have been systematically studied, but in general, a reduction in cell deformability is observed, often because of membrane loss.

3. Normal red cells can be induced to undergo shape changes following ATP depletion, or by drug binding. Deuticke (1968) noted many years ago that shape changes induced by amphipathic compounds exist on a continuum between echinocyte and stomatocyte, with the normal discocyte in the middle. Within limits (Chailley et al., 1973), a combination of stomatocytogenic and echinocytogenic agents appear to cancel one another, leaving the cell discocytic. Many, if not all, of these observations can be rationalized by the bilayer couple hypothesis (Sheetz and Singer, 1974) which states that stomatocytic agents are positively charged and prefer to intercalate into the inner leaflet which is enriched in negatively charged phosphatidylserine, while echinocytic agents are negatively charged and partition mainly into the outer leaflet of the membrane. Since little of the membrane's ability to resist mechanical forces depends on the lipid component, it is not clear that these drug-induced shape changes would affect the mechanical properties of the membrane. The existing evidence, although fragmentary, suggests that they do not. Ramp assays of red cells by Bessis and Mohandas (1975b) indicated that red cells with the amphiphile-induced echinocyte II or stomatocyte II morphology have unaltered membrane deformability. (At higher concentrations, vesiculation occurs and cell deformability drops because of geometrical factors.)

Echinocytic shape changes also occur when internal ATP declines. Among the earliest applications of the ektacytometer was the demonstration by Féo and Mohandas (1977) that the shape change lags ATP depletion by many hours. Normal cell deformability is maintained for an even longer period, but after 18 hr, the cell becomes less deformable because of membrane rigidity (Mohandas et al., 1978). The normal deformability of membranes rendered echinocytic by ATP depletion has been confirmed in other work (Meiselman and Baker, 1977; Clark et al., 1981). The echinocytosis and eventual loss of deformability that follows ATP depletion has an unknown mechanism, although a number of hypotheses have been proposed.

The behavior of ghosts is more complex. Ordinary hypotonic ghosts are discoidal in the low-ionic-strength media used for lysis. On resealing in isotonic media, they become echinocytic, unless ATP is present, which converts them to discs. Schrier (1987) observed a loss of membrane flexibility (ramp assay) when hypotonic ghosts were resealed with ATP. That is, discocytic ghost membranes were less flexible than echinocytic ghosts. In addition, resealed ghosts containing antibodies to spectrin were also discoidal and more rigid. The underlying mechanism of these interesting results is not known. Cross-linking the spectrin network with antibodies might reasonably be expected to make the membrane less flexible, but the reason for the discocytic morphology is mysterious. Many skeletal proteins are phosphorylated when ghosts are exposed to ATP, but the connection with membrane shape and deformability is unknown. While the bilayer couple hypothesis is successful in rationalizing lipid-associated effects on shape, skeletal proteins must play the leading role in certain circumstances.

9. STABILITY AND RIGIDITY

In a striking use of induced and naturally occurring membrane alterations, Chasis and Mohandas (1986) showed that membrane flexibility (as determined by the ramp assay) and stability (determined by the ghost fragmentation assay) are independently regulated by skeletal associations. While most treatments reduced membrane flexibility, membrane stability could be increased or decreased. For example, exposure to diamide increased stability while 2,3-DPG decreased it. Chasis and Mohandas also pointed out that HS membranes were unstable, without any noticeable loss of flexibility, while HE membranes were abnormal in both properties. Schrier and Mohandas (1992) have demonstrated the opposing effects of phenylhydrazine and methylhydrazine on stability.

10. RELATIONSHIP TO OTHER TECHNIQUES

Valuable reviews of methods in blood rheology are found in Bull et al. (1986), Shiga et al., (1990), and Stuart and Nash (1990). For whole blood, the main factors affecting the macroscopic viscosity are the hematocrit, rouleau formation (mediated primarily by fibrinogen), and the deformability of the erythrocyte (Chien et al., 1970; Mohandas et al., 1980; Chien, 1987). In the ektacytometer, the dilution of the cells and the applied shear forces both inhibit aggregation. Hence, cell deformability is the only factor seen by the ektacytometer.

In the rheoscope (Fischer et al., 1978; Shiga et al., 1990), cells are observed by microscopy while under shear in a cone-plate viscometer. This has the great advantage of allowing direct observation of the effects of shear on individual erythrocytes. It was used, for example, to validate the interpretation of DI as equivalent to the elongation of sheared cells (Mohandas et al., 1980). The technique is difficult to master, and does not lend itself to continuous measurements, since flow through the observation chamber is not possible. There is no method of automated image analysis. Nevertheless, the rheoscope remains a primary reference method for determinations of erythrocyte deformability, since it yields direct visual information about erythrocyte response to shear.

The deformability of populations of intact erythrocytes can be measured by other techniques as well. The filtration rate through narrow pores has been frequently used (reviewed in Stäubli et al., 1986). Filtration through pores of 5 μm depends on cytoplasmic viscosity (MCHC), while narrower (3 μm) pores are sensitive to cell volume. Neither size is particularly sensitive to membrane rigidity.

The cell transit analyzer (CTA) measures the time required for individual cells to pass through the micrometer pores, with less deformable cells requiring longer times (Koutsouris et al., 1988; Stuart and Nash, 1990). A constant current is maintained across the porous membrane, and transit times are seen as increases in voltage as resistance rises during pore occlusion. The instrument has been modified to report the shape of the voltage peak as well (Fisher et al., 1992). There is not enough experience with this instrument to know if the various influences on erythrocyte deformability can be determined separately, or if membrane properties can be evaluated.

A new development of the ektacytometer principle uses high-frequency electric fields (1 MHz) to deform erythrocytes suspended in low-conductivity media, while their elongation is observed by laser light scattering at a four-quadrant diode (Kage et al., 1990). This is reported to have the advantage of detecting membrane material properties independent of cell geometry and MCHC.

While there are numerous alternatives to ektacytometry for measuring aspects of cell and membrane deformability, the ability of the ektacytometer to measure the membrane fragmentation rate in a large number of cells is unique and valuable. Palek's group has used the rate of fragmentation of isolated membrane skeletons shaken on a hematological pipet shaker to identify spectrin defects (Liu et al., 1981), but this has not received as wide an application as the ghost $t_{1/2}$ assay. The yield strength of the membrane in individual erythrocytes can be determined by either extension (Berk et al., 1988). The membrane stability measured in the ghost fragmentation assay is clearly related to anemia in HE, HS, and similar conditions with genetic variants of membrane skeletal proteins. Many studies (reviewed in Mentzer et al., 1984; Dhermy et al., 1986; Delaunay et al., 1990) have demonstrated that $t_{1/2}$ is correlated with clinical severity and is a valuable test for the presence of an underlying skeletal defect.

11. RELATIONSHIP TO BLOOD PHYSIOLOGY

Rigidity and stability are also relevant to broader issues of cardiovascular circulation and physiology. Increased erythrocyte rigidity is a common occurrence in clinical situations, because of dehydration, membrane loss, or membrane rigidity. In turn, a loss of red cell deformability can increase blood viscosity, since the main determining factors are hematocrit, rouleau formation (aggregation), and cell deformability. There have been many proposals that the increased viscosity may add to the pathology of these clinical states, but there is convincing evidence that this is not so. One informative condition, discussed by Clark (1989b), is the presence of hemoglobin C, which, by an unknown mechanism, causes the erythrocyte to dehydrate and thus to become rigid. Despite this significant loss of deformability, neither homolysis nor circulatory flow is affected. Moreover, the presence of very nondeformable Malayan ovalocytes is not associated with circulatory difficulties (Mohandas *et al.*, 1984; Saul *et al.*, 1984). These negative findings are probably related to the fact that cell deformability affects the macroscopic viscosity of the blood, which is not the most significant factor affecting circulatory flow (Chien, 1987; Stuart and Nash, 1990; Nash, 1991). The macroscopic viscosity is likely to be relevant only in large vessels. In the small vessels of the circulation, where blood cells make direct contact with the endothelial wall, the properties of white cells are likely to be of major significance (Nash, 1991, 1992; Shields and Sarin, 1992). Leukocytes can become extremely rigid after activation, and they are also capable of adhesion to the endothelium. Either of these events can occlude the capillary. It is also clear that in some situations where red cells are known to cause vessel blockage, it is their abnormal adhesiveness that is relevant. For example, in *Plasmodium* malaria, the appearance of adhesive knobs on the infected erythrocyte is the pathologic event that allows the cell to adhere to vessel walls and block blood flow. The adhesive cells of sickle-cell anemia (Hebbel, 1991) may be another example. There may, however, be subtle interactions between red cell deformability and adhesion. A close contact between the vessel wall and the red cell membrane probably requires that the cell have sufficient flexibility to follow the contours of the vessel. More rigid red cells are known to adhere less well to each other (Nash *et al.*, 1992) and to cultured endothelium (Hebbel, 1991). In this way, rigidity might actually improve the flow properties of otherwise adhesive cells in the capillaries. This has been proposed as an explanation of the paradoxical finding that sickle-cell anemia is less severe when the DI is low and large numbers of rigid cells are found in the circulation (Ballas *et al.*, 1988; Lande *et al.*, 1988; 1991).

Finally, in sickle-cell anemia, the rigidity of deoxygenated sickle cells far exceeds the loss of red cell deformability seen in other clinical conditions, and these extremely rigid cells are likely to participate in the vasoocclusion that is a prominent part of this disease. The relation between rigidity and circulatory physiology is therefore likely to be particularly complex in sickle-cell disease.

12. MOLECULAR BASIS FOR MEMBRANE MECHANICAL PROPERTIES

What is the molecular basis for the membrane properties of deformability and mechanical stability observed in the ektacytometer? The stability of the membrane is reduced if components of the membrane skeleton are genetically defective, as has been shown for many variants of spectrin and protein 4.1. Oxidative damage to spectrin or protein 4.1 that lowers their binding affinities also lowers the fragmentation time. Weakening the spectrin–actin junction with DPG destabilizes the membrane. Since the membrane skeleton maintains the structure of the red cell membrane, it is readily understood that erythrocytes in which skeletal interactions are defective would have diminished shear resistance. The molecular changes in membrane with increased stability (in α-thalassemia or after methylhydrazine treatment) are undefined as yet. The hypothesis that elements of the skeleton become covalently cross-linked naturally suggests itself, but experimental evidence one way or the other is lacking. It is more difficult to rationalize the destabilizing effect of weakened linkages between the skeleton and the bilayer. Ektacytometric fragmentation tests have shown that the membrane is less stable when glycophorin–protein 4.1 linkages are absent (Reid *et al.*, 1987), or ankyrin–band 3 linkages are disrupted by elevated pH (Low *et al.*, 1991). Low *et al.* (1991) have suggested that the phenomenon of "load sharing" might explain these results.

With regard to membrane rigidity, ektacytometric assays indicate that neither aging, ATP depletion, storage, nor amphipathic compounds have an effect on membrane deformability. Heat treatment (Mohandas *et al.*, 1978) or exposure to high shear forces (Wolf *et al.*, 1992) diminishes deformability, and membrane rigidity is also increased in all cases where the membrane has been exposed to oxidative damage. Indeed, it is difficult to find an instance where experimental manipulations of the erythrocyte membrane result in an increase in flexibility, suggesting that the membrane structure is optimized for maximal deformability. The invariable increase in membrane rigidity after experimental manipulations contrasts with the opposing effects on membrane stability (fragmentation time under shear) seen after treatment with hydrazines and other oxidants. Rigidity is reduced below normal values in HS, where spectrin density on the inner surface is reduced, but the spectrin deficiency diminishes the strength of the membrane as well. The spectrin–actin network appears to be constructed to have the greatest flexibility consistent with a mechanically stable membrane.

Although membrane deformability is often regarded as a property of the membrane skeleton, it is clear that bilayer–skeleton interactions can also play a role. The naturally occurring genetic variants of hereditary ovalocytosis, where band 3 is mutated, and glycophorin C and D deficiency are both associated with a less flexible membrane. Ligand binding to glycophorin A can also bring about a loss in membrane deformability. It is remarkable that binding to the external domain of an integral protein alters the flexibility of the membrane. These observations led Chasis and co-workers to propose that a conformational change in glycophorin A is transmitted through the bilayer to the skeleton.

The glycophorin studies also provide a link between ektacytometry and the material properties of the membrane. In micropipet studies, small concentrations of lectin (0.1 μg/ml) increased surface viscosity (Smith and Hochmuth, 1982), while higher concentrations (up to 2 μg/ml) increased the shear modulus μ by a factor of ten (Evans and Leung, 1984). Although these experiments, done in different laboratories, may not be entirely comparable, the implication is that ektacytometric rigidity is primarily influenced by increases in the elastic shear modulus, since Chasis et al. (1985) used glycophorin in the concentration range of 0.5 to 2.5 μg/ml. The bending modulus of normal erythrocyte membranes is small, but this may not be true in chemically altered membranes. In principle, the bending modulus can also be anisotropic, differing for inward and outward curvature. This possibility is suggested by the observation that lectin binding to glycophorin A inhibits crenation (an outward curvature) but not invagination (Chasis and Schrier, 1989).

While biochemical models have been proposed to explain membrane elasticity (see, e.g., Berk et al., 1988; Shen, 1988), they have not been extensively tested for their consistency with experimental findings in membranes with altered rigidity. Electron micrographs of the erythrocyte membrane skeleton in a relatively unperturbed state show a condensed network and compact spectrin. Fully extended spectrin tetramers are 200 nm in length, but their length in the normal membrane is 70 nm. The folded spectrin hypothesis (Shen, 1988) suggests that spectrin is held in the compact state by specific weak intramolecular interactions between the repeated domains of spectrin, and that membrane elongation dissociates these interactions. Alternative models include proposals that conformational changes within spectrin occur during elongation (McGough and Josephs, 1990), that oligomerization of spectrin is involved, or that purely physical ionic interactions are the basis of elasticity (Stokke et al., 1986). Contractile proteins related to muscle proteins are present in the erythrocyte membrane and these may confer elastic properties (Bennett, 1989). The folded spectrin hypothesis (Shen, 1988) predicts that membranes can lose elasticity through the formation of covalent bonds between otherwise weakly associated domains. Many oxidizing agents increase the membrane's resistance to shear. It has been suggested that this is a consequence of covalent cross-linking between components of the membrane skeleton, but direct evidence for such cross-linking is sparse. More detailed biochemical analysis of membranes whose mechanical properties have been altered by various treatments would be of great value in elucidating the underlying molecular basis. Because of the ease and rapidity of ektacytometric measurements of membrane mechanical properties, the ektacytometer should play a role in the clarification of the biochemical changes that control the unusual deformability of the erythrocyte membrane.

ACKNOWLEDGMENTS. The work in the author's laboratory has been funded by the National Institutes of Health, Grant HL-16008 to the Detroit Comprehensive Sickle Cell Center. I am also grateful for a Fogarty Fellowship from the

National Institutes of Health that allowed me to learn ektacytometry at the Institut de Pathologie Cellulaire.

13. REFERENCES

Abraham, D. J., Mehanna, A. S., Wireko, F. C., Whitney, J., Thomas, R. P., and Orringer, E. P., 1991, Vanillin, a potential agent for the treatment of sickle cell anemia, *Blood* **77**:1334–1341.

Acquaye, C. T., Carter, T. L., Morson, T. T., Zahoor, M., Johnson, R. M., Mizukami, H., and Bustamente, C., 1988, 2,6-Di(isobutylamino)hexanoic acid as a potential therapeutic agent for the treatment of sickle cell disease, *Ann. N. Y. Acad. Sci.* **565**:353–355.

Advani, R., Sorenson, S., Shinar, E., Lande, W., Rachmilewitz, E., and Schrier, S. L., 1992a, Characterization and comparison of the red blood cell membrane damage in severe human alpha-and beta-thalassemia, *Blood* **79**:1058–1063.

Advani, R., Rubin, E., Mohandas, N., and Schrier, S. L., 1992b, Oxidative red blood cell membrane injury in the pathophysiology of severe mouse beta-thalassemia, *Blood* **79**:1064–1067.

Agre, P., Asimos, A., Casella, J. F., and McMillan, C., 1986, Inheritance pattern and clinical response to splenectomy as a reflection of erythrocyte spectrin deficiency in hereditary spherocytosis, *N. Engl. J. Med.* **315**:1579–1583.

Allard, C. F., Mohandas, N., Wacjman, H., and Krisnamoorthy, R., 1976, Un cas d'instabilité majeure de l'hémoglobine. L'hémoglobine Bicêtre, *Nouv. Rev. Fr. Hematol.* **16**:23–36.

Baklouti, F., Maréchal, J., Wilmotte, R., Alloisio, N., Morlé, L., Ducluzeau, M. T., Denoroy, L., Mrad, A., Ben Aribia, M. H., Kastally, R., and Delaunay, J. 1992, Elliptocytogenic alpha I/36 spectrin Sfax lacks nine amino acids in helix 3 of repeat 4. Evidence for the activation of a cryptic 5'-splice site in exon 8 of spectrin alpha-gene, *Blood* **79**:2464–2470.

Ballas, S. K., 1991, Sickle cell anemia with few painful crises is characterized by decreased red cell deformability and increased number of dense cells, *Am. J. Hematol.* **36**:122–130.

Ballas, S. K., Clark, M. R., Mohandas, N., Colfer, H. F., Caswell, M. S., Bergren, M. O., Perkins, H. A., and Shohet, S. B., 1984, Red cell membrane and cation deficiency in Rh null syndrome, *Blood* **63**:1046–1055.

Ballas, S. K., Tabbara, K. F., Murphy, D. L., Mohandas, N., Clark, M. R., and Shohet, S. B., 1985, Erythrocyte deformability changes in autoimmune hemolytic anemia during development of NZB mice and their (NZB/NZW)F1 hybrid, *J. Clin. Lab. Immunol.* **16**:217–222.

Ballas, S. K., Larner, J., Smith, E. D., Surrey, S., Schwartz, E., and Rappaport, E. F., 1987, The xerocytosis of Hb SC disease, *Blood* **69**:124–130.

Ballas, S. K., Larner, J., Smith, E. D., Surrey, S., Schwartz, E., and Rappaport, E. F., 1988, Rheologic predictors of the severity of the painful sickle cell crisis, *Blood* **72**:1216–1223.

Ballas, S. K., Dover, G. J., and Charache, S., 1989, Effect of hydroxyurea on the rheological properties of sickle erythrocytes in vivo, *Am. J. Hematol.* **32**:104–111.

Ballas, S. K., Bator, S. M., Aubuchon, J. P., Marsh, W. L., Sharp, D. E., and Toy, E. M., 1990, Abnormal membrane physical properties of red cells in McLeod syndrome, *Transfusion* **30**:722–727.

Bareford, D., Stone, P., Caldwell, N., Meiselman, H., and Stuart, J., 1985, Comparison of instruments for measurement of erythrocyte deformability, *Clin. Hemorheol.* **5**:311–322.

Bareford, D., Jennings, P. E., Stone, P. C., Baar, S., Barnett, A. H., and Stuart, J., 1986, Effects of hyperglycaemia and sorbitol accumulation on erythrocyte deformability in diabetes mellitus, *J. Clin. Pathol.* **39**:722–727.

Becker, P. S., Morrow, J. S., and Lux, S. E., 1987, Abnormal oxidant sensitivity and beta-chain structure of spectrin in hereditary spherocytosis associated with defective spectrin–protein 4.1 binding, *J. Clin. Invest.* **80**:557–565.

Bennett, V., 1989, The spectrin–actin junction of erythrocyte membrane skeletons. *Biochim. Biophys. Acta* **988**:107–121.

Berk, D., Hochmuth, R., and Waugh, R., 1988, Viscoelastic properties and rheology, in: *Red Blood Cell Membranes: Structure, Function, Clinical Implications* (P. Agre and J. Parker, eds.), pp. 423–454, Dekker, New York.

Bessis, M., and Mohandas, N., 1974, Mesure continue de la deformabilité cellulaire par une méthode diffractométrique, *C. R. Acad. Sci. Ser. D* **278**:3263–3265.

Bessis M., and Mohandas, N., 1975a, A diffractometric method for the measurement of cellular deformability, *Blood Cells* **1**:307–313.

Bessis, M., and Mohandas, N., 1975b, Deformability of normal, shape-altered and pathological red cells, *Blood Cells* **1**:315–321.

Bessis, M., and Mohandas, N., 1976, Deformation et orientation des globules rouges falciformes soumis à des forces de cisaillement, *C. R. Acad. Sci. Ser. D* **282**:1567–1570.

Bessis, M., and Mohandas, N., 1977, Laser diffraction patterns in sickle cells in fluid shear fields, *Blood Cells* **3**:229–239.

Bessis, M., Mohandas, N., and Féo, C., 1980, Automated ektacytometry: A new method of measuring red cell deformability and red cell indices, *Blood Cells* **6**:315–327.

Bessis, M., Féo, C., and Jones, E., 1982, Quantitation of red cell deformability during progressive deoxygenation and oxygenation in sickling disorders (the use of an automated ektacytometer), *Blood Cells* **8**:17–28.

Bessis, M., Féo, C., Jones, E., and Nossal, M., 1983, Adaptation of the ektacytometer to automated continuous pO$_2$ changes: Determination of erythrocyte deformability in sickling disorders, *Cytometry* **3**:296–299.

Bull, B., Chien, S., Dormandy, J., Kiesewetter, H., Lewis, S., Lowe, G., Meiselman, H., Shohet, S., Stoltz, J., Stuart, J., and Teitel, P., 1986, Guidelines for measurement of blood viscosity and erythrocyte deformability, *Clin. Hemorheol.* **6**:439–453.

Card, R. T., Mohandas, N., Perkins, H. A., and Shohet, S. B., 1982, Deformability of stored red blood cells. Relationship to degree of packing, *Transfusion* **22**:96–101.

Cavadini, B., 1975, Réalisation d'un prototype de viscomètre-diffractomètre (ektacytomètre) permettant l'étude de la deformabilité des cellules du sang, Diplome de l'Ecole Pratique des Hautes Etudes, Paris.

Chailley, B., Weed, R., Leblond, P., and Maigne, J., 1973, Formes echinocytaires et stomatocytaires du globule rouge. Leur reversibilité et leur convertibilité *Nouv. Rev. Fran. Hematol.* **13**:71–88.

Chasis, J. A., and Mohandas, N., 1986, Erythrocyte membrane deformability and stability: Two distinct membrane properties that are independently regulated by skeletal protein associations. *J. Cell Biol.* **103**:343–350.

Chasis, J. A., and Schrier, S. L., 1989, Membrane deformability and the capacity for shape change in the erythrocyte, *Blood* **74**:2562–2568.

Chasis, J. A., Mohandas, N., and Shohet, S. B., 1985, Erythrocyte membrane rigidity induced by glycophorin A–ligand interaction. Evidence for a ligand-induced association between glycophorin A and skeletal proteins, *J. Clin. Invest.* **75**:1919–1926.

Chasis, J. A., Reid, M. E., Jensen, R. H., and Mohandas, N., 1988a, Signal transduction by glycophorin A: Role of extracellular and cytoplasmic domains in a modulatable process, *J. Cell Biol.* **107**:1351–1357.

Chasis, J. A., Agre, P., and Mohandas, N., 1988b, Decreased membrane mechanical stability and in vivo loss of surface area reflect spectrin deficiencies in hereditary spherocytosis, *J. Clin. Invest.* **82**:617–623.

Chien, S., 1987, Red cell deformability and its relevance to blood flow, *Annu. Rev. Physiol.* **49**:177–192.

Chien, S., Usami, S., and Bertles, J., 1970, Abnormal rheology of oxygenated blood of sickle cell anemia, *J. Clin. Invest.* **49:**623–634.

Clark, M. R., 1989a, Computation of the average shear-induced deformation of red blood cells as a function of osmolality, *Blood Cells* **15:**427–439.

Clark, M. R., 1989b, Mean corpuscular hemoglobin concentration and cell deformability, *Ann. N.Y. Acad. Sci.* **565:**284–294.

Clark, M. R., Mohandas, N., Caggiano, V., and Shohet, S. B., 1978, Effect of abnormal cation transport on deformability of desiccytes, *J. Supramol. Struct.* **8:**521–532.

Clark, M. R., Mohandas, N., Féo, C., Jacobs, M. S., and Shohet, S. B., 1981, Separate mechanisms of deformability loss in ATP-depleted and Ca-loaded erythrocytes, *J. Clin. Invest.* **67:**531–539.

Clark, M. R., Mohandas, N., and Shohet, S. B., 1982, Hydration of sickle cells using the sodium ionophore monensin: A model for therapy, *J. Clin. Invest.* **70:**1074–1080.

Clark, M. R., Mohandas, N., and Shohet, S. B., 1983, Osmotic gradient ektacytometry: Comprehensive characterization of red cell volume and surface maintenance, *Blood* **61:**899–910.

Clark, M. R., Aminoff, M. J., Chiu, D. T., Kuypers, F. A., and Friend, D. S., 1989, Red cell deformability and lipid composition in two forms of acanthocytosis: Enrichment of acanthocytic populations by density gradient centrifugation, *J. Lab. Clin. Med.* **113:**469–481.

Conboy, J. G., Shitamoto, R., Parra, M., Winardi, R., Kabra, A., Smith, J., and Mohandas, N., 1991, Hereditary elliptocytosis due to both qualitative and quantitative defects in membrane skeletal protein 4.1, *Blood* **78:**2438–2443.

Cummings, D. M., and Ballas, S. K., 1990, Effects of pentoxifylline and metabolite on red blood cell deformability as measured by ektacytometry, *Angiology* **41:**118–123.

Delaunay, J., Alloisio, N., Morlé, L., and Pothier, B., 1990, The red cell skeleton and its genetic disorders, *Mol. Aspects Med.* **11:**161–241.

Deuticke, B., 1968, Transformation and restoration of biconcave shape of human erythrocytes induced by amphiphilic agents and changes of ionic environment, *Biochim. Biophys. Acta* **163:**494–500.

Dhermy, D., Garbarz, M., Lecomte, M. C., Féo, C., Bournier, O., Chaveroche, I., Gautero, H., Galand, C., and Boivin, P., 1986, Hereditary elliptocytosis: Clinical, morphological and biochemical studies of 38 cases, *Nouv. Rev. Fr. Hematol.* **28:**129–140.

Discher, D., Parra, M., Conboy, J., and Mohandas, N., 1992, Physiologic function of the alternatively-spliced spectrin–actin binding domain of protein 4.1, *Blood* **80:**143a.

Evans, E., 1988, Deformability and adhesivity properties of blood cells and membrane vesicles: Direct methods, in: *Red Cell Membranes* (S. Shohet and N. Mohandas, eds.), pp. 271–297, Churchill Livingstone, Edinburgh.

Evans, E., and Leung, A., 1984, Adhesivity and rigidity of erythrocyte membrane in relation to wheat germ agglutinin binding, *J. Cell Biol.* **98:**1201–1208.

Evans, E., and Mohandas, N., 1986, Developments in red cell rheology at the Institut de Pathologie Cellulaire, *Blood Cells* **12:**43–56.

Evans, E. A., and Mohandas, N., 1987, Membrane-associated sickle hemoglobin: A major determinant of sickle erythrocyte rigidity, *Blood* **70:**1443–1449.

Evans, E., and Skalak, R., 1980, *Mechanics and Thermodynamics of Biomembranes*, CRC Press, Boca Raton, Fla.

Féo, C., and Mohandas, N., 1977, Clarification of role of ATP in red-cell morphology and function, *Nature* **265:**166–168.

Féo, C. J., Fischer, S., Piau, J. P., Grange, M. J., and Tchernia, G., 1980, Première observation de l'absence d'une protéine de la membrane érythocytaire (bande 4.1) dans un cas d'anémie elliptocytaire familiale [1st instance of the absence of an erythrocyte membrane protein (band 4.1) in a case of familial elliptocytic anemia], *Nouv. Rev. Fr. Hematol.* **22:**315–325.

Féo, C., Jones, E., and Bessis, M., 1981, Mesure de la deformabilité cellulaire en fonction de la pression d'oxygène dans la drépanocytose. Resultants donnés par une méthode visco-diffractométrique (ektacytomètre) [Measurement of sickle cell deformability during deoxygenation–reoxygenation experiments in the ektacytometer], *C. R. Acad. Sci.* **293**(III):57–62.

Féo, C. J., Nossal, M., Jones, E., and Bessis, M., 1982, Une nouvelle technique d'étude de la physiologie des globules rouges: la mesure de leur deformabilité en fonction de l'osmolarité. Résultats obtenus par un ektacytomètre automatisé sur du sang normal et dans différentes anémies hemolytiques [A new technique for the study of the physiology of erythrocytes: measurement of their deformability as a function of osmolarity. Results obtained by an automated ektacytometer in normal blood and in various hemolytic anemias], *C. R. Acad. Sci.* **295**(III):687–691.

Féo, C. J., Johnson, R. M., and Nossal, M., 1983, Action de trois variétés de substances antidrépanocytaires mesurée par ektacytometrie en fonction de la pression d'oxygène, *C. R. Acad. Sci.* **296**(III):911–916.

Fischer,T. M., Stohr, M., and Schmid-Schönbein, H., 1978, Red blood cell (RBC) microrheology: Comparison of the behavior of single RBC and liquid droplets in shear flow, *A IChE Symp. Ser. 182* **74**:38–45.

Fisher, T., Wenby, R., and Meiselman, H., 1992, Pulse shape-analysis of RBC micropore flow via new software for the cell transit analyzer (CTA), *Biorheology* **29**:185–201.

Fortier, N., Snyder, L. M., Garver, F., Kiefer, C., McKenney, J., and Mohandas, N., 1988, The relationship between in vivo generated hemoglobin skeletal protein complex and increased red cell membrane rigidity, *Blood* **71**:1427–1431.

Gao, S. J., 1989, [A study of erythrocyte deformability using an ektacytometer in normal adults and in patients with acute cerebral infarction], *Chung Hua Shen Ching Ching Shen Ko Tsa Chih* **22**:155–156.

Gardner, K., and Bennett, V., 1987, Modulation of spectrin–actin assembly by erythrocyte adducin, *Nature* **328**:359–362.

Ghanem, A., Pothier, B., Marechal, J., Ducluzeau, M. T., Morle, L., Alloisio, N., Féo, C., Ben, A. A., Fattoum, S., and Delaunay, J., 1990, A haemolytic syndrome associated with the complete absence of red cell membrane protein 4.2 in two Tunisian siblings, *Br. J. Haematol.* **75**:414–420.

Groner, W., Mohandas, N., and Bessis, M., 1980, New optical technique for measuring erythrocyte deformability with the ektacytometer, *Clin. Chem.* **26**:1435–1442.

Guetarni, D., Roux, A. F., Alloisio, N., Morle, F., Ducluzeau, M. T., Forget, B. G., Colonna, P., Delaunay, J., and Godet, J., 1990, Evidence that expression of Sp alpha I/65 hereditary elliptocytosis is compounded by a genetic factor that is linked to the homologous alpha-spectrin allele, *Hum. Genet.* **85**:627–630.

Gulley, M. L., Ross, D. W., Féo, C., and Orringer, E. P., 1982, The effect of cell hydration on the deformability of normal and sickle erythrocytes, *Am. J. Hematol.* **13**:283–291.

Hardeman, M., Goedhart, P., and Breederveld, D., 1987, Laser diffraction ellipsometry of erythrocytes under controlled shear stress using a rotational viscosimeter, *Clin. Chim. Acta* **165**:227–234.

Hardeman, M., Bauersachs, R., and Meiselman, H., 1988, RBC laser diffractometry and RBC aggregation with a rotational viscometer: Comparison with rheoscope and Myrenne aggregometer, *Clin. Hemorheol.* **8**:581–593.

Heath, B. P., Mohandas, N., Wyatt, J. L., and Shohet, S. B., 1982, Deformability of isolated red blood cell membranes, *Biochim. Biophys. Acta* **691**:211–219.

Hebbel, R. P., 1991, Beyond hemoglobin polymerization: The red blood cell membrane and sickle disease pathophysiology, *Blood* **77**:214–237.

Jain, S. K., Mohandas, N., Sensabaugh, G., Shojania, A. M., and Shohet, S. B., 1982, Hereditary plasma lecithin–cholesterol acyl transferase deficiency. A heterozygous variant with erythrocyte membrane abnormalities, *J. Lab. Clin. Med.* **99:**816–826.

Jarolim, P., Palek, J., Amato, D., Hassan, K., Sapak, P., Nurse, G. T., Rubin, H. L., Zhai, S., Sahr, K. E., and Liu, S. C., 1991, Deletion in erythrocyte band 3 gene in malaria-resistant Southeast Asian ovalocytosis, *Proc. Natl. Acad. Sci. USA* **88:**11022–11026.

Johnson, R. M., 1985, pH effects on red cell deformability, *Blood Cells* **11:**317–321.

Johnson, R. M., 1989, Ektacytometry of red blood cells, *Methods Enzymol.* **173:**35–54.

Johnson, R. M., Féo, C. J., Nossal, M., and Dobo, I., 1985, Evaluation of covalent antisickling compounds by PO_2-scan ektacytometry, *Blood* **66:**432–438.

Kage, H., Engelhardt, H., and Sackmann, E., 1990, A precision method to measure average viscoelastic parameters of erythrocyte populations, *Biorheology* **27:**67–78.

Koutsouris, D., Guillet, R., Wenby, R. B., and Meiselman, H. J., 1988, Determination of erythrocyte transit times through micropores. II—Influence of experimental and physicochemical factors, *Biorheology* **25:**773–790.

Kuypers, F. A., Scott, M. D., Schott, M. A., Lubin, B., and Chiu, D. T., 1990, Use of ektacytometry to determine red cell susceptibility to oxidative stress, *J. Lab. Clin. Med.* **116:**535–545.

Lande, W. M., Andrews, D. L., Clark, M. R., Braham, N. V., Black, D. M., Embury, S. H., and Mentzer, W. C., 1988, The incidence of painful crisis in homozygous sickle cell disease: Correlation with red cell deformability, *Blood* **72:**2056–2059.

Liu, S.-C., Palek, J., Prchal, J., and Castleberry, R., 1981, Altered spectrin dimer–dimer association and instability of erythrocyte membrane skeletons in hereditary pyropoikilocytosis, *J. Clin. Invest.* **68:**597–605.

Liu, S.-C., Palek, J., and Prchal, J., 1982, Defective spectrin dimer–dimer association in hereditary elliptocytosis, *Proc. Natl. Acad. Sci. USA* **79:**2072–2076.

Liu, S.-C., Zhai, S., Palek, J., Golan, D. E., Amato, D., Hassan, K., Nurse, G. T., Babona, D., Coetzer, T., Jarolim, P., Zaik, M., and Borwein, S., 1990, Molecular defect of the band 3 protein in Southeast Asian ovalocytosis, *N. Engl. J. Med.* **323:**1530–1538.

Low, P. S., Willardson, B. M., Mohandas, N., Rossi, M., and Shohet, S., 1991, Contribution of the band 3–ankyrin interaction to erythrocyte membrane mechanical stability, *Blood* **77:**1581–1586.

McGough, A., and Josephs, R., 1990, On the structure of erythrocyte spectrin in partially expanded membrane skeleton, *Proc. Natl. Acad. Sci. USA* **87:**5208–5212.

Meiselman, H., and Baker, R., 1977, Flow behavior of ATP-depleted erythrocytes, *Biorheology* **14:**111–126.

Mentzer, W., Turetsky, T., Mohandas, N., Schrier, S., and Wu, C.-S. C., 1984, Identification of the hereditary pyropoikilocytosis carrier state, *Blood* **63:**1439–1446.

Mentzer, W., Iarocci, T. A., Mohandas, N., Lane, P. A., Smith, B., Lazerson, J., and Hays, T., 1987, Modulation of erythrocyte membrane mechanical stability by 2,3-diphosphoglycerate in the neonatal poikilocytosis/elliptocytosis syndrome, *J. Clin. Invest.* **79:**943–949.

Messmann, R., Gannon, S., Sarnaik, S., and Johnson, R. M., 1990, Mechanical properties of sickle cell membranes, *Blood* **75:**1711–1717.

Mohandas, N., 1988, Measurement of cellular deformability and membrane material properties of red cells by ektacytometry, in: *Red Cell Membranes* (S.Shohet and N.Mohandas, eds.), pp. 299–320, Churchill Livingstone, Edinburgh.

Mohandas, N., and De Boisfleury, A., 1977, Antibody-induced spherocytic anemia I: Changes in red cell deformability, *Blood Cells* **3:**187–196.

Mohandas, N., Greenquist, A. C., and Shohet, S. B., 1978, Effects of heat and metabolic depletion on erythrocyte deformability, spectrin extractibility and phosphorylation, *Prog. Clin. Biol. Res.* **21:**453–472.

Mohandas, N., Phillips, W. M., and Bessis, M., 1979, Red blood cell deformability and hemolytic anemias, *Semin. Hematol.* **16**:95–114.

Mohandas, N., Clark, M. R., Jacobs, M. S., and Shohet, S. B., 1980, Analysis of factors regulating erythrocyte deformability, *J. Clin. Invest.* **66**:563–573.

Mohandas, N., Clark, M. R., Health, B. P. Rossi, M., Wolfe, L. C., Lux, S. E., and Shohet, S. B., 1982, A technique to detect reduced mechanical stability of red cell membranes: Relevance to elliptocytic disorders, *Blood* **59**:768–774.

Mohandas, N., Lie, I. L., Friedman, M., and Mak, J. W., 1984, Rigid membranes of Malayan ovalocytes: A likely genetic barrier against malaria, *Blood* **63**:1385–1392.

Mohandas, N., Rossi, M. E., and Clark, M. R., 1986, Association between morphologic distortion of sickle cells and deoxygenation-induced cation permeability increase, *Blood* **68**:450–454.

Mohandas, N., Winardi, R., Knowles, D., Leung, A., Parra, M., George, E., Conboy, J., and Chasis, J., 1992, Molecular basis for membrane rigidity of hereditary ovalocytosis. A novel mechanism involving the cytoplasmic domain of band 3, *J. Clin. Invest.* **89**:686–692.

Morlé, L., Pothier, B., Alloisio, N., Féo, C., Garay, R., Bost, M., and Delaunay, J., 1989, Reduction of membrane band 7 and activation of volume stimulated (K^+, Cl^-)-cotransport in a case of congenital stomatocytosis, *Br. J. Haematol.* **71**:141–146.

Mrad, A., Kister, J., Féo, C., Poyart, C., Kastally, R., Blibech, R., Galacteros, F., and Wajcman, H., 1989, Hemoglobin Athens-Georgia in association with β0-thalassemia in Tunisia, *Am. J. Hematol.* **32**:117–122.

Nash, G. B., 1991, Red cell mechanics: What changes are needed to adversely affect in vivo circulation, *Biorheology* **28**:231–239.

Nash, G., 1992, White blood cell rheology and atherosclerotic ischemic disease, *Clin. Hemorheol.* *12*, Suppl. 1:57–69.

Nash, G. B., Parmar, J., and Reid, M. E., 1990, Effects of deficiencies of glycophorins C and D on the physical properties of the red cell, *Br. J. Haematol.* **76**:282–287.

Nash, G. B., Cooke, B. M., Marsh, K., Berendt, A., Newbold, C., and Stuart, J., 1992, Rheological analysis of the adhesive interactions of red blood cells parasitized by *Plasmodium falciparum*, *Blood* **79**:798–807.

Noguchi, C., Torchia, D., and Schechter, A., 1980, Determination of deoxyhemoglobin S polymer in sickle erythrocytes upon deoxygenation, *Proc. Natl. Acad. Sci. USA* **77**:5487–5491.

Noguchi, C. T., Mohandas, N., Blanchette, M. J., Mackie, S., Raik, E., and Charache, S., 1991, Hemoglobin aggregation and pseudosickling in vitro of hemoglobin Setif-containing erythrocytes, *Am. J. Hematol.* **36**:131–139.

Orringer, E. P., Blythe, D. S., Whitney, J. A., Brockenbrough, S., and Abraham, D. J., 1992, Physiologic and rheologic effects of the antisickling agent ethacrynic acid and its N-butylated derivative on normal and sickle erythrocytes, *Am. J. Hematol.* **39**:39–44.

Palek, J., and Sahr, K. E., 1992, Mutations of the red blood cell membrane proteins: From clinical evaluation to detection of the underlying genetic defect, *Blood* **80**:308–330.

Pasvol, G., Chasis, J. A., Mohandas, N., Anstee, D. J., Tanner, M. J., and Merry, A. H., 1989, Inhibition of malarial parasite invasion by monoclonal antibodies against glycophorin A correlates with reduction in red cell membrane deformability, *Blood* **74**:1836–1843.

Pinder, J., 1991, Red cell membrane cytoskeleton and the control of membrane properties, *Biochem. Soc. Trans.* **19**:1039–1041.

Reid, M. E., Chasis, J. A., and Mohandas, N., 1987, Identification of a functional role for human erythrocyte sialoglycoproteins beta and gamma, *Blood* **69**:1068–1072.

Saul, A., Lamont, G., Sawyer, W., and Kidson, C., 1984, Decreased membrane deformability in Melanesian ovalocytes from Papua New Guinea, *J. Cell Biol.* **98**:1348–1354.

Scheven, C., 1989, Ektazytometrie–ein Verfahren zur Charakterisierung der Deformierbarkeit von

Erythrorozyten [Ektacytometry: A method for characterizing erythrocyte deformability], *Folia Haematol. (Leipzig)* **116**:653–669.

Schofield, A., Tanner, M., Pinder, J., Clough, B., Bayley, P., Nash, G., Dluzewski, A., Reardon, D., Cox, T., Wilson, R., and Gratzer, W., 1992, Basis of unique red cell membrane properties in hereditary ovalocytosis, *J. Mol. Biol.* **223**:949–958.

Schrier, S. L., 1987, Shape changes and deformability in human erythrocyte membranes, *J. Lab. Clin. Med.* **110**:791–797.

Schrier, S. L., and Mohandas, N., 1992, Globin-chain specificity of oxidation-induced changes in red blood cell membrane properties, *Blood* **79**:1586–1592.

Schrier, S. L., Rachmilewitz, E., and Mohandas, N., 1989, Cellular and membrane properties of alpha and beta thalassemic erythrocytes are different: Implication for differences in clinical manifestations, *Blood* **74**:2194–2202.

Schwartz, R. S., Madsen, J. W., Rybicki, A. C., and Nagel, R. L., 1991, Oxidation of spectrin and deformability defects in diabetic erythrocytes, *Diabetes* **40**:701–708.

Scott, M., Rouyer-Fessard, P., Soda Ba, M., Lubin, B., and Beuzard, Y., 1992, Alpha- and beta-hemoglobin chain induced changes in normal erythrocyte deformability: Comparison to beta thalassemia intermedia and Hb H disease, *Br. J. Haematol.* **80**:519–526.

Sheetz, M., and Singer, S., 1974, Biological membranes as bilayer couples. A molecular mechanism of drug–erythrocyte interactions, *Proc. Natl. Acad. Sci. USA* **71**:4457–4461.

Shen, B., 1988, Ultrastructure and function of membrane skeleton, in: *Red Blood Cell Membranes: Structure, Function, Clinical Implications* (P. Agre and J. Parker, eds.), pp. 261–297, Dekker, New York.

Shields, D., and Sarin, S., 1992, Haemorheological treatment of ischaemia—facts or fancy? *J. R. Soc. Med.* **85**:365.

Shiga, T., Maeda, N., and Kon, K., 1990, Erythrocyte rheology, *Crit. Rev. Oncol. Hematol.* **10**:9–48.

Smith, J. E., Mohandas, N., Clark, M. R., Greenquist, A. C., and Shohet, S. B., 1980, Deformability and spectrin properties in three types of elongated red cells, *Am. J. Hematol.* **8**:1–13.

Smith, L., and Hochmuth, R., 1982, Effect of wheat germ agglutinin on the viscoelastic properties of erythrocyte membrane, *J. Cell Biol.* **94**:7–11.

Snyder, L. M., Fortier, N. L., Trainor, J., Jacobs, J., Leb, L., Lubin, B., Chiu, D., Shohet, S., and Mohandas, N., 1985, Effect of hydrogen peroxide exposure on normal human erythrocyte deformability, morphology, surface characteristics, and spectrin–hemoglobin cross-linking, *J. Clin. Invest.* **76**:1971–1977.

Snyder, L. M., Fortier, N. L., Leb, L., McKenney, J., Trainor, J., Sheerin, H., and Mohandas, N., 1988, The role of membrane protein sulfhydryl groups in hydrogen peroxide-mediated membrane damage in human erythrocytes, *Biochim. Biophys. Acta* **937**:229–240.

Sobue, K., Muramoto, Y., Fujita, M., and Kakiuchi, S., 1981, Calmodulin-binding protein of erythrocyte cytoskeleton, *Biochem. Biophys. Res. Commun.* **100**:1063–1070.

Sorette, M. P., Lavenant, M. G., and Clark, M. R., 1987, Ektacytometric measurement of sickle cell deformability as a continuous function of oxygen tension, *Blood* **69**:316–323.

Stäubli, M., Stone, P., Straub, P., and Stuart, J., 1986, Evaluation of methods for measuring erythrocyte deformability, *Clin. Hemorheol.* **6**:589–602.

Stokke, B. T., Mikkelsen, A., and Elgsaeter, A., 1986, The human erythrocyte membrane skeleton may be an ionic gel. I. Membrane mechanochemical properties, *Eur. Biophys. J.* **13**:203–218.

Stolitz, J. F., Ravey, J. C., Larcan, A., Mazeron, P., and Lucius, M., 1981, Deformation and orientation of red blood cells in a simple shear flow, *Scand. J. Clin. Lab. Invest.* **41**(Suppl. 156):67–75.

Stozicky, F., Blazek, V., and Muzik, J., 1979, Deformabilita Cervenych Krvinek. Difraktometricka

Metoda Mereni [Deformability of erythrocytes. A diffractometric method of measurement], *Cas. Lek. Cesk.* **118**:362–368.

Stuart, J., and Nash, G., 1990, Technological advances in blood rheology, *Crit. Rev. Clin. Lab. Sci.* **28**:61–93.

Takakuwa, Y., and Mohandas, N., 1988, Modulation of erythrocyte membrane material properties by Ca²⁺ and calmodulin. Implications for their role in regulation of skeletal protein interactions, *J. Clin. Invest.* **82**:394–400.

Takakuwa, Y., Tchernia, G., Rossi, M., Benabadji, M., and Mohandas, N., 1986, Restoration of normal membrane stability to unstable protein 4.1-deficient erythrocyte membranes by incorporation of purified protein 4.1 *J. Clin. Invest.* **78**:80–85.

Tchernia, G., Mohandas, N., and Shohet, S. B., 1981, Deficiency of skeletal membrane protein band 4.1 in homozygous hereditary elliptocytosis. Implications for erythrocyte membrane stability, *J. Clin. Invest.* **68**:454–460.

Tzanck, A., and Bessis, M., 1947, Un nouvel hémo-diffractomètre, *Sang* **18**:71–76.

Wajcman, H., Baudin-Chich, V., Kister, J., Féo, C., Gombaud-Saintonge, G., Bohn, B., Marden, M., Pagnier, J., Poyart, C.. Dodé, C., Galacteros, F., Blouquit, Y., Cynober, T., and Tchernia, G., 1988, Hemoglobin J Guantanamo [alpha2 beta2 128 (H6) Ala→Asp] in association with hemoglobin C and alpha-thalassemia in a family from Benin, *Am. J. Hematol.* **28**:170–175.

Waugh, R. E., and Agre, P., 1988, Reductions of erythrocyte membrane viscoelastic coefficients reflect spectrin deficiencies in hereditary spherocytosis, *J. Clin. Invest.* **81**:133–141.

Waugh, R. E., Narla, M., Jackson, C. W., Mueller, T. J., Suzuki, T., and Dale, G. L., 1992, Rheologic properties of senescent erythrocytes: Loss of surface area and volume with red blood cell age, *Blood* **79**:1351–1358.

Wolf, G., Bayer, R., and Ostuni, D., 1992, Stress-induced rigidfication of erythrocytes as determined by laser diffraction and image analysis, *Opt. Eng.* **31**:1475–1481.

Yip, R., Mohandas, N., Clark, M. R., Jain, S., Shohet, S. B., and Dallman, P. R., 1983, Red cell membrane stiffness in iron deficiency, *Blood* **62**:99–106.

Chapter 6

Spin-Label ESR Study of Molecular Dynamics of Lipid/Protein Association in Membranes

László I. Horváth

1. INTRODUCTION

The basic structural unit of membranes is the lipid bilayer which, as a permeability barrier, separates two aqueous phases. Solute transport between these two phases is maintained via special transmembrane proteins which are partially embedded into the hydrophobic region and partially extended into the interfacial regions. Since each physiological function, as a rule, is associated with different protein species, and very often with a self-regulatory unit of several proteins, and the optimal lipid environment of proteins shows great variation, the composition of membranes is usually very complex and is probably optimized for its specific task (Sandermann, 1978).

From the point of view of molecular dynamics the major components of biological membranes, namely proteins and lipids, have rather different motional properties owing to their different hydrodynamic sizes (Saffman and Delbrück, 1975; Cherry, 1979; Clegg and Vaz, 1985). Since these molecules are located in

László I. Horváth   Institute of Biophysics, Biological Research Centre, H6701 Szeged, Hungary.

Subcellular Biochemistry, Volume 23: Physicochemical Methods in the Study of Biomembranes, edited by Herwig J. Hilderson and Gregory B. Ralston. Plenum Press, New York, 1994.

close proximity in the lipid bilayer, their different mobilities inevitably give rise to frictional forces at the colliding molecular interfaces. The motional gradient between proteins and lipids leads to the formation of a solvation shell of lipids which is motionally coupled to proteins, on the one hand, and consequently motionally restricted with respect to the fluid bulk lipids, on the other hand (Watts and de Pont, 1985, 1986). The great difference in dynamic properties is rather useful for maintaining tight molecular packing in the membrane and, hence, a continuous permeability barrier since fluid lipids can match any packing irregularities arising during the lateral movement and conformational rearrangements of proteins practically instantaneously.

These structural and dynamical features can only be resolved by spectroscopic techniques. Among the approaches that can highlight different aspects of membrane organization, magnetic resonance spectroscopy, including electron spin resonance (ESR)* and nuclear magnetic resonance (NMR) spectroscopy, has proved to be particularly powerful (Seelig et al., 1982; Bloom and Smith, 1985; Marsh, 1985; Devaux and Seigneuret, 1985; Knowles and Marsh, 1991). ESR spectroscopy can detect the presence of paramagnetic species and, since most membranes are devoid of or contain only small amounts of intrinsic paramagnetic molecules, a special branch of ESR spectroscopy, namely spin-labeling ESR, has become the method of choice (Berliner, 1976, 1979; Schreier et al., 1978). Spin labeling consists of the intercalation of a nitroxyl analogue of a membrane constituent, usually a spin-labeled lipid, into the membrane or alternatively the covalent modification of a membrane-bound macromolecule, usually a protein, by a reactive nitroxyl group. It is useful to make a fine distinction between these two strategies: in the former case, which is strictly speaking molecular probing rather than labeling, the nitroxyls can freely diffuse like all the noncovalently attached lipids in the plane of the membrane and, hence, undergo rapid reorientational motion in the liquid-crystalline phase (Griffith and Jost, 1976). Covalent attachment to proteins, on the other hand, leads to motional restriction and, hence, the mobility of the nitroxyl is more characteristic of the motion of macromolecules (Berliner, 1976).

NMR spectroscopy can detect magnetic nuclei, such as ^{1}H, ^{2}H, ^{13}C, ^{31}P, and ^{19}F, to mention a few that are most frequently studied (Seelig, 1977, 1978; Bloom and Smith, 1985). The intrinsic ^{31}P moiety of phospholipid head groups and the ^{2}H nuclei of selectively deuterated lipids are two alternatives for the

---

*Abbreviations used in this chapter: DMPC, 1,2-dimyristoyl-sn-glycero-3-phosphocholine; DMPC-d_9, 1,2-dimyristoyl-sn-glycero-3-phosphocholine deuterated in the phosphocholine methyl groups [-$(CD_3)_3$]; ESR, electron spin resonance; NMR, nuclear magnetic resonance; PC, phosphatidylcholine; PLP, myelin proteolipid protein; 14-PASL, -PCSL, -PGSL, -PSSL, 1-acyl-2-[14-(4,4-dimethyl-oxazolidine-N-oxyl) stearoyl]-sn=glycero-3-phosphatidic acid, -phosphocholine, -phosphoglycerol, -phosphoserine; 14-SASL, [14-(4,4-dimethyl-oxazolidine-N-oxyl) stearic acid.

above-mentioned spin labeling with the obvious advantage of having no or very little membrane perturbations.

When the two branches of magnetic resonance spectroscopy were first applied to the study of lipid/protein systems, conflicting results were obtained (Paddy et al., 1981; Bienvenue et al., 1982). In spin-labeling ESR experiments on mitochondrial cytochrome oxidase/dimyristoylphosphatidylcholine (DMPC) complexes, two populations of lipids could be resolved in a rather consistent manner in both lipid/protein titration and lipid selectivity experiments (Jost et al., 1973; Knowles et al., 1979). In ^{2}H-NMR experiments, on the other hand, no motionally restricted spectral component could be detected: with the addition of protein a gradual line broadening was observed (Seelig et al., 1982; Tamm and Seelig, 1983). The reason for this apparent contradiction is the time-scale paradox. As shown more rigorously below, the magnetic resonance (both ESR and NMR) spectrum of randomly oriented molecules consists of a set of overlapping spectral lines corresponding to different orientations providing the molecular reorientational motion is sufficiently slow. As molecular tumbling rates increase, this anisotropic line shape collapses into a single motionally averaged (nearly) isotropic spectrum and a rather well-defined limiting frequently can be ascribed to the transition from anisotropic to isotropic line shape. The limiting frequency of spin-labeling ESR spectroscopy is 3×10^8 sec^{-1} (McConnell, 1976) and the motional rates of fluid and motionally restricted lipids lie above and below this limit, respectively (Marsh and Watts, 1982). Hence, two spectral components are observed (the reciprocal of the limiting frequency is the time scale of the method). ^{2}H-NMR has a significantly lower limiting frequency of 2×10^3 sec^{-1} and so both populations will exhibit motionally averaged spectra: the observed experimental spectrum will be a weighted average of these two line shapes and usually exhibits no obvious two-component character (Seelig et al., 1982; Bloom and Smith, 1985).

Clearly, spin-labeling ESR spectroscopy has an optimal time scale to resolve motionally restricted lipids in the solvation layer of integral membrane proteins. The major emphasis of this chapter will be on the spin-labeling ESR method. A special feature of magnetic resonance spectroscopy is that because the rotational and translational motions have significant contribution to spin relaxation their rate can be determined from the magnetic resonance spectra. In particular, the molecular dynamics of fluid-to-solvation site exchange will be quantified by the modified Bloch equations; this analysis will lead to a correlation between lipid selectivity and exchange dynamics. The special problem of slowly exchanging cardiolipins at the interface of a mitochondrial protein highlights the role of hydrophobic interactions in lipid/protein association. In such an extreme case the fluid-to-solvation exchange may become so slow that it is possible to resolve two lipid populations by ^{2}H-NMR spectroscopy. Finally, the cross-relaxation effects of two-site exchange will be explored in spin-lattice relaxation experiments using spin-labeling ESR spectroscopy.

2. THERMODYNAMIC MODEL OF LIPID/PROTEIN ASSOCIATION

In this section the motional sensitivity of nitroxyl spectra will be addressed first and special emphasis will be paid to spectral characteristics in the fast- and slow-motion regimes. Then, the thermodynamic model of lipid/protein association will be discussed. Finally, the stoichiometry and lipid specificity will be determined by quantitative analysis of two-component spectra ubiquitously observed for spin-labeled lipids in protein/lipid complexes.

2.1. Spectral Line Shapes of Nitroxyls in Different Motional Regimes

The ESR absorption of nitroxyls exhibits an angular dependence with respect to the spectrometer magnetic field because of the anisotropic distribution of spin density of the unpaired electron. To good approximation, the spin Hamiltonian (assuming axial symmetry) can be written as (McConnell and McFarland, 1970; Berliner, 1976; Marsh and Horváth, 1989)

$$\mathcal{H} = g_o\beta B_z S_z + a_o I_z S_z + \frac{1}{3}(3\cos^2\vartheta - 1)(\Delta g\beta S_z + \Delta A I_z S_z) \qquad (1)$$

where S and I denote the electron spin and nuclear angular momenta, respectively; $\mathbf{B} = (0, 0, B_z)$ is the magnetic induction vector, β is the Bohr magneton, and ϑ is the angle between the magnetic field direction and the orientation of the p_z orbital (the nitroxyl z-axis). The spin Hamiltonian tensor values are $\mathbf{g} = (2.0088, 2.0061, 2.0026)$ and $\mathbf{A} = (0.625 \text{ mT}, 0.580 \text{ mT}, 3.350 \text{ mT})$ and have been separated into an isotropic (g_o and a_o) and an anisotropic part (Δg and ΔA).

In the case of no molecular motion, the ESR spectrum of randomly oriented nitroxyls with different orientations relative to the magnetic field is a superposition of absorptions defined by the g and hyperfine tensor anisotropies (Δg and ΔA). This is the so-called powder spectrum (McConnell and McFarland, 1970). At the other extreme of fast molecular reorientational motion the last two terms in Equation (1) become time dependent and in isotropic liquids rapidly average to zero; the resulting spectrum shows no anisotropy and the g value (e.g., the spectral position) and hyperfine splitting are determined by the isotropic values (g_o and a_o). As a result, with the onset of molecular motion the powder spectrum gradually collapses into the isotropic three-line spectrum as illustrated in Figure 1. The line shape change is the most pronounced as the motional rate becomes comparable with the spectral anisotropies (in frequency units):

$$\tau_{\text{lim}}^{-1} = \frac{\Delta g\beta B}{h} + m_I\Delta A = 1 \times 10^8 \text{sec}^{-1} \qquad (2)$$

FIGURE 1. Effect of molecular rotation on the ESR line shape of nitroxyl radicals. These simulated line shapes are calculated by solving the exchange-coupled Bloch equations (see also Section 4); molecular rotation is treated as exchange between different orientations with respect to the magnetic field direction. The rotational correlation time, τ_R, decreases from top to bottom: (1) rigid limit line shape ($\tau_R \geq 10^{-7}$ sec); (2) slow motion line shape ($\tau_R \approx 10^{-8}$ sec) shown together with the rigid limit line shape (dashed line) to illustrate the incipient line broadening and decreasing anisotropy; (3) intermediate line shape for motional rates close to the limiting frequency of the spin label ESR method; (4) motionally averaged line shape ($\tau_R \approx 10^{-9}$ sec) shown together with the fast motion limit line shape (dashed line); in this case motional rates can be determined from linewidth (or root amplitude) measurements; (5) fast motion limit line shape ($\tau_R \leq 10^{-10}$ sec).

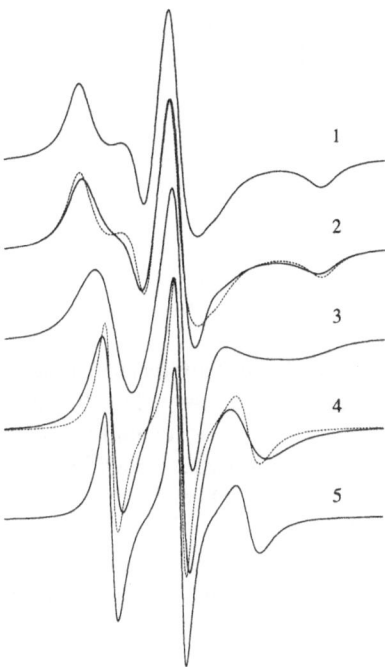

Since the g and hyperfine tensor anisotropies have opposite signs, different limiting frequencies (reciprocal time scales) can be ascribed to each nuclear manifold ($m_I = +1, 0, -1$). The motional sensitivity of the nitroxyl ESR spectra is confined to an interval about this limiting frequency $[\tau_{\lim}^{-1}/100; 100\tau_{\lim}^{-1}]$. The region between the lower bound and approximately half the limiting frequency $[\tau_{\lim}^{-1}/100; \tau_{\lim}^{-1}/2]$ is referred to as the slow motion regime, while the region between the double limiting frequency and the upper bound $[2\tau_{\lim}^{-1}; 100\tau_{\lim}^{-1}]$ is referred to as the fast motion regime.

The observed spectral changes shown in Figure 1 can be described by the two-site exchange model to be discussed later in detail assuming that reorientational motion brings about spectral packet exchanges between different orientations in the interval of g-tensor and hyperfine anisotropy (Marsh, 1986). (The three nuclear manifolds can be treated separately.) At slow exchange rates the effect of two-site exchange is an incipient line broadening of the spectral packets and a slight inward shift toward the center of the anisotropy interval (Carrington and McLachlan, 1967). Hence the first spectral change in the slow motion regime is a gradual line broadening and decreasing anisotropy. This is illustrated in Figure 1 overlaying a slow motion spectrum (trace 2) with the rigid limit spectrum (trace 1). Both effects can be utilized for measuring a motional parameter, the rotational correlation time, τ_R (Goldman *et al.*, 1972; Mason and Freed, 1974; Freed, 1976):

$$\tau_R = a'_m \left[\frac{\Delta W_m}{\Delta W_m^{ref}} - 1 \right]^{b'_m} \qquad (3)$$

and

$$\tau_R = a \left[1 - \frac{A_{zz}}{A_{zz}^{ref}} \right]^{b_m} \qquad (4)$$

where A_{zz} is half the separation of the outer extrema and ΔW_m denote their linewidths ($m_I = \pm 1$), all values normalized with respect to the rigid limit reference values. It should be noted that the linewidths of the outer extrema give the spectral packet linewidths in the $\vartheta = 0^0$ orientation (McConnell and McFarland, 1970). Other constants, assuming Brownian diffusion, are: $a'_m = 1.15 \times 10^{-8}$ sec, $b'_m = -0.143$ and $a = 1.10 \times 10^{-9}$ sec, $b = -1.01$.

As the motional rate exceeds the limiting frequency defined by Equation (2), the individual spectral packets merge into a single narrow, motionally averaged line, the width of which is determined by the motional rate. Accordingly, the motional rate can be determined from linewidth (or root amplitude) measurements; a typical motionally averaged spectrum is shown together with the fast limit line shape in Figure 1 (traces 4 and 5). The rotational correlation time can be determined in this fast motion regime according to the formula (Kivelson, 1960; Schreier *et al.*, 1978)

$$\tau_R = kW_o \left[\sqrt{\frac{h_{-1}}{h_0}} - \sqrt{\frac{h_{+1}}{h_0}} \right] \qquad (5)$$

where h_{+1}, h_0, and h_{-1} denote the amplitudes of the three hyperfine lines, respectively, W_o is the linewidth of the central line, and $k = 6.5 \times 10^{-8}$ sec/mT.

The center of the sensitive range about τ_{lim}^{-1} is the so-called intermediate regime where the line-shape changes are very rapid and complicated. None of the above approximate methods can be applied; the quantitative analysis of such line shapes requires the solution of the stochastic Liouville equation in conjunction with line-shape simulations. A detailed treatise of this method is given by Freed (1976).

2.2. Two-Component ESR Spectra of Protein/Lipid Complexes

Lipid/protein association in biological membranes can be studied by isolating and purifying different integral membrane proteins and subsequently reassembling with a single phospholipid species in lipid vesicles (Abney and Owicki,

1985). On incorporating freely diffusible spin-labeled lipids into such protein/lipid complexes, two-component ESR spectra are usually obtained (Marsh and Watts, 1982; Marsh and Horváth, 1989). The reconstitution process allows the variation of the molar ratio of proteins and lipids over a wide range. A typical pattern of ESR spectral variation with lipid/protein ratio is shown in Figures 2A. In the case of a high protein/lipid ratio, corresponding to approximately a single shell of lipid sites at the hydrophobic interface of proteins, an almost rigid limit line shape is obtained. On adding more lipid molecules, shells of lipids not directly interacting with the protein interface are occupied and a second isotropic spectrum component will dominate the line shape which is very similar to the ESR spectra of freely diffusible labels in protein-free, pure lipid vesicles.

The association can be modeled as an exchange equilibrium between labeled (L^*) and unlabeled (L) lipids competing for N_{solv} solvation layer sites at the protein interface (PL_i) [a detailed discussion is given by Griffith et al. (1982)]:

$$PL_1 \ldots L_i L_{i+1}^* \ldots L_N^* \rightleftharpoons PL_1 \ldots L_{i-1} L_i^* \ldots L_N^* \tag{6}$$

We assume that the N_{solv} sites are occupied independently and so the equilibrium equation for the ith site can be written as

$$P' L_i + L^* \rightleftharpoons P' L_i^* + L \tag{7}$$

The relative binding constant is defined as

$$K_r = \frac{[L][PL_i^*]}{[L^*][PL_i]} \tag{8}$$

where the concentrations are given in mole fractions of lipids in the membrane. Since solvation sites are occupied independently and the total amount of labeled and unlabeled lipids remains unchanged during the exchange event, the mole fractions can be replaced by the respective moles of the two species. If the amount of labeled lipids is $\leq 1\%$ and all of the solvation sites have an identical relative binding constant, from the mass balance equations the following expression can be derived for the ratio of labeled lipids at fluid and solvation sites as a function of lipid-to-protein ratio (Brotherus et al., 1981; Griffith et al., 1982):

$$\frac{[L^*]}{[PL^*]} = \left[\frac{[L]/[P]}{N_{solv}} - 1 \right] \frac{1}{K_r} \tag{9}$$

The construction of the titration diagram shown in Figure 2B requires the measurement of the total lipid/protein molar ratio and the relative fractions of labeled lipids in fluid and solvation sites. According to Equation (9) the y-intercept of the

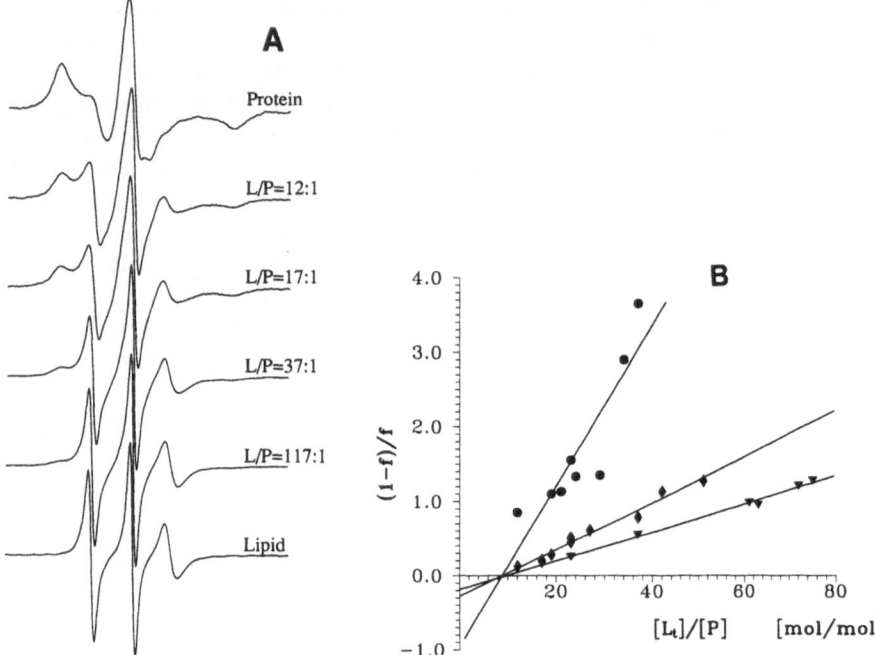

FIGURE 2. Lipid/protein titration experiments with spin-labeled phosphatidic acid (14-PASL) in myelin proteolipid protein/dimyristoylphosphatidylcholine (DMPC) complexes.

(A) 9-GHz ESR spectra as a function of lipid/protein ratio (in mol/mol). For comparison, single-component spectra recorded in a protein-alone and in a fluid lipid sample are shown together with two-component line shapes. The decreasing intensity of the motionally restricted component with decreasing lipid/protein ratio should be noted.

(B) The titration diagram with different spin-labeled lipids in myelin proteolipid protein/dimyristoylphosphatidylcholine complexes. The spectra shown in A were evaluated by difference spectroscopy as described in Section 2. f is the fractional (integrated) ESR intensity of the motionally restricted spectral component [see Equation (10)]. From the x-axis intercept the number of solvation sites, N_{solv}, and from the y-axis intercept the relative association constant, $1/K_r$, can be determined. ▼, spin-labeled stearic acid, 14-SASL; ◆, spin-labeled phosphatidic acid, 14-PASL; ▲, spin-labeled phosphatidylserine, 14-PSSL; ●, spin-labeled phosphatidylcholine, 14-PCSL.

titration curve will yield $-1/K_r$, the x-intercept gives the number of solvation sites N_{solv}, and the slope is $1/(N_{solv}K_r)$.

The two-component ESR spectra of lipid spin labels ubiquitously observed in protein/lipid complexes are quantitatively evaluated by difference spectroscopy (Marsh and Watts, 1982). The basic idea of such analyses is that the fractional (integrated) intensity of each component is proportional to its mole fraction (Wertz and Bolton, 1972):

$$f = \frac{[PL^*]}{[L^*] + [PL^*]} = \frac{I_{\text{solv}}}{I_{\text{tot}}} = \frac{\iint v_{1,\text{solv}}(B)d^2B}{\iint v_{1,\text{tot}}(B)d^2B} \qquad (10a)$$

$$1 - f = \frac{[L^*]}{[L^*] + [PL^*]} = \frac{I_{\text{fluid}}}{I_{\text{tot}}} = \frac{\iint v_{1,\text{fluid}}(B)d^2B}{\iint v_{1,\text{tot}}(B)d^2B} \qquad (10b)$$

where v_1 (B) denotes the first-derivative ESR line-shape function.

If there is no exchange between the two populations, the component spectra can be generated by difference spectroscopy for any fractional distribution, although digital resolution sets a practical limit in the range of $0.1 \leq f \leq 0.9$. This is illustrated in Figure 3A with a computer-simulated two-component spectrum and its simulated single components; a well-defined rigid limit end point can be

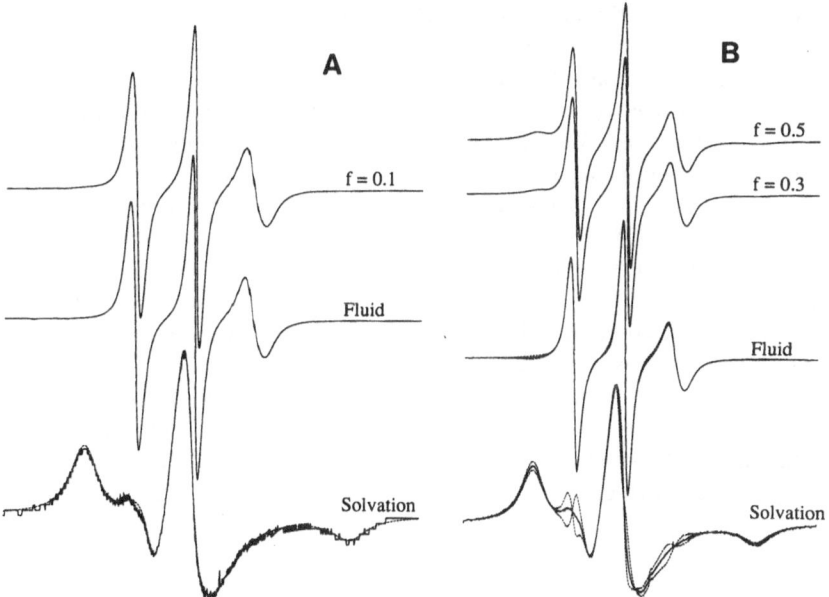

FIGURE 3. Spectral subtractions with simulated two-component ESR spectra using simulated single-component slow motion and motionally averaged fluid lipid reference spectra. f is the fractional (integrated) ESR intensity of the motionally restricted spectral component [see Equation (10)].

(A) The fraction of the solvation component is set to $f = 0.1$. Subtracting an approximate slow motion (e.g., protein-alone) spectrum a fluid lipid end point, or vice versa, subtracting a fluid lipid spectrum a solvation end point is obtained. The respective (integrated) intensities are used to quantify the lipid/protein titration and the lipid selectivity experiments.

(B) The above fluid and solvation end points are obtained by pairwise intersubtractions of two spectra with $f = 0.3$ and 0.5. The fractions of solvation lipids are calculated according to Equations (11) and (12). Typical subtraction errors caused by over/undersubtraction are shown by dashed lines.

obtained by subtracting as much as 90% of the integrated intensity although the rigid limit difference spectrum has a poor digital resolution assuming 12-bit analogue-to-digital acquisition. In the case of experimental spectra the instrumental noise sets a further practical limit unless a signal-to-noise ratio of $100:1$ or better is attained.

A decisive step in spectral subtractions is to what extent the available single-component spectrum (either fluid lipid or rigid limit reference) should be subtracted in order to generate a difference spectrum, which closely matches the line shape of the other (often not available) component. Griffith and Jost (1976) formulated a general end-point criterion: the integrated intensity of the difference spectrum must be nonnegative throughout the scan range. Another difficulty arises because of the ubiquitous exchange between the two lipid populations. As shown below, slow two-site exchange between the two populations will result in line broadening and so, even if a pure lipid spectrum is available, its linewidths do not match exactly those of the fluid component. A commonly accepted solution to this problem is the use of a series of pure lipid spectra recorded at progressively decreasing temperatures in order to match the line broadening related to exchange with that related to slower motional relaxation (Brotherus *et al.*, 1980; Marsh and Watts, 1982). When the two-component spectra are recorded in the fluid phase (e.g., at 30°C in the case of DMPC or at 10°C in the case of egg yolk PC), a temperature correction of 2–5°C yields significantly better rigid limit difference spectra than subtraction of lipid spectra recorded at the same temperature without temperature correction. Needless to say, this temperature correction is only applicable to slow lipid exchanges and its general success in various protein/lipid systems gives a first indication that the frequency of lipid exchanges is in the range of $1-10 \times 10^6 \ \text{sec}^{-1}$.

Two-component spectra can also be analyzed by pairwise intersubtraction provided the fractional intensities of the two components are sufficiently different (Brotherus *et al.*, 1980; Knowles *et al.*, 1981). If the spectral intensities of the two spectra are written as

$$I'_{1,\text{tot}} = f' \, I_{\text{solv}} + (1 - f')I_{\text{fluid}} \tag{11a}$$

$$I''_{2,\text{tot}} = f'' \, I_{\text{solv}} + (1 - f'')I_{\text{fluid}} \tag{11b}$$

The solvation fractions f' and f'' can be determined by subtracting the two spectra from each other as illustrated in Figure 3B. Both fluid and rigid limit end points can be generated with intersubtraction factors f_{12} and f_{21}, respectively, and the solvation fractions can be calculated as follows:

$$f' = \frac{1 - f_{21}}{1 - f_{21}f_{12}} \tag{12a}$$

$$f'' = f'f_{12} \qquad (12b)$$

As in the case of spectral subtractions, digital resolution and instrumental noise will set a practical limit of $(f' - f'') \geq 0.05$.

2.3. Stoichiometry of Lipid/Protein Association

Lipid/protein titration experiments with spin-labeled lipids yield rather detailed insight into the stoichiometry and specificity of lipid association with the hydrophobic interface of integral membrane proteins (Marsh, 1985). Of the two parameters determined from such experiments, namely, the number of solvation sites N_{solv} and the relative association constant K_r, the former will be discussed in the present paragraph, while the latter will be the subject of the next section. A typical lipid/protein titration diagram was shown in Figure 2B. With three different spin-labeled lipids a linear relationship was obtained for the ratio of fluid to solvation fractions as a function of lipid-to-protein ratio conforming to Equation (9). The number of lipid solvation sites is obtained from the intercept on the abscissa giving $N_{solv} = 10$ in the case of the myelin proteolipid protein (Brophy et al., 1984). Other integral membrane proteins that were studied by similar lipid/protein titration experiments include Na$^+$,K$^+$-ATPase (Brotherus et al., 1980; Esmann et al., 1987), Ca^{2+}-ATPase (Thomas et al., 1982; Silvius et al., 1984; East et al., 1985), cytochrome c oxidase (Knowles et al., 1979), acetylcholine receptors (Ellena et al., 1983), rhodopsin (Watts et al., 1979; Pates et al., 1985; Ryba et al., 1987), ADP/ATP carrier (Horváth et al., 1990b), and M13 coat protein (Datema et al., 1987; Wolfs et al., 1989). These results were recently reviewed by Marsh (1985) and Knowles and Marsh (1991).

The solvation shell stoichiometries were interpreted with a geometrical model (Marsh, 1985). As a rule the number of solvation sites correlates with the protein molecular mass M, being greater for the larger proteins and scaling with the square root of the molecular mass $\approx \sqrt{M}$. The correlation diagram constructed from spin-labeling ESR data obtained for the above integral membrane proteins is shown in Figure 4. The experimental points are fitted with a straight line assuming linear dependence as predicted by the geometric model whereby the proteins are approximated as smooth cylinders partly embedded in the lipid bilayer and partly protruding into the aqueous phase and surrounded by a single shell of lipids. Assuming a lipid interchain distance of 4.8 Å and an even distribution of the protein mass between the intramembranous and extramembranous phase, the slope of the correlation curve $N_{solv}/\sqrt{M} \approx 0.11$.

Each protein/lipid complex is seen to conform to the predicted linear dependence of the model, although the experimental points clearly scatter around two distinct lines. Two opposing effects can significantly alter the number of solvation sites at the protein interface as compared with the predictions of the model. First, a ragged surface contour can accommodate more lipid molecules in the

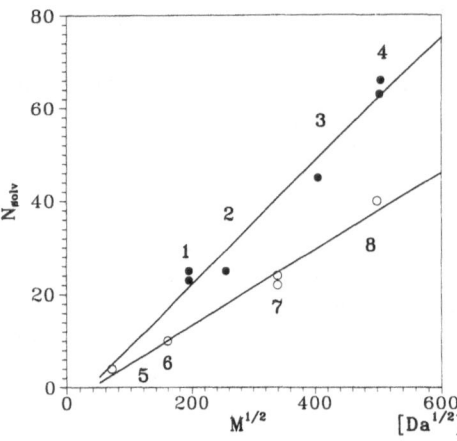

FIGURE 4. Correlation diagram between the number of solvation sites, N_{solv}, and the root of the molecular mass, $M^{1/2}$, for several integral membrane proteins. The experimental points are fitted with straight lines predicted by the geometric model described in Section 2. 1, rod outer segment rhodopsin (Watts *et al.*, 1979); 2, mitochondrial ADP/ATP carrier (Horváth *et al.*, 1990b); 3, acetylcholine receptor (Ellena *et al.* 1983); 4, Na$^+$,K$^+$-ATPase (Esmann and Marsh, 1985); 5, viral M13 coat protein (Wolfs *et al.*, 1989); 6, myelin proteolipid protein (Brophy *et al.*, 1984); 7, Ca^{2+}-ATPase (East *et al.*, 1985); 8, mitochondrial cytochrome oxidase (Knowles *et al.*, 1979).

solvation shell than predicted for a smooth cylinder and there is no compelling evidence that solvation lipids cover homogeneously the entire hydrophobic surface. Second, protein/protein interaction, whenever the interaction energy is greater between the hydrophobic interfaces of proteins in direct contact than that between the protein interface and solvation lipids, can displace a certain fraction of solvation lipids thereby lowering the accessible hydrophobic surface.

Rod outer segment rhodopsin, the mitochondrial nucleotide carrier and cytochrome oxidase, and rectal gland Na$^+$,K$^+$-ATPase belong to the first group of integral membrane proteins whose hydrophobic interface can be described accurately as a smooth cylinder. Of these the nucleotide carrier and the Na$^+$, K$^+$-ATPase (Esmann *et al.*, 1987; Horváth *et al.*, 1989) were shown to form protein dimers and, hence, the good agreement between the measured values and the predicted numbers of solvation sites must be a fortuitous cancellation of the above two effects.

The other four proteins, namely, the viral M13 coat protein, the myelin proteolipid, the Ca^{2+}-ATPase, and the acetylcholine receptor, are all associated with significantly fewer solvation lipids. In the case of the first two proteins this is probably related to the higher degree of protein aggregation; the myelin proteolipid forms hexameric aggregates (Smith *et al.*, 1983), whereas the M13 coat protein, according to freeze-fracture evidence (van Gorkom *et al.*, 1990), forms high-molecular-weight linear arrays. The Ca^{2+}-ATPase and the acetylcholine receptor are representatives of another group of integral membrane proteins for which the overall molecular mass, assuming an even distribution between the hydrophobic and the aqueous phase, grossly overestimates the effective molar volume in the lipid bilayer because of the large extramembranous domains (MacLennan *et al.*, 1985; Toyoshima and Unwin, 1988).

Lipid/protein titration experiments using spin-labeled lipids in conjunction with difference ESR spectroscopy can only be performed in a rather limited range of lipid/protein molar ratios. At high lipid/protein ratios the signal-to-noise ratio of the spectra and occasionally the digital resolution will limit the applicability of the method to the range $f/(1 - f) \leq 6$. At very low lipid/protein ratios, on the other hand, the titration curve saturates at $f/(1 - f) \neq 0$ indicating that at a certain fraction of solvation sites lipids are displaced by the contacting (solvated or unsolvated) hydrophobic interface of another protein molecule. This will set a lower limit of $f/(1 - f) \geq 0.15$ to the applicability of the geometric model. In between these upper and lower bounds the lipid/protein association can be described with a fixed stoichiometry. At very low lipid/protein ratios the stoichiometry gradually falls off as a result of random protein/protein contacts which displace solvation lipids from the protein interface. In this extreme case the concept of fixed stoichiometry is no longer applicable and very often the protein also loses its activity. Molecular dynamics simulation of the lateral distribution of proteins and lipids can quantitatively describe the off-lying data points in the low lipid-to-protein range taking the interaction energy of protein/protein interaction into account (Laidlaw and Pink, 1985).

3.   ELECTROSTATIC ORIGIN OF LIPID SELECTIVITY

The central theme of the present section will be the analysis of lipid selectivity. The role of the charge state of lipid head groups will be illustrated with pH and salt titration experiments. Finally, the reciprocal nature of electrostatic interactions will be highlighted by the anomalous association of negatively charged lipids to the interface of a mutant protein which has fewer positively charged amino acid side chains.

3.1.   Thermodynamic Analysis of Lipid Selectivity

From a lipid/protein titration diagram such as shown in Figure 2B both the number of solvation sites N_{solv} and the relative association constant K_r can be determined. Comparing the association of three different spin-labeled lipids with the myelin proteolipid protein it can be concluded that all three lipids compete for the same set of solvation sites since the number of solvation sites is the same in each case (Brophy et al., 1984). It should be noted that there is one further solvation site on the surface of myelin proteolipid protein which is occupied by a covalently attached fatty acid (for a review see Schulkz et al., 1988). However, the above three lipids exhibit rather different y-axis intercepts and slopes in the titration diagram. Essentially identical results were obtained in every other case studied so far: the preferential association of certain lipids at the hydrophobic interface of integral membrane proteins is a result of changes in the association constant K_r rather than in the number of solvation sites N_{solv} (Brotherus *et al.*, 1980; Knowles *et al.*, 1981; Ellena *et al.*, 1983; Esmann *et al.*, 1987; Wolfs *et al.*, 1989; Horváth *et al.*, 1990b).

An important corollary of this observation is that the relative association constants of various lipids, besides recording a complete lipid/protein titration diagram, can also be determined from the ratio of the solvation/fluid factors, $f/(1 - f)$, measured at a fixed lipid/protein ratio ($[L_t]/[P]$ = const.) as shown in Figure 5A. Then for any lipid (L) the solvation-to-fluid ratio can be written with respect to the phosphatidylcholine (PC) reference as (Marsh, 1985)

$$\frac{[f/(1 - f)]^L}{[f/(1 - f)]^{PC}} = \frac{K_r^{PC}}{K_r^L} \tag{13}$$

The significantly lower amount of purified membrane protein needed for lipid selectivity measurements at a fixed lipid/protein ratio allowed a rather extensive mapping of the different proteins. The selectivity profiles of different lipids at the interface of various integral membrane proteins are summarized in Figure 5B. The selectivity of association is characterized by the effective free energy ΔG given by

$$\Delta G^L - \Delta G^{PC} = - RT\ln \frac{K_r^L}{K_r^{PC}} \tag{14}$$

In the first group of proteins, including the myelin proteolipid protein (Brophy *et al.*, 1984), rectal gland Na^+,K^+-ATPase (Esmann and Marsh, 1985), the mitochondrial ADP/ATP carrier (Horváth *et al.*, 1990b), cytochrome oxidase (Knowles *et al.*, 1981), and the acetylcholine receptor (Ellena *et al.*, 1983), there is a clear selectivity for negatively charged lipids, although their relative association constants vary between $K_r = 7$ and 2. In the second group of proteins, on the other hand, including the Ca^{2+}-ATPase (Silvius *et al.*, 1984) and rhodopsin (Watts *et al.*, 1979), there is significantly less or no selectivity among the various lipids.

The selectivity for negatively charged lipids must arise because of electrostatic interactions between surface charges of the protein interface and lipid head groups. It is not solely determined by lipids; the accessibility of (positively) charged amino acid side chains at the protein interface also plays an important role as seen from the different selectivity profiles of the proteins studied. In some cases the highest selectivity is observed for spin-labeled cardiolipin. The integral membrane proteins of the mitochondrial inner membrane, like cytochrome oxidase and the adenine nucleotide carrier, clearly fall into this category. In other

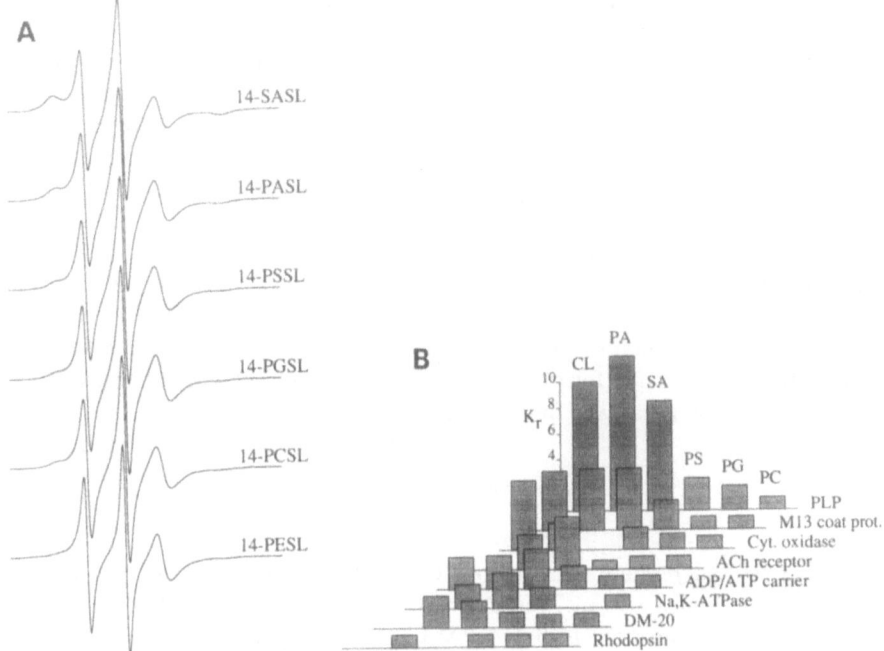

FIGURE 5. Lipid selectivity experiments using different spin-labeled lipids in various lipid/protein complexes.

(A) ESR spectra of different spin-labeled lipids in myelin proteolipid protein/dimyristoylphosphatidylcholine complexes of lipid/protein ration 23:1 mol/mol. The lipid selectivity is quantified by the relative association constant [for definition see Equation (13)] derived from spectral subtraction data.

(B) The selectivity pattern of different spin-labeled lipids at the hydrophobic interface of various integral membrane proteins as determined by single point experiments. CL, cardiolipin; PA, phosphatidic acid; SA, stearic acid; PS, phosphatidylserine; PG, phosphatidylglycerol; PC, phosphatidylcholine. References to the different proteins are given in Figure 4.

cases, however, the selectivity for cardiolipin is more modest and other negatively charged lipids, like phosphatidic acid and stearic acid, are preferred in the solvation shell. The reciprocity in electrostatic interactions between proteins and lipids in the interfacial region will be demonstrated in two different ways. First, the lipid head group charge will be altered by pH and salt titration experiments demonstrating that protonation at low pH or electrostatic screening by monovalent cations leads to a (partial) loss of selectivity. Second, the role of charged groups at the protein interface will be discussed by comparing two partially homologous proteins of myelin which differ in their amino acid sequences.

3.2.   Influence of Lipid Head Group Charge on the Selectivity

By changing the pH or monovalent cation concentration it is possible to vary the relative fractions of fluid and solvation lipids, essentially on the same sample without changing either protein/lipid ratio or spin label (Horváth et al., 1988b). From quantitative difference spectroscopy by the relative fractions were determined in the pH range of 6–9 and in the salt concentration range of 0–1 M; the results are shown in Figure 6 for three different negatively charged lipids and the phosphatidylcholine control. Rather similar titration curves were obtained for spin-labeled lipids at the interface of Na^+,K^+-ATPase (Esmann and Marsh, 1985).

With the variation of pH a very clear titration curve was obtained in each case (the titration curve for phosphatidylserine being only partial because of the limited pH range) that was fitted to the titration equation:

$$f = f^{min} + \frac{f^{max} - f^{min}}{1 + [H^+]/K_a} \tag{15}$$

where f^{max} and f^{min} are the fractions of solvation component for the deprotonated and protonated forms, respectively, and $pK_a = -\log K_a$. For stearic acid (14-SASL), phosphatidic acid (14-PASL), and phosphatidylserine (14-PSSL), the following limiting values were obtained: $f^{max} = 0.74, 0.75$, and 0.74; $f^{min} = 0.46, 0.64$, and 0.48; and $pK_a = 7.7, 7.6$, and 9.4, respectively. Because phosphatidylserine is only partially deprotonated at pH 9.6, the above values should be regarded as somewhat approximate.

The salt dependence of the selectivity shown in Figure 6B can be interpreted, at least semiquantitatively, by using the Debye–Hückel theory of electrolytes (Esmann and Marsh, 1985; Horváth et al., 1988b). The activity coefficients of the various charged lipids and proteins are given by

$$\ln\gamma_i = \frac{-Z_i^2 e^2}{8\pi\epsilon_0\epsilon kT} \frac{\kappa}{1 + \kappa a_i} \tag{16}$$

where Z_i is the charge on species i, a_i is the effective interaction distance of species i with counterions, and the inverse Debye screening length is defined as

$$\kappa^2 = \frac{Ne^2I}{\epsilon_0\epsilon kT} \tag{17}$$

I being the ionic strength of the electrolyte, and other symbols have their usual meanings. In terms of the relative association constant K_r^0, defined by the activ-

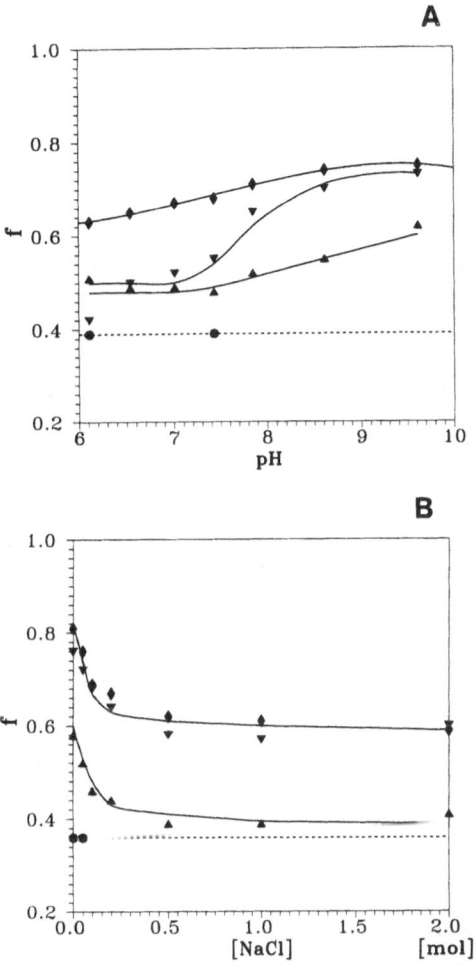

FIGURE 6. The pH and salt dependence of the fraction of different spin-labeled lipids in myelin proteolipid protein/dimyristoylphosphatidycholine complexes of protein/lipid ratio 1:37 mol/mol (Horváth *et al.*, 1988b). ▼, spin-labeled stearic acid, 14-SASL; ◆, spin-labeled phosphatidic acid, 14-PASL; ▲, spin-labeled phosphatidylserine, 14-PSSL; ●, spin-labeled phosphatidylcholine, 14-PCSL.

(A) The fraction of the solvation component as a function of pH. The pH dependence is fitted with the titration equation [Equation (15)].

(B) The fraction of the solvation component as a function of monovalent salt concentration. The salt titration is semiquantitatively described by the Debye–Hückel theory of electrolytes (Section 3).

ities rather than concentrations [see Equation (8)], the ratio of relative association constants in the presence and absence of salt is given by

$$\ln \frac{K_r}{K_r^o} = - \frac{e^2\kappa}{8\pi\epsilon_o\epsilon kT} \left[\frac{Z_L^2}{1 + \kappa a_L} + \frac{Z_P^2}{1 + \kappa a_P} + \frac{(Z_P + Z_L)^2}{1 + \kappa a_L a_P} \right] \quad (18)$$

For the sake of simplicity it is assumed that the charge on the lipid and the net charge on the protein approximately cancel and so the last term in Equation (18) can be neglected. The effective protein radius can be taken to be $a_p = 1.65$ nm for myelin proteolipid protein and the effective lipid interaction radius a_L is treated as an adjustable parameter to fit the observed salt dependence. For the above three negatively charged lipids the following value was obtained: $a_L = 0.15$ nm (Horváth et al., 1988b), in good agreement with previous estimates from salt titration data for the same spin labels in Na^+,K^+-ATPase (Esmann and Marsh, 1985) and reconstituted cytochrome oxidase complexes (Powell et al., 1987).

3.3. Influence of Polar Residue Deletions at the Protein Surface

The pH and salt titration experiments clearly demonstrate that the origin of lipid selectivity is electrostatic interactions between lipid head group charges and polar amino acids at the protein interface. A direct approach to the problem of identifying the protein residues involved and, hence, proving the reciprocity of electrostatic interactions would be to investigate the effects of site-directed mutagenesis on the lipid selectivity pattern. The myelin proteolipid protein is ideally suited for such experiments, since a naturally occurring alternative form is available in myelin (Stoffel et al., 1984; Nave et al., 1987; Simons et al., 1987). The native protein is 276 residues in length, and its alternative form, the so-called DM-20 protein, is a homologous integral membrane protein which is 241 residues in length. The two proteins are expressed from the same nucleic acid region: the proteolipid protein expression consists of seven exons, while DM-20 expression derives from the same seven but one exons and so the difference between these two proteins consists of a long deletion of 35 consecutive amino acids corresponding to residue positions 116–150 in the native protein. According to the currently accepted structural model of the proteolipid protein (Stoffel et al., 1984), this sequence is contained in the largest extracytosolic loop and is highly polar in nature (3 Arg, 2 Lys, 1 Glu, and 1 Asp groups). A comparison of the lipid selectivity patterns of the proteolipid protein and DM-20 allowed determination of the effects of this highly basic portion (Horváth et al., 1990a).

The lipid selectivities of the myelin DM-20 and proteolipid protein were studied by the different spin-labeled lipids in complexes of fixed protein/lipid ratio. As shown in Figure 7 there is a great degree of difference between the two

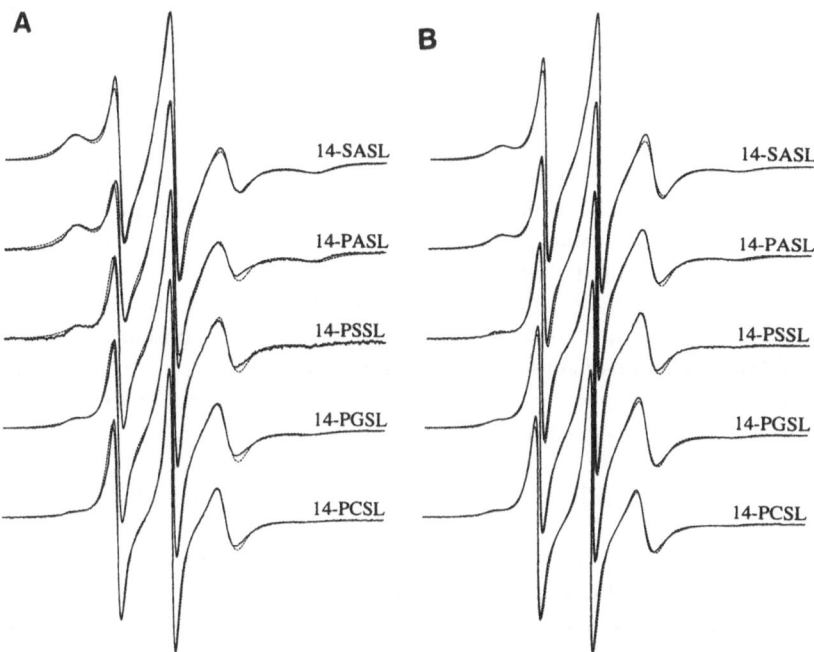

FIGURE 7. ESR spectra of different spin-labeled lipids in myelin protein/lipid complexes. The lipid/protein ration is 28:1 mol/mol in both cases and the spectra are recorded at T = 30°C (from Horváth *et al.*, 1990a). The experimental spectra (solid lines) are shown together with best-fitting two-component simulations (dashed lines). The off-rates derived from spectral simulations are shown in Figure 9.

(A) 9-GHz ESR spectra of various spin-labeled lipids incorporated into myelin (proteolipid protein + DM-20) /dimyristoylphosphatidylcholine complexes. The ratio of the two protein components is 3.5:1 mol/mol.

(B) 9-GHz ESR spectra of various spin-labeled lipids incorporated into purified DM-20 /dimyristoylphosphatidylcholine complexes.

proteins particularly in the case of acidic lipids (14-SASL and 14-PASL), which show the highest selectivity for both proteins, while the associations of lipids of low selectivity (14-PGSL and 14-PCSL) are practically identical. The observed change in selectivity is consistent with the structural alterations since the positively charged residues are exposed on the membrane interface and their deletion is expected to decrease the relative association constants of acidic lipids.

In the lipid selectivity experiments described in the previous section the natural mixture of the two proteins was investigated and so the protein associated fraction in the mixed (proteolipid + DM-20/lipid complexes is a weighted average of the solvation components from these two proteins. Assuming that the proteolipid and DM-20 protein molecules are solvated independently, the observed fraction can be given in terms of the fractions associated with DM-20 and the proteolipid protein (PLP), f_{DM-20} and f_{PLP}:

$$f_{obs} = X_{DM-20}f_{DM-20} + X_{PLP}f_{PLP} \qquad (19)$$

where the mole fractions of the two proteins are $X_{DM-20} = 0.29$ and $X_{PLP} = 0.71$. From subtraction analyses of the two spectrum series shown in Figure 7 the relative association constants of both proteins can be determined and for the two negatively charged lipids (14-SASL and 14-PASL) the following values were found: $K_r^{SA} = 2.2$ and 6.5, and $K_r^{PA} = 2.5$ and 10.4, respectively (Horváth et al., 1990a). This difference between the relative association constants of the proteolipid protein and its natural alternative form (the DM-20 protein) corresponds to a decrease in free energy of interaction of ≈ 1 kJ/mol per charge.

4. LINE SHAPE EFFECTS OF SOLVATION-TO-FLUID EXCHANGE

The motional effects of exchange dynamics on the ESR line shape will be described by the modified Bloch equations. An optimized spectrum simulation algorithm will be applied to the analysis of two-component ESR spectra. As a result a quantitative relationship will be derived for the correlation between lipid selectivity and exchange dynamics.

4.1. Modified Bloch Equations

The spectral effects of two-site exchange can be described by incorporating the rate equations into the phenomenological Bloch equations of spin magnetization (McConnell, 1958; Davoust and Devaux, 1982; Marsh and Horváth, 1989). The rate equations for the exchange of spin magnetization between the solvation and fluid populations are given by

$$\frac{dM_{solv}}{dt} = -\frac{M_{solv}}{\tau_{off}} + \frac{M_{fluid}}{\tau_{on}} \qquad (20a)$$

$$\frac{dM_{fluid}}{dt} = \frac{M_{solv}}{\tau_{off}} - \frac{M_{fluid}}{\tau_{on}} \qquad (20b)$$

where τ_{off}^{-1} and τ_{on}^{-1} are the probabilities per unit time of transfer from a solvation site to a fluid site and vice versa, respectively. The material balance [see Equation (7)] can be rewritten as

$$\frac{\tau_{off}^{-1}}{\tau_{on}^{-1}} = \frac{M_{fluid}}{M_{solv}} = \frac{1-f}{f} \qquad (21)$$

At exchange equilibrium $dM_{solv}/dt = dM_{fluid}/dt = 0$ and so the steady-state Bloch equations for the complex transverse magnetizations, M_{solv} and M_{fluid}, can be written from Equation (20) as

$$[(\omega - \omega_{solv}) + i\tau_{off}^{-1}]M_{solv} - i\tau_{on}^{-1}M_{fluid} = \gamma B_1 M_o f \qquad (22a)$$

$$[(\omega - \omega_{fluid}) + i\tau_{on}^{-1}]M_{fluid} - i\tau_{off}^{-1}M_{solv} = -\gamma B_1 M_o(1 - f) \qquad (22b)$$

where ω_{solv} and ω_{fluid} are the complex angular resonance frequencies of the two components, respectively: $\omega_{solv} = \omega_{o,solv} - iT_{2,solv}^{-1}$ and $\omega_{fluid} = \omega_{o,fluid} - iT_{2,fluid}^{-1}$. Other notations are the following: M_o is the equilibrium value of magnetization, B_1 is a small circularly polarized magnetic field which is applied perpendicular to the static magnetic field and rotating at the angular resonance frequency ω, and γ is the gyromagnetic ratio. From Equation (21) the total complex magnetization, $M = M_{solv} + M_{fluid}$, can then be calculated:

$$M = M_{solv} + M_{fluid}$$

$$= -\gamma B_1 M_o \frac{fL_{solv} + (1 - f)L_{fluid} + L_{solv}L_{fluid}[f\tau_{off}^{-1} + (1 - f)\tau_{on}^{-1}]}{1 - L_{solv}L_{fluid}\tau_{off}^{-1}\tau_{on}^{-1}} \qquad (23a)$$

$$L_{solv} = [(\omega - \omega_{solv}) + i(T_{2,solv}^{-1} + \tau_{off}^{-1})]^{-1} \qquad (23b)$$

$$L_{fluid} = [(\omega - \omega_{fluid}) + i(T_{2,fluid}^{-1} + \tau_{on}^{-1})]^{-1} \qquad (23c)$$

The intensity of the ESR signal is proportional to the imaginary part of M. In the case of very slow exchange the spectra consist of a weighted sum of the solvation and fluid components, each centered at its resonance frequency with a line shape corresponding to the intrinsic linewidth. For slow rates of exchange ($\tau_{off}^{-1}, \tau_{on}^{-1} \ll \omega_{o,solv} - \omega_{o,fluid}$) the linewidths are increased by the exchange lifetime:

$$T_{2,eff}^{-1}(solv) = T_{2,solv}^{-1} + \tau_{off}^{-1} \qquad (24a)$$

$$T_{2,eff}^{-1}(fluid) = T_{2,fluid}^{-1} + \tau_{on}^{-1} \qquad (24b)$$

and the lines begin to move together. Because of the incipient line broadening caused by slow exchange an equivalent temperature correction was applied to the reference spectra in all of the subtractions described in Section 2. As the exchange frequency is further increased the spectral lines collapse to a single, broad line centered about the mean position. In the case of fast exchange ($\tau_{off}^{-1}, \tau_{on}^{-1} \gg \omega_{o,solv} - \omega_{o,fluid}$) the linewidth is the mean of the two linewidths, but with an additional contribution from lifetime broadening related to exchange.

The observed ESR spectra are more complex than predicted by Equation (23) since, in protein/lipid vesicles, there is an isotropic distribution of director orientations with respect to the static spectrometer magnetic field. Hence, the ESR spectrum is a weighted sum of spectra corresponding to different orientations, the weighing factor being $p(\vartheta)d\vartheta = \sin\vartheta d\vartheta$. To simulate the entire line shape, a summation must be made over angular orientations; the line shapes corresponding to different nuclear manifolds ($m_I = -1, 0, +1$) are calculated separately. Depending on the nature of the reorientation of the molecular axes during the exchange process, two cases can be distinguished: the correlated and the uncorrelated model (Davoust and Devaux, 1982).

In the correlated model the exchange event has no effect on the instantaneous molecular orientation; the angular orientation of the magnetic tensors remains correlated during the solvation-to-fluid jump. In this instance, only transitions corresponding to the same orientations are connected by exchange $[\omega_{fluid}(\vartheta_i)$ and $\omega_{solv}(\vartheta_i)]$. The overall spectrum is then a weighted sum of the exchange averaged line shapes given by Equation (23), with an isotropic distribution of orientations. In the uncorrelated model the angular orientation of the spin label is uncorrelated before and after the exchange event, i.e., there is a random reorientation of the lipid chains on collision with the ragged protein interface. In this instance, the exchange process takes place between transitions in the fluid component, with a resonant frequency $\omega_{fluid}(\vartheta_i)$, and a distribution of transitions in the solvation component, with resonant frequencies $\Sigma\omega_{solv}(\vartheta_j)$. Again, the summation must be made over an isotropic distribution of orientations ϑ_i.

4.2.   Two-Site Exchange as a Function of the Protein/Lipid Ratio

The spectral parameters needed for exchange simulations can be adjusted in two stages (Horváth et al., 1988a; Marsh and Horváth, 1989). First, fluid lipid and solvation lipid reference spectra are compared with simulated single-component spectra in order to set the principal values of the g and A tensors and to find the best matching linewidth parameters ($T_{2,m}^{-1}$). Motionally restricted end points obtained from spectral subtraction as outlined in Section 2 are useful for adjusting the line shape of the solvation component, whenever delipidated protein-alone spectra are not available. The fluid lipid spectrum used to establish the best-fitting simulation parameters is recorded at the same temperature as the two-component spectrum of the protein/lipid system. A typical pair of best-fitting solvation and fluid line shapes is shown together with (pure) fluid lipid and delipidated protein-alone spectra in Figure 8A.

In the second stage of simulation, only the fraction of the solvation component and the exchange frequency are varied. The effect of these two parameters is illustrated with simulated two-component spectra and an experimental spectrum for a myelin proteolipid protein/lipid recombinant in Figure 8B. An approximate best fit is obtained with an exchange rate in the region of $f \approx 0.7$ and $\tau_{of}^{-1} \approx 5 \times$

10^6 sec^{-1}; at 0 and 10^7 sec^{-1} less satisfactory agreement is obtained. A precise estimate of the exchange requires an objective assessment of the goodness of fit and so a steepest descent algorithm has been used to minimize the sums of squares of the differences between simulated (Y_{sim}) and experimental lines shapes (Y_{exptl}):

$$\Delta Y = \frac{\Sigma_i (Y_{sim,i} - Y_{exptl,i})^2}{\Sigma_i Y^2_{exptl,i}} \tag{25}$$

A typical root mean square (rms) error contour diagram, as a function of the parameters f and τ^{-1}_{off}, is shown in the inset of Figure 8B. According to this contour plot a well-defined minimum is obtained, in the above case, for the following parameter values: $f = 0.71$ and $\tau^{-1}_{off} = 4.8 \times 10^6$ sec^{-1}.

The complete spectrum series of lipid spin labels in myelin proteolipid protein/lipid complexes of varying protein/lipid ratio (see Figure 2A) could be simulated with the same single-component spectra and the exchange off-rate constant τ^{-1}_{off} showed only insignificant variations (Horváth et al., 1988a; Wolfs et al., 1989). Spectral simulations for different lipid spin labels in various integral membrane protein/lipid recombinants of varying protein/lipid ratio have led to the general conclusion that the off-rate τ^{-1}_{off} for the solvation lipid population is constant, independent of the total protein/lipid ratio (Marsh and Horváth, 1989). It is found, however, that the lipid selectivity is rather significantly reflected in the different off-rate constants as discussed in the next section.

The off-rate τ^{-1}_{off} of solvation lipids in all of the protein/lipid systems studied so far lies in the range $\tau^{-1}_{off} \approx 0.8$–$1.5 \times 10^7$ sec^{-1} at 30°C for nonselective lipids. These values are significantly lower than the translational diffusion rates of lipids in protein-free lipid bilayers ($\tau^{-1}_{diff} \approx 7.5 \times 10^7$ sec^{-1}; Vaz et al., 1985; Clegg and Vaz, 1985).

4.3. The Lipid Selectivity of Two-Site Exchange

According to exchange dynamics data derived from spectral simulations, all of the solvation lipids are exchanged at a rate of 10^6–10^7 sec^{-1} by the laterally diffusing lipids. The off-rates τ^{-1}_{off} are modulated by the lipid selectivity; negatively charged lipids of high selectivity are exchanged more slowly. This is illustrated with the off-rate data for various spin-labeled lipids in myelin proteolipid protein/DMPC recombinants of varying protein/lipid ratio (Figure 9). As the selectivity for negatively charged lipids was suppressed by pH and salt titration the exchange dynamics of these lipids (notably stearic acid, phosphatidic acid, and phosphatidylserine) became considerably faster (Horváth et al., 1988b). Essentially the same trends were deduced from the respective selectivity and exchange dynamics of spin-labeled lipids at the interface of myelin proteolipid protein and that of its natural mutant (DM-20; Horváth et al., 1990a).

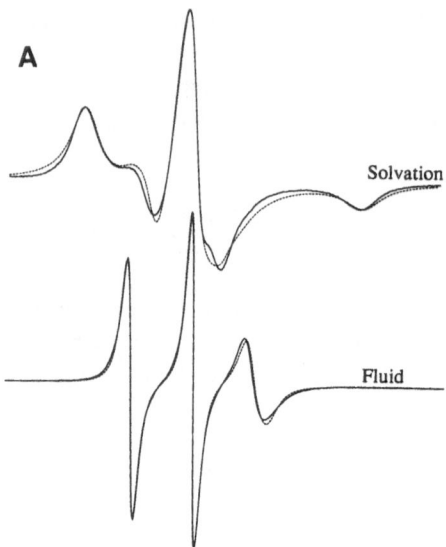

A

Solvation

Fluid

FIGURE 8. Strategy of two-component line-shape simulations using the exchange-coupled Bloch equation algorithm. The best-fitting simulated line shapes are selected in two stages: first, the line shapes of the two spectral components and, then, their relative weights are adjusted by least-squares fitting algorithm.

(A) ESR spectra of spin-labeled phosphatidic acid, 14-PASL, bound to myelin proteolipid protein aggregates and incorporated into fluid dimyristoylphosphatidylcholine bilayers. Both ESR spectra (solid lines) are shown together with best-fitting single-component simulated spectra (dashed lines). All subsequent two-component line shapes shown in Figures 7 and 8, are fitted with various proportions of these two components.

The two variables that can be determined from two-component line-shape simulations are the solvation-to-fluid exchange rate τ_{off}^{-1} and the fraction of the solvation component f. The connection between these two quantities is determined by the material balance equation [Equation (21)]. Substituting this condition into the equation for equilibrium association [Equation (9)], the following relationship is obtained (Marsh and Horváth, 1989):

$$\frac{\tau_{off}^{-1}}{\tau_{on}^{-1}} = \left[\frac{[L_t]/[P]}{N_{solv}} - 1 \right] \frac{1}{K_r} \tag{26}$$

This general equation predicts a linear relationship between the off-rate τ_{off}^{-1} and the reciprocal relative association constant $1/K_r$ whenever the on-rate τ_{on}^{-1} is constant and the exchange dynamics data are compared at fixed protein/lipid ratio. The constancy of the on-rate was verified in the pH and salt titration experiments (Horváth *et al.*, 1988b). In addition, measurements of the collision rates between diffusible ^{14}N-labeled lipids and ^{15}N-labeled chains covalently

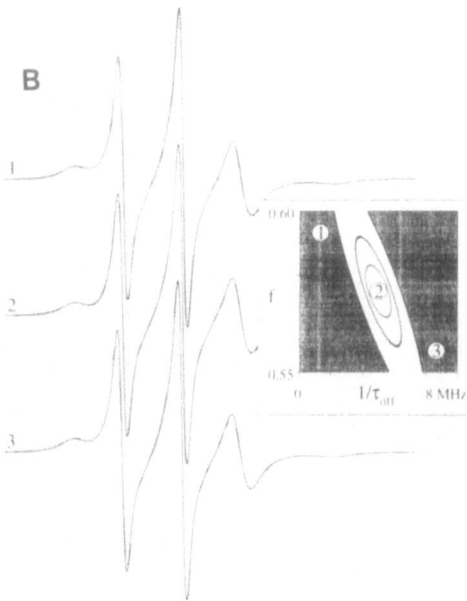

FIGURE 8. (*Continued*)

(B) Comparison of experimental (solid lines) and simulated two-component ESR spectra (dashed lines) for various values of the off-rate. In trace 1 the exchange rate is set to 0 MHz; the line shape of the fluid component and, hence, the resolution between the two components does not match the experimental curve. In trace 3 the exchange rate is set to 8.4 MHz and, as judged from the linewidths, this is an overestimation of the exchange rate. In trace 2 the best-fitting two-component simulation is shown together with the experimental spectrum which has been selected by least-squares optimization. In the inset a contour plot of the error function is shown to illustrate the global minimum which defines an objective criterion for best fitting.

linked to the protein, have demonstrated directly that the lipid on-rate is diffusion controlled (Davoust and Devaux, 1982; Davoust *et al.*, 1983).

This correlation plot between $1/K_r$ and τ_{off}^{-1} is shown in Figure 9 for different spin-labeled lipids in DM-20/DMPC and mixed (proteolipid protein + DM-20)/DMPC complexes, and the data are seen to conform to the inverse relation predicted by Equation (26).

5. SLOWLY EXCHANGING SOLVATION LIPIDS

This section will highlight three exceptional cases in which certain lipids of the solvation layer are exchanged anomalously slowly. The contribution of lipid unsaturation in the hydrophobic region to selectivity will be demonstrated by parallel biochemical and spin-labeling ESR experiments. In the last part, slowly

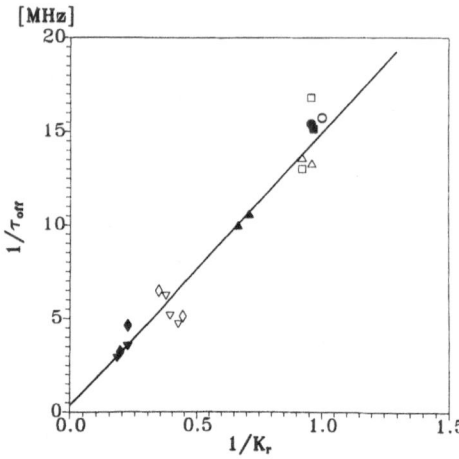

FIGURE 9. Correlation diagram between the off-rate constants, τ_{off}^{-1} and the relative association constants, K_r, for different spin-labeled lipids in myelin (proteolipid protein + DM-20)/dimyristoylphosphatidylcholine (closed symbols) and DM-20/dimyristoyl-phosphatidylcholine (open symbols) complexes of lipid/protein ration 28:1 mol/mol. ∇, $\blacktriangledown$, spin-labeled stearic acid, 14-SASL; $\diamond$, $\blacklozenge$, spin-labeled phosphatidic acid, 14-PASL; $\triangle$, $\blacktriangle$, spin-labeled phosphatidyl-serine, 14-PSSL; $\blacksquare$, spin-labeled phosphatidylglycerol, 14-PGSL; $\bigcirc$, $\bullet$, spin-labeled phosphatidylcholine, 14-PCSL.

exchanging lipids will be identified by ^{2}H-NMR experiments and their exchange rate will be measured by NMR line-shape simulations.

5.1. Cardiolipin Exchange at the Mitochondrial Nucleotide Carrier

Two-component spectral simulations for different spin-labeled lipids in various integral membrane protein/lipid recombinants led to the general conclusion that solvation lipids at the hydrophobic interface of proteins are rapidly exchanged by the laterally diffusing fluid lipids on the submicrosecond time scale (Horváth *et al.*, 1988a,b; Marsh and Horváth, 1989). However, there must be exceptions since, in certain cases, a few moles of endogenous lipids are very difficult to remove from the protein interface on solubilizing the membrane. One such example is the mitochondrial inner membrane which is rich in cardiolipin. This diphosphatidylglycerol copurifies with several integral membrane proteins and its total removal very often leads to a drastic decrease or complete irreversible loss of enzymatic activity (Awasthi *et al.*, 1971; Sandermann, 1978; Kadenbach *et al.*, 1982). In the present section three lines of spectroscopic evidence are presented for the presence of a few moles of slowly exchanging solvation lipids (mostly cardiolipins) among the other rapidly exchanging solvation lipids.

The ADP/ATP carrier, the most abundant integral protein of the inner mitochondrial membrane, shows high selectivity for the negatively charged lipids as shown in Figure 5B. The exchange rates at the lipid/protein interface display the expected dependence on protein/lipid ratio, selectivity, and ionic strength (Horváth *et al.*, 1990b). The off-rates of solvation lipids τ_{off}^{-1} are modulated by the lipid selectivity as demonstrated above for the myelin proteolipid

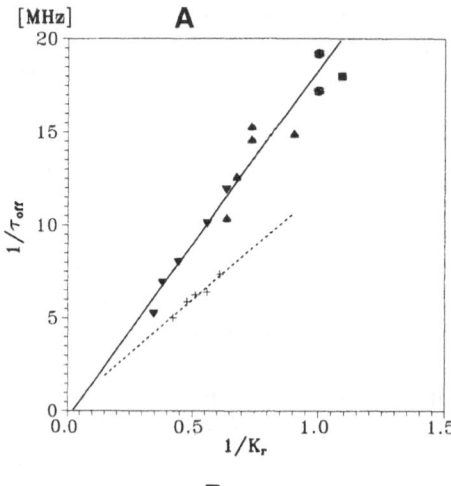

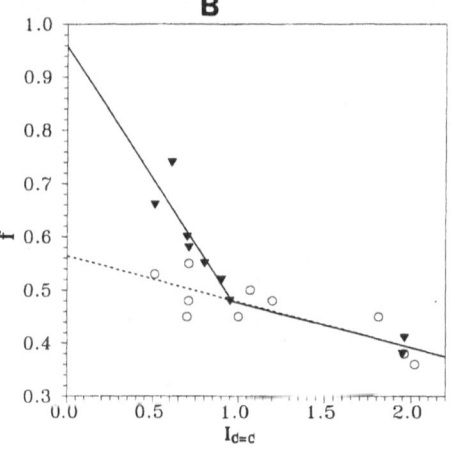

FIGURE 10. Two examples for slowly exchanging lipids in the solvation layer of mitochondrial proteins.

(A) Correlation diagram between the off-rate constants, τ_{off}^{-1}, and the relative association constants, K_r, for different spin-labeled lipids in mitochondrial ADP/ATP carrier/egg phosphatidylcholine complexes of lipid/protein ratio 106:1 mol/mol. This diagram is compiled from the results of two-component simulations; the different selectivities for a particular label correspond to different salt concentrations (Horváth et al., 1990b). ▼, spin-labeled stearic acid, 14-SASL; ♦, spin-labeled phosphatidic acid, 14-PASL; ▲, spin-labeled phosphatidylserine, 14-PSSL; ■, spin-labeled phosphatidylglycerol, 14-PGSL; ●, spin-labeled phosphatidylcholine, 14-PCSL; +, spin-labeled cardiolipin, 14-CLSL.

(B) The variation of fractions, f, of solvation lipids in native mitochondria after various degrees of lipid hydrogenation using spin-labeled phosphatidylcholine (14-PCSL, ○) and stearic acid (14-SASL, ▼). The double bond index $I_{C=C}$ is calculated from the fatty acid patterns of the major phospholipids and the fraction of solvation lipids is determined by difference ESR spectroscopy (Schlame et al., 1990).

protein. From the two-component simulation data of the lipid selectivity series and from the salt titration with four different spin-labeled lipids, a τ_{off}^{-1} versus $1/K_r$ correlation plot was constructed and all lipids were found to conform to the inverse relation predicted by Equation (26), except spin-labeled cardiolipin (Figure 10A). This lipid displayed significantly slower off-rates for any given value of $1/K_r$. This indicates that part of the cardiolipin is associated at more highly selective sites for which the exchange rate is much slower than at other solvation sites. Assuming that the exchange rate at these specific sites is too low to produce any detectable line broadening of the ESR spectrum ($\tau_{off}^{-1} > 10^6$ sec^{-1}), it can be estimated that approximately 30% of the spin-labeled cardiolipin (corresponding to approximately 16 moles cardiolipin/protein monomer) is associated at such sites. These spin-labeling ESR results are consistent with ^{31}P-NMR studies which have indicated that 6 moles of cardiolipin are tightly associated with the detergent-solubilized ADP/ATP carrier while other phospholipids are in fast exchange (Beyer and Klingenberg, 1983).

5.2. Selectivity for Unsaturated Cardiolipin in Mitochondria

Like the adenine nucleotide carrier, several other mitochondrial proteins were shown to copurify with cardiolipin or to possess a functional requirement for this lipid. Of these proteins, as discussed in Section 3, cytochrome oxidase and the ADP/ATP carrier are associated with high specificity with negatively charged lipids and, in particular, with cardiolipin according to spin-labeling ESR results. On incorporating spin-labeled lipids into the native mitochondrial membrane a rather similar lipid selectivity pattern is obtained, the average relative association constants being dominated by the most abundant adenine nucleotide carrier. The origin of the high selectivity for cardiolipin is related to, in part, the negative head-group charge and, in part, the high degree of unsaturation in the acyl chain region (Schlame et al., 1990). It is demonstrated that a certain fraction of cardiolipin, but not the other phospholipids, is tightly associated with integral proteins because of its unsaturated acyl chains.

The major phospholipids of the mitochondrial inner membrane, namely phosphatidylcholine, phosphatidylethanolamine, and cardiolipin, are highly unsaturated; catalytic hydrogenation (Chapman and Quinn, 1976; Quinn et al., 1989) affects these lipids to different extents according to a rather specific pattern (Schlame et al., 1990). First, phosphatidylcholine and phosphatidylethanolamine are hydrogenated and, then, cardiolipin is saturated significantly later. The same pattern cannot be observed while hydrogenating extracted lipids in protein-free vesicles, indicating that cardiolipin is not accessible to the catalyst in the native membrane under mild hydrogenation conditions.

In parallel ESR experiments the distribution of spin-labeled lipids between solvation and fluid sites was followed and a gradually altering lipid selectivity could be demonstrated (Schlame et al., 1990). As shown in Figure 10B, stearic acid and phosphatidylcholine display nearly identical specificity under mild hydrogenation conditions (double bond index decreased from 2 to 1), while more extensive hydrogenation, which also affects cardiolipin, leads to an augmentation of stearic acid in the solvation layer (double bond index $I_{C=C} < 1$). The lipid composition and spin-labeling ESR data were interpreted assuming that cardiolipin (1) is associated with high selectivity with integral proteins and (2) is not accessible to catalytic hydrogenation while in the solvation layer. Since no significant alteration could be detected in the number of acyl double bonds of cardiolipin under mild hydrogenation conditions, it was concluded that the majority of the cardiolipin molecules were not taking part in the rapid exchange of other phospholipids. However, more extensive hydrogenation displaced this residual cardiolipin pool from the protein interface. Hydrogenation resulted in a reduced protein selectivity and consequently in an increase in proportion of the (fully saturated) stearic acid that was found to be associated with protein in these membranes. A corollary of this assumption is that the slowly exchanging cardiolipin pool could also be displaced from the protein interface and became prone to hydrogenation in the fluid regions under extensive hydrogenation conditions. Solvation sites, in such a case, were occupied by stearic acid with steady medium selectivity in contrast to the hydrogenated host lipids of gradually decreasing selectivity. This is indirect evidence that the high selectivity for cardiolipin must arise mostly because of its unsaturated acyl chains and only to a smaller extent because of the negative head-group charge.

5.3. NMR Study of Slowly Exchanging Trapped Solvation Lipids

The coat protein of the bacteriophage M13 represents another example of the occurrence of slowly exchanging solvation lipids. During the infection period, parental and newly synthesized coat protein are stored as an integral membrane protein in the cytoplasmic membrane with a strong tendency to form elongated protein aggregates (van Gorkom et al., 1990). One possible packing arrangement for aggregation which is consistent with the above freeze fracture results is a quasilinear aggregate. Lipid molecules can now be situated in one of three different environments: at the outside or between the closely packed parallel strands of linear protein/lipid aggregates, or in the bulk fluid lipid phase. Implicit in this distinction between solvation lipids at the perimeter of the protein aggregate and those trapped between the parallel strands is that nearest-neighbor protein monomers must share a single layer of solvation lipids entrapping thereby a certain fraction of solvation sites between the parallel strands.

The exchange dynamics of lipids in these two kinds of solvation sites is rather different but, on the spin-labeling ESR time scale, both types of protein-associated lipids give rise to a similar solvation component without discrimination between either type of motional restriction. The sole indication for the presence of slowly exchanging lipids is the unusually slow off-rate of the nonspecific spin-labeled phosphatidylcholine ($\tau_{\text{solv}}^{-1} \approx 4\text{--}7 \times 10^6 \text{ sec}^{-1}$). However, in deuterium NMR spectroscopy, the time scale as defined by the quadrupole splitting anisotropy [see Equation (2)] is sensitive to much slower motions $\approx 10^{-3}$ sec. Using DMPC deuterated in the trimethyl segments of the choline head group (DMPC-d_9), two-component ^{2}H-NMR spectra were obtained over a wide range of protein/lipid ratios (Figure 11A). The first component is a motionally averaged axially symmetric spectrum of a quadrupole splitting $\Delta\nu_q = 1150$ Hz typical for lipid systems in the fluid phase. The second isotropic spectral component ($\Delta\nu_q = 0$ Hz) increases with increasing protein concentration and is very similar in shape at all protein concentrations.

Comparing the lipid/protein titration diagrams deduced from spin-labeling

A

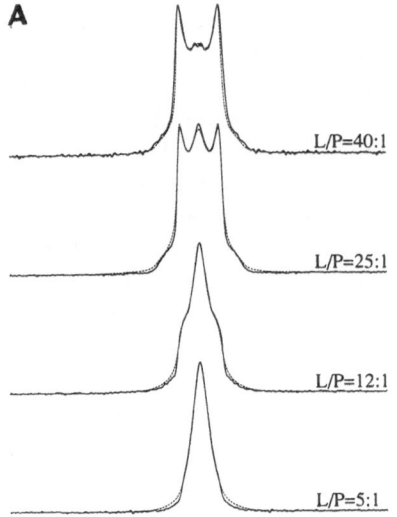

L/P=40:1

L/P=25:1

L/P=12:1

L/P=5:1

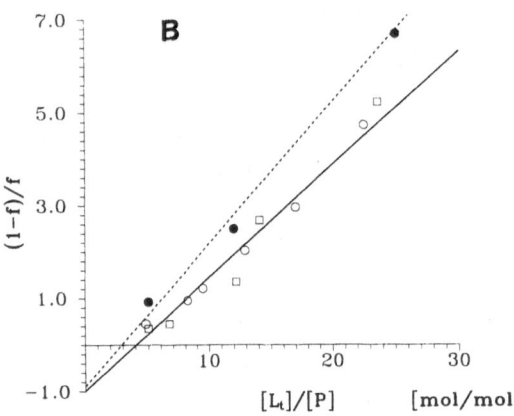

FIGURE 11. Resolution of fast and slowly exchanging lipids in the solvation layer of viral M13 coat protein.

(A) Quadrupole echo ^{2}H-NMR spectra of viral M13 coat protein/dimyristoylphosphatidylcholine-d_9 (DMPC-d_9) complexes as a function of lipid/protein ratio (in mol/mol). The experimental ^{2}H-NMR spectra (solid lines) are shown together with best-fitting two-component simulated line shapes (dashed lines); all spectra recorded at 30°C (van Gorkom *et al.*, 1990).

(B) The ESR and ^{2}H-NMR titration diagrams of spin-labeled phospholipids and head-group-deuterated phosphatidylcholine (DMPC-d_g) in viral M13 coat protein/dimyristoylphosphatidycholine complexes. The different x-axis intercepts of the two titration diagrams should be noted which is a result of the different time scales of these two methods as described in Section 5.3. □, spin-labeled phosphatidylglycerol, 14-PGSL; ○, spin-labeled phosphatidylcholine, 14-PCSL; ●, head-group-deuterated phosphatidylcholine, DMPC-d_g.

ESR and ^{2}H-NMR measurements (Figure 11B), different numbers of solvation sites were obtained because of the different time scales of these two methods. In spin-labeling ESR experiments, all of the solvation lipids are readily resolved from the fluid lipids, while the two kinds of solvation lipids, both exchanging much slower than the limiting frequency of the method ($\tau_{lim}^{-1} = 1 \times 10^8$ sec^{-1}), cannot be further resolved. In ^{2}H-NMR experiments, on the other hand, fast exchanging solvation lipids are averaged with fluid lipids as in the case of most of the integral membrane proteins (Seelig *et al.*, 1982; Bloom and Smith, 1985; Knowles and Marsh, 1991), while slowly exchanging lipids are resolved as a distinct component.

The good resolution in ^{2}H-NMR spectra between fast and slowly exchanging solvation lipids sets an upper limit for the exchange rate between these two populations: $\tau_{slow}^{-1} = 10^3$ sec^{-1}. A more accurate estimate can be obtained from two-component line-shape simulations (van Gorkom et al., 1990). The decreasing quadrupole splitting and the accompanying line broadening effects can be accounted for by two-site exchange between the average of the fluid and the fast exchanging solvation lipids and the slowly exchanging solvation lipids. Slow exchange can also successfully explain the apparent contradiction between the observed line broadening and the increasing spin–spin T_{2e} relaxation times. By solving the steady-state Bloch equations, the best-fitting line shapes, shown in Figure 11A together with the respective experimental NMR spectra, were obtained and for the off-rate of the slowly exchanging solvation lipids a value of $\tau_{slow}^{-1} = 230$ Hz was calculated, as expected, independent of the protein/lipid ratio.

The steady-state Bloch equations describe the line shapes in the frequency domain. A complete description of the pulse experiments including the effects of the pulse delay times requires the solution of the time-dependent Bloch equations (Woessner, 1961). This method is particularly powerful as the exchange rate becomes comparable with the frequency separation of the two components.

6.   NOVEL EVIDENCE FOR THE TWO-SITE EXCHANGE MODEL

In this last section, two further experiments will be described which shed light on hitherto undiscussed aspects of the two-site exchange model. First, the line broadening related to exchange will be compared at two different microwave frequencies and, then, two-site exchange will be shown to be an effective spin-lattice cross-relaxation mechanism.

6.1.   Exchange Effects at Different Microwave Frequencies

Spin-label ESR spectra are usually recorded at a microwave frequency of 9 GHz where the spectral anisotropy is dominated by the hyperfine term [see Equation (2)]. At 34 GHz the field-dependent g-tensor term becomes comparable with the hyperfine term and so the spectral resolution between the two components and, hence, the sensitivity to two-site exchange will greatly differ providing thereby a rather critical test of the two-component nature of the spectra (Horváth et al., 1994).

The 34-GHz ESR spectra of spin-labeled phosphatidic acid in myelin proteolipid protein/DMPC complexes of various protein/lipid ratios are shown in Figure 12A together with best-fitting simulated line shapes. There are several points to note in the simulations of 34-GHz spectra; the corresponding 9-GHz

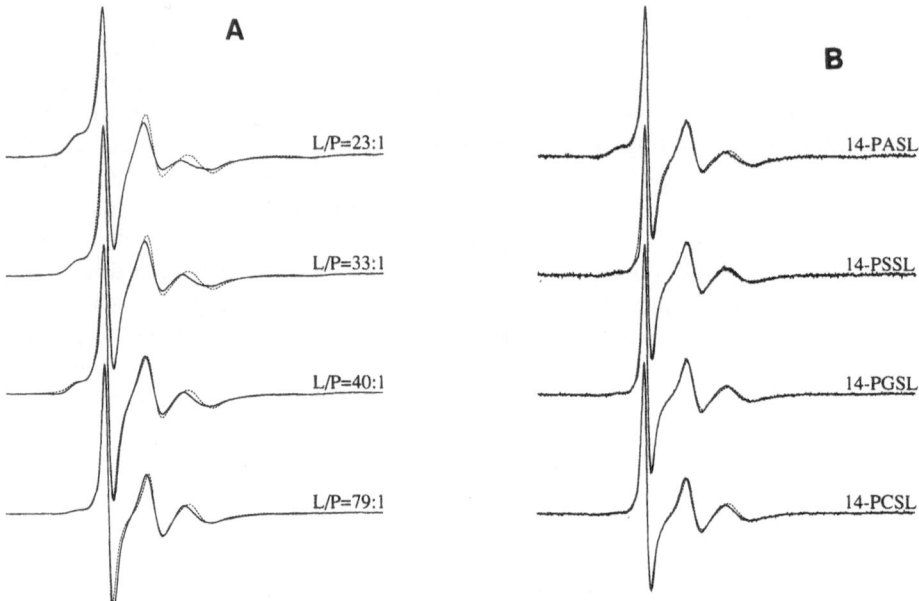

FIGURE 12. 34-GHz ESR spectra of various spin-labeled lipids in myelin proteolipid protein/ dimyristoylphosphatidylcholine complexes. The effect of two-site exchange is illustrated by overlaying two-component 34-GHz ESR spectra (solid lines) and best-fitting line-shape simulations using the two-site model (dashed lines). Total scan width is 12 mT about a center field of 1.218 T and the microwave frequency is 34.15 GHz (Horváth et al., 1994).

(A) 34-GHz ESR spectra of spin-labeled phosphatidic acid, 14-PASL, in complexes of various lipid/protein ratios (given in mol/mol next to the spectra).

(B) The lipid selectivity pattern of different spin-labeled lipids in complexes of fixed lipid/protein ratio (22:1 mol/mol).

spectra were presented in Figure 2A. First, no motional averaging takes place while lipids occupy solvation sites and so, following McConnell (1976), for the rates of rotational isomerization and long axis rotation of acyl chains the approximate lower bounds can be given as

$$\tau_{\text{off-axis}} \geq \frac{h}{\beta B_0[(g_{xx} + g_{yy})/2 - g_{zz}]} = 2.0 \times 10^{-9}\text{sec} \qquad (27a)$$

$$\tau_{\text{axial}} \geq \frac{h}{\beta B_0(g_{xx} - g_{yy})} = 4.0 \times 10^{-9}\text{sec} \qquad (27b)$$

This latter inequality relies on the presence of well-resolved x and y extrema in the 34-GHz spectra. Incipient (x,y) averaging leads to an axial line shape in the 9-GHz spectrum at rates of $\tau_{\text{axial}}^{-1} \leq 8.9 \times 10^7 \text{ sec}^{-1}$; at higher microwave frequency the expanded spectral anisotropy sets a smaller lower bound. The spectral anisotropies at the three nuclear manifolds (m_I = $+1,0,-1$) are 388, 180, 665 MHz and 168, 652, 1135 MHz in the case of 9- and 34-GHz spectra, respectively. Clearly, the dominant low-field band at 34 GHz must be more sensitive to exchange effects than the central band at 9 GHz; the better resolution at high-field may, in principle, be useful in spectral subtractions. A detailed discussion of several special problems is given in Pasenkiewicz-Gierula *et al.* (1983), Ovchinnikov and Konstantinov (1978), and Horváth *et al.* (1994).

These data allow a comparison of exchange rates obtained at two different microwave frequencies. The pronounced influence of lipid specificity on exchange has been previously discussed in light of the 9-GHz spectra (Figure 5A); the corresponding 34-GHz spectra are shown in Figure 12B. The ESR spectra recorded at two microwave frequencies leave little doubt regarding the two-component nature of the spectra. A quantitative agreement has been found between the exchange frequency data deduced from 9- and 34-GHz spectral simulations substantiating the two-site exchange model.

6.2. Two-Site Exchange as a Spin-Lattice Relaxation Mechanism

Inherent in the exchange averaging mechanism is that best sensitivity to exchange is anticipated when the exchange rate is comparable to the spectral resolution (in MHz) between the two components (Carrington and McLachlan, 1967). In the case of slow exchange ($\tau_{ex}^{-1} < 10^7$ sec^{-1}) T_1 relaxation measurements can extend the sensitivity of linewidth (T_2 relaxation) measurements. Such a T_1 experiment relies on two assumptions: (1) the spin-lattice relaxation rates of the two lipid populations are different and (2) the two-site exchange is an effective cross-relaxation mechanism. Consequently, exchange rates can be determined from the cross-relaxation term. T_1 relaxation measurements can be performed, additionally to pulsed ESR methods, by progressive saturation (c.w.) and saturation transfer ESR techniques (Horváth *et al.*, 1993).

A typical series of progressive saturation data for spin-labeled phosphatidylcholine (14-PCSL) in protein-alone and fluid lipid vesicles, all recorded at 30°C, is shown in Figure 13A. As seen with increasing microwave power the rigid limit spectrum of solvation lipids saturates rather steeply with a half-saturation power of $\sqrt{P} = 6.6$ mW, whereas the motionally averaged spectrum of fluid lipids saturates beyond the maximum attainable power, $\sqrt{P} \sim 18$ mW. Since the reciprocal spin-lattice relaxation time is, in good approximation, proportional to the motional rate (Carrington and McLachlan, 1967), this saturation pattern is consistent with the different mobilities of the two lipid populations. A further point to note is that the spectral effects of saturation are different in the three regions of the spectrum (m_I = $+1, 0, -1$); subsequently we shall

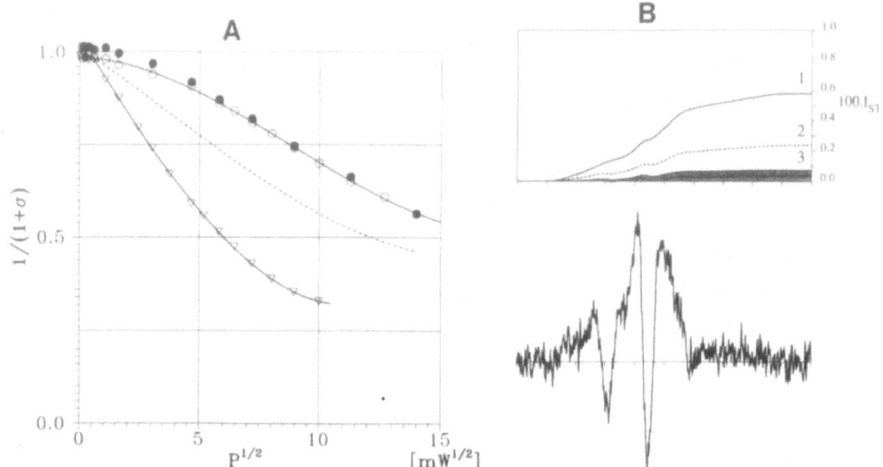

FIGURE 13. Effect of cross-relaxation related to two-site exchange on progressive saturation (c.w.) and integrated intensity of saturation transfer ESR spectra (Horváth *et al.*, 1993).

(A) The progressive saturation curve of spin-labeled phosphatidylcholine (solid line) is shifted to higher microwave power relative to the weighted-average saturation of the solvation and fluid populations (dashed line). ∇, spin-labeled stearic acid, 14-SASL, in myelin proteolipid protein aggregates; ○, spin-labeled phosphatidylcholine, 14-PCSL, in fluid dimyristoylphosphatidylcholine vesicles; ●, spin-labeled phosphatidylcholine, 14-CSL, in myelin proteolipid protein/dimyristoyl-phosphatidylcholine complexes.

(B) The saturation transfer ESR spectrum of spin-labeled phosphatidylcholine in myelin proteolipid protein/dimyristoylphosphatidylcholine complexes of lipid/protein ratio 22:1 mol/mol. The second-harmoic, out-of-phase (v'_2) spectrum is shown together with the integral spectrum which has less intensity than the weighted average of the two components (dashed line). 1, spin-labeled stearic acid, 14-SASL, in myelin proteolipid protein aggregates; 2, the weighted average of the integrated intensities of the two components; 3, spin-labeled phosphatidylcholine, 14-PCSL, in lipid/protein complexes. The integrated intensity of the saturation transfer ESR spectrum of spin-labeled 14-PCSL in dimyristoylphosphatidylcholine vesicles is ≈ 5 × 10^{-2} (unmarked bottom curve).

focus our attention on the most easily saturable central region of the nitroxyl spectrum.

The anomalous saturation properties of the two-component spectrum of spin-labeled phosphatidylcholine in myelin proteolipid protein/lipid complexes at 30°C are illustrated in Figure 13A. The experimental points are shown together with the weighted-average saturation of the solvation and fluid components and are seen to be shifted toward higher microwave fields; no such shift could be observed at 4°C below the phase transition (data not shown).

An alternative T_1 experiment for studying the spectral effects of slow two-site exchange on spin-lattice relaxation is provided by saturation transfer ESR measurements (Thomas *et al.*, 1976; Hyde and Dalton, 1979). A typical two-component second-harmonic, out-of-phase (v_2') spectrum of spin-labeled lipids in protein-alone, protein/lipid complex, fluid lipid vesicles is shown in Figure 13B, together with its normalized integral curve (Horváth *et al.*, 1993). The saturation transfer ESR spectra show a similar pattern to that observed in progressive saturation experiments. Again, the integrated intensities of the two components are markedly different and the experimental saturation transfer ESR spectrum of the protein/lipid complex has significantly less intensity than the weighted average of those of the two components.

The normalized integrated intensity of the STESR spectrum is proportional to the effective spin-lattice relaxation time: $I_{ST} \sim T_1^{eff}$ (Thomas *et al.*, 1976) and the exchange rate can be determined from the difference between the (reciprocal) relaxation time in the presence and absence of exchange (Subczynski and Hyde, 1981; Horváth *et al.*, 1993):

$$\frac{1}{T_{1,solv}^{obs}} - \frac{1}{T_{1,solv}^{o}} = \frac{(1-f)\tau_{solv}^{-1}}{(1-f) + fT_{1,fluid}^{o}\tau_{1,solv}^{-1}} \tag{28}$$

From the observed reduced value of I_{ST} (Figure 13B) and, hence, of T_{1solv}^{eff}, and with reasonable assumptions regarding the unmodified relaxation time $T_{1,solv}^{o}$, off-rates in the range of $1-5 \times 10^6 \, sec^{-1}$ could be deduced in good order-of-magnitude agreement with values obtained previously from two-site exchange simulations (Section 4). Furthermore, the exchange rates deduced from the STESR spectra were consistent with lipid selectivity data. The saturation data of the conventional two-component ESR spectrum in the progressive saturation (c.w.) experiments could be evaluated with the same model, assuming that the half-saturation power was proportional to the relaxation time: $\sqrt{P_{1/2}} \sim T_1$, and good quantitative agreement was obtained (Horváth *et al.*, 1993).

7. CONCLUSIONS

Lipid molecules at the interface of membrane proteins have a specialized role in interfacing the two-dimensional fluid lipid bilayer with slowly fluctuating macromolecules. As a rule at least one layer of lipids, the so-called solvation layer, is motionally restricted with respect to the bulk fluid lipids because of motional coupling between proteins and lipids. Using spin-labeled lipids these two populations can be resolved by ESR spectroscopy and the number of solvation lipids can be determined by protein/lipid titrations.

The protein interface exhibits a pronounced lipid selectivity toward negatively charged lipids indicating the role of electrostatic interactions in lipid/protein interactions. This lipid specificity can be modulated by pH and (monovalent cation) salt titration of lipid head groups. The reciprocity of electrostatic interactions was demonstrated by comparing the selectivities of myelin proteolipid protein and its natural mutant which consists of a long deletion including several positively charged amino acid groups.

The molecular dynamics of lipid/protein association can be quantified by spectrum simulations based on the exchange-coupled Bloch equations. It is found that solvation sites at the protein interface are continuously and rapidly replaced by the laterally diffusing lipids and these sites are occupied by the oncoming lipids for various lengths of time depending on their selectivities. The off-rate of negatively charged lipids of high selectivity is varied in the range of $2-8 \times 10^6 \, sec^{-1}$, whereas phosphatidylcholine and other lipids of low selectivity are exchanged significantly faster ($\tau_{off}^{-1} = 1.5 \times 10^7 \, sec^{-1}$). Analyzing the lipid selectivity observed in solvation-to-fluid exchange of spin-labeled lipids, a linear correlation was found between the off-rate τ_{off}^{-1} and the reciprocal association constant $1/K_r$.

Of the different protein/lipid systems studied, in three cases anomalously slowly exchanging lipids were identified in the off-rate versus selectivity correlation diagram. Approximately 30% of cardiolipin was seen not to be exchanged in the solvation layer along with other lipids, and while in the solvation shell this population of cardiolipins was inaccessible to lipid hydrogenation. The off-rate of such slowly exchanging lipids was determined by ^{2}H-NMR spectroscopy by resolving fast and slowly exchanging solvation lipids at the interface of M13 coat protein.

The solvation-to-fluid exchange was shown to be an effective T_1 cross-relaxation mechanism by progression saturation (c.w.) and saturation transfer ESR experiments. Good agreement was found between the exchange data deduced from linewidth (T_2 relaxation) measurements and saturation (T_1 relaxation) experiments.

ACKNOWLEDGMENTS. I thank Dr. Derek Marsh (Göttingen) for a long-standing cooperation during which the fundamental aspects of this work have been elaborated. I am grateful to Drs. T. Szörényki (Szeged) and Nicholas F. P. Ryba (N.I.H. Bethesda) for critically reading the manuscript. This work was supported in part by the Hungarian National Science Foundation (OTKA 175/1988 and 913/1991).

8. REFERENCES

Abney, J. R., and Owicki, J. C., 1985, Theories of protein–lipid and protein–protein interactions in membranes, in: *Progress in Protein–Lipid Interactions* (A. Watts and J.J.H.H.M. de Pont, eds.), Vol. 1, pp. 1–60, Elsevier, Amsterdam.

Awasthi, Y. C., Chuang, T. F., Keenan, T. W., and Crane, F. L., 1971, Tightly bound cardiolipin in cytochrome oxidase, *Biochim. Biophys. Acta* **226**:42–52.

Berliner, L. J., ed., 1976, 1979, *Spin Labeling: Theory and Applications*, Vols. 1 and 2, Academic Press, New York.

Beyer, K., and Klingenberg, M., 1983, Reincorporation of ADP/ATP carrier into phospholipid membranes. Phospholipid–protein interaction as studied by ^{31}P NMR and electron microscopy, *Biochemistry* 22:639–645.

Bienvenue, A., Bloom, M., Davis, J. H., and Devaux, P. F., 1982, Evidence for protein-associated lipid from deuterium nuclear magnetic resonance studies of rhodopsin–dimyristoylphosphatidyl choline recombinants, *J. Biol. Chem.* 257:3032–3038.

Bloom, M., and Smith, I.C.P., 1985, Manifestations of lipid–protein interactions in deuterium NMR, in *Progress in Protein–Lipid Interactions* (A. Watts and J.J.H.H.M. de Pont, eds.), Vol. 1, pp. 61–88, Elsevier, Amsterdam.

Brophy, P. J., Horváth, L. I., and Marsh, D., 1984, Stoichiometry and specificity of lipid–protein interaction with myelin proteolipid protein studied by spin label electron spin resonance, *Biochemistry* 23:860–865.

Brotherus, J. R., Jost, P. C., Griffith, O. H., Keana, J.F.W., and Hokin, L. E., 1980, Charge selectivity at the lipid–protein interface of membranous Na,K-ATPase, *Proc. Natl. Acad. Sci. USA* 77:272–276.

Brotherus, J. R., Griffith, O. H., Brotherus, M. O., Jost, P. C., Silvius, J. R., and Hokin, L. E., 1981, Lipid–protein multiple binding equilibria in membranes, *Biochemistry* 20:5261–5267.

Carrington, A., and McLachlan, A. D., 1967, *Introduction to Magnetic Resonance*, Harper & Row, New York.

Chapman, D., and Quinn, P. J., 1976, A method for the modulation of membrane fluidity: Homogeneous catalytic hydrogenation of phospholipid and phospholipid–water model biomembranes, *Proc. Natl. Acad. Sci. USA* 73:3971–3975.

Cherry, R. J., 1979, Rotational and lateral diffusion of membrane proteins, *Biochim. Biophys. Acta* 559:289–327.

Clegg, R. M., and Vaz, W.L.C., 1985, Translational diffusion of proteins and lipids in artificial lipid bilayer membranes. A comparison of experiment with theory, in: *Progress in Protein–Lipid Interactions* (A. Watts and J.J.H.H.M. de Pont, eds.), Vol. 1, pp. 173–229, Elsevier, Amsterdam.

Datema, K. P., Wolfs, C.J.A., Marsh, D., Watts, A., and Hemminga, M. A., 1987, Spin-label electron spin resonance study of bacteriophage M13 coat protein incorporation into mixed lipid bilayers, *Biochemistry* 26:7571–7574.

Davoust, J., and Devaux, P. F., 1982, Simulation of electron spin resonance spectra of spin-labelled fatty acids covalently attached to the boundary of an intrinsic membrane protein. A chemical exchange model, *J. Magn. Reson.* 48:475–494.

Davoust, J., Seigneuret, M., Hervé, P., and Devaux, P. F., 1983, Collisions between nitrogen-14 and nitrogen-15 spin labels. Investigations on the specificity of the lipid environment of rhodopsin, *Biochemistry* 22:3146–3151.

Devaux, P. F., and Seigneuret, M., 1985, Specificity and lipid–protein interaction as determined by spectroscopic techniques, *Biochim. Biophys. Acta* 822:63–125.

Devaux, P. F., Houtson, G. L., Favre, E., Fellmann, P., Farsen, B., MacKay, A. L., and Bloom, M., 1986, Interaction of cytochrome *c* with mixed dimyristoyl phosphatidylcholine–dimyristoyl phosphatidylserine bilayers: A deuterium nuclear magnetic resonance study, *Biochemistry* 25:3804–3812.

East, M., Melville, D., and Lee, A. G., 1985, Exchange rates and numbers of annular lipids for the calcium and magnesium ion dependent adenosinetriphosphatase, *Biochemistry* 24:2615–2623.

Ellena, J. F., Blazing, M. A., and McNamee, M. G., 1983, Lipid–protein interactions in reconstituted membranes containing acetylcholine receptor, *Biochemistry* 22:5523–5535.

Esmann, M., and Marsh, D., 1985, Spin label studies on the origin of the specificity of lipid–protein interactions in Na+,K+-ATPase membranes from *Squalus acanthias, Biochemistry* **24:**3572–3578.

Esmann, M., Horváth, L. I., and Marsh, D., 1987, Saturation transfer electron spin resonance studies on the mobility of spin-labeled sodium and potassium activated adenosinetriphosphatase in membranes from *Squalus acanthias, Biochemistry* **26:**8675–8683.

Freed, J. H., 1976, Theory of slowly tumbling ESR spectra for nitroxides, in: *Spin Labeling: Theory and Applications* (L. J. Berliner, ed.), Vol. 1, pp. 53–132, Academic Press, New York.

Goldman, S. A., Bruno, G. V., and Freed, J. H., 1972, Estimating slow-motional rotational correlation times for nitroxides by electron spin resonance, *J. Phys. Chem.* **76:**1858–1860.

Griffith, O. H., and Jost, P. C., 1976, Lipid spin labels in biological membranes, in: *Spin Labeling: Theory and Applications* (L. J. Berliner, ed.), Vol. 1, pp. 453–523, Academic Press, New York.

Griffith, O. H., Brotherus, J. R., and Jost, P. C., 1982, Equilibrium constants and number of binding sites for lipid–protein interactions in membranes, in: *Lipid–Protein Interactions* (P. C. Jost and O. H. Griffith, eds.), Vol. 2, pp. 225–237, Wiley, New York.

Horváth, L. I., Brophy, P. J., and Marsh, D., 1988a, Exchange rates at the lipid–protein interface of myelin proteolipid protein studied by spin label electron spin resonance, *Biochemistry* **27:**46–52.

Horváth, L. I., Brophy, P. J., and Marsh, D., 1988b, Influence of lipid headgroup on the specificity and exchange dynamics in lipid–protein interactions. A spin label study of myelin proteolipid apoprotein–phospholipid complexes, *Biochemistry* **27:**5296–5304.

Horváth, L. I., Münding, A., Beyer, K., Klingenberg, M., and Marsh, D., 1989, Rotational diffusion of mitochondrial ADP/ATP carrier studied by saturation-transfer electron spin resonance, *Biochemistry* **28:**407–414.

Horváth, L. I., Brophy, P. J., and Marsh, D., 1990a, Influence of polar residue on lipid–protein interactions with the myelin proteolipid protein. Spin label ESR studies with DM-20/lipid recombinants, *Biochemistry* **29:**2635–2638.

Horváth, L. I., Drees, M., Beyer, K., Klingenberg, M., and Marsh, D., 1990b, Lipid–protein interactions in ADP–ATP carrier/egg phosphatidylcholine recombinants studied by spin label ESR spectroscopy, *Biochemistry* **29:**10664–10669.

Horváth, L. I., Brophy, P. J., and Marsh, D., 1993, Exchange rates at the lipid–protein interface of the myelin proteolipid protein determined by saturation transfer electron spin resonance and continuous wave saturation studies, *Biophys. J.* **62:**622–631.

Horváth, L. I., Brophy, P. J., and Marsh, D., 1994, Microwave frequency dependence of ESR spectra from spin labels undergoing two-site exchange myelin proteolipid complexes, *J. Magn. Reson.*, in press.

Hyde, J. S., and Dalton, L. R., 1979, Saturation transfer spectroscopy, in: *Spin Labeling: Theory and Applications* (L. J. Berliner, ed.), Vol. 2, pp. 1–70, Academic Press, New York.

Jost, P. C., Griffith, O. H., Capaldi, R. A., and Vanderkooi, G. A., 1973, Evidence for boundary lipid in membranes, *Proc. Natl. Acad. Sci. USA* **70:**4756–4763.

Kadenbach, B., Mende, P., Kolbe, H.J.V., Stipani, I., and Palmieri, F., 1982, The mitochondrial phosphate carrier has an essential requirement for cardiolipin, *FEBS Lett.* **139:**109–112.

Kivelson, D., 1960, Theory of ESR linewidth of free radicals, *J. Chem. Phys.* **33:**1094–1106.

Knowles, P. F., and Marsh, D., 1991, Magnetic resonance of membranes, *Biochem. J.* **274:**625–641.

Knowles, P. F., Watts, A., and Marsh, D., 1979, Lipid immobilization in dimyristoyl phosphatidylcholine-substituted cytochrome oxidase, *Biochemistry* **18:**4480–4487.

Knowles, P. F., Watts, A., and Marsh, D., 1981, Spin label studies of head-group specificity in the interaction of phospholipids with yeast cytochrome oxidase, *Biochemistry* **20:**5888–5894.

Laidlaw, D. J., and Pink, D. A., 1985, Protein lateral distribution in lipid bilayer membranes. Applications to ESR studies, *Eur. Biophys. J.* **12**:143–151.

McConnell, H. M., 1958, Reaction rates by nuclear magnetic resonance, *J. Chem. Phys.* **28**:430–431.

McConnell, H. M., 1976, Molecular motion in biological membranes, in: *Spin Labeling: Theory and Applications* (L. J. Berliner, ed.), Vol. 1, pp. 525–560, Academic Press, New York.

McConnell, H. M., and McFarland, B. G., 1970, Physics and chemistry of spin labels, *Q. Rev. Biophys.* **3**:91–136.

MacLennan, D. H., Brandl, C. J., Korczak, B., and Green, N. M., 1985, Amino-acid sequence of Ca^{2+} + Mg^{2+}-dependent ATPase from rabbit muscle sarcoplasmic reticulum, deduced from its complementary DNA sequence, *Nature* **316**:696–700.

Marsh, D., 1985, ESR spin label studies of lipid–protein interactions, in: *Progress in Protein–Lipid Interactions* (A. Watts and J.J.H.H.M. de Pont, eds.), Vol. 1, pp. 143–172, Elsevier, Amsterdam.

Marsh, D., 1986, Spin label ESR spectroscopy and molecular mobility in biological systems, in: *Supramolecular Structure and Function* (G. Pifat-Mrzlijak, ed.), pp. 48–62, Springer-Verlag, Berlin.

Marsh, D., and Horváth, L. I., 1989, Spin label studies of the structure and dynamics of lipids and proteins in membranes, in: *Advanced EPR in Biology and Biochemistry* (A. J. Hoff, ed.), pp. 707–772, Elsevier, Amsterdam.

Marsh, D., and Watts, A., 1982, Spin labeling and lipid–protein interactions in membranes, in: *Lipid–Protein Interactions* (P. C. Jost and O. H. Griffith, eds.), Vol. 2, pp. 53–126, Wiley, New York.

Mason, R., and Freed, J. H., 1974, Estimating microsecond rotational correlation times from lifetime broadening of nitroxide ESR spectra near the rigid limit, *J. Phys. Chem.* **78**:1321–1323.

Nave, K.-A., Lai, C., Bloom, F. E., and Milner, R. J., 1987, Splice site selection in the proteolipid protein (PLP) gene transcript and primary structure of the DM-20 protein of central nervous system myelin, *Proc. Natl. Acad. Sci. USA* **84**:5665–5669.

Ovchinnikov, I. V., and Konstantinov, V. N., 1978, Extra absorption peaks in EPR spectra of systems with anisotropic *g*-tensor and hyperfine structure in powders and glasses, *J. Magn. Reson.* **32**:179–190.

Paddy, M. R., Dahlquist, F. W., Davis, J. H., and Bloom, M., 1981, Dynamical and temperature dependent effects of lipid–protein interactions. Application of deuterium nuclear magnetic resonance and electron paramagnetic resonance spectroscopy to the same reconstitutions of cytochrome *c* oxidase, *Biochemistry* **20**:3152–3162.

Pasenkiewicz-Gierula, M., Hyde, J. S., and Pilbrow, J. R., 1983, Simulation of Q-band ESR spectra of immobilized spin labels, *J. Magn. Reson.* **55**:255–265.

Pates, R. D., Watts, A., Uhl, R., and Marsh, D., 1985, Lipid–protein interactions in frog rod outer segment disc membranes. Characterization by spin labels, *Biochim. Biophys. Acta* **814**:389–397.

Powell, G. L., Knowles, P. F., and Marsh, D., 1987, Spin-label studies on the specificity of interaction of cardiolipin with beef heart cytochrome oxidase, *Biochemistry* **26**:8138–8145.

Quinn, P. J., Joó, F., and Vígh, L., 1989, The role of unsaturated lipids in membrane structure and stability, *Prog. Biophys. Mol. Biol.* **53**:71–103.

Ryba, N., Horváth, L. I., Watts, A., and Marsh, D., 1987, Exchange at the lipid–protein interface. Spin label ESR studies of rhodopsin–dimyristoyl phosphatidylcholine recombinants, *Biochemistry* **26**:3234–3240.

Saffman, P. G., and Delbrück, M., 1975, Brownian motion in biological membranes, *Proc. Natl. Acad. Sci. USA* **72:**3111–3113.

Sandermann, H., 1978, Regulation of membrane enzymes by lipids, *Biochim. Biophys. Acta* **515:**209–237.

Schlame, M., Horváth, L. I., and Vígh, L., 1990, Relationship between lipid saturation and lipid–protein interaction in liver mitochondria modified by catalytic hydrogenation with reference to cardiolipin molecular species, *Biochem. J.* **265:**79–85.

Schreier, S., Polnaszek, C. F., and Smith, I.C.P., 1978, Spin labels in membranes. Problems in practice, *Biochim. Biophys. Acta* **515:**375–436.

Schultz, A. M., Henderson, L. E., and Oroszlan, S., 1988, Fatty acylation of proteins, *Annu. Rev. Cell Biol.* **4:**611–647.

Seelig, J., 1977, Deuterium magnetic resonance: Theory and application to lipid membranes, *Q. Rev. Biophys.* **10:**353–418.

Seelig, J., 1978, ^{31}P nuclear magnetic resonance and the head group structure of phospholipid membranes, *Biochim. Biophys. Acta* **515:**105–140.

Seelig, J., Seelig, A., and Tamm, L., 1982, Nuclear magnetic resonance and lipid–protein interactions, in: *Lipid–Protein Interactions* (P. C. Jost and O. H. Griffith, eds.), Vol. 2, pp. 127–148, Wiley, New York.

Silvius, J. R., McMillen, D. A., Saley, N. D., Jost, P. C., and Griffith, O. H., 1984, Competition between cholesterol and phosphatidylcholine for the hydrophobic surface of sarcoplasmic reticulum Ca^{2+}-ATPase, *Biochemistry* **23:**538–547.

Simons, R., Alon, N., and Riordan, J. R., 1987, Human myelin DM-20 proteolipid protein deletion defined by cDNA sequence, *Biochem. Biophys. Res. Commun.* **146:**666–671.

Smith, R., Cook, J., and Dickens, P. A., 1983, Structure of the proteolipid protein extracted from bovine central nervous system myelin and nondenaturating detergents, *J. Neurochem.* **42:**306–313.

Stoffel, W., Hillen, H., and Giersiefen, H., 1984, Structure and molecular arrangement of proteolipid protein of central nervous system myelin, *Proc. Natl. Acad. Sci. USA* **81:**5012–5016.

Subczynski, W. K., and Hyde, J. S., 1981, The diffusion–concentration product of oxygen in lipid bilayers using the spin-label T_1 method, *Biochim. Biophys. Acta* **643:**283–291.

Tamm, L. K., and Seelig, J., 1983, Lipid solvation of cytochrome *c* oxidase. Deuterium, nitrogen-14 and phosphorus-31 nuclear magnetic resonance studies on the phosphocholine headgroup and on cis-unsaturated fatty acyl chains, *Biochemistry* **22:**1474–1483.

Thomas, D. D., Dalton, L. R., and Hyde, J. S., 1976, Rotational diffusion studied by passage saturation transfer electron paramagnetic resonance, *J. Chem. Phys.* **65:**3006–3024.

Thomas, D. D., Bigelow, D. J., Squier, T. C., and Hidalgo, C., 1982, Rotational dynamics of protein and boundary lipid in sarcoplasmic reticulum membrane, *Biophys. J.* **37:**217–225.

Toyoshima, C., and Unwin, N., 1988, Ion channel of acetylcholine receptor reconstructed from images of postsynaptic membranes, *Nature* **336:**247–249.

Van Gorkom, L.C.M., Horváth, L. I., Hemminga, M. A., Sternberg, B., and Watts, A., 1990, Identification of trapped and boundary lipid binding sites in M13 coat protein/lipid complexes by deuterium NMR spectroscopy, *Biochemistry* **29:**3828–3834.

Vaz, W.L.C., Clegg, R. M., and Hallmann, D., 1985, Translational diffusion of lipids in liquid crystalline phase phosphatidylcholine multibilayers. A comparison of experiment with theory, *Biochemistry* **24:**781–787.

Watts, A., and de Pont, J.J.H.H.M., eds., 1985, 1986, *Protein–Lipid Interactions,* vols. 1 and 2, Elsevier, Amsterdam.

Watts, A., Volotovski, I. D., and Marsh, D., 1979, Rhodopsin/lipid associations in bovine rod outer segment membranes. Identification of immobilized lipid by spin labels, *Biochemistry* **18:**5006–5013.

Wertz, J. E., and Bolton, J. R., 1972, *Electron Spin Resonance*, McGraw–Hill, New York.

Woessner, D. E., 1961, Nuclear transfer effects in nuclear magnetic resonance pulse experiments, *J. Chem. Phys.* **35**:41–48.

Wolfs, C.J.A.M., Horváth, L. I., Marsh, D., Watts, A., and Hemminga, M. A., 1989, Spin label ESR of bacteriophage M13 coat protein in mixed lipid bilayers. Characterization of molecular selectivity of charged phospholipids for the bacteriophage M13 coat protein in lipid bilayers, *Biochemistry* **28**:9995–10001.

Chapter 7

NMR Methods for Measuring Membrane Transport

Philip W. Kuchel, Kiaran Kirk, and Glenn F. King

1. INTRODUCTION

1.1. Scope

The aim of this chapter is to outline the reasons why certain items of information are sought for a description of membrane transport, to describe the various types of NMR experiment that can be used to obtain estimates of the rates of membrane transport, and to illustrate the various NMR procedures with biological examples. The major appeal of the NMR method in this context lies in the fact that the measurements do not usually require the physical separation of the cells or vesicles from their suspending solution. The work described here is a subset of the large number of NMR studies that have been carried out in recent years on living systems (for reviews see Shulman, 1979; Gadian and Radda, 1981; Kuchel, 1981, 1989; Gadian, 1982; Avison *et al.*, 1986; Cerdan and Seelig, 1990; Lundberg *et al.*, 1990). NMR studies of membrane transport have recently been reviewed (Kirk, 1990; Kuchel, 1990; King and Boyd, 1991); the first two reviews are directed mostly at "NMR audiences" and the latter to "biological

Philip W. Kuchel and Glenn F. King   Department of Biochemistry, University of Sydney, Sydney, NSW, 2006, Australia.   **Kiaran Kirk**   University Laboratory of Physiology, University of Oxford, OX1 3PT, England.

Subcellular Biochemistry, Volume 23: Physicochemical Methods in the Study of Biomembranes, edited by Herwig J. Hilderson and Gregory B. Ralston. Plenum Press, New York, 1994.

audiences." It is the intention that this chapter might be useful to readers who are NMR experts as well as those who are not; the latter may be encouraged to try the NMR counterparts of more conventional experiments and the former may find something of interest in these newer applications of NMR spectroscopy.

1.2. History

"Biological NMR" has been said to have begun when Felix Bloch placed his finger into the probe-coil of the first nuclear induction apparatus (Bloch *et al.*, 1946) and obtained a signal from the protons in his blood, connective tissues, and lipid (Andrew, 1990). However, the "modern era" of biological NMR spectroscopy was initiated about 25 years later when Eakin *et al.* (1972) used ^{13}C NMR to follow the metabolism of [1-^{13}C]glucose by *Candida utilis* (yeast) cells, and Moon and Richards (1973) showed that ^{31}P NMR could be used to follow the metabolism of 2,3-bisphosphoglycerate to orthophosphate in a suspension of rabbit erythrocytes. Furthermore, in the latter study, the positions of the ^{31}P NMR peaks in the spectra (i.e., the resonance frequencies) varied systematically with changes in pH, so that after the appropriate calibration of the frequency shift with respect to pH, the pH of the intracellular compartment could be read from a spectrum.

From the perspective of this chapter, the measurement by Conlon and Outhred (1972) of the rate of exchange of water across the human erythrocyte membrane by ^{1}H NMR spectroscopy is a "landmark." Conlon and Outhred estimated that the residence lifetime of a water molecule in a human red cell at 37°C was $\sim$ 10 msec. The ability of NMR spectroscopy to yield estimates of rates of such fast transmembrane exchange reactions, occurring in the subsecond time scale, awakened the interest of many biologists. Thus, there is now a vast literature devoted to the measurement of water transport into red cells (Section 5.1.2; for recent reviews see Benga, 1988, 1989a). The application of NMR spectroscopy to measuring the rapid exchange of molecules other than water is a much more recent occurrence (Kirk, 1990; Kuchel, 1990; King and Boyd, 1991).

1.3. Two-Site Exchange Theory

The canonical model of membrane transport is that of two solute populations (referred to as *sites*) separated by a permeability barrier, and with the rate of exchange between the sites described by first-order kinetics (Figure 1). The *flux* (flow of mass per unit time) of the solute across the membrane is described by the equations (e.g., Pirkle *et al.*, 1979)

$$dS_i/dt = -k_{-1}S_i + k_1S_o \tag{1}$$

$$dS_o/dt = k_{-1}S_i - k_1S_o \tag{2}$$

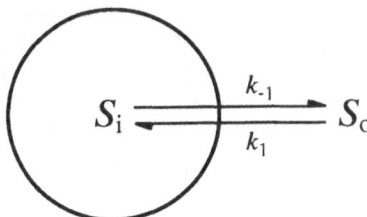

FIGURE 1. Diagram of a two-site membrane transport system. S_i and S_o denote the inside and outside solutes that are in exchange across the membrane, and k_1 and k_{-1} are the unitary rate constants that characterize the influx and efflux, respectively.

where t denotes time, S_i and S_o denote the total amounts (mass) of the solute inside and outside the compartment enclosed by the membrane, respectively, k_{-1} is the unitary efflux rate constant and is equal to the reciprocal of the mean residence lifetime $(1/\tau_i)$ of the solute inside, and k_1 is the unitary rate constant for influx and is equal to the reciprocal of the mean residence lifetime $(1/\tau_o)$ outside.

Note that for a cell suspension k_1 is an *adjustable parameter* since reducing the cytocrit (the fraction of the sample volume that is occupied by cells) will increase the extracellular volume; thus, the mean residence lifetime of molecules outside the cells will be increased. This implies that the "time window" over which a transport process can be measured can be altered by the simple expedient of altering the cytocrit. For example, in measurements of the rate of exchange of the small nonelectrolyte dimethyl methylphosphonate (DMMP; see Section 5.6.5) across the membranes of intact human red cells, the transport was found to be too fast to measure precisely when the usual hematocrit of blood, 0.45, was employed; however, the measurements were made with high precision by using a hematocrit of 0.05 (Kirk and Kuchel, 1986a; Potts *et al.*, 1989).

1.4. Permeability Coefficient

The "ease" with which a solute can permeate a membrane is often indicated by the permeability coefficient $(P;$ cm/sec) of the membrane for the solute. It is a normalized value that is independent of both the membrane surface area and the cell volume used in the experiment. In relation to the reaction scheme in Figure 1,

$$P = k_{-1}V_i/A = k_1 V_o/A \qquad (3)$$

where V_i and V_o are the total volumes accessible to S inside and outside the cells, respectively, and A is the total surface area of the membranes.

The permeability is determined by both the ease with which the solute *enters,* and the ease with which it *crosses,* the membrane. The former is determined by the buffer/membrane partition coefficient and the latter by the diffusion coefficient of the solute in the lipid bilayer. Permeability theory was largely developed with a view to explaining the transport characteristics of amphipathic molecules with various molecular volumes and lipid/water partition coefficients (Stein, 1986). It was not initially intended for saturable transport systems, such as membrane transport proteins, but it transpires that the concepts can be usefully applied to these systems; however, the apparent P value in such cases will vary with the extent of saturation of the transporter and may also depend on factors such as the membrane potential.

While it is often possible to determine experimentally the values of the unitary rate constants in the above equations, it may not be a straightforward matter to measure either V_i or A. When large quantities of cells are available, V_i may be estimated from the cytocrit (commonly measured by capillary centrifugation), provided that the fraction of the cell volume that is accessible to solutes, as well as the volume of extracellular solution that is trapped in the pellet during centrifugation, are known. In the case of normovolumic human erythrocytes, water occupies ~0.71 of the total cell volume (Savitz *et al.*, 1964; Raftos *et al.*, 1990) and the trapping of extracellular solution in the cell pellet normally accounts for 2–3% of the packed cell volume (Dacie and Lewis, 1975). If direct measurement of the cytocrit in this manner is not possible (because the cells are either scarce or not amenable to dense packing), V_i may be estimated gravimetrically (Gary-Bobo and Solomon, 1968), or by using a radioisotope distribution method, or NMR-based procedures (Kuchel, 1989).

Estimating the mean surface area of a cell, A, is often more difficult. Many isolated cells in suspension are approximately spherical so the estimation by microscopy is simple; but red cells from various animals have stable shapes which depart dramatically from spherical (e.g., Benga *et al.*, 1993). Morphometric analysis under light microscopy can be used to measure the dimensions of avian erythrocytes which are flattened ellipsoidal shapes and contain a nucleus which often appears to bulge from the two flattened surfaces. The value of V can be obtained from the mathematical expression for the volume of an ellipsoid, $V = 4\pi abc/3$, where a, b, and c are the semiaxis lengths. It is surprising to note, however, that no mathematical handbooks appear to contain an expression for the surface area of an ellipsoid. A recently derived expression involves elliptic functions and although it is an infinite series it is rapidly convergent. Furthermore, if the infinite series is truncated at the third term it fortuitously yields the correct expression for the surface area of a sphere, when $a = b = c$ (Bulliman and Kuchel, 1988). Therefore, it may be possible to determine more accurately the P values of the membranes of avian red cells than has hitherto been the case. In the case of chicken erythrocytes, the surface area was estimated in this way to be 1.24–1.27×10^{-10} m^2, which contrasts with the value of 1.75×10^{-10} m^2 used on the basis of a cylindrical approximation to the cell shape; the former value is more similar to the surface area of normal human red cells (1.43×10^{-10} m^2; Grimes, 1980).

2. GENERAL CONCEPTS OF NMR RELEVANT TO TRANSPORT ANALYSIS

2.1. Basic Concepts of NMR

The technique of NMR spectroscopy is a natural development from Wolfgang Pauli's discovery in 1924 that certain nuclei have angular momentum (i.e, they spin). The spinning of these positively charged nuclei causes them to behave like a current loop and consequently they have an associated magnetic dipole moment, μ (see Figure 2A). Thus, the nuclei can be considered analogous to tiny bar magnets possessing angular momentum.

When these spinning nuclei are placed in an external magnetic field, B_0, the nuclei align with the external field and additionally, because of the torque resulting from the interaction between B_0 and μ, they will *precess* (i.e., rotate) at a small angle around the axis of the external field (see Figure 2A). This nuclear precession is completely analogous to the precession of a spinning top in the earth's gravitational field, and is described by the Larmor equation:

$$\omega_0 = -\gamma \, B_0 \qquad (4)$$

where ω_0 is the angular frequency of the precession, B_0 is the strength of the external magnetic field (which is conventionally assigned to be along the z axis), and γ is the so-called magnetogyric ratio, a fundamental constant which varies from one type of nucleus to another. The convention used by NMR spectroscopists to describe an NMR spectrometer is to indicate the Larmor frequency of ^1H nuclei at the magnetic field strength of the spectrometer magnet; thus, a 600-MHz NMR spectrometer (B_0 = 14.1 T) is one in which ^1H nuclei resonate at 600 MHz.

It transpires that more than one orientation is possible for a nucleus under the influence of an external magnetic field, and these correspond to the discrete nuclear energy (eigen) states in the quantum-mechanical description of NMR. For a spin-$^1/_2$ nucleus (i.e., a nucleus with spin quantum number $I = {}^1/_2$) there are two possible orientations: the lower-energy state corresponds to nuclei whose magnetic dipoles are aligned with B_0 and the high-energy state corresponds to nuclei with their dipoles aligned against B_0 (see Figure 2B). For an ensemble of like nuclei, the ratio of the number of low-energy spin (n_1) to high energy spins (n_h) is given by the Boltzmann equation (Shaw, 1984):

$$n_1/n_h = \exp\left(\Delta E/kT\right) \qquad (5)$$

where ΔE is the energy difference between the two eigenstates, k is Boltzmann's constant, and T is the absolute temperature of the system. At a magnetic field

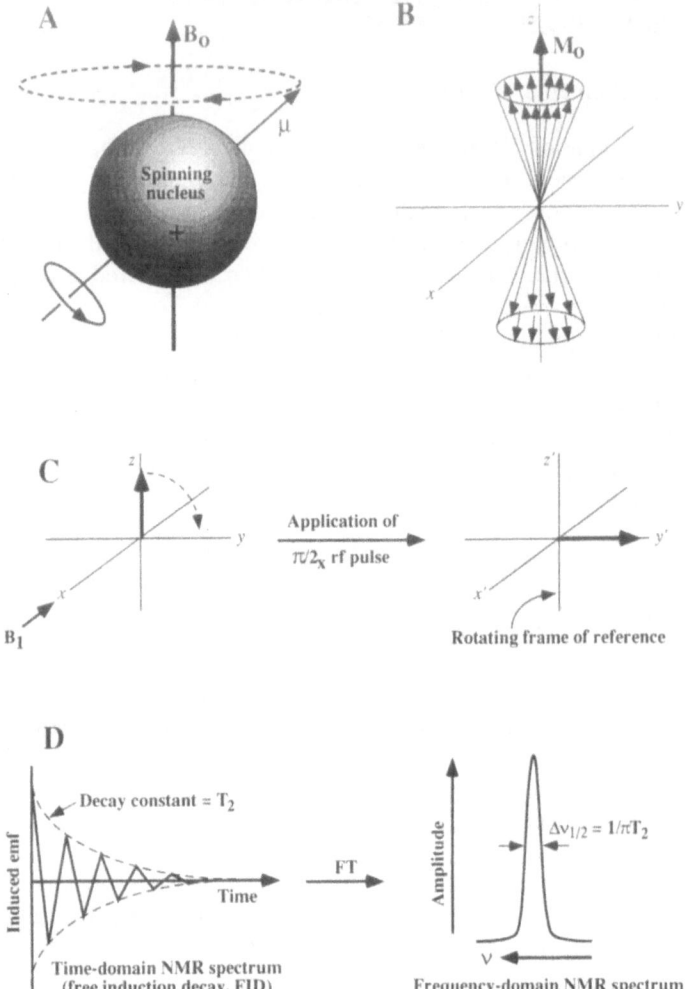

FIGURE 2. (A) A spinning nucleus with associated magnetic dipole moment, μ, precessing around the axis of an external magnetic field, B_0. (B) An ensemble of nuclei in an external magnetic field. The spins distribute between the low-energy state (aligned with B_0) and the high-energy state (aligned against B_0) according to the Boltzmann equation. The small excess of spins in the low-energy state (which is grossly exaggerated in this diagram) causes the sample to develop a net macroscopic magnetization vector, M_0. (C) Application of a second magnetic field in the form of a short-duration r.f. pulse, in this case along the x axis, causes the net macroscopic magnetization to nutate onto the y axis in the rotating frame of reference. (D) The Larmor precession of the magnetization induces an emf in an appropriately located receiver coil; this emf, which is known as the time-domain NMR spectrum or free induction decay (FID), oscillates cosinusoidally and decays with a time constant T_2. Fourier transformation of the FID gives the familiar frequency-domain NMR spectrum in which the peak linewidth at half-height ($\Delta\nu_{1/2}$) is equal to $1/\pi T_2$.

strength of 14.1 T (i.e., a 600-MHz spectrometer), ΔE for ^1H nuclei corresponds to $\sim 4 \times 10^{-25}$ J, and consequently at room temperature (298 K) there are approximately 1,000,097 nuclei in the lower-energy state for every million nuclei in the high-energy state; in other words the nuclear ensemble has a very small excess of spins aligned with B_0. Thus, if the nuclear dipoles are vectorially added, the resultant is a macroscopic magnetization vector (M_0) aligned along the z axis (i.e., the direction of B_0).

From a simplistic, but often very convenient, classical mechanical viewpoint, NMR experiments involve the manipulation and subsequent detection of the net macroscopic magnetization vector, M_0. The inherent insensitivity of the NMR technique arises from the fact that M_0 reflects the excess of nuclear spins in the low-energy state, and this excess corresponds to a very small fraction of the total number of nuclei in the ensemble (typically several thousandths of 1% of the total population of spins at a field strength of 14.1 T).

In pulsed Fourier transform NMR spectroscopy, M_0 is manipulated by the application of a second magnetic field (B_1) in the form of a short-duration (typically 5–100 μsec) radiofrequency (r.f.) pulse. When B_1 rotates at the Larmor frequency of the nuclei, resonance will occur: the low-energy spins will absorb the requisite energy for transition to the high-energy state. It would be difficult in practice to apply a purely rotating B_1 field, and consequently it is applied in the form of an r.f. pulse which can be decomposed into two counterrotating magnetic fields, only one of which will be of the correct sense relative to the nuclear precession to cause resonance.

Resonance causes M_0 to precess around the direction of the applied field (B_1) while still continuing its precession around the main field at the Larmor frequency (see Figure 2C); M_0 is said to *nutate* around B_1. The angle through which M_0 nutates is determined by the strength and duration of the r.f. pulse; in most NMR experiments, the r.f. pulses are applied along arbitrary x, y, −x, and −y axes. A $\pi/2_x$ pulse is taken to indicate the application of an r.f. pulse along the x axis for sufficient time to cause nutation of the magnetization by 90°: application of this pulse to the equilibrium magnetization, M_0, will cause it to nutate onto the y axis (see Figure 2C). In viewing the behavior of M_0 in this Cartesian coordinate system, it is convenient to use the "trick" of imagining that the axes rotate at the Larmor precession frequency of the nucleus; in this socalled *rotating frame of reference* (in which the axes are marked with primes, i.e., x', y', and z'), M_0 appears to be stationary in the transverse plane. However, it is important to remember that M_0 continues to precess around B_0 in the *laboratory frame of reference*.

As M_0 precesses in the transverse (xy) plane, it induces an electromotive force (emf) in an appropriately arranged receiver coil. In fact, most modern spectrometers are designed such that both the x and y components of M_0 can be independently detected; thus, the induced emf will reflect the time dependence of the x and y components of the nuclear magnetization. If we imagine the time dependence of the y component of the magnetization after a $\pi/2_x$ pulse, it will start with maximum amplitude (as it is initially nutated onto the y axis) and its amplitude will periodically increase and decrease as it precesses around B_0 (i.e., the z-axis) at its Larmor frequency; the amplitude of the y component will be zero when the magnetization is aligned along the $-x$ or x axis, and it will have a maximal negative amplitude when aligned along the $-y$ axis. Thus, it can be seen that the induced emf will oscillate cosinusoidally at the Larmor frequency of the nucleus.

Because of a variety of relaxation phenomena (see Section 2.4), M_o will slowly return to its equilibrium orientation along the z axis; in other words, the equilibrium distribution of spins between the various eigenstates will be restored. The transverse component of the nuclear magnetization decays with a time constant T_2 (see Section 2.4.2) and consequently the induced emf will decay with this time constant in addition to its cosinusoidal oscillation. This leads to an NMR signal of the type shown in Figure 2D, known as a free induction decay (FID) or time-domain NMR spectrum.

It is more conventional (and certainly much more useful!) to convert the FID into a frequency-domain spectrum for display and analysis. This can be done by means of a mathematical transformation known as the Fourier transform (FT) after its inventor, Jean-Baptiste Fourier (Fourier, 1822). Fortunately for NMR spectroscopists, the Larmor frequency of a given nucleus is very dependent on its immediate chemical environment, which means that the resonance frequencies of like nuclei can vary considerably. The beauty of the FT is that although all of these frequencies with their associated amplitudes will be summed in a single FID in the NMR experiment, the mathematical transformation is able to decompose the FID into its component amplitudes and frequencies to give a conventional frequency-domain NMR spectrum, which is really a plot of the amplitude of the induced emfs against their frequencies of oscillation. Thus, the frequency of a peak in a conventional NMR spectrum is an indication of the chemical environment in the vicinity of the nucleus, and its intensity (or area) is an indication of the number of nuclei in the sample contributing to the resonance. The linewidth of the peak is related to the relaxation time constant T_2; specifically, the linewidth at half the peak height is equivalent to $1/\pi T_2$.

2.2. Features of the NMR Spectrum

The five characteristics of a peak (also called a *line*) in a frequency-domain NMR spectrum (see Figure 3) are (e.g., Günther, 1980; Shaw, 1984; Homans, 1989): (1) the *chemical shift* [this is the normalized absorption (Larmor) frequency, which is expressed in parts per million and is denoted by δ]; (2) the *linewidth*,

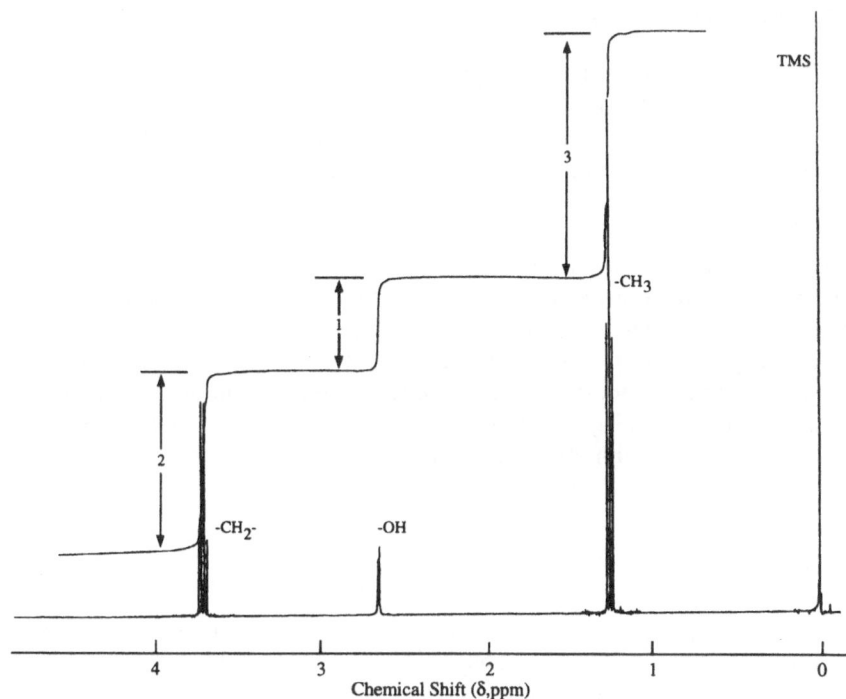

FIGURE 3. ^{1}H NMR spectrum of pure ethanol acquired at 400 MHz. The singlet at 0.000 ppm is from the reference compound tetramethylsilane (TMS) contained in a coaxial capillary in the sample tube. Note the peak multiplicity in the case of the resonances assigned to the ethanol CH_3 (triplet) and CH_2 (quartet) moieties; the coupling constant that characterizes this splitting is denoted $^3J_{HH}$ and it has a value of ~7 Hz. The continuous line offset from the spectrum represents a continuous integration of the spectral intensity starting from the left-hand (high frequency) end of the spectrum, the ratios of the peak integrals correspond to the number of protons giving rise to the peak [i.e., 3($-CH_3$) : 2($-CH_2$-) : 1($-OH$)].

generally given as the width of the peak at half its height ($\Delta\nu_{1/2}$, measured in Hz); (3) the *area* of the peak (measured in arbitrary units and often called the integral or intensity) which is usually proportional to the number of nuclei contributing to the resonance; (4) the *multiplicity*, or splitting into fine structure of the peak, that occurs principally as a result of the effects of local magnetic fields from adjacent nuclei in the molecule (so-called *J*-coupling); and (5) the *phase* of the peak relative to that of other peaks in the spectrum; i.e., whether it is upright (as is usually the case), inverted, or dispersive. The latter two situations occur in the spectra obtained by applying some multiple-pulse sequences (e.g., Section 4.1).

2.3. Elements Amenable to NMR Spectroscopy

Every element of the periodic table, except for argon, has at least one isotope for which the nuclei have a magnetic moment and are therefore NMR receptive (Harris and Mann, 1978). Nuclides which have been detected in NMR studies of transport have, however, been largely confined to those found in natural or fluorinated bio-organic solutes; these include ^1H, ^7Li, ^{13}C, ^{19}F, ^{23}Na, ^{31}P, ^{35}Cl, ^{37}Cl, ^{39}K, ^{87}Rb, and ^{133}Cs. Most of the commonly utilized NMR-receptive nuclei have a spin quantum number, I, of $1/2$; examples include ^1H, ^{31}P, ^{19}F, and ^{13}C. However, some have $I > 1$, and in these cases the nuclei are no longer simple dipoles (classically "spherical") as described in Section 2.1, but are quadrupoles ($I = 1$) or octopoles ($I \geq 3/2$); for example, the ^2H nucleus has a nuclear quadrupole moment. Consequently, the interaction of these nuclei with B_0 and other local magnetic fields is more complex and, additionally, these nuclei also interact with electrostatic field gradients.

2.4. NMR Relaxation Times

2.4.1. T_1

After excitation by r.f. irradiation a nuclear ensemble loses its energy to the surrounding "lattice" at a rate which is characterized by the rate constant $1/T_1$, where T_1 is called the *longitudinal relaxation time*. In recognition of the primary mechanism of the relaxation it is also referred to as the spin-lattice relaxation time (Abragam, 1978).

2.4.2. T_2

As detailed in Section 2.1, the net magnetization of a sample (M_o) undergoes reorientation in the magnetic field of the NMR spectrometer when r.f. energy is absorbed by the sample nuclei. The precession of this magnetization is detected as the emf induced in the receiver coil whose axis of symmetry is orthogonal to the direction of the main magnetic field. The rate constant characterizing this loss of NMR signal from the transverse plane is $1/T_2$; hence, T_2 is called the *transverse relaxation time*. The relaxation process is an adiabatic one; mechanistically for simple nuclei ($I = 1/2$) it involves the coupled flipping of two nuclei in opposite directions between the high and one low energy states. Thus, T_2 is also called the spin–spin relaxation time (Günther, 1980; Shaw, 1984; Homans, 1989).

2.4.3. $T_{1\rho}$

By the use of an appropriate r.f. pulse sequence (e.g., Hennig and Limbach, 1982), it is possible to "lock" the magnetization of a nuclear ensemble in the transverse plane and measure the rate of spin-lattice relaxation in this new orientation. Since the strength of the r.f. (rotating) magnetic field is much less than that of the main field (B_0), and since the rate of relaxation depends on the field strength, the relaxation time is usually less than T_1. The mechanism of the relaxation is the same as for T_1 but it takes place in the rotating magnetic field; hence it is called the *rotating frame* T_1 and is denoted by $T_{1\rho}$ (e.g., Shaw, 1984; Homans, 1989).

2.4.4. Enhancement of Relaxation Rates

All three of the above relaxation rate constants are able to be made larger by various means; this change can be the basis of the discrimination between solute molecules inside and outside membrane-enclosed spaces. A common means to enhance this rate in the extracellular compartment is the addition of a membrane-impermeant paramagnetic species such as Mn^{2+} and certain complexes of it (Chapman *et al.*, 1973; Section 5.1.2).

2.5. Suppressing the Water Signal in ^1H NMR Spectra

2.5.1. General

When using ^1H NMR spectroscopy to measure solute transport, a major problem is coping with the dynamic range of the signals. Specifically, it is often necessary to observe signals from solutes at a concentration below 1 mM and yet in most cases there is a potentially interfering signal that is much more intense, from the ~ 80 M ^1H_2O protons of the solvent water, and the ~ 30 M total concentration of protein protons, in cells. Even in modern NMR spectrometers the peak intensities in spectra, acquired serially, vary by $\sim 1\%$ so that difference spectra cannot be used for isolating a small peak next to a large one ~ 100 times its size (Redfield, 1985). Thus, it is advantageous to reduce the size of the ^1H_2O signal *before* it reaches the digitizer of the spectrometer; there are basically three categories of procedure for achieving this.

2.5.2. Chemical Methods

The most obvious way to reduce the ^1H_2O resonance in spectra of cells is to wash the cells centrifugally in a medium made with ^2H_2O (Kuchel, 1989); the ^2H_2O also provides an internal field/frequency lock for the spectrometer. The ^2H_2O may produce subtle isotope effects which can alter protein–protein association and conformations, and the rates of enzyme-catalyzed reactions (Kuchel, 1989); however, the method has been used often since the early NMR studies of water transport in nerves (Fritz and Swift, 1967) and erythrocyte metabolism (F. F. Brown *et al.*, 1977).

2.5.3. Radio-frequency Irradiation of ^1H_2O Nuclei

The equilibrium nuclear magnetization (M_o) of ^1H_2O can be reduced to zero by selectively irradiating the sample at the absorption frequency of the water in order to equalize the number of nuclei in the high- and low-energy states (Jesson *et al.*, 1973; Campbell *et al.*, 1974). This method can be combined with many of the other solvent-suppression techniques to increase the overall effect (Kuchel, 1989).

2.5.4. Selective Nonexcitation of ^1H_2O Nuclei

The ^1H_2O resonance intensity can be reduced in spectra by simply avoiding the excitation of water when acquiring the FID. There are fundamentally four types of method used to achieve this.

Correlation spectroscopy, which although useful in earlier studies of bacterial metabolism, has not found general use in studies of membrane transport (Dadok and Sprecher, 1974; Gupta *et al.*, 1974; Ogino *et al.*, 1978).

Soft-pulse sequences such as the "2–1–4" sequence encompass a relatively broad spectral region over which ^1H_2O is selectively not excited (Redfield, 1978). The technique has, however, been largely superseded by hard-pulse procedures because of their ease of implementation in contemporary NMR spectrometers.

Hard-pulse sequences are those which are designed to nutate the nuclear magnetization in some spectral regions by a large angle, while not perturbing others, such as that of ^1H_2O. The simplest of the procedures are the 1–1 (Clore *et al.*, 1983), 1–$\bar{1}$ (also called the jump and return, or JR sequence; Plateau *et al.*, 1983), and 1–$\bar{2}$–1 (Sklenár and Starcuk, 1982) sequences. The integers denote the relative durations of nonselective (hard) r.f. pulses and the overbars denote a 180° phase shift of the pulse. The rationale of the design of these pulse sequences, and even more elaborate ones, is given in detail elsewhere (e.g., Kuchel, 1989).

Selective excitation of protons by polarization transfer from one nucleus to another can be used to avoid excitation of the protons of ^1H_2O. Thus, the "reverse" polarization transfer pulse sequences, such as reverse-INEPT (Freeman *et al.*, 1981; Bendall *et al.*, 1981), reverse-DEPT (Bendall *et al.*, 1983), and reverse-POMMIE (Bulsing *et al.*, 1984), have the potential to achieve ^1H_2O suppression in heteronuclear experiments that employ ^1H detection. The homonuclear polarization transfer (HPT) experiment is a homonuclear modification of the refocused-INEPT experiment (Burum and Ernst, 1980; Dumoulin and Williams, 1986); the pulse sequence discriminates between coupled and uncoupled resonances by polarization transfer within a homonuclear spin system. Pulse-phase alternation is used to cancel resonances from uncoupled spins and to add signals from those which are coupled. Although the sequence can successfully suppress solvent singlets (Dumoulin and Williams, 1986), it does not avoid the dynamic range problem, alluded to above (Section 2.4.1), and relies on precise signal subtraction to achieve selectivity. While these techniques could in principle be applied to studies of solute transport in cell suspensions, there appear not to be any relevant reports.

Spin-lock purge pulses can be used to obliterate the net magnetization of the ^1H_2O, while leaving desirable magnetization unperturbed. This approach has been used in inverse heteronuclear experiments that employ proton detection (Messerle *et al.*, 1989). These experiments are designed so that an r.f. pulse of relatively high intensity (1–2 msec duration) is applied along either the x or y axis of the rotating frame of reference such that it is coaxial with the desired magnetization, but not with that of the water protons. This pulse effectively locks the desired magnetization along the axis of the r.f. pulse but leads to the randomization of the ^1H_2O magnetization because of the inhomogeneity of the r.f. magnetic field. More than one spin-lock field can be used in a pulse sequence, thus enhancing the extent of suppression of the ^1H_2O signal. While this technique has found widespread use in studies of proteins, it appears not to have been applied in NMR studies on membrane transport.

3. CAUSES OF TRANSMEMBRANE CHEMICAL SHIFT DIFFERENCES

3.1. General

Any procedure, including NMR, which is used to measure the transport of a solute across a membrane depends on some property of the solute that distinguishes it on either side of the partition. In NMR experiments the properties are those of one or more of the nuclei in the solute and include their relaxation times (Section 2.4) and, in particular, various phenomena which give rise to transmembrane differences in the chemical shifts of the nuclei. The various phenomena that give rise to these transmembrane chemical shift differences are specifically discussed in the following sections.

3.2. Magnetic Susceptibility

The resonance frequency (v) of a population of NMR-receptive nuclei is directly proportional to the strength of the magnetic field B_0 (strictly this is called the *magnetic induction field;* Bleaney and Bleaney, 1983; Dobbs, 1984) in their immediate environment. This outcome is expressed in the Larmor equation [Equation (4), Section 2.1]. In turn, the magnitude of the magnetic induction created by a magnet is dependent on a bulk property of the medium, called the magnetic susceptibility, χ (Dobbs, 1984). A major complication with this apparently simple picture of magnetizable material arises because, if an NMR sample consists of two compartments (denoted i for inside and o for outside) both of which contain a solute with an NMR-receptive nuclide, and the respective magnetic susceptibilities χ_i and χ_o of the two compartments are different, then an NMR spectrum will contain one or two separate peaks depending on the *shape* of the inner and outer compartments. The reason for this shape dependence is that the applied magnetic field causes the average alignment of all of the molecular magnets in the sample such that at the surface of the compartment (body) there will be unpaired north poles at one end and unpaired south poles at the other. This polar distribution yields a so-called depolarizing field which opposes the main field inside the compartment.

An early deduction from the theory of fields of force, developed by Poisson in the early 19th century (Hele-Shaw and Hay, 1901; Maxwell, 1954; see also Kuchel and Bulliman, 1989; Mendz *et al.*, 1989), was that the magnetic field *inside* an isolated sphere, placed in a previously uniform imposed magnetic field, is uniform, and its magnitude is *independent* of the radius of the sphere. The greater the difference between χ_i and χ_o, the greater will be the inhomogeneity brought about in the imposed field in the region *outside* the sphere. An important result for practical NMR spectroscopy is that if a reference compound is placed in a small spherical bulb inside the sample (so that the radius of the glass bulb is less than $\sim 1/3$ of the radius of the NMR tube), then the chemical shift of the peak will be the same as that for the compound if it were in the outer solution. This somewhat curious result is accounted for by assuming that there exist so-called "spheres of Lorentz" around each solute molecule (Chu *et al.*, 1990); thus, inside the spheres of Lorentz the magnetic field is uniform, as is the *average* field inside the whole glass sphere. The depolarizing field from the large sphere and the "sphere of Lorentz" cancel each other exactly [because, as stated above, the field strengths do not depend on the radius of a sphere but merely on the polarization density at the surface of each (Bleaney and Bleaney, 1983)]; in other words the field intensity is the same, but of opposite sign, for the glass sphere and the "sphere of Lorentz." Thus, in practice, chemical shift reference compounds should be placed in a small spherical bulb, coaxial with the sample tube; also a small glass sphere with a cylindrical capillary stem is used in one of the most reliable methods for measuring the relative magnetic susceptibility values of solutions (Frei and Bernstein, 1962; Batley and Redmond, 1982).

With human erythrocytes in suspension the intra- and extracellular magnetic fields will be nonuniform for the following reasons. (1) Even an isolated biconcave disc has a nonuniform internal field. (2) The close proximity of other cells, even if they are spheres, imposes a nonuniform outer field thus contributing further to a nonuniform "overlap" field inside the cells (Brown, 1983; Brown *et al.*, 1983). (3) As shown in Figure 4 there is an additional effect of orientation of

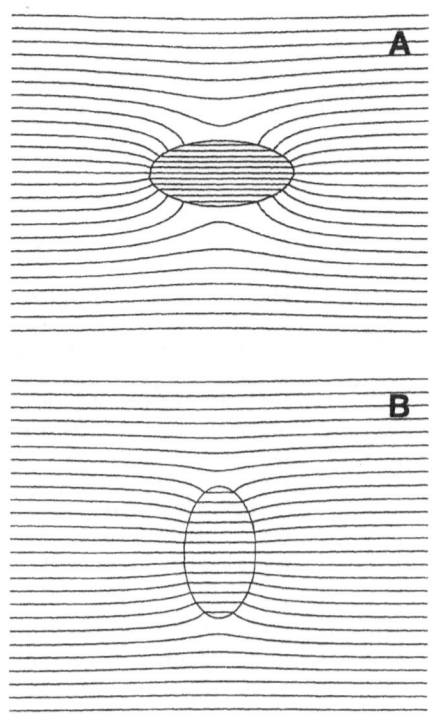

FIGURE 4. Representation of the lines of force that characterize the space- and time-averaged magnetic field in and around a prolate spheroid that serves as a model of a cell. (A) Spheroid oriented with the long axis in the direction of the field and (B) spheroid orthogonal to the field. Note the higher inner field strength, as indicated by the closeness of the lines of force, in A. The axial ratio of the cross-sectional ellipse was 1:2. For the sake of illustrating the uniformity of the inner field the ratio between the magnetic permeabilities inside and outside was 100:1, which is vastly more than would normally be encountered with cells in suspension. Based on Kuchel and Bulliman (1989).

a nonspherical cell on the strength of the field inside an isolated cell. In the case of a spheroidal cell, note that the field is uniform inside the spheroid but the field strength, which is proportional to the separation distance of the lines of force, is dependent on the orientation of the spheroid relative to the imposed field. (4) In addition, the field in the central region of a *series* of confocal spheroids is also always uniform if the imposed field is uniform (Figure 5; Kuchel and Bulliman, 1989). Thus, from these theoretical analyses it is seen that the mean magnetic field, in nonisolated and/or nonspherical cells with high packing density, will be different inside and outside and thus may give rise to different mean resonance frequencies from a solute that is in both compartments.

3.3. pH

Nuclei that are proximal to a titratable group in a molecule often resonate at different NMR frequencies that depend on the protonation state of the group. If the protonation reaction is fast on the NMR time scale the chemical shift of the resonance will be pH-dependent (Moon and Richards, 1973); this is because the chemical shift of the single peak in the spectrum is the average of the chemical shifts of each of the protonated and unprotonated species, weighted according to

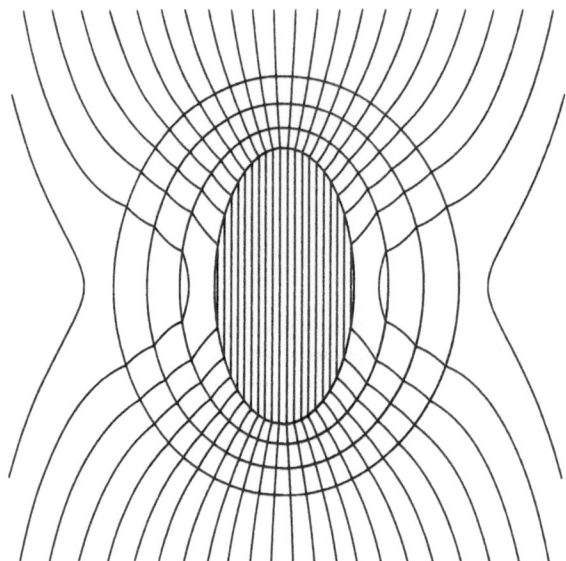

FIGURE 5. Lines of magnetic force in and around four confocal oblate spheroids that serve as a model of a cell with a nucleus and cytoplasmic regions of different magnetic susceptibility. In calculating the field lines the relative magnetic permeabilities were set to 10,000, 1000, 100, 10, and 1 from the "nucleus" to the outside. Note the uniformity of the field in the "nucleus." Adapted from Kuchel and Bulliman (1989).

the relative amounts of the two species. The *NMR time scale* in this context is the mean lifetime of the high-energy nuclear magnetic state, which is $\sim T_1$. This phenomenon of chemical-shift-averaging forms the basis of the widely used procedure for measuring pH in cellular systems (e.g., Iles, 1981; Okerlund and Gillies, 1988; Kuchel, 1989).

In whole human blood the pH of the plasma is normally 7.4 while inside the erythrocytes it is ~ 7.2. In many media this ΔpH of 0.2 is maintained and it leads to the ^{31}P NMR spectrum of red cells having separate peaks from the orthophosphate inside and outside. In one study the pH outside was reduced to 6.4 so the chemical shift difference was made even greater and the uptake of orthophosphate was able to be recorded over many minutes (Brauer *et al.*, 1985). Methylphosphonate has been well investigated as a pH-probe molecule for cellular systems (Slonczewski *et al.*, 1981; Labotka and Kleps, 1983; Labotka, 1984; Stewart *et al.*, 1986); it has a pK_a of ~ 7.6 which varies only slightly with ionic strength, and possibly also in the presence of divalent cations, although the effect is much less than with orthophosphate (Stewart *et al.*, 1986). The uptake into human erythrocytes can be readily monitored under physiological conditions because the intracellular resonance is shifted by ~ 0.2 ppm to high frequency as it enters the lower pH environment. At 37°C the transport occurs on a time scale of many minutes so it can be recorded with sequential NMR spectra. The uptake is blocked by the stilbene derivatives 4,4′-diisothiocyanatostilbene-2,2′-disulfonic acid (DIDS) and 4,4′-dinitrostilbene-2,2′-disulfonic acid (DNDS) (Stewart *et al.*, 1986; Kuchel *et al.*, 1992) which are band 3 (capnophorin) inhibitors.

Imidazole, which has a pK_a of ~7, yields ^{1}H NMR resonances with different chemical shifts inside and outside erythrocytes when added to a suspension of these cells (Rabenstein and Isab, 1982). The extremes of the chemical shift of imidazole at low and high pH differ by ~1.0 ppm, and a ΔpH of 0.2 under physiological conditions gives a chemical shift difference of ~0.04 ppm. Since the transmembrane exchange of imidazole is relatively slow, sequential spectra can be recorded in order to monitor its transport into or out of the cells. The range of compounds with pH-sensitive ^{1}H NMR chemical shifts that are sufficiently large to give resolution of inside and outside peaks is small in comparison with, say, ^{31}P-containing compounds. The problems lie with: (1) the small chemical-shift range (the *dispersion*) of only ~12 ppm for hydrogen nuclei overall, although this range does not include some extreme examples of paramagnetic and ring-current induced shifts of protons adjacent to heme rings in hemoglobin (Han *et al.*, 1989); and (2) there are few titratable groups, with adjacent covalently bonded protons, with pK_a values in the physiological pH range; imidazole and its derivatives are among the more useful ones (Kuchel, 1981, 1989; Ferrige *et al.*, 1979).

Synthetic phospholipid vesicles are more stable than most cells, under extremes of both pH and ionic strength. Thus, whereas the intra- and extravesicular vinyl proton resonances of fumarate (or maleate) are coincident in a suspension of small unilamellar vesicles prepared with an internal pH of 7.0, the resonances from the two compartments can be discriminated when the extravesicular pH is adjusted to be close to the pK_a of the ionizable carboxyl group (4.0–5.5). The sensitivity of the vinyl proton chemical shifts to the protonation state of the adjacent carboxyl group enables the transport of these compounds to be monitored at appropriate extravesicular pH values using sequential ^{1}H NMR spectra (Cramer and Prestegard, 1977; Prestegard *et al.*, 1979).

The chemical shift range in ^{19}F NMR is large (~400 ppm; Bruker Almanac, 1993), so ^{19}F atoms proximal to titratable groups can be used to measure intracellular pH (Deutsch and Taylor, 1987). A number of titratable fluorinated compounds (including trifluoroethylamine and fluorinated α-methylamino acids) have been shown to give separate intra- and extracellular ^{19}F NMR resonances when added to a variety of different cell types (Taylor *et al.*, 1981; Taylor and Deutsch, 1983). The transmembrane chemical shift difference is related primarily to transmembrane pH gradients, though it is likely that hydrogen-bonding effects also contribute to this effect.

3.4. Intra- or Extracellular Binding

When a compound or ion binds to another one, the change in the nuclear magnetic environment may be such as to lead to a change in the corresponding NMR chemical shift (Dwek, 1973). For example, fast-exchange averaging of chemical shifts occurs when Mg^{2+} binds to ATP, and this leads to a decrease in the separation between the β and γ resonances up to 2.33 ppm (Gupta *et al.*, 1978a). This phenomenon can be used to measure the intracellular free Mg^{2+} concentration in human erythrocytes (Gupta *et al.*, 1984). In contrast, when ATP binds to hemoglobin, there is no change in the respective ^{31}P chemical shifts (Gupta *et al.*, 1978b). When 2,3-bisphosphoglycerate binds to hemoglobin, a readily measured increase occurs in both the 2-P and 3-P chemical shifts (Marshall *et al.*, 1977). The binding occurs more avidly to deoxyhemoglobin than oxyhemoglobin (Labotka and Schwab, 1990) so the chemical shifts can, in principle, be used to estimate the oxygen partial pressure in a system of perfused erythrocytes (Lundberg *et al.*, 1992).

An ionic or hydrophobic interaction may take place between a solute and macromolecules or membranes in much the same way as a solute will partition between solvent phases. The interaction could appear like a nonspecific nonsaturable binding. For example, the intraerythrocyte ^{31}P NMR resonances of phenylphosphinate and diphenylphosphinate are shifted and broadened relative to the extracellular resonances; the mechanism probably involves interaction of the benzyl moieties with hydrophobic amino acid residues of hemoglobin (Kirk and Kuchel, 1988b). The interation also occurs with lysozyme and it is expected to occur with most proteins since nearly all possess hydrophobic regions (Kirk and Kuchel, 1988b).

3.5. Shift Reagents

The unpaired electrons in the valence orbitals of the rare earth metals in molecular complexes are often anisotropically distributed about the nucleus and so lead to a paramagnetic dipolar interaction (through space) with an adjacent nucleus, thus shifting its resonance frequency (Williams, 1982; Springer, 1987). The shift can be to higher or lower frequency depending on the geometry of the complex (Günther, 1980; Williams, 1982). Shift reagents have been very successfully applied to move particular peaks in unresolved regions of NMR spectra, thus aiding in the assignment of the peaks. The reagents function best in hydrophobic solvents and few of them have been discovered that can be applied in physiological media. Nevertheless, there are currently three main classes of shift reagents that can be used successfully in physiological experiments.

3.5.1. Dysprosium Complexes

The lanthanide ion Dy^{3+} forms a $2:1$ complex with tripolyphosphate $[Dy(PPP)_2^{7-}]$ that is highly stable in physiological media. $^{23}Na^+$ ions in the medium are in fast exchange binding with the complex and thus experience an altered time-averaged nuclear magnetic field that causes a change in their chemical shift (to low frequency); this is a so-called pseudocontact shift. The tripolyphosphate and the complex appear not to enter erythrocytes or phospholipid vesicles. Hence, the ^{23}Na NMR spectrum obtained from a suspension of erythrocytes, after the addition of ~5.0 mM of the complex, shows separate ^{23}Na$^+$ resonances from the intra- and extracellular species (Gupta and Gupta, 1982; Ogino et al., 1983; Pettegrew et al., 1984; Boulanger et al., 1985; Anderson et al., 1988). A serious limitation in the use of the complex is the insolubility of Ca^{2+} and Mg^{2+} adducts of tripolyphosphate; these not only precipitate and reduce the free concentration of the metabolically important ions, but the decrease in tripolyphosphate concentration also decreases the effect on the ^{23}Na$^+$ shift (Chu et al., 1984).

The 1:1 complex of Dy^{3+} and triethylenetetraaminehexaacetic acid [Dy(TTHA)$^{3-}$] also shifts the ^{23}Na$^+$ resonance in spectra from aqueous samples (Chu et al., 1984) although not to the same extent, at a given concentration, as Dy(PPP)$_2$$^{7-}$. However, the shift reagent is less affected by interactions with Ca^{2+} and Mg^{2+} than is Dy(PPP)$_2$$^{7-}$, so it may be more useful in some applications.

The shift effects induced on ^{23}Na$^+$ by the above dysprosium complexes are also observed in cellular or vesicular systems with other NMR-receptive alkali metal cations including ^{7}Li$^+$ (Pettegrew et al., 1987; Espanol and Mota De Freitas, 1987; Partridge et al., 1988; Hughes et al., 1988a,b; Thomas et al., 1988; Espanol et al., 1989), ^{39}K$^+$ (Brophy et al., 1983; Riddell and Arumugam, 1989), and ^{89}Rb$^+$ (Helpern et al., 1987, 1989; Allis et al., 1989; Endre et al., 1989).

3.5.2. Thulium and Lutetium Complexes

While Dy^{3+} has been the most frequently used lanthanide in the present context, the complex of thulium (Tm^{3+}) and tripolyphosphate, which has a less dramatic paramagnetic effect, can also be used as a shift reagent (Chu et al., 1984). Furthermore, if a shift of the resonance in the opposite direction is required, then the most potent diamagnetic ion that can be used is lutetium (Lu^{3+}).

3.5.3. Noncomplexed Ions

An example of a noncomplexed diamagnetic ion is Co^{2+}; at 40 mM it causes the intracellular ^{1}H$_2$O resonance, in the ^{1}H NMR spectra of frog sciatic nerve, to be moved to high frequency of the extracellular resonance (Fritz and Swift, 1967). This enabled an estimate of the exchange rate of water across the cell membranes by measuring the change in linewidth of the water as a function of temperature. The data were interpreted as indicating that the pseudo-first-order rate constant for proton exchange between the inside of the nerves and the outside decreases on depolarization of the nerves, and also that the transverse relaxation time of the intracellular water decreases significantly on depolarization (Fritz and Swift, 1967).

The paramagnetic species, Pr^{3+}, has been used as a noncomplexed ion to separate the intra- and extravesicular resonances of acetic acid in a study of the kinetics of the uptake of this carboxylic acid by vesicles (Alger and Prestegard, 1979; Section 6.4). The use of free ions of this type, however, appears to be restricted to only a few situations where complexes with biological molecules are not required.

3.6. Ionic Composition of the Media

3.6.1. $^{133}Cs^+$

The chemical shifts of ionic species are influenced by several factors including, obviously, ion–ion interactions. For example, $^{133}Cs^+$ NMR resonances in cesium halide solutions show large shifts which vary with concentration, temperature, and nature of the solvent; the effects arise from short-range ion–solvent and ion–ion interactions which can be described quantitatively in terms of temperature and concentration (Halliday *et al.*, 1969). The variation of chemical shift with salt concentration is markedly nonlinear down to 0.2 M. Counterions also form a sequence in terms of their effectiveness to induce the shifts. Nitrate causes a shift to low frequency, and hydroxide and the halides to high frequency; for the halides the effectiveness is $F^- < Cl^- < Br^- < I^-$. It is noteworthy that the order of the halide sequence is reversed for Na^+ and Li^+ (Halliday *et al.*, 1969).

3.6.2. $^{19}F^-$

The physical basis of the separation of the resonances of $^{19}F^-$ in suspensions of human erythrocytes (Chapman and Kuchel, 1990) is possibly the same as for $^{133}Cs^+$ (Halliday *et al.*, 1969; Davis *et al.*, 1988; Wittenkeller *et al.*, 1992); the $^{19}F^-$ linewidth decreases with an increase in F^- concentration, suggesting that binding to macromolecules contributes to the effect. The peak separation of $\sim$1.7 ppm (650 Hz at 376.47 MHz) occurs at a concentration of $\sim$50 mM, with the intracellular resonance at higher frequency. The exchange, which takes place via band 3, is sufficiently fast to be measured using saturation transfer (Section 5.6) and 1D EXSY analysis (Section 5.8).

3.7. Hydrogen Bonding

3.7.1. ^{31}P Compounds

When the highly water-soluble uncharged liquid, dimethyl methylphosphonate (DMMP), is added to a suspension of human erythrocytes, the ^1H-decoupled ^{31}P NMR spectrum contains two resonances (Kirk and Kuchel, 1985, 1986a,b).

Dilution of the suspension with isotonic medium leads to a relative diminution of the intensity of the low-frequency peak. Lysis of the cells by freezing and thawing the sample yields one resonance at a frequency intermediate between the two previous ones. Therefore, it was deduced that it is the intracellular population of solute nuclei that resonate at lower frequency (Kirk and Kuchel, 1985). Reducing the cell volume by the addition of concentrated NaCl to the suspension causes an increase in the peak separation; the cell volume dependence of the separation frequency can be calibrated, thus enabling cell-volume determinations to be made during ^{31}P NMR spectral time courses (Kirk and Kuchel, 1985; Raftos et al., 1988).

DMMP is not alone in showing this unusual chemical shift behavior; a similar effect is seen for a number of other phosphoryl compounds when added to erythrocyte suspensions (Kirk and Kuchel, 1988b,c). Many of these compounds do not have pK_a values in the physiological pH range, so the "split peak phenomenon" cannot be the result of transmembrane ΔpH. The transmembrane susceptibility gradient probably contributes to the chemical shift difference (Fabry and San George, 1983; Kirk and Kuchel, 1988a); however, it seems likely that in most cases the peak separation is largely a consequence of the disruption by intracellular hemoglobin of the hydrogen bonding of the phosphoryl oxygen to intracellular water. Hydrogen bonding at the oxygen atom has the effect of decreasing the electron shielding at the phosphorus nucleus. Thus, a disruption of hydrogen bonding increases the shielding and hence decreases the ^{31}P NMR chemical shift.

In experiments with a range of different solvents (that differ in their tendency to form hydrogen bonds, as assessed for example by infrared spectroscopy), it was shown that there is a good correlation between the extent to which the ^{31}P NMR chemical shifts of the phosphoryl compounds vary with changes in the hydrogen-bonding conditions and the extent to which hemoglobin influences the chemical shifts (Kirk and Kuchel, 1988b). Furthermore, it was demonstrated that lysozyme, a protein that is structurally unrelated to hemoglobin, causes similar variations in the ^{31}P NMR chemical shifts, indicating that the phenomenon is a property of proteins in general rather than being unique to hemoglobin (Kirk and Kuchel, 1988b).

3.7.2. ^{19}F Compounds

Halogenated compounds are known to form medium to strong hydrogen bonds with water and other compounds (Günther, 1980). Thus, when 3-fluoro-3-deoxy-D-glucose is added to a suspension of erythrocytes, separate intra- and extracellular resonances are evident in the ^{19}F NMR spectrum; in contrast to the phosphoryl compounds the intracellular resonance is shifted to high frequency relative to the extracellular one (Potts et al., 1990; Potts and Kuchel, 1992). The explanation for the different responses is as follows: The phosphoryl compounds form hydrogen bonds between the oxygen atoms that are vicinal to the phosphorus, whereas in the fluorinated compounds the hydrogen bond forms directly with the atom containing the "reporter" nucleus. The increase in electron density on formation of a hydrogen bond shields the ^{19}F nucleus and reduces its resonance frequency. Inside the cells the extent of hydrogen bonding of the ^{19}F to water is reduced because the water interacts with proteins, so the ^{19}F resonance moves to higher frequency.

The split peak effect also occurs with trifluoracetate, a membrane-permeable anion with a $pK_a \sim 0.3$ (Merck Index, 1989), thus making it a useful probe of membrane potential (London and Gabel, 1989). The experiment is the ^{19}F NMR counterpart of the ^{31}P NMR one which uses hypophosphite (Kirk *et al.*, 1988). Another related compound that can be used as a membrane potential probe in either ^{19}F or ^{31}P NMR experiments is difluorophosphate (Xu and Kuchel, 1991; Xu *et al.*, 1991).

Split peaks are evident in the ^{19}F NMR spectra obtained from the brains of dogs *in vivo* that had been infused with 3-fluoro-3-deoxy-D-glucose (Nakada *et al.*, 1988); although the authors did not comment on this, the likely explanation for the splitting is the differential hydrogen bonding in the cerebral compartments.

3.7.3. ^{13}C Compounds

The chemical shift of the carbonyl-^{13}C of acetone varies in solvents that can be arranged in sequence according to their tendency to form hydrogen bonds with the carbonyl oxygen (Maciel and Ruben, 1963; Maciel and Natterstad, 1965). ^{13}C NMR spectra of human erythrocytes to which D-[1-^{13}C]glucose is added reveal an unresolved splitting of the resonances of the β- but not the α-anomer (Kuchel *et al.*, 1987b; Section 5.9.2). The most probable explanation for the phenomenon is the reduced extent of hydrogen bonding between solvent water and the glucose hydroxyls that occurs inside the cells because of the high protein concentration. A similar effect is seen with [^{13}C]urea in suspensions of these cells (Potts *et al.*, 1992; Section 5.10.3), probably as a result of a combination of magnetic susceptibility and hydrogen-bonding effects.

3.8.   Covalent Chemical Modification

3.8.1.   General

The chemical shift of a nucleus in a transported solute will probably change if, on crossing the membrane, the solute undergoes conversion to a different compound. This phenomenon has been used, for example, to monitor the uptake of arginine by human erythrocytes (Kuchel *et al.*, 1984): endogenous arginase in the erythrocytes rapidly catalyzes the hydrolysis of arginine to ornithine. The hydrolysis can be monitored by means of ^{1}H spin-echo NMR, from the decreasing intensity of the arginine δ-proton resonance (δ 2.99) and the concomitant increase in that of ornithine δ-protons (δ 2.81). Since the rate of formation of ornithine is ~ten times faster in a hemolysate than in intact cells, it was concluded that the rate-limiting step in the overall reaction sequence is the transport of arginine across the membranes. The rate, inferred from a time course of spectra, is similar to that obtained by the more conventional procedure of forming cell extracts (Young and Ellory, 1977; Kuchel *et al.*, 1984).

3.8.2. Caveats

Several important caveats must be mentioned in relation to this method: First, if even a small fraction of the cells undergo lysis during the time course of the reaction, the otherwise intracellular enzyme(s) will be released into the extracellular fluid and will catalyze the formation of the product which is supposed to have been formed inside the cells. Such an outcome has led to an artifactually high estimate of peptide transport rates in human erythrocytes (Young *et al.*, 1987; Kuchel *et al.*, 1987a; Odoom *et al.*, 1990). Second, in addition to avoiding lysis, it is critical to ensure that the metabolic conversion, which is presumed to be intracellular, is not actually catalyzed by enzymes on the exoplasmic surface of the plasma membrane.

4. NMR STUDIES OF SLOW MEMBRANE TRANSPORT PROCESSES

4.1. Studies Exploiting Endogenous Magnetic Field Gradients

The spin-echo pulse sequence ($\pi/2-\tau-\pi-\tau$–acquire; see Section 6.2 for more details) is commonly used to acquire ^{1}H NMR spectra of various cell types because of its ability to filter out NMR signals from large, relatively immobile molecules such as proteins and membrane phospholipids, thus yielding a spectrum consisting predominantly of resonances from small metabolites (F. F. Brown *et al.*, 1977). Under certain circumstances, the NMR spin-echo signal from a small molecule increases as it enters a cell from the extracellular medium. This suggests that the nuclei in the different compartments have different T_2 (T_2^*) relaxation times.

The intensity of a resonance in a spin-echo spectrum, obtained from a sample with isotropic unbounded diffusion, is given by Equation (18) in Section 6.3. It can be seen from this equation that the attenuation of the spin-echo signal becomes larger as the values of the diffusion coefficient, D, and the magnetic field gradient, g, are increased. If the magnetic susceptibility differs between the inside and outside of cells in a suspension, then microscopic magnetic field gradients arise, which are usually greater outside the convex surface of cells than inside (Glasel and Lee, 1973; Brindle *et al.*, 1979; Brown, 1983; Brown *et al.*, 1983; Endre *et al.*, 1983b). As an aside it is useful to note that since carbonmonoxy hemoglobin is diamagnetic (Cerdonio *et al.*, 1985) the magnetic field gradients in suspensions of human erythrocytes can be stabilized by gassing the cells with carbon monoxide before NMR experiments.

Brindle *et al.* (1979) enhanced the endogenous field gradients in a suspension of erythrocytes by using the inert paramagnetic complex of Dy^{3+} and diethylenetriaminepentaacetic acid $[Dy(DTPA)^{2-}]$ and thereby reduced the contribution of the external solute signal to the intensity of the ^1H NMR spin-echo signal from the solute. Under the conditions used, the intra- and extracellular resonances of the compounds were superimposed on one another. However, because the T_2^* in the extracellular compartment was less than that in the intracellular compartment, the net movement of the compounds into the cells gave rise to an increase in the intensity of the corresponding NMR resonances. In analyzing the time-dependent intensity changes, Brindle *et al.* (1979) defined a ratio of peak heights that varied in relation to the concentration of the solute inside the cells; the data were analyzed to yield a membrane permeability coefficient from which an initial transport rate was calculated. The same NMR approach was used by Brown *et al.* (1982) to measure the transport of glycerol into porcine erythrocytes, and into the alga *Dunaliella salina*. Jones and Kuchel (1980) used Fe^{3+} ligated to the ferric iron chelator desferrioxamine to cause suppression of the extracellular choline signal in a suspension of human erythrocytes and thus were able to monitor the transport of this organic cation into the cells.

The ability to use endogenous magnetic field gradients to differentiate between the solute molecules in each compartment is greatest for ^1H because of its high γ value; in fact there appear to be no reports of the use of this method with any other type of nuclide.

4.2. Studies Exploiting Transmembrane Chemical Shift Differences

4.2.1. ^{31}P

Since the ^{31}P NMR resonance of orthophosphate shifts with pH (Moon and Richards, 1973) the transport of this anion into cells can be recorded if a transmembrane ΔpH exists. With human erythrocytes, the extracellular pH rises while intracellular pH falls during orthophosphate transport into the cells because the protonated monoanionic $H_2PO_4^-$ form enters rather than the deprotonated dianionic HPO_4^{2-} form (Brauer et al., 1985).

The transport of other ions, that bring about a change in pH, can in principle also be measured indirectly from the time-dependent changes in the intensities of the orthophosphate resonances.

Methylphosphonate ($H_3CPO_3^-$) is a nonphysiological compound which can be used to measure the transmembrane ΔpH in human erythrocytes under a range of experimental conditions (Labotka and Kleps, 1983; Stewart et al., 1986). The pK_a of methyl phosphonate is $\sim$7.6 under physiological conditions and its ^{31}P NMR resonance moves between two extreme values which are separated by 3.6 ppm, with the shift being to higher frequency as the pH is lowered. The chemical shift of methylphosphonate does not change as much, with changes in the intracellular composition, as orthophosphate (Stewart et al., 1986). The ionic strength dependence of the chemical shift is, however, comparable in magnitude, but opposite in sign, to that of orthophosphate. The entry of this compound into human erythrocytes occurs with a half-life of $\sim$20 min with an initial extracellular concentration of 30 mM. The time course can be readily recorded from a series of ^{31}P NMR spectra because the intra- and extracellular resonance of methylphosphonate are resolved as a result of a transmembrane ΔpH of $\sim$0.2. The influx is inhibited by DIDS and increases as the pH is lowered below 7 (Stewart et al., 1986; Kuchel et al., 1992), which suggests that, as with inorganic phosphate, the monovalent anionic form is transported by band 3.

As a result of the operation of several factors that bring about the separation of intra- and extracellular ^{19}F and ^{31}P NMR resonances of solutes, the rate of influx of a series of phosphate analogues (fluorophosphate, thiophosphate, phosphite, phosphate, methylphosphonate, hypophosphite, and dimethyl phosphinate) into human erythrocytes has been measured and shown to be saturable (Labotka and Omachi, 1987a,b). The factors that caused the peak separation were deemed to include ΔpH and differences in magnetic susceptibility, but it is now recognized that the most important contribution comes from difference in the extent of hydrogen bonding (Section 3.7). Interestingly, it was concluded that the major determinant of the rate was neither the molecular weight of the analogue nor the net charge on the anion, but rather the "shape" of the anion. Phosphite (HPO_3^{2-}), which most closely resembles bicarbonate, has the highest influx rate of the compounds listed above (Labotka and Omachi, 1987a,b).

The net accumulation of phosphoenolpyruvate (PEP) in human erythrocytes, from extracellular medium containing 65 mM of the compound at pH 6.0 in a citrate buffer, has been monitored with sequential ^{31}P NMR spectra over $\sim$1 hr (Hamasaki et al., 1981). The extracellular signal of PEP was suppressed by using 1 mM Mn^{2+}, and in order to prevent the rapid metabolism of PEP once it entered the cells, glycolysis was inhibited with 10 mM NaF.

4.2.2. $^{23}Na^+$

Although it is a quadrupolar nucleus ($I = {}^3/_2$), ^{23}Na has a moderately high relative receptivity (9.25 $\times$ 10^{-2} compared with 1.0 for ^1H; Harris and Mann, 1978; Bruker Almanac, 1993). Also, because of its high extracellular concentration in most living systems ($\sim$140 mM), and relatively low concentration inside cells, a study of its NMR behavior can yield information on the intracellular environment. An extensive literature has developed on the study of $^{23}Na^+$ transport into vesicles and cells. These studies have largely been possible because of the introduction of shift reagents. The most widely used has been the membrane-impermeant paramagnetic complex of tripolyphosphate and the lanthanide dysprosium [$Dy(PPP)_2{}^{7-}$] which was introduced by Gupta and Gupta (1982); however, other reagents (Section 3.5) have been introduced in an attempt to avoid some of the cytotoxicity that attends the $Dy(PPP)_2{}^{7-}$ complex (Springer, 1987). At a concentration of $\sim$5 mM the complex imparts a transmembrane chemical shift separation of $\sim$10 ppm, the shift outside the cells being to low frequency (Ogino *et al.*, 1985). The earliest measurements of $^{23}Na^+$ transport made using $Dy(PPP)_2{}^{7-}$ were carried out by Pike *et al.* (1982); they monitored the uptake of $^{23}Na^+$ into synthetic phospholipid vesicles mediated by the antibiotic gramicidin. The first reported application to the study of Na^+ transport in erythrocytes also revealed an enhancement of uptake of the cation by gramicidin (Ogino *et al.*, 1985).

There have now been numerous applications of the previous methodology to biological systems, including *Escherichia coli* (Castle *et al.*, 1986) and lasalocid A-mediated Na^+ exchange across erythrocyte membranes (Fernandez *et al.*, 1987). Other examples are listed by Grandjean and Laszlo (1987). Studies on vesicles with ionophore mediation of $^7Li^+$, $^{23}Na^+$, and $^{39}K^+$ transport are exemplified by an extensive series of experiments carried out by Riddell and colleagues (Riddell and Hayer, 1985; Riddell *et al.*, 1988a,b; Riddell and Arumugam, 1988, 1989).

Altered cation exchange in tissues is of medical importance and considerable opportunities exist for investigating Na^+ transport using NMR. For example, erythrocytes show enhanced Na^+ influx in essential hypertension (Duhm and Behr, 1987; Kojima *et al.*, 1989) and Na^+ exchange is increased in lymphocytes stimulated to proliferate by a mitogen (Prasad *et al.*, 1987).

Volume regulation is an important response in many cells; it occurs on exposure of the cells to media of different osmolalities. The kinetics of volume regulation in the erythrocytes of the giant salamander *Amphiuma* have been investigated using ^{23}Na NMR and the shift reagent $Dy(TTHA)^{3-}$; this shift reagent appears to be less toxic to these cells than $Dy(PPP)_2{}^{7-}$ (Anderson *et al.*, 1988).

4.2.3. ^{87}Rb$^+$

Radioactive-tracer studies with the rubidium cation, ^{86}Rb$^+$, have established it as a congener of K$^+$ and the energy dependence of its transport into intact HeLa cells has been studied (e.g., Ikehara *et al.*, 1984). The corresponding NMR-receptive isotopes are ^{85}Rb and ^{87}Rb. ^{85}Rb has a natural abundance of 72.15% but its spin quantum number $I = \frac{5}{2}$ ensures very broad resonances and thus renders it much less useful for biological studies than ^{87}Rb which has a natural abundance of 27.95% and a spin quantum number of $I = \frac{3}{2}$ (Harris and Mann, 1978; Bruker Almanac, 1993). The natural receptivity of ^{87}Rb is only half that of ^{23}Na, but 19 times more receptive than ^{39}K (Allis *et al.*, 1989). Like ^{23}Na$^+$, ^{87}Rb$^+$, in the presence of macromolecules, gives a resonance which can be deconvoluted into broad and narrow components (Endre *et al.*, 1989). In isolated perfused kidneys the assignment of the broad and narrow peak components to the intra- and extracellular compartments, respectively, could not be confirmed by the use of the complex of Dy^{3+} and triethylenetetraaminehexaacetic acid [Dy(TTHA)$^{3-}$]; the already-broad natural linewidth of the ^{87}Rb$^+$ peak was merely further broadened by the shift reagent (Endre *et al.*, 1989). An additional complication to the spectral analysis is the fact that high concentrations of the shift reagent are required to shift the ^{87}Rb$^+$ peak in the presence of other cations and protein. Notwithstanding these difficulties it is possible to demonstrate that the resonance from intracellular ^{87}Rb$^+$ is significantly broadened relative to the ion outside and thus it is possible to monitor the accumulation of the ion over a time course of several minutes; plateau levels are attained after ~20 min and persist for over an hour. As with ^{23}Na$^+$, the extent to which the intracellular resonance of ^{87}Rb$^+$ is visible is still a contentious issue (Grandjean and Laszlo, 1987; Endre *et al.*, 1989).

Human erythrocytes, with 10 mM RbCl in the presence of ~10 mM Dy(TTHA)$^{3-}$ shift reagent, give ^{87}Rb NMR spectra in which the intra- and extracellular resonances are resolved. This enables an estimate of the unidirectional influx of the ion to be measured. In one report the rate of influx from a 15 mM solution was 1.8 ± 0.3 mmol/liter cells per hr, which is similar to, but less than, the corresponding rate of K$^+$ influx (Allis *et al.*, 1989).

4.2.4. ^{133}Cs$^+$

^{133}Cs$^+$ is readily detected with NMR spectroscopy because, although the nucleus is quadrupolar (spin quantum number $I = \frac{7}{2}$), the quadrupolar moment is small (Harris and Mann, 1978; Bruker Almanac, 1993). ^{133}Cs$^+$ is 100% naturally abundant and its chemical shift is highly sensitive to the nature and concentrations of counterions in solution (Halliday *et al.*, 1969). This leads to conditions under which the intra- and extracellular $^{133}Cs^+$ pools give resonances which are resolved without the further requirement to introduce extracellular shift reagents. This property of the ion has been used in studies of suspensions of human erythrocytes and perfused rat hearts (Davis *et al.*, 1988; see Section 3.5).

Longitudinal relaxation of $^{133}Cs^+$ in biological media, in all cases, appears to be monoexponential and the rate is 3–4 times faster for the intracellular component compared with the extracellular one. All of the $^{133}Cs^+$ is "visible" in the NMR spectrum, implying total observability of intracellular pools of the ion. Sequential spectra acquired over a period of hours show that the uptake of Cs^+ by human erythrocytes occurs at approximately one third the rate reported for K^+, and this rate is reduced by a factor of two on addition of the Na,K-ATPase inhibitor ouabain (Davis *et al.*, 1988). The shift of the intracellular resonance to *high* frequency relative to that of the resonance from the outside, despite the fact that the Cl^- concentration inside is less (as mentioned in Section 3.6.1, Cl^- causes a shift to high frequency), was thought to be an effect of phosphate ions and, possibly, the nonideality of intracellular solvent water (Davis *et al.*, 1988).

As stated in Section 3.6.1, the basis of the separate resonances from $^{133}Cs^+$ is related to that seen with various phosphoryl and halogenated compounds and has recently been studied in detail (Wittenkeller *et al.*, 1992).

4.2.5. ^{1}H

^{1}H NMR spectroscopy has been used to monitor the uptake of a number of compounds by human erythrocytes by exploiting the fact that the permeant is covalently modified within the cell to give a metabolite with chemical shifts differing from the parent compound. In addition to the example of arginine transport given in Section 3.8.1, the uptake of dipeptides (King *et al.*, 1983; King and Kuchel, 1984, 1985), tripeptides (Vandenberg *et al.*, 1985), and γ-glutamylalanine (York *et al.*, 1984) by human erythrocytes have also been monitored by exploiting the hydrolysis of these solutes that takes place only inside the cells; the enzymes involved in the particular reactions are diglycinase (EC 3.4.13.11), several possible tripeptidases (EC 3.4.16.x), and γ-glutamyl amino acid cyclotransferase (EC 2.3.2.4), respectively.

4.3. Results Based on T_2 Differences

^{35}Cl NMR

^{35}Cl NMR has been used to study the binding of Cl^- to a wide variety of proteins (Forsén and Lindman, 1981). The binding of Cl^- to band 3 of human erythrocytes was studied by Shami *et al.* (1977) who showed that the broadening of the $^{35}Cl^-$ resonance in the presence of erythrocyte ghosts, or their Triton

X-100 extracts, was decreased in the presence of the band 3 inhibitor DIDS. These studies were extended by Falke *et al.* (1984a,b) who demonstrated two types of Cl^- binding sites on the erythrocyte membrane: a high-affinity "transport" site, and a low-affinity one which is possibly involved in "substrate inhibition" of transport. More detailed relaxation analyses have been carried out using ^{35}Cl and ^{37}Cl NMR with Cl^- in the presence of erythrocyte ghosts (Price *et al.*, 1991). DIDS brought about a significant alteration in the relaxation time and the general conclusion was that transmembrane flux of Cl^- is not limited by the rate of chloride binding to the external chloride binding site(s) of band 3.

The first report of the use of ^{35}Cl NMR to measure the net transport of Cl^- in human erythrocytes was that by Brauer *et al.* (1985). They exploited the finding that the ^{35}Cl NMR resonance from intracellular Cl^- is broad (linewidth > 200 Hz) thus rendering the resonance virtually undetectable, while that of the extracellular Cl^- is relatively narrow (linewidth ~ 30 Hz). When phosphate is added to a suspension of human erythrocytes, the one-for-one exchange of phosphate for chloride results in the progressive increase in the intensity of the $^{35}Cl^-$ resonance; initially it is virtually undetectable in the spectrum (derived from 1000 transients acquired in just over 2 min), but there is gradual appearance of a resonance, with a signal-to-noise ratio of $\sim 35 : 1$, over a period of 1.8 hr (Brauer *et al.*, 1985).

4.4.   Indirect Detection of Cation Transport

The transport of a cation across a membrane can alter the NMR signal from another nucleus. An example of this is the change in proton concentration in large unilamellar vesicles that can be inferred from alterations in the resonance frequency of orthophosphate (see Section 3.3); proton transport mediated by valinomycin, gramicidin D, and amphotericin B has been investigated using this phenomenon (Hervé *et al.*, 1985). With the "channel-forming" ionophores (gramicidin D and amphotericin B) two internal phosphate signals were observed suggesting that two vesicle populations coexisted in the suspension. A ^{31}P NMR signal at the initial chemical shift therefore corresponded to those vesicles which were devoid of ionophore molecules, and the other peak was from vesicles with incorporated gramicidin D or amphotericin B. On the other hand, in the presence of the "carrier" ionophore, valinomycin, the intravesicular resonance was not split. The different behavior of the carrier- and channel-forming ionophores is therefore consistent with the latter transferring much more slowly between vesicles (Grandjean and Laszlo, 1987).

Another means of measuring ion translocation is to employ paramagnetic cations such as Pr^{3+} or Eu^{3+}. These ions, when localized in a compartment, interact with the phospholipid head groups on the same side of the membrane and bring about a chemical shift difference between internal and external resonances.

The shifts have been observed in ^{1}H (Hunt, 1975), ^{31}P (Bystrov *et al.*, 1971), and ^{13}C (Shapiro *et al.*, 1975) NMR spectra. Since phospholipid bilayers are essentially impermeable to the lanthanide cations, the incorporation of an ionophore into the membrane induces selective cation translocation. Thus, transport of the ion across the membrane causes a shift of the resonances from the reporter molecules in the compartment into which the transport occurs.

Quantitative measurements using this technique have been carried out by Hunt (1975) in studies of the transport of Pr^{3+} into unilamellar vesicles via the ionophore A23187, which is normally used as a Ca^{2+} translocator. In subsequent studies, different channel-forming agents were used (Hunt and Jones, 1982, 1983; Hunt *et al.*, 1984). There have now been numerous applications of this basic procedure to the study of transport processes in vesicles (Grandjean and Laszlo, 1987).

Rapid relaxation of the magnetization of protons on the lipid head groups in contact with added Mn^{2+} causes a decrease in the signal intensity of the lipid head groups of phospholipid vesicles. Therefore, as Mn^{2+} ions pass into vesicles, the signal declines (Degani, 1978). Further studies have been carried out to monitor Mn^{2+} transport mediated by the ionophore X-547A (Degani, 1978; Degani *et al.*, 1981) and the hormone angiotensin II (Degani and Lenkinski, 1980).

4.5. Multiple Quantum Detection

4.5.1. Multiple Quantum Coherence

An explanation of multiple quantum coherence is beyond the scope of this chapter, but some aspects of the behavior can be readily visualized by analogy with the more familiar single-quantum coherence. An isolated spin-half nucleus has available to it two energy levels brought about by the imposed magnetic field; these are called the Zeeman energy levels or eigenstates (see Section 2.1). The difference in energy, ΔE, between these eigen states is related to the absorption frequency v by the Planck equation ($\Delta E = hv$, where h is Planck's constant). In a description of NMR involving populations of high- and low-energy nuclei a single-quantum transition occurs when a nuclear spin is promoted from the lower to the upper eigenstate by absorption of an appropriate quantum of energy. Experimentally we do not detect that transition directly; instead we detect the evolution of the magnetization in the x,y plane as described in Section 2.1. Thus, we detect what is called the single-quantum coherence which is brought about by the r.f. pulse.

Analogously, two nuclear spins which are within three or four chemical bonds of one another, and are thus *J*-coupled (see Section 2.2), have available four energy levels. In addition to the set of possible single-quantum transitions, there are also double-quantum transitions which connect the eigenstate in which both nuclear spins have low energy to that in which they both have high energy. A zero-quantum transition corresponds to the mutual swapping of low- and high-energy eigenstates of the paired spins. It is obvious that these transitions, which cannot be detected in conventional NMR experiments, are only present in J-coupled systems. Magnetization can, however, be made to evolve at the double-quantum frequency during the period that follows an appropriate sequence of r.f. preparation pulses. This evolution of states is called double-quantum coherence. The evolution is not directly detected but its effects can be manifest after suitable additional r.f. pulses followed by acquisition of the FID.

Nuclei with $I > 1/2$ have more than two eigenstates even in the absence of J-coupling, and, under special circumstances, multiple-quantum coherences can be brought about by sequences of r.f. pulses that "expose" the evolution of eigenstates via multiple quantum coherence. Coherence, unlike simple magnetization (or polarization), is oscillatory in nature and really refers to the ordering or phase of a nuclear spin system. It is a more complex concept than magnetization and concerns the relationships between the wave functions which describe a spin system. There is no simple physical model which can be applied to the concept but we can conveniently state that coherence can be *transferred* between energy levels by r.f. fields in much the same way that transitions can occur between populations.

4.5.2. Multiple Quantum Detection of ^{23}Na$^+$

Since ^{23}Na has spin quantum number $I = 3/2$, multiple quantum coherences can be brought about under special conditions. Thus, an interesting development for the selective detection of intracellular ^{23}Na$^+$ is double-quantum filtering (Pekar and Leigh, 1986; Pekar et al., 1987; Jelicks and Gupta, 1989a,b). The electric quadrupole moment of the ^{23}Na nucleus interacts with fluctuating electric field gradients at sites on macromolecules which bind Na$^+$, such as proteins in cells; this causes a lack of redundancy in the energy differences between the four possible eigenstates, with the result that double-quantum transitions can be observed. In the presence of Na$^+$-binding macromolecules the "outer" eigenstate transitions ($3/2 \rightarrow 1/2$ and $-1/2 \rightarrow -3/2$) occur faster during relaxation than the "inner" eigenstate transitions ($-1/2 \rightarrow 1/2$), even though only a small fraction of the Na$^+$ may be bound to the macromolecule; this leads to a biexponential NMR relaxation time course (Hubbard, 1970). It can be shown that the biexponentially relaxing ^{23}Na$^+$ can be "passed" through a state of double-quantum coherence while that which relaxes monoexponentially cannot (Pekar and Leigh, 1986).

Quantitative distinction between bi- and monoexponential relaxation is usually imprecise if the two relaxation times do not differ by a factor of more than 5. Rapidly relaxing transitions decay so rapidly that the signal resulting from the outer transitions in $^{23}Na^+$ may be so fast as to be effectively "invisible." The measurement of the two relaxation times is, however, aided by using a one-dimensional multiple-quantum filter during signal acquisition (Bax, 1984). The requisite pulse sequence causes the nuclear spin system to pass through a state of multiple-quantum coherence, and since the response of a spin I = $^3/_2$ system can be quantitatively predicted, it is possible to estimate the relaxation times (Bax, 1984). Thus, by measuring $^{23}Na^+$ peak intensity as a function of a double-quantum "creation time" (one of the delays in the pulse sequence), the two relaxation times of $^{23}Na^+$ in an albumin solution can be measured even though the relaxation rates differ by a factor of only ~4 (Pekar and Leigh, 1986).

More importantly, in the context of membrane transport studies, is the finding that in a cell suspension, of which only intracellular $^{23}Na^+$ may relax biexponentially, selective detection of the intracellular ion can be achieved using a pulse sequence that acts as a double-quantum filter. Pekar *et al.* (1987) showed that intracellular sodium ions can be detected selectively from the extracellular ions. Adding $Dy(PPP)_2^{7-}$ to a suspension of washed canine erythrocytes shifts the extracellular $^{23}Na^+$ resonance, thus giving two peaks in the "unfiltered" spectrum, but only the intracellular $^{23}Na^+$ is evident in the double-quantum-filtered spectrum. Although double-quantum filtering can in principle be used to detect selectively $^{23}Na^+$ in cell compartments, in which the ion has different binding and relaxation behavior, the "sensitivity penalty" is probably too large for many applications. The method, though, has the advantage of avoiding the use of exogenous chemical shift reagents which are potentially cytotoxic (Section 3.5).

In principle, any nucleus with spin quantum number not equal to $^1/_2$ can be detected by multiple-quantum NMR procedures. For example, $^{87}Rb^+$ has been studied in agarose gels using triple-quantum filtration (Allis *et al.*, 1990) and the results have been used to infer the level of NMR "visibility" in tissues. The 100% visibility of the ion in erythrocytes (Allis *et al.*, 1989) is consistent with the results of this multiple-quantum analysis. There appear to be no reports yet of the use of triple-quantum ^{87}Rb NMR to select intra- and extracellular $^{87}Rb^+$ signals in transport experiments.

5. NMR STUDIES OF FAST MEMBRANE TRANSPORT PROCESSES

5.1. Studies Exploiting Transmembrane Differences in Relaxation Rates

5.1.1. NMR Studies of the Properties of Water

The molecular organization of water probably plays a key role in the function of all cells (Mathur-De Vre, 1979). Water influences the structure of organelles and the rate of enzyme-catalyzed reactions by binding to proteins and changing their conformations, by binding to ions, and simply by being the medium through which most reactants must diffuse. The most prominent of all available NMR signals in biological samples is that of the ~80 M water hydrogen nuclei (Section 2.5.1). Therefore, it is not surprising that the first NMR studies of various tissues, using the low-resolution spectrometers of the time, were directed at gaining insight into the properties of "biological" water (Oderblad *et al.*, 1956). The most extensively studied membrane transport process has been that of water in human erythrocytes (Benga, 1989a,b). The ready availability of mammalian erythrocytes and their lack of internal membranes makes them ideally suited for these investigations.

The physical properties of water can be studied using the NMR of four different nuclides: the three isotopes of hydrogen (^1H, ^2H, ^3H) and one of oxygen (^{17}O). Understandably, ^1H has been the most extensively employed with only limited use being made of ^2H and ^{17}O (Mathur-De Vre, 1979; Kuchel, 1981). There do not appear to have been any reports of the use of ^3H NMR to measure the transport of tritiated water into cells. The high receptivity of ^3H (the highest of all nuclides, being 1.21 times that of ^1H; Harris and Mann, 1978; Bruker Almanac, 1993) does not offset the disadvantage of it being radioactive with the attendant practical problems associated with handling such a sample.

5.1.2. Water Transport in Erythrocytes

The ^1H nuclei of the water in a sample, in the magnetic field of the spectrometer, acquire a short-lived magnetic "label" after the application of an r.f. pulse. The label decays at a rate characterized by one of the two relaxation times, T_1 or T_2 (Section 2.4), depending on the type of experiment being performed. The T_2 type of experiment is favored now so it alone will be dealt with in detail here.

In a simple system consisting of two compartments, A and B (Section 1.3), two transverse relaxation times, denoted T_{2A} and T_{2B}, characterize the decay of the magnetization of the respective fractional populations P_A and P_B ($P_B = 1 - P_A$), which have mean residence times of τ_A and τ_B. In a T_2 type of experiment the NMR signal decays in a manner that is determined by four independent parameters, T_{2A}, T_{2B}, P_A (or P_B), and τ_A (or τ_B) (Fabry and Eisenstadt, 1975; Herbst and Goldstein, 1984). Data analysis is relatively complicated using this approach, so the preferred NMR method for measuring water transport across erythrocyte membranes involves a minor modification of the original paramagnetic "doping" method of Conlon and Outhred (Conlon and Outhred, 1972, 1978; Morariu and Benga, 1977; Benga, 1988). The method is analogous to radioactive tracer diffusion methods (see also Section 5.3) and it uses the relatively impermeable paramagnetic Mn^{2+} ion, added in high concentrations (12–20 mM) to the cell suspension. Dextran–magnetite has also been used as a relaxation reagent

(Ohgushi *et al.*, 1978; Ashley and Goldstein, 1980; Renshaw *et al.*, 1986). The paramagnetic atoms or ions dramatically increase the rate of relaxation of the high-energy nuclear magnetic states outside the cells. The intracellular nuclei undergo relaxation at a relatively slow rate characterized by the rate constant $1/T_{2,in}$, but some of the water molecules pass outside and relax at a much faster rate characterized by the rate constant $1/T_{2,out}$. The rate of the reverse transfer of magnetization is much less than the rate of relaxation due to the external Mn^{2+}, so no coherent magnetization (signal) enters the cells from outside. As a result, the time course of the observable NMR signal from the water has an apparent relaxation time $T'_{2,in}$ which is shorter than $T_{2,in}$; the faster the exchange, the smaller will be the value of $T'_{2,in}$. Thus, the latter value can be used to determine τ_{in} by nonlinear regression of a single-exponential function onto the relaxation data. Separate measurements of the value of $T_{2,in}$ using a suspension of packed cells enables the resonance lifetime (τ_{in}) of a water molecule inside the cell to be calculated using the expression (Benga, 1988)

$$\tau_{in} = 1/\{1/T'_{2,in} - 1/T_{2,in}\} \tag{6}$$

5.2. Magnetization Transfer Studies

An important outcome of the application of pulsed NMR spectroscopy to the study of living systems has been the measurement of the rates of some rapid protein-catalyzed exchange reactions in living cells (Alger and Shulman, 1984; Brindle and Campbell, 1987; Kuchel, 1990). Furthermore, the kinetics of coupled three-site enzymatic reactions have been studied with NMR spectroscopy (Mendz *et al.*, 1986; Chapman *et al.*, 1988). The first explicit suggestion that NMR might be used to measure chemical exchange rates was made by Gutowsky and Saika (1953) and was implemented in 1956 (Arnold, 1956; Gutowsky and Holm, 1956). The procedure they employed is called "total lineshape analysis" and the data were acquired in the field-sweep mode. The method used with a modern high-field NMR spectrometer has only recently been applied to the measurement of the rates of membrane transport in cells, but in some respects it is more difficult to apply than other NMR procedures (Section 5.10). It is the NMR magnetization-transfer techniques (also called spin transfer; Sections 5.6–5.9) which have enabled the relatively facile measurement of rapid transport rates of solutes (Kuchel, 1990).

For the NMR magnetization-transfer experiments there is the almost universal requirement that there be separate resonances for at least one nucleus in the solute when the solute is inside and outside the compartment. Separation of the resonances can be induced with shift reagents, or can be an intrinsic property of the system, but in some circumstances this restriction can be circumvented (Section 5.9).

5.3. Tracer Exchange

Conventionally, the rapid rates of exchange between solutes in a chemical reaction at equilibrium have been measured by sampling the reaction mixture at short times after the addition of a small amount of radioactively labeled reactant; these tracer-exchange procedures have been used extensively in enzyme kinetic analyses (Morales et al., 1962; references in Kuchel, 1987). Other approaches are "temperature jump" and "pressure jump" methods, in which the equilibrium position of a reaction is perturbed by a rapid change in temperature or pressure, respectively. The change in concentration of the reactants as the new equilibrium is established is monitored, usually spectrophotometrically (Eigen and De Maeyer, 1963). However, it is difficult to conceive of how these methods might be applied to living systems including suspensions of whole cells.

5.4. NMR and Exchange Reactions

The theory relating the shapes of lines in an NMR spectrum to the rates of chemical exchange processes occurring in a sample (McConnell and Thompson, 1957; McConnell, 1958; Woessner, 1961; Forsén and Hoffman, 1963, 1964a,b; Section 5.10) preceded the application of pulsed NMR to biological systems by over 10 years. Proteins were the subject of the first pulsed NMR magnetization-transfer experiments (Gupta and Redfield, 1970; Campbell et al., 1974, 1977). The transfer of magnetic saturation between ^1H nuclei in a range of peptides and proteins (Gupta and Redfield, 1970; Campbell et al., 1977) and also between the protein protons and the solvent water was studied (Glickson et al., 1974), but the quantification of the rates was only obtained in the latter work.

The first reported use of the method to actually measure the rate of exchange in an enzyme-catalyzed reaction was by Brown and Ogawa (1977) using inversion transfer (Section 5.7), although Navon (G. Navon, personal communication) had already demonstrated the utility of the technique for measuring the catalase-catalyzed hydration/dehydration of acetaldehyde (Cheshnovsky and Navon, 1978, 1980). The first reported experiment on an enzyme system in whole cells was the study of ATPase in *Escherichia coli*, using saturation transfer (Section 5.6; T. R. Brown et al., 1977); it is worth noting, though, that it has now been shown that most of the saturation transfer occurs via the substrate-level phosphorylation reactions centered around cytoplasmic phosphoglycerate kinase (Brindle and Campbell, 1987). In many respects saturation transfer is the simplest of the spin-transfer procedures, because evidence of rapid chemical exchange is obtained from only two spectra, one with and one without the selective saturating r.f. field. However, the measurement of the rates of the exchange requires an estimate of the relaxation time of the nonsaturated nuclei and it is in the determination of this value that many pitfalls can be encountered.

The first quantitative estimates of the rate of *transmembrane* exchange of a solute across a plasma membrane in whole cells, made using a magnetization-transfer method, was carried out on the nonelectrolyte DMMP (Kirk and Kuchel, 1986b; Section 3.7). Magnetization transfer analysis had, however, already been shown to be applicable to studying the exchange of solvent between the two environments inside and outside gel polymer beads (Ford *et al.*, 1985).

Before discussing the methods and results of magnetization-transfer experiments, it is relevant to consider some fundamental definitions and concepts used in these studies.

5.5. Bloch–McConnell Equations

The magnetizations of the nuclei of S_i and S_o, in a reaction such as that shown in Figure 1, are detected in the NMR experiment as resonance lines with well-specified areas; the areas are proportional to the mass of the corresponding solute in the "detected volume" of the sample. Selective r.f. irradiation at the resonance frequency of the S_i nuclear population is applied to the sample, thus altering the Boltzmann distribution of spins and hence the magnetization of S_i. The rate of change of the z component of the magnetization of S_o and S_i, after this r.f.-induced perturbation, is described by an extension of Equations (1) and (2) called the Bloch–McConnell equations (McConnell, 1958):

$$dM_i/dt = -\{(M_i - M_i^e)/T_{1,i}\} - k_{-1}M_i + k_1M_o \tag{7}$$

$$dM_o/dt = -\{(M_o - M_o^e)/T_{1,o}\} + k_{-1}M_i - k_1M_o \tag{8}$$

where the magnetizations of S_i and S_o, at spin–temperature equilibrium, are denoted by M_i^e and M_o^e, and at other times as M_i and M_o. $T_{1,i}$ and $T_{1,o}$ are the corresponding longitudinal relaxation times. Bloch–McConnell equations that describe the rates of change of magnetization in systems with multiple reacting species are all of the same form as above, and since they are linear differential equations the only real increase in complexity in these equations is their number. The analytical solution of Equations (7) and (8) can be obtained by a variety of mathematical techniques but the general solution can only be obtained for a maximum of four sites. Beyond four simultaneous differential equations there is no general solution because in the analysis a quintic, or higher-degree polynomial, must be solved for its roots, and there is no general solution for the roots of a polynomial of degree five or above (Kuchel and Chapman, 1983; Kuchel, 1990). There are some special initial conditions for which the solution of Equations (7) and (8) are very simple; one such situation occurs with the saturation transfer experiment.

5.6. Saturation Transfer

5.6.1. The Experiment

Selective irradiation of a nuclear population S_i, if sufficiently intense and of sufficient duration, will reduce its magnetization to zero (see Section 2.5.3). The NMR signal intensity recorded from S_o in the presence of this additional r.f. field is proportional to its magnetization, so we denote it by M_o^{sat}, while that in its absence is denoted by M_o^e. Thus, in Equation (8) $M_i = 0$, and the rate of change of the magnetization of S_o, prior to obtaining the NMR spectrum, is zero, so $dM_o/dt = 0$. Hence, after some rearrangement of Equation (8), an expression is obtained for k_{-1} in terms of the longitudinal relaxation time of M_o (T_{1,S_o}^{sat}) measured in the presence of the exchange and the M_i-saturating field (Brown, 1980):

$$T_{1,S_o}^{sat} = T_{1,S_o}/(1 + k_{-1}T_{1,S_o}) \qquad (9)$$

$$k_{-1} = \{(M_o^e - M_o^S)/M_o^e\}\, 1/T_{1,S_o}^{sat} \qquad (10)$$

where the bracketed term in Equation (10) is called the *degree of saturation transfer*.

The selective r.f. field is commonly generated using the main transmitter of the spectrometer in conjunction with a DANTE pulse sequence (Morris and Freeman, 1978) or one of the more recent top-hat series (Geen and Freeman, 1989; Geen et al., 1989); alternatively, a separate frequency synthesizer could be used.

In situations where there are several sites of exchange, it is possible to selectively saturate more than one site and thus elegantly simplify the data analysis (Ugurbil, 1985). However, there are no examples in the literature where this has been done on a transport system. In recent studies on ionophore-mediated $^{23}Na^+$ exchange between two populations of vesicles (i.e., a three-site system), saturation transfer was not used for the quantitative analysis which was carried out using 1D EXSY (Section 5.8) as this method appeared to give more reliable results (Waldeck and Kuchel, 1993; Figure 6).

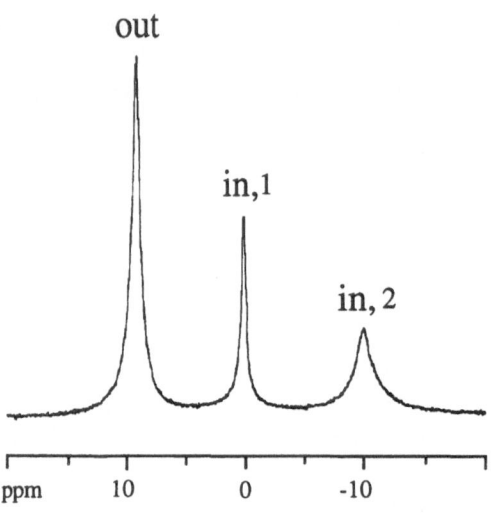

out

in,1

in, 2

ppm   10   0   -10

FIGURE 6. ^{23}Na NMR spectrum of a suspension of large unilamellar vesicles containing two different compartments. The suspending medium contained 8 mM Tm(PPP)$_2$$^{7-}$ (see Section 3.5.2) and consequently the extravesicular ^{23}Na$^+$ resonance (labeled "out") is upfield-shifted. One population of vesicles contained 2.5 mM Dy(PPP)$_2$$^{7-}$ inside (see Section 3.5.1), and these correspond to the lowest-frequency ^{23}Na$^+$ peak (labeled "in,2"). The other population of vesicles contained no shift reagents, and the chemical shift of the resulting ^{23}Na$^+$ resonance (labeled "in,1") was arbitrarily set to 0.000 ppm. The percentage encapsulated volume of each of the internal compartments was ~20%. A line broadening factor of 5 Hz was applied to the spectrum. Adapted from Waldeck and Kuchel (1993).

5.6.2. Experimental Problems

Five main problems are associated with measuring rate constants using saturation transfer experiments:

1. The selective r.f. irradiation, by virtue of its power and/or duration (it must be applied for at least 5 times T_1), may not fully saturate the designated resonance. In this case Equation (10) cannot validly be applied to the data. If the signal-to-noise ratio is poor and the resonances are broad, it may be difficult to determine if indeed $M_i = 0$ (e.g., as has occurred with some *in vivo* enzyme kinetic studies; Gadian and Radda, 1981).

2. The resonance frequencies of S_o and S_i may be so close that the supposedly selective radiation is not selective enough and power "spills over" from the region of M_i to partially saturate M_o.

3. The "control" spectrum for which M_i is not saturated, nevertheless, must have the selective r.f. field directed at another part of the spectrum in order to control for "spillover" of r.f. power; the frequency used is generally that on the opposite side of M_o from M_i and separated from M_o by the frequency difference $\Delta\nu = |\nu_i - \nu_o|$. This protocol assumes a symmetrical power-versus-frequency "envelope" for the selective r.f. field, and yet this is sometimes not realized and more elaborate control experiments must be performed (Potts *et al.*, 1989).

4. In the conventional saturation transfer analysis T_{1,S_o}^{sat} is measured, but in some experiments it may be more convenient to measure the T_1 of M_o in the

absence of exchange, for example, by using a transport inhibitor or by simply making the measurement on the supernatant taken from the cell suspension. We then use Equation (9) with Equation (10) to evaluate k_{-1}. However, as was pointed out by Kuhn et al. (1986), an off-resonance r.f. field (such as the saturating r.f. field near M_i) yields an apparent T_1 that is less than the real value for S_o. Therefore, the measurement of T_1, even if exchange is inhibited, should be carried out in the presence of the M_i-selective r.f. field. This approach was also used for a supernatant sample containing DMMP as the transported solute in erythrocytes (Potts et al., 1989).

5. Finally, the selective r.f. irradiation which must be applied continuously to saturate M_i, except perhaps during the acquisition of the free induction decays, can give rise to significant sample heating. For example, in some of our earlier work using a DANTE pulse train in a typical ^{31}P NMR experiment on erythrocytes, a temperature rise of ~9°C occurred (Bubb et al., 1988). The temperature rise for a given decoupler power and duration is an increasing function of the ionic strength of the solution and is related to its electrical conductivity. The increase in the rate of an enzymatic reaction for a 10°C rise in temperature, the so-called Q_{10}, is ~2.5 for many enzymes (e.g., Beutler, 1984) and it may be higher for some membrane transport processes. Therefore, when using NMR to study the temperature dependence of the rate of a transport reaction, the temperature of the sample in the spectrometer must be measured at each new setting of the thermostat. A standard NMR "thermometer," such as ethylene glycol contained in a capillary, can be placed in the sample; although neat ethylene glycol (1,2-ethanediol) is generally considered to be an ^1H NMR thermometer, it can be readily adapted for use in X-nucleus experiments (Bubb et al., 1988).

5.6.3. Estimating Errors in Rate Constants

It is now accepted practice in enzyme (and transport protein) kinetic analyses to give estimates of the precision of the measured value of a rate constant. The analytical rationale of this procedure is given elsewhere (Kuchel, 1990), but we give here the expression for the standard deviation (s.d.) of k_{-1} in Equation (10), as derived from the formula for the standard deviation of a function of several variables (Kendall and Stuart, 1977):

$$\text{s.d.}\{k_{-1}(x)\} = [M_o^s/\{(M_o^e)^2 T_{1,s_o}^{sat}\}]^2 \{\text{s.d.}(M_o^e)\}^2$$

$$+ [-1/(M_o^e T_{1,s_o}^{sat})]^2 \{\text{s.d.}(M_o^s)\}^2$$

$$+ \{[(M_o^s - M_o^e)/\{M_o^e(T_{1,s_o}^{sat})^2\}]^2 [\text{s.d.}(T_{1,s_o}^{sat})]^2\}^{1/2} \qquad (11)$$

where $x = (M_o^e, M_o^s, T_{1,s_o}^{sat})$, and it is assumed that there is no "statistical interaction" between the three parameters. This assumption appears to be valid in practice since each parameter is measured from a separate experiment. The standard deviations of the parameter values can be obtained as follows: (1) The peak areas in the spectra are measured in arbitrary units using the same overall spectral scaling factor. The signal-to-noise ratio, r, is determined for the region of the spectrum containing the S_o peak. Thence,

$$\text{s.d.}(M_o^e) = M_o/r \tag{12}$$

(2) The value of s.d.(T_{1,S_o}^{sat}) is obtained from the nonlinear regression of the relevant exponential expression (Shaw, 1984) onto data from an inversion recovery experiment; this fitting can usually be done automatically on modern spectrometers.

In summary, Equation (11) is a formula for calculating the standard deviation of the estimated value of the rate constant in a saturation transfer experiment. It should probably be used routinely when reporting the rate constant values obtained from all such experiments. Indeed, when comparisons are drawn between the values of rate constants obtained under different experimental conditions it is essential to know whether the differences are statistically significant. A first step in assessing the significance is to see if the parameters differ by more than the sum of their standard deviations.

5.6.4. Water Transport in Chloroplasts

The ^{1}H NMR spectra of the leaves of some plants show two or three broad unresolved ^1H_2O resonances, which are most distinguishable when stacks of discs cut from the leaves are arranged with their planes perpendicular to the direction of the magnetic field in the NMR spectrometer. For leaves of the tulip tree (*Liriodendron tulipifera*) the resonance with the largest amplitude had a linewidth of ~1.7 ppm and it was assigned to overlapping signals from ^1H_2O in the extracellular space, the vacuole, and the cell cytoplasm. The associated low-frequency shoulder, of similar linewidth, was assigned to the ^1H_2O in the chloroplasts (McCain and Markley, 1985).

The particular saturation transfer procedure used for the chloroplast experiments involved the partial saturation of the chloroplast ^1H_2O signal and the detection of the time evolution of the resonance intensity, i.e., partial-saturation recovery (Campbell *et al.*, 1978). Because the water peaks in the spectra were not resolved, their intensities were obtained by regressing a bi-Lorentzian onto each spectrum in the time course. The estimated permeability coefficient of water in chloroplast membranes, 9.2×10^{-4} cm/sec, is similar to that for human red blood cell membranes (Benga, 1989a).

5.6.5. Phosphoryl Compound Transport

The first quantitative estimates of the rate of transmembrane exchange of a solute across a plasma membrane in whole cells, made using a magnetization-transfer method, were made possible by the observation of separate intra- and extracellular ^{31}P NMR resonances from the nonelectrolyte DMMP (Kirk and Kuchel, 1985, 1986a,b; Section 3.6). In osmotically shrunken cells the ^{31}P NMR resonances from the intra- and extracellular DMMP are separated by ~0.43 ppm and have a linewidth of ~10 Hz; it is therefore possible to measure the trans-membrane exchange of this molecule using a saturation transfer experiment. The influx rate constant for the transmembrane exchange of DMMP is inversely proportional to the mean residence lifetime in the extracellular space; the domain of values of the rate constant can therefore be varied by adjusting the hematocrit of the cell suspension (see Section 1.3). Using the saturation transfer technique it was shown that the transport of DMMP across the human erythrocyte membrane was nonsaturable up to a concentration of 600 mM, not inhibited by a range of known transport blockers, and enhanced by butanol and phloretin (Potts et al., 1989). The results imply that DMMP crosses the membrane by a process of simple diffusion through the lipid of the membrane bilayer (Potts et al., 1989). Hypophosphite ($H_2PO_2^-$), which is a ^{31}P NMR probe of membrane potential (Kirk et al., 1988), has also been investigated in relation to its rate and means of entry in erythrocytes by using magnetization-transfer experiments; it enters the erythrocyte via band 3 (Price and Kuchel, 1990).

5.6.6. Fluoro-glucose Transport

3-Fluoro-3-deoxy-D-glucose (fluoro-glucose) gives separate ^{19}F NMR reso-nances from the species inside and outside erythrocytes and, because of this property, its rate of exchange across the membrane can be measured using a saturation transfer ^{19}F NMR experiment (Potts et al., 1990; Potts and Kuchel, 1992). Because of its ease of use, this method should supplant the one developed earlier to measure glucose transport into erythrocytes using D-[1-^{13}C]glucose and ^{13}C NMR saturation transfer (Kuchel et al., 1987b; Section 5.9.2). In turn, the saturation transfer approach has given way to 1D EXSY or inversion transfer analysis (Potts and Kuchel, 1992). Among the results of the latter studies of human erythrocytes was the finding that the equilibrium exchange Michaelis constant of the α-anomer of fluoro-glucose is less than that of the β-anomer; the values at 34°C were 6.0 ± 0.9 and 8.8 ± 1.3 mM, respectively, and the corre-sponding maximal velocity was 28 ± 3 mmol/liter cells per sec (Potts and Kuchel, 1992). The permeability coefficients of the human erythrocyte for fluoro-glucose, at concentrations similar to those encountered for glucose in

A   B   C

Control   With DNDS   Without DNDS

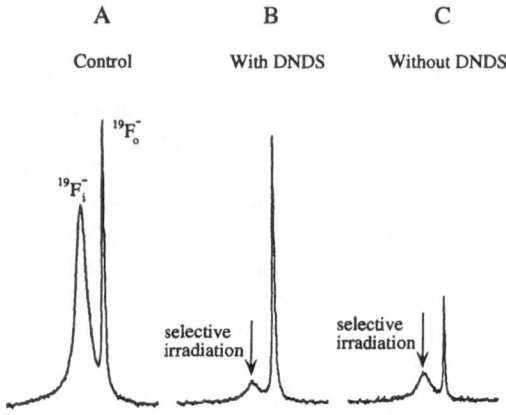

FIGURE 7. ^{19}F NMR spectra obtained at 376 MHz from human erythrocytes in 40 mM NaF at 37°C. The narrow resonance at low frequency and the broad resonance at high frequency are from the extra- and intracellular populations of ^{19}F$^-$, respectively. The control (A) is a "fully relaxed" equilibrium spectrum with no selective irradiation applied. B was obtained with selective irradiation applied at the frequency indicated by the arrow, namely that of the intracellular ^{19}F$^-$ population, in the presence of the band 3 inhibitor DNDS. C shows the result of selective irradiation of the intracellular ^{19}F$^-$ population in the absence of the band 3 inhibitor. Adapted from Chapman and Kuchel (1990).

plasma, are $\sim 3 \times 10^{-5}$ cm/sec, which implies a residence lifetime of a fluoro-glucose molecule in the cell of ~ 1 sec.

5.6.7. ^{19}F$^-$ Transport

The rate of exchange of F$^-$ across the membranes of the human erythrocyte is high; at 37°C the maximum rate is 0.26 of that of Cl$^-$, a value that was determined using magnetization transfer analysis with ^{19}F NMR. Figure 7 displays the "split peak" phenomenon operating on the F$^-$ ion, and the spectra provide qualitative evidence of the rapid exchange; however, as was discussed with regard to fluoro-glucose the quantification of the exchange rate constants was carried out with an "overdetermined" 1D EXSY analysis (Chapman and Kuchel, 1990).

5.7. Inversion Transfer

5.7.1. The Experiment

For a reaction scheme like that in Figure 1 a selective r.f. pulse (usually a π pulse) is used to perturb the orientation of the z component of the net magnetization of S_i or S_o. The return of both of the magnetizations to their thermal equilibrium orientations is measured in a manner similar to the inversion recovery experiment used to measure T_1 (Shaw, 1984). The selective irradiation can be delivered by a separate transmitter (Brown and Ogawa, 1977) or the main transmitter can be used with special pulse trains such as the DANTE sequence (Morris and Freeman, 1978). Another r.f. pulse sequence that is particularly useful is the so-called "δ-ordered inversion transfer" sequence (Robinson *et al.*, 1985):

$$\pi/2_x - t_1 - \pi/2_x - t_m - \pi/2_{x,y,-x,-y} - \text{acquire}$$

where the x and y subscripts denote the axis in the rotating frame along which the r.f. pulse is applied, and t_1 is called the *evolution time* and is set to the value $1/(2|\nu_i - \nu_o|)$. After this time the magnetization vectors of S_i and S_o are disposed at 180° to each other in the x,y plane, and then the second $\pi/2$ pulse is applied. This pulse aligns one of the vectors along the $+z$ axis and the other along the $-z$ axis. The time during which the z components of the magnetizations of S_i and S_o can interchange is called the *mixing time* (t_m) and it is varied from one spectrum to the next in a series of spectra; it is generally varied through at least ten different values, ranging from 0 to 5 times the T_1 value of the slowest relaxing of the two species (Robinson *et al.*, 1984). The pulse sequence is given its name because the magnetization vectors, which lie in the x,y plane after the first r.f. pulse, dephase during t_1 relative to the y axis in a spatial order that depends on their Larmor frequencies, or chemical shifts (δ). The transmitter frequency is assigned to the value of the resonance which is ultimately inverted through the angle π after the first two pulses.

Magnetization transfer between nuclear populations, in which the magnetization has been locked in the transverse plane, can be measured with a special "spin locking" pulse sequence (Section 2.4.3; Hennig and Limbach, 1982). However, there appear not to have been any studies of biological systems using this sequence.

5.7.2. Data Analysis

The general solutions of the two-site Bloch–McConnell equations [Equations (7) and (8)] have been given by several authors (Campbell *et al.*, 1978; Robinson *et al.*, 1984; Degani *et al.*, 1985; Kuchel, 1990). The mathematical expressions describing the time dependence of M_i and M_o may be regressed onto the inversion transfer data. Nonlinear regression algorithms must be used in the computer programs for this analysis as the magnetization functions are nonlinear with respect to the parameters being estimated, namely the T_1, k_1, and k_{-1} values.

A convenient method for obtaining an initial estimate of k_{-1} for a two-site reaction from inversion transfer data is to note that the signal from the uninverted magnetization, say it is M_o, decreases to a minimum with increasing t_m and then rises to its equilibrium value for long values of t_m. At the minimum the derivative on the left-hand side of Equation (8) is zero, and the corresponding values of M_o and M_i at the minimum, denoted by M_o^{min} and M_i^{min}, respectively, can be read from the graphs of the magnitude of the z components of magnetization (NMR signal) versus time. Thus, Equation (8) can be rearranged to yield an expression for k_{-1} (King et al., 1986):

$$k_{-1} = (M_o^{min} - M_o^e)/\{T_{1,o}(M_i^{min} - K_e M_o^{min})\} \tag{13}$$

where $K_e = M_i^e/M_o^e = k_1/k_{-1}$, and $T_{1,o}$ must be obtained from a separate experiment.

An alternative approach to fitting the analytical solution of the Bloch–McConnell equations is to employ a computer program that uses the differential equations per se and integrates them numerically (Kuchel et al., 1988a). The method is more numerically intensive, but it is operationally more direct as it obviates the need to obtain analytical solutions of the differential equations. Like all nonlinear regression programs, it requires initial estimates of the relevant parameters. It then computes the sum of squares of residuals that exist between the data and the numerically integrated time course. The parameter values are altered according to a particular algorithm and the integration is repeated yielding a new sum of squares of residuals. Iteration of the procedure continues until a minimum of the sum of squares of the residuals is obtained. The parameter values which give this minimum are called best fit values.

Led and Gesmar (1982; Gesmar and Led, 1986) stress the importance of using "complementary experiments" in order to maximize the precision of parameter estimates: Inversion transfer experiments should be performed with, say, the S_i magnetization inverted (e.g., in the δ-ordered inversion transfer sequence the transmitter frequency would be set to that of the S_i nuclei) for the series of t_m values. Then the experiment should be repeated with the magnetization of S_o inverted. The data would then be fitted as a complete set in a single nonlinear regression process.

If at all possible, independent estimates of some of the parameters of the exchanging species should be obtained. For example, it is often possible to obtain independent estimates of the T_1 values for the nonexchanging solute, such as that in the supernatant separated from the cells or in a cell lysate. This reduces the dimensionality of the parameter space in the regression analysis and thus enhances the prospects of obtaining unique estimates of the set of parameter values (Kuchel, 1990).

5.7.3. Experimental Problems

A potential problem with the δ-ordered inversion transfer experiment (Robinson et al., 1984) is that T_2 and T_1 may be short relative to the evolution time, t_1, which depends on the difference between the absorption frequency of the solute inside and outside. Hence, the magnitude of the magnetization inverted onto the

$-z$ axis is less than required for a precise analysis. In other words, the problem arises when $\Delta\nu = |\nu_i - \nu_o|$ is small, because of the closeness of the peaks in the spectrum. This is not a problem with $^{23}Na^+$ because the frequency difference between the resonances for the inside and outside species can be made very large (several ppm) by the addition of $Dy(PPP)_2^{7-}$ to the suspension medium (Section 3.4). Thus, the monensin-mediated exchange of Na^+ across the membranes of phospholipid vesicles has been measured in an experiment in which the intra- and extravesicular $^{23}Na^+$ populations have separate resonances. The experiment was successful even though the relaxation times in the system were 20–40 msec (Shungu and Briggs, 1988).

5.7.4. Acetic Acid Transport in Vesicles

NMR was first used quantitatively to measure the rapid permeation by a solute of the lipid bilayers of synthetic phospholipid vesicles by Alger and Prestegard (1979). They employed both band-shape analysis (Section 5.10) and magnetization-transfer procedures (Section 5.2). Membrane-impermeant $Pr(NO_3)_3$ was added to suspensions of vesicles at 7°C, to yield separate ^1H NMR resonances from the methyl groups of acetic acid/acetate inside and outside the vesicles. The Pr^{3+} ion (5 mM) induced a shift of the extravesicular methyl resonance to 2.8 ppm while the intravesicular resonance remained at 1.9 ppm, although it was broadened. Despite the large frequency separation the two peaks were very broad and were not completely resolved. Therefore, in analyzing an inversion transfer experiment, the "true" line intensities had to be estimated by nonlinear regression of a bi-Lorentzian onto each spectrum in the series (e.g., as was used by McCain and Markley, 1985; Section 5.6.4). When the samples were progressively heated to 35, 45, and 55°C, the two resonances increasingly broadened and were seen to be almost merged at the highest temperature. Thus, the spectra were suitable for line shape analysis (Section 5.10) and an estimate of the exchange rate constant for the permeation at 35°C was 200 sec^{-1}. The broadness of the peaks at the high temperatures, of course, precluded quantitative analysis of the rates by a magnetization-transfer procedure but at 7.5°C the well-resolved peaks allowed selective inversion of either peak by use of a separate r.f. transmitter in the NMR spectrometer. The efflux rate constant under the conditions employed was ~1 sec^{-1}. Consideration of the permeation rate as a function of pH led to the conclusion that it is the acid form of the solute which accounts for virtually all of the transmembrane exchange, and from the temperature dependence of the rate constant the activation energy for the transmembrane exchange was estimated to be 41.8 kJ/mol.

Alger and Prestegard (1979) concluded that with acetic acid permeation, it is likely that the slowest step is the one in which the acid crosses the interfacial region between the aqueous solution and the hydrocarbon of the lipid, rather than diffusion within the hydrocarbon. Crossing the interface may involve one or more dehydration steps or there may be steric hindrance by the lipid head groups. These characteristics of the system may be manipulated by changing the lipid composition of the vesicles and also that of the suspension medium. Surprisingly, but probably because of the relative complexity of the data analysis, many of the fundamental questions relating to the kinetics of permeation of phospholipid membranes have not yet been addressed, and yet NMR line shape analysis and other techniques are among the few methods amenable to such analysis.

5.8. 1D EXSY

This is a recent procedure for measuring rate constants in n-site exchanging systems: the r.f. pulse sequence is that of Section 5.7.1 but it is used to obtain a limited series of 1D spectra. In its simplest application only five different spectra are acquired; two different values of t_1 are used, each with $t_m = 0$, and again with t_m a fixed nonzero value, plus a "fully relaxed" spectrum (Engler *et al.*, 1988). One way of viewing the analysis is that it constitutes the fitting of an exponential function to a data set consisting of only two time points; these data are the magnetizations (signal intensities) of the exchanging species when t_m is zero and the other value. In the form of the experiment as originally described (Engler *et al.*, 1988) the mathematical system of equations describing the results is said to be "totally specified" so that no statistical information on the precision of the parameters is obtained; it would require repeated experiments to obtain several estimates of the parameter values in order to get a mean value and its associated standard error.

The analysis is derived from that of the 2D *exchange* spectroscopy (EXSY) experiment (Jeener *et al.*, 1979; Macura and Ernst, 1980; Macura *et al.*, 1981; Abel *et al.*, 1986; Baine, 1986) and entails a computational procedure known as "back-transformation" (Bremer *et al.*, 1984; Perrin and Gipe, 1984; Kuchel *et al.*, 1988b). It has been applied, for example, to spectra acquired from solutions of D-[1-^{13}C]glucose undergoing rapid exchange between the α- and β-anomers, catalyzed by mutarotase. Some spectra, which superficially appeared to be suitable for rate constant determination, could not be used in the analysis because of a negative eigenvalue which appeared during the numerical processing. We therefore introduced the procedure called "overdetermined" 1D exchange analysis (Bulliman *et al.*, 1989) for which, in order to analyze an n-site system, a series of spectra is acquired with m different t_1 values (each with $t_m = 0$, and t_m = a fixed nonzero value), where $m > n$. This apparently minor extension to the experimental design, when coupled with the new analytical procedure, has the important result of enabling estimates of the precision of the rate constants and of circumventing most of the problems, alluded to above, that arise from the sensitivity of the previous analysis to noisy data. Thus, in the previous analysis, n

parameter values were estimated from n data elements, whereas in the newer procedure m data elements are used to estimate a smaller number of parameter values; thus, the problem is said to be "overdetermined."

When planning one of these experiments it is important to make a choice of t_m that will yield the most precise estimate of the various rate constants in the reaction scheme. Perrin (1989) concluded that to a good approximation the optimum mixing time is given by

$$t_m = 1/(R + k) \tag{14}$$

where R and k are the mean values of the respective relaxation and exchange rate constants. Thus, it is this value, determined from preliminary experiments, that is used in the 1D EXSY experiment.

Figure 8 is an example of the spectra from an overdetermined 1D EXSY experiment that was carried out to study the rapid exchange of fluoro-glucose in human erythrocytes (Potts and Kuchel, 1992); it was this experiment and not saturation transfer which provided the most precise estimates of the exchange rate constants (see Section 5.6.6). Note the separate resonances from the α- and β-anomers of the fluoro-glucose and that each is "split," as discussed in Section 3.7.2.

Perrin and Engler (1990) recently introduced a related procedure, which uses a series of t_1 and t_m values. The analysis employs the back-transformation procedure mentioned above, but its statistical advantage over the overdetermined 1D EXSY experiment (Bulliman et al., 1989) has not yet been fully explored.

5.9. Differential Saturation Transfer

5.9.1. Bicarbonate Exchange

The resonances of intra- and extracellular $H^{13}CO_3^-$ are not resolved in the ^{13}C NMR spectra of erythrocyte suspensions, so magnetization-transfer analyses of the type discussed in Sections 5.6–5.8 cannot be used to measure the rates of transmembrane exchange of this ion. However, $H^{13}CO_3^-$ is in rapid exchange with $^{13}CO_2$ inside the cells, in a reaction catalyzed by carbonic anhydrase. The activity of the enzyme is low in plasma where the spontaneous hydration rate of CO_2 is characterized by first-order rate constant of $\sim 1 \times 10^{-4}$ sec^{-1} (Itada and Forster, 1977; Weith et al., 1982). Both bicarbonate and carbon dioxide exchange rapidly across the cell membrane; the dissolved gas passes via the lipid bilayer and the ion traverses the membrane via band 3. Thus, Figure 9 is a representation of the interconversion and transmembrane exchange of the solutes B (bicarbonate) and C (CO_2) inside (i) and outside (o) a red blood cell. The values of the rate constants k_2, k_{-2}, k_4, and k_{-4} are the reciprocals of the mean residence lifetimes of the respective nuclei in the compartment from which

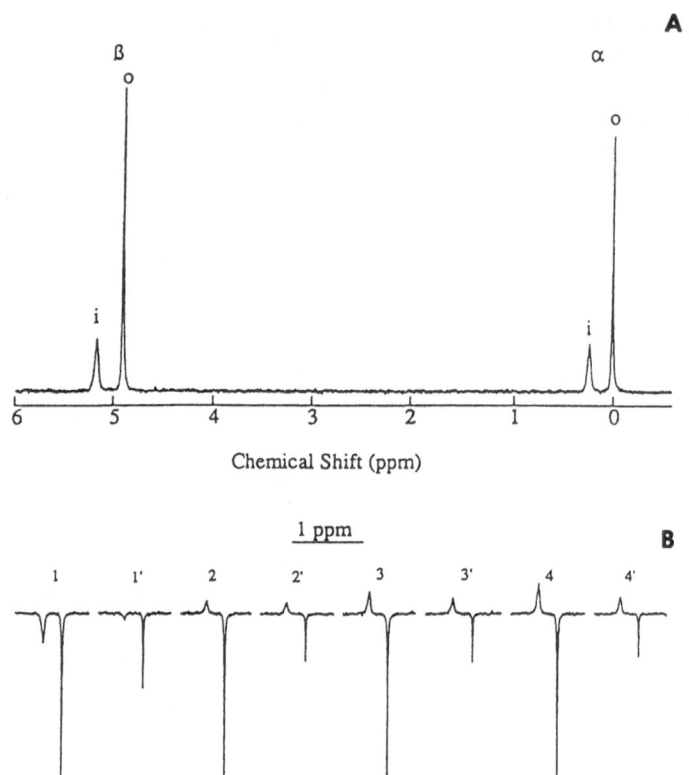

FIGURE 8. "Overdetermined" 1D exchange analysis of the exchange of fluoro-glucose across the membranes of human erythrocytes. (A) "fully relaxed" ^{19}F NMR spectrum of fluoro-glucose (17 mM) in a suspension of human erythrocytes of hematocrit 0.40. For each of the α- and β-anomers there is an intra- (i) and an extra- (o) cellular population of the solute. The spectrum was acquired at 376.43 MHz with 16 transients and an intertransient delay of 12 sec. (B) Four pairs of spectra (β-anomer region of the spectrum only). Each pair of spectra arose from a single evolution time with 0 sec (unprimed numbers above spectrum) and 0.5 (primed numbers above spectrum) mixing times. The evolution times for the pairs of spectra were, from left to right, 3 μsec, ~3.3 msec, ~4 msec, ~4.7 msec, respectively. Each spectrum was acquired from eight transients and a 12 sec intertransient delay. Adapted from Potts and Kuchel (1992).

exchange takes place. Some analysis enables an estimation of these rate constants; it is based on the suppositions that: (1) the NMR spectrum of the suspension of cells with B and C present has only two resonances; in other words, there is no resolution of peaks due to B_o and B_i, and C_o and C_i; (2) B and C are in chemical exchange that is slow on the NMR time scale but still fast enough to be amenable to study by saturation transfer experiments; the extremely rapid carbonic anhydrase reaction is subject to inhibition (e.g., by methazolamide) thus leaving the relatively slow spontaneous reaction; (3) the rate constants for the

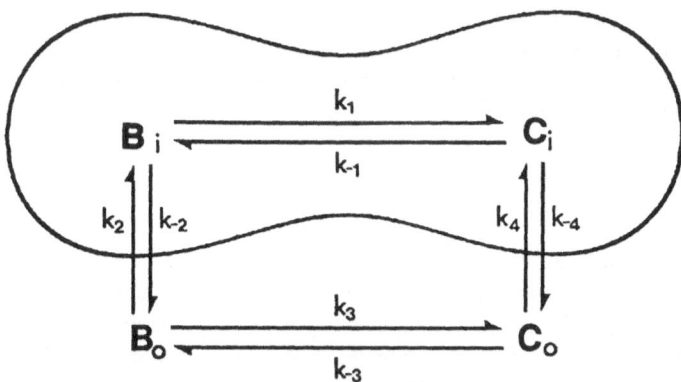

FIGURE 9. Reaction scheme depicting the chemical interconversion and the transmembrane exchange of: B, bicarbonate; and C, carbon dioxide; i, inside; and o, outside, an erythrocyte. Adapted from Kuchel *et al.* (1987c).

exchange between B and C differ inside and outside the cells; (4) a significant fraction of B_o exchanges with B_i and C_o (and B_i with B_o and C_i) in the lifetime of the high-energy nuclear magnetic state. Specifically, B_i in the presence of the saturating field is measured while the resonance corresponding to C (i.e., C_o + C_i) is selectively irradiated and its magnetization is fully saturated. Bloch-McConnell equations can be written to describe the reaction scheme shown in Figure 9 and can be solved to yield an expression for the value of k_2 (Kuchel *et al.*, 1987c,d). In its simplest form the analysis requires three saturation transfer experiments together with their controls to yield estimates of the degree of saturation transfer in the supernatant alone, in the packed cells alone, and in a cell suspension in which the intra- and extracellular compartments are of approximately equal volume.

A qualitative explanation of the overall experiment is as follows: The rate of the carbonic anhydrase reaction inside the erythrocytes is slowed by an inhibitor (methazolamide) to such an extent that the chemical-exchange lifetimes are of the same order of magnitude as the T_1 of the $^{13}CO_2$ and $H^{13}CO_3^-$. When the $^{13}CO_2$ is irradiated at its absorption frequency, with sufficient power to saturate the magnetization of both the intra- and extracellular populations, then saturation of magnetization can be transferred to the extracellular $H^{13}CO_3^-$ by two routes: (1) by direct hydration, which is very slow, so an insignificant amount takes place by this means; (2) by translocation of the $^{13}CO_2$ to the inside of the cell, followed by carbonic anhydrase-catalyzed hydration to yield intracellular $H^{13}CO_3^-$, then export via band 3 to give extracellular $H^{13}CO_3^-$. In the absence of this final step the degree of saturation transfer to the overall population of $H^{13}CO_3^-$ will be less. Therefore, the extent of suppression of the bicarbonate resonance by saturation transfer from $^{13}CO_2$ will be less in a dilute suspension of cells than in packed cells. An alternative to using packed cells is to add an inhibitor of band 3 to the dilute suspension, in which case the degree of saturation transfer to the bicarbonate resonance will be less in the presence of the inhibitor than in its absence (Kuchel *et al.*, 1987c,d).

5.9.2. Glucose Exchange

Differential saturation transfer has also been used to measure the rate of transmembrane exchange of D-[1-^{13}C]glucose in a suspension of human erythrocytes (Kuchel *et al.*, 1987b). When added to a suspension of erythrocytes the labeled glucose yields two resonances, one each for the α- and β-anomers. Additionally, the resonances are split into two peaks which are unresolved at 400 MHz, with the partial separation being smaller for the α-peaks. For both anomers the high-frequency component of the resonance is from the intracellular glucose. Because the intra- and extracellular peaks are not resolved, a direct saturation transfer analysis cannot be used to measure the glucose transport. However, the method outlined above for determining the rate of H^{13}CO$_3^-$ exchange can be adapted to the study of D-[1-^{13}C]glucose exchange across erythrocytes (Kuchel *et al.*, 1987b).

The exchange between the two glucose anomers outside the cells is enhanced by the addition of purified mutarotase, which catalyzes the anomerization reaction. The rate of anomerization is made arbitrarily large by increasing the enzyme concentration. The rate constant of the mutarotase-catalyzed reaction can be measured by saturation or inversion transfer experiments (Sections 5.6–5.8). There is no mutarotase in human erythrocytes so the scheme depicted in Figure 9 applies to the present system except B and C refer to the β- and α-anomers and k_1, $k_{-1} = 0$. The mathematical expression relating the transmembrane exchange rate constants to the degree of saturation transfer in the experiments on cell-free supernatant, packed cells, and a moderately dilute cell suspension is evaluated. In one series of experiments with a D-[1-^{13}C]glucose concentration of 30 mM, the rate constants for efflux from the cells were 0.71 $\pm$ 0.30 and 1.2 $\pm$ 0.40 sec^{-1} for the β- and α-anomers, respectively. These values compare favorably with those determined (albeit more accurately) using ^{19}F magnetization transfer with fluoro-glucose (Potts and Kuchel, 1992; Section 5.6.6).

5.10. Band-Shape Analysis

5.10.1. Background

If two nuclear populations are in exchange that is slow on the NMR time scale, then two resonances will probably be visible in the spectrum, and the shape of each line will not be significantly altered as a result of the exchange. On the other hand, if exchange between the two populations is fast, then the spectrum distinguishes only an average chemical (or physical) environment of the nuclei and only a single resonance will be observed in the NMR spectrum (Gutowsky, 1975; Akitt, 1983). In the intermediate region of exchange rates (i.e., between the above two extremes), an increase in the rate of exchange between the populations results in a broadening of the two NMR spectral lines; this arises from the presence in each population of spins of some from the other population which have just entered it and retain the characteristics of their original population. Thus, there exist those spins with characteristics intermediate between the two populations (Dwek, 1973). For a given rate of exchange, whether two populations of spins are in slow, intermediate, or fast exchange on the NMR time scale, will depend on the difference in absorption frequency between the resonances that exists in the absence of exchange; this difference depends on the field strength (B_0) of the spectrometer magnet.

In the presence of intermediate rates of exchange, a change in the rate causes a change in the shapes of the resonances in the spectrum. Analysis of these shapes, to estimate the rates of exchange between two populations of spins, is known as "dynamic" NMR spectroscopic analysis. In the past dynamic NMR spectroscopy was used primarily to determine the activation energies of various chemical exchange processes (Gutowsky, 1975; Kaplan and Fraenkel, 1980; Sandström, 1982; Oki, 1985). The estimates of rates were often obtained using approximations to rather complicated mathematical expressions (see Section 5.10.2), so the analysis depended on changes in linewidth, or chemical shift separation, between the resonances that occurred with a change in temperature. With the advent of high-speed computers, nonlinear least-squares regression analysis of so-called "total line shapes" has become a much more realistic proposition (Potts et al., 1992).

There are very few reports in the literature of the use of total line shape analysis for measuring exchange rate constants in vesicles, and only one so far for cells. The first report appears to have been that of Alger and Prestegard (1979) who studied the temperature dependence of the permeation of acetic acid through the membranes of large unilamellar vesicles. They observed the change in the ^{1}H NMR spectrum of acetic acid, in a suspension of vesicles to which an impermeable paramagnetic shift reagent had been added. Later, Hoffman and Henkens (1987) used ^{13}C NMR line shape analysis to measure CO_2 transport across erythrocyte membranes (Section 5.2); however, the experiment relies on the exchange between CO_2 and a second species ($H^{13}CO_3^-$) rather than simply exchange between the intra- and extracellular CO_2 populations. Most recently, total line shape analysis has been applied to characterizing the transmembrane exchange of ^{13}C-labeled urea in human erythrocytes (Potts et al., 1992).

5.10.2.   Theory for Two Sites

The NMR line shape (v) of two exchange-broadened Lorentzian resonances (sites A and B) is derived from the steady-state solution of the Bloch–McConnell equations [Section 5.5; Equations (7) and (8)] and is given by the following equation (Rogers and Woodbrey, 1962; Sandström, 1982):

$$v = \{C_o\,[S(1 + V) + QR]/(S^2 + R^2)\} + b_c \tag{15}$$

where

$$\Delta v = v_A - v_B$$

$$B = 0.5\,(2v_A - \Delta v) - v$$

$$A = \pi^2\,(\Delta v_{1/2}^A)(\Delta v_{1/2}^B) - 4\pi^2 B^2$$

$$S = (P_B/k_1)\,[A + \pi^2(\Delta v)^2] + \pi[(1-P_B)(\Delta v_{1/2}^A) + P_B(\Delta v_{1/2}^B)]$$

$$Q = (\pi P_B/k_1)\,[2B - \Delta v(1 - 2P_B)]$$

$$T = 1 + (\pi P_B/k_1)(\Delta v_{1/2}^A + \Delta v_{1/2}^B)$$

$$R = 2\pi BT + [(\pi^2 P_B \Delta v/k_1)(\Delta v_{1/2}^A - \Delta v_{1/2}^B)] + \pi\Delta v(1 - 2P_B)$$

$$V = (\pi P_B/k_1)[P_B(\Delta v_{1/2}^B) + (1 - P_B)(\Delta v_{1/2}^A)]$$

$$\pi\Delta v_{1/2}^A = 1/T_{2,A}$$

$$\pi\Delta v_{1/2}^B = 1/T_{2,B}$$

These expressions are almost identical to those of Sandström (1982) except for the replacement of the terms in $1/T_{2,A}$ and $1/T_{2,B}$ by the final two expressions; in other words, the linewidths at half-height are more conveniently used, being part of the raw data, rather than the $1/T_{2,A/B}$ values which are imperfectly *derived* from the data. The parameters v_A and v_B denote the resonance frequencies of the extra- and intracellular populations of solute, respectively, in the absence of exchange, and v is the so-called spectral "offset" frequency. P_A and P_B, as in the previous sections, are proportional to the relative accessible volumes (amounts) of the extra-and intracellular compartments, respectively, and the values are normalized so that $P_A + P_B = 1$. $\Delta v_{1/2}^A$ and $\Delta v_{1/2}^B$ are the linewidths at half-height of the extra- and intracellular resonances, respectively, in the absence of exchange; as has been stated above, they are related to the transverse relaxation rate constants $1/T_{2,A}$ and $1/T_{2,B}$. The first-order influx rate constant for solute exchange is k_1. The parameter b_c is included to account for any baseline variation in the spectra but, in a study of [^{13}C]urea exchange, adequate fits were generally obtained when this parameter was held constant at zero (Potts et al., 1992).

The value of C_0 in the above expression is proportional to the concentration of the exchanging solute. In other words the assumption is made that the concentration of the solute on either side of the membrane is the same. However, for a charged solute in a vesicle or cellular system, where there are membrane potentials, care must be taken to refer to the populations in terms of amounts rather than concentrations (Waldeck and Kuchel, 1993).

5.10.3. [^{13}C]urea Transport

The essential prerequisite for total line shape analysis is that there be partially resolved resonances from the solute on either side of the membrane; this phenomenon is observed with [^{13}C]urea in a suspension of human erythrocytes (Potts et al., 1992). The human erythrocyte membrane is highly permeable to urea (Brahm, 1983; Mayrand and Levitt, 1983) and it has been suggested that this is required in order for the cell to maintain its osmotic stability when passing through the renal medulla where urea concentrations are ~1 M (Macey, 1984; Macey and Yousef, 1988). The exchange appears to be protein-mediated; it is saturable (Hunter, 1970; Brahm, 1983; Mayrand and Levitt, 1983; Potts et al., 1992), inhibited by the glucose transport inhibitor phloretin (Macey and Farmer, 1970; Brahm, 1983), and by the sulfhydryl reagent pCMBS (Macey and Farmer, 1970; Naccache and Sha'afi, 1974; Brahm, 1983; Toon and Solomon, 1991). Urea exchange in erythrocytes is also competitively inhibited by the urea derivatives thiourea and dimethylurea (Mayrand and Levitt, 1983).

The dotted line in Figure 10 shows a ^{13}C NMR spectrum of [^{13}C]urea at a concentration of 29 mM in an erythrocyte suspension of hematocrit ~0.4. The high-frequency component is from the extracellular solute; this was ascertained by diluting the cell suspension and noting which resonance diminished in relative intensity, thus identifying it as the intracellular one, and by the addition of the paramagnetic ion Mn^{2+} which enhances the rate of relaxation of the extracellular nuclei and thus broadens the extracellular resonance. Mn^{2+} only permeates the cell membrane slowly at low millimolar concentrations (Pirkle et al., 1979) so it does not affect the amplitude of the intracellular resonance, at least over several minutes (Potts et al., 1992). The separation between the maxima of the two unresolved peaks can be increased if the mean cell volume is reduced by some means such as increasing the solution osmolarity by the addition of NaCl. This causes an elevation of the protein concentration which increases the extent of

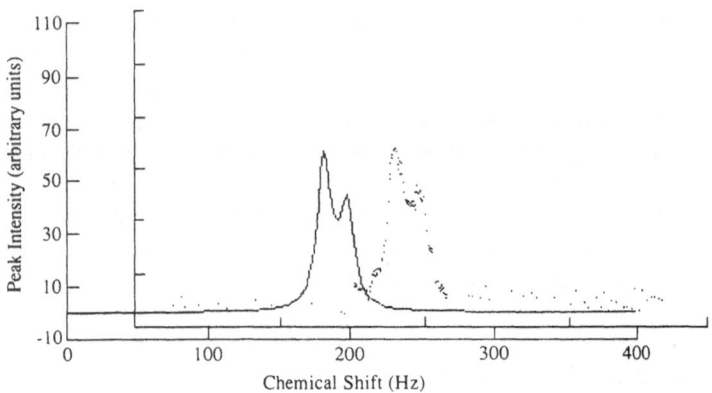

FIGURE 10. The dots represent the ^{13}C NMR spectrum of [^{13}C]urea (29 mM) in a suspension of human erythrocytes (hematocrit ~0.4). These data were subjected to band-shape analysis to determine the equilibrium exchange rate constants that characterized the membrane transport of urea; the solid line slightly offset from the spectrum is the best fit of Equation (25) to the data. NMR parameters: spectrometer frequency, 100.62 MHz; intertransient delay of 200 sec; eight transients per spectrum; temperature, 25°C. Adapted from Potts *et al.* (1992).

disruption of hydrogen bonding between the urea and water (Section 3.6). Other factors, including alterations in the intracellular magnetic susceptibility (Section 3.2), may contribute to this effect. The solid line in Figure 10 (offset from the data for the sake of clarity) was obtained by using nonlinear least-squares regression to fit Equation (25) to the NMR spectrum. Among the parameter values obtained is the apparent first-order rate constant for urea influx. This has a value in the range 0.5 to 4 sec^{-1} depending on experimental conditions; most notable is the fact that the value is reduced as higher urea concentrations are used (Potts *et al.*, 1992). This outcome is consistent with a smaller fraction of the total urea population exchanging, during the time in which the molecules are in their high-energy state, as the urea concentration is increased. Furthermore, estimation of k_1 for a range of urea concentrations enables the construction of a graph relating apparent permeability coefficient (Section 1.4) of erythrocytes for urea. From data such as these an estimate of the maximal velocity (V_{max}^{ee}) and Michaelis constant (K_m^{ee}) under the equilibrium exchange (ee) conditions were shown to be $3.1 \pm 0.6 \times 10^{-8}$ mol/cm^2 per sec and 44 ± 18 mM (Potts *et al.*, 1992).

5.10.4. Transmembrane Carbon Dioxide Exchange

Carbonic anhydrase in the erythrocyte catalyzes the reversible hydration of CO_2:

$$CO_2 + H_2O \leftrightarrow HCO_3^- + H^+$$

In blood plasma this reaction is uncatalyzed and slow (Section 5.9.1). Addition of millimolar concentrations of $MnCl_2$ causes very rapid relaxation of the magnetization of the extracellular $H^{13}CO_3^-$ added to erythrocytes but affects little that of $^{13}CO_2$ (Chapman et al., 1986; Hoffman and Henkens, 1987). The paramagnetic ion brings about broadening of the $^{13}CO_2$ and $H^{13}CO_3^-$ resonances in the ^{13}C NMR spectrum of such a cell suspension and the extent of line broadening increases when the temperature of the sample is raised from $\sim 13°C$ to $\sim 41°C$ (Hoffman and Henkens, 1987). These authors fitted the four-site counterpart of Equation (15) to the spectra (Binsch, 1975; Sandström, 1982) to obtain estimates of the residence lifetime of $^{13}CO_2$ and $H^{13}CO_3^-$ in human erythrocytes; for $^{13}CO_2$ the lifetime was 1.1–1.9 msec while that of $H^{13}CO_3^-$ was only imprecisely determined to be 200 msec. Also, the lifetime of the uncatalyzed $^{13}CO_2$ hydration was measured to be 16 sec which is considerably shorter than expected on the basis of the previously mentioned value (Section 5.9.1). Hoffman and Henkens (1987) assumed, probably unreasonably, that the inner space of the cells was totally accessible to $^{13}CO_2$ and that the concentration of $H^{13}CO_3^-$ was the same inside and outside the cells. Thus, from their line shape analysis, which yielded relative population sizes at each of the four sites, they also estimated the relative concentrations of $^{13}CO_2$ and $H^{13}CO_3^-$. By using the previously reported pK_a of the protonation reaction of the latter ion with the Henderson–Hasselbalch equation, they obtained an estimate of the pH inside the cells; this is an alternative method for measuring the intracellular pH using ^{13}C NMR spectroscopy.

6. COMPARTMENTAL DISCRIMINATION USING DIFFUSION RATES

6.1. The Nature of Diffusion

6.1.1. Three-Dimensional Random Walk

The physical process which underlies the diffusion of a molecule from the interstitial medium between cells, across the plasma membrane, and into the cytoplasm, is *thermal* motion. The diffusion rate of a molecule in the various environments of a tissue is an important determinant of the overall rate of its distribution in the tissue or even the whole body.

A generally accepted view of diffusion is one in which there is random jostling of molecules so that after a period of time net movement is apparent (Crank, 1975; Tyrrell and Harris, 1984; Berg and von Hippel, 1985; Cussler, 1986). But the *net* distance moved is much less than the sum of the lengths of all of the small "jumps" which have occurred in that time. For gases there are substantial mean free paths of moving particles and well-defined average particle velocities for the linear portions of the trajectories between collisions. However, diffusion in liquids is a very different matter. Collisions between molecules (where typically a solute molecule is several times larger than a solvent molecule) ensure that the solute molecule moves no farther than the diameter of a solvent molecule, before the next collision. A molecule exchanges momentum continuously with surrounding molecules as it collides with them. Frequently a macroscopic particle will undergo short excursions that result from the summation of several solvent encounters that provide a net vectorial momentum component. This is the basis of "Brownian" motion that was first observed under a microscope as the irregular motion of minute pollen grains on the surface of water (Brown, 1828).

The translation of an individual molecule in a bulk solution is described as a three-dimensional random walk, with a mean free path much less than its diameter. The diffusional path of a molecule can also be described as a fractal (Mandelbrot, 1977), since in a particular time interval a molecule will come close to its starting position on a large number of times, prior to achieving appreciable separation from its initial position. Thus, in considering the rate of collision between a solute molecule and a transport protein, it is important to distinguish between a diffusional "macrocollision," which is the coming together of the two molecules from afar ("macro" implies a large *distance*) and the intervening "microcollisions" with the solvent molecules.

6.1.2. Diffusion Control of Reactions

The time for a diffusional macrocollision (i.e., the mean time the molecules remain in proximity to one another and experience the multiple microcollisions) is $\sim \sqrt{r/D}$ where r is the collision radius and D is the diffusion coefficient; thus, the rate at which macrocollisions occur constitutes the "diffusion limit" to the rate of a chemical reaction between two molecules.

The volume-filling behavior of diffusion is important to keep in mind in order to make intuitive sense of the concept of diffusion control of enzymatic and membrane transport reactions. The rates of all association processes in solution are ultimately limited by the time it takes to bring reactants together by diffusion. Additionally, most macromolecular interactions require that the molecules attain a correct mutual orientation so that the relevant binding sites are properly aligned. This reorientation is also a diffusional process (Tanford, 1966) and if subsequent covalent chemical interactions take place, which are slower than the rate of association of the molecules, then the process is said to be "reaction controlled" rather than "diffusion controlled" (Berg and von Hippel, 1985).

In principle, a diffusion-limited association reaction is distinguished by the fact that the concentrations of the reactants, or the spatial distribution of these, will be inhomogeneous during the course of the reaction. Because the chemical step is fast in this limit, the region around each molecule will be depleted of reaction partners and the diffusion of unreacted molecules, into the depleted regions, will limit the overall rate of the reaction. Conversely, a diffusion-limited dissociation reaction is one in which the rate-limiting step will be the diffusing apart of the reaction products and concentration inhomogeneities will therefore result. On the other hand, if the chemical-reaction step is slow, all concentration inhomogeneities will have time to relax and the reaction will proceed in the presence of a homogeneous spatial distribution of reactants (Berg and von Hippel, 1985).

For a spherical molecule the Stokes–Einstein relationship relates the diffusion coefficient (D) of a molecule to the viscosity (η) of the medium (e.g., Tanford, 1966):

$$D = RT/Nf \tag{16}$$

$$f = 6\pi\eta r \tag{17}$$

where R is the universal gas constant, N is Avogadro's number, T is the absolute temperature, and r is the radius of the molecule. Thus, in this ideal case a diffusion-limited association rate constant will be proportional to the diffusion coefficient and therefore inversely proportional to the viscosity of the solution. This implies that the most direct experimental way of distinguishing between diffusion-controlled and reaction-controlled processes is to examine the viscosity dependence of the reaction. Furthermore, the temperature dependence of most diffusion coefficients is weak (e.g., the activation energy for the self-diffusion of water, in the temperature range 1–45°C, is ~20 kJ/mol; Mills, 1973) with the activation energy for association being determined by the temperature dependence of the solvent viscosity; on the other hand, the Arrhenius relationship for a chemical reaction shows a much more marked temperature dependence (Moore, 1981).

In a suspension of cells or a tissue the viscosity of the medium inside the cells will be different from that outside the cells (Endre et al., 1983a; Endre and Kuchel, 1986), thus giving rise to different diffusion coefficients for the same type of solute molecule. Furthermore, penetrable and impenetrable barriers, such as are afforded by the selective entry characteristics of membrane transport proteins, lead to a confining of the motion of some molecules (Stein, 1986). When the average net distance moved by solute molecules in a specified time is shofter than the distance moved by them in free solution, the molecules are said to be undergoing *restricted diffusion*. Experimentally, this may be detected by the demonstration of a time dependence of the measured diffusion coefficient; in other words, the experimental estimate of the diffusion coefficient decreases to a limiting value as the time interval over which the molecular motion is measured is increased. Thus, even if the viscosity inside and outside cells in a suspension is similar, the *apparent* diffusion coefficients may well differ as a result of this confining effect (Kärger *et al.*, 1988; Price and Kuchel, 1990). Since it was shown by Hahn in 1950 that the intensity of the signal recorded in a so-called spin-echo experiment is sensitive to diffusion, it has emerged that NMR methods are among the most useful for measuring diffusion coefficients. These NMR methods are almost unique in their ability to measure translational motion of molecules in heterogeneous systems such as cell suspensions and tissues (Callaghan, 1991).

6.2. Principles of NMR Diffusion Measurements

6.2.1. Gradient-Induced Echo

The "ancestor" of the contemporary NMR methods for measuring diffusion is the Hahn spin-echo experiment which was developed for the measurement of the transverse relaxation time, T_2 (Hahn, 1950). This experiment uses the pulse sequence $\pi/2-\tau-\pi-\tau-$acquire (Figure 11B, upper frame), where τ is the delay between the two r.f. pulses. However, the basis of the NMR-diffusion experiment is probably best visualized by considering a modification of Hahn's original experiment, which was first suggested by McCall *et al.* (1963) and then implemented by Stejskal and Tanner (1965) (see Figure 11A). In this experiment the uniform and static magnetic field B_0 is superimposed on, for *two* short periods of duration δ, by a linearly nonuniform field $\Delta B_0 = (\pm) \, g \, z$. These so-called field-gradient pulses are thus applied first in one direction, and then after the period Δ, in the opposite direction. During the application of these two pulses the rate of precession of the transverse magnetization in the sample depends on its z coordinate. Thus, the first field-gradient pulse causes a dephasing of the transverse magnetization vectors (also called spin-isochromats), relative to those at different z positions in the sample, and thus there is a decay in the *vector* sum of the magnetization. If an individual nucleus does not change its position with respect to the z coordinate in the interval Δ, then the effect of the first pulse is exactly compensated by the second. This comes about through the inversion of the direction of the pulsed field gradient, so that the magnetization vectors now precess in the *opposite* direction, but at the previous rate. In other words, the dephasing magnetization vectors retain their rate of dephasing, but the direction is reversed. Provided the duration of the two field gradient pulses is exactly the same, all of the magnetization vectors should return into *coherence* after the second pulse, thus forming an "echo" in the signal along the $-y$ axis of the rotating reference frame.

If, during the second field-gradient pulse, the z coordinates of the individual nuclei differ from those during the first pulse (i.e., the nuclei have moved), the

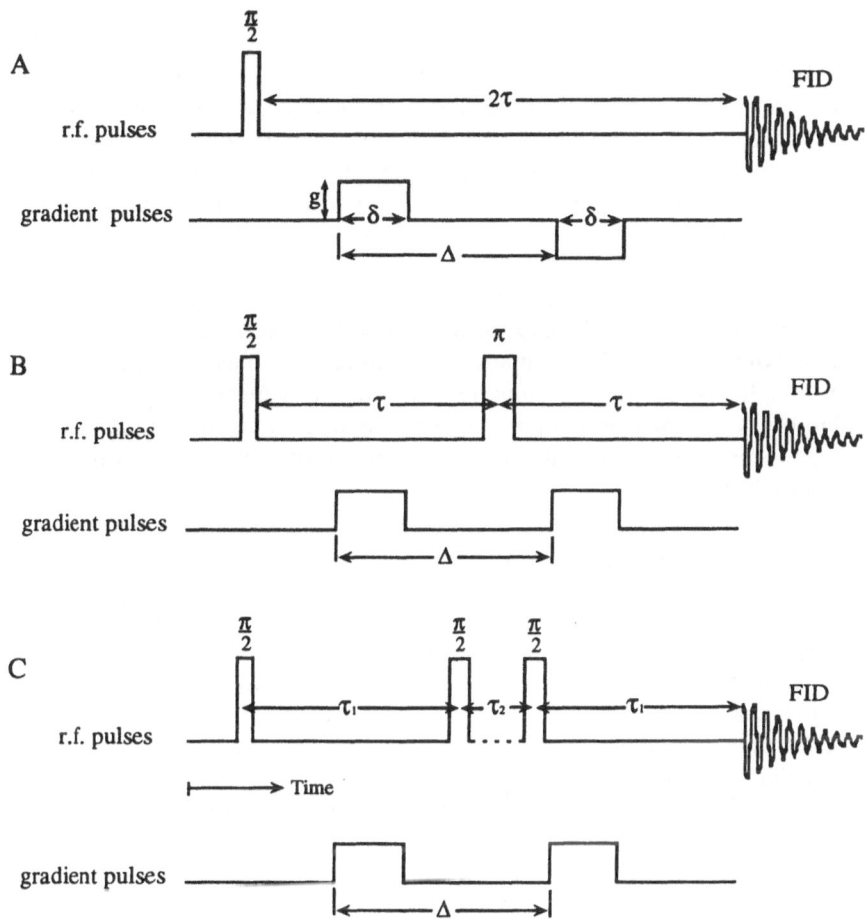

FIGURE 11. Schematic representation of the three basic types of NMR pulse sequences used to measure diffusion coefficients. (A) The gradient echo pulse sequence. (B) The "classical" pulsed field gradient spin-echo pulse sequence as introduced by Stejskal and Tanner (1965). (C) The stimulated echo pulse sequence; the magnetic field gradient pulses associated with this sequence are applied at the times indicated for B.

rate of dephasing of the magnetization vectors during the second pulse will be different, so that all of the magnetization vectors will not be superimposed along the $-y$ axis at any time after the second pulse. This will lead to a decrease in the net magnetization that is projected onto the $-y$ axis of the NMR receiver coil and thus will lead to a decrease in the NMR echo signal. The intensity of the NMR signal (i.e., the FID) will be decreased more when a larger z distance is traversed by the nuclei during the time interval between the two field-gradient pulses.

6.2.2. Hahn Spin-Echo

In the original Hahn spin-echo experiment (Hahn, 1950), it is the π r.f. pulse which reverses the sign of the phase change of the dephasing magnetization vectors (see Figure 11B); thus, at the end of the second τ interval the magnetization vectors are realigned (refocused) to give the signal-echo. If no molecular diffusion occurred, T_2 could be evaluated by repeating the experiment and measuring the amplitude of the refocused echo for a range of τ values (e.g., Shaw, 1984). A graph of echo amplitude (or what is equivalent, the area of the corresponding peak in the frequency-domain NMR spectrum) versus τ is described by an exponential decay, with an exponent $-\tau/T_2$. However, molecular diffusion is a source of error in the T_2 measurement; it manifests itself as an attenuation of the echo amplitude arising from imperfect refocusing of magnetization. Thus, the value of T_2 measured in the presence of a significant amount of diffusion and magnetic field imhomogeneity will be systematically low and the value is denoted by T_2^*.

6.3. Experimental Procedures

Following from the previous two sections it transpires that there are basically two types of NMR experiment for measuring diffusion coefficients: (1) the steady field-gradient experiment and (2) the pulsed field-gradient experiment. The steady field-gradient experiment involves measuring spin-echo amplitudes in the presence of a known steady linear magnetic field gradient for a range of gradient strengths, or τ times. The expression relating echo intensity to these parameters is (Carr and Purcell, 1954; Kärger *et al.*, 1988)

$$A(2\tau) = A(0) \exp(-2\tau/T_2 - 2\gamma^2 g^2 D \tau^3 / 3) \tag{18}$$

where $A(2\tau)$ is the echo amplitude for a given τ time in the presence of the field gradient, $A(0)$ is the amplitude of the FID at time $\tau = 0$, and the other symbols are as defined previously.

The pulsed field-gradient spin-echo (PFGSE) technique extends the range of applicability of the NMR procedure to smaller diffusion coefficients, multicomponent systems, and is also better suited for measuring restricted, or spatially dependent, diffusion (Callaghan, 1984; Stilbs, 1987; Kärger *et al.*, 1988). The pulse sequence which is most commonly employed is depicted in Figure 11B and the relationship between echo intensity, the magnitude of the field gradient, and the timing of the pulses is given by (Stejskal and Tanner, 1965)

$$A(2\tau) = A(0) \exp[-2\tau/T_2 - (\gamma g \Delta)^2 D(\Delta - \delta/3)] \tag{19}$$

where δ is the duration of the magnetic field-gradient pulses and Δ is the time interval between their leading edges. For this equation to apply it is assumed that there are no static magnetic field gradients when the gradient pulses are off, and the diffusion is unrestricted, isotropic, and a simple three-dimensional random walk (Section 6.1.1).

If significant static field gradients exist in the z direction, Equation (19) must be replaced by the modified expression (Stejskal and Tanner, 1965):

$$A(2\tau) = A(0) \exp[-2\tau/T_2 - (\gamma g \delta)^2 D(\Delta - \delta/3)]$$
$$\times \exp (2\gamma^2 g_0^2 D\tau^3/3) \times \exp \{\gamma^2 g_0 D\, g\delta[t_1^2 + t_2^2 + \delta(T_1 + t_2) + 2\delta^2/3 - 2\tau^2]\} \tag{20}$$

where t_1 is the time between the first r.f. pulse and the leading edge of the first gradient pulse, t_2 is the time between the trailing edge of the second gradient pulse to the echo which occurs at time 2τ, and g_0 is the magnitude of the static field gradient, taken here to be linear, but this is generally not the case. In practice, for accurate estimates of D, g should exceed g_0 by at least an order of magnitude and when $g_0\tau$ is $<< g\delta$ the steady gradient is deemed to be "insignificant."

There exist various modifications of the basic PFGSE sequence which include alternating or sinusoidal gradient pulses. Some of these variations are applicable to the study of high-molecular-weight species such as proteins. Of particular importance is the *stimulated echo* sequence originally proposed by Hahn (1950) and modified in various ways since then (Gibbs and Johnson, 1991; Waldeck et al., 1993); it enables the measurement of D in systems in which the T_1 is large relative to T_2, as often occurs in biological systems. The pulse sequence consists of three $\pi/2_x$ r.f. pulses in which the second and third pulses are delivered at times τ_1 and τ_2, respectively (Figue 11C). The first echo arises at time $2\tau_1 + \tau_2$. Field gradient pulses are applied between the first and second r.f. pulses and after the third. The intensity of the spin-echo amplitude is given by (Kärger et al., 1988)

$$A(\tau_1 + \tau_2) = (1/2) A(0) \exp[-(\tau_2 - \tau_1)/T_1 - 2\tau_1/T_2 - (\gamma g \Delta)^2 D(\Delta - \delta/3)] \tag{21}$$

Pulse sequences that generate multiple-quantum coherences and spin-echoes have also been developed (Kuchel and Chapman, 1993; Chapman and Kuchel, 1993). A major advantage of these experiments is that the extent of dispersion of the phases of the magnetization vectors, elicited by the field gradient pulses, is directly proportional to the order of the multiple-quantum coherence. Thus, for coherences that are of order two or above, the magnitude of g

required to bring about a specified echo attenuation is substantially reduced. Alternatively, it can be shown that for a given value of g, slower diffusional processes can be monitored without recourse to very high magnetic-field-gradient strengths.

6.4. Restricted Diffusion

The phenomenon of restricted diffusion can be used in PFGSE experiments to measure both slow *and* rapid membrane transport processes. Fist, for un-bounded diffusion in an isotopic medium, it was shown by Einstein (1906) that the root mean square displacement ($\bar{x}$) of a particle (with a diffusion coefficient D) in such a medium is given by

$$\bar{x} = (6\,D\,t)^{1/2} \tag{22}$$

where t is the time over which the displacement is measured. Thus, by rearranging this equation, it can be seen that

$$D = \bar{x}^2/6t \tag{23}$$

If the diffusion is now considered to take place in a sample in which there are impermeable barriers, separated by distances of the order of $\bar{x}^2$, then the *apparent* value of the diffusion coefficient, D_{app}, asymptotically approaches a smaller value as the time allowed between the start and finish of the experiment increases. The influence of molecular diffusion on the NMR spin-echo ampli-tude, in the case of a heterogeneous system, in which the motion of the mole-cules can be described by two different diffusion coefficients, has been investi-gated theoretically by Kärger and co-workers (Kärger, 1971, 1985; Kärger *et al.*, 1988). If a molecule undergoes exchange between two regions, with one region, A, characterized by free diffusion (or "obstructed" diffusion which may still be described by a diffusion coefficient which is independent of Δ) and the other region, B, by restricted diffusion, then an equation relating the mean lifetime at region B to the observed diffusion coefficient D_B may be applied to PFGSE data. The theory does not involve a consideration of the geometry of the compartments but merely supposes that the echo attenuation can be described by the superposi-tion of two exponentials:

$$R = P_1\exp(-KD_1\Delta) + P_2\exp(-KD_2\Delta) \tag{24}$$

where

$$D_{1,2} = 1/2(D_A + D_B + 1/K(1/\tau_A + 1/\tau_B)$$

$$\pm \sqrt{[D_B - D_A + 1/K(1/\tau_B - 1/\tau_A)]^2 + 4/K^2\tau_A\tau_B}\}$$

$$P_2 = [1/(D_2 - D_1)](P_A D_A + P_B D_B - D_1)$$

$$P_1 = 1 - P_2$$

$$K = \gamma^2\delta^2 g^2$$

where τ_A and τ_B are the mean lifetimes of the molecules at regions A and B, characterized by the diffusion coefficients D_A and D_B, and P_A and P_B are the fractions of the molecules at regions A and B, respectively. The underlying assumptions used in developing this equation are that the exchange rates are much faster than the transverse relaxation rates of the nucleus observed (i.e., $1/\tau_{A,B} >> 1/T_{2A,B}$) and that the time spacing between the pulses, Δ, is much larger than their duration, δ. The complicated expression may be simplified in two limiting cases: (1) When there is very rapid exchange ($K\tau_{A,B}D_{A,B} >> 1$), in which case the equation becomes a single exponential with $D_1 = P_A D_A$ and $D_2 = P_B D_B$; and (2) when there is slow exchange ($K\tau_{A,B}D_{A,B} << 1$), $D_1 = D_A$, $D_2 = D_B$, $P_1 = P_A$, and $P_2 = P_B$. However, in these special cases, the values of $\tau_{A,B}$ cannot be obtained from diffusion measurements. Regression of Equation (24) onto spin-echo data obtained in experiments in which Δ, δ, or g are varied, yields estimates for D_1, D_2, P_1, and P_2. Alternatively, nonlinear regression can be employed in which estimates of D_A, D_B, τ_A, and τ_B, together with the P_1 and P_2, are obtained.

Human erythrocytes, like many other cells, are characterized by dimensions which lead to restriction of the motion of molecules undergoing diffusion measured on a time scale like that of PFGSE experimens (viz., 10–100 msec). A molecule diffusing inside a cell will move freely until it reaches the membrane, whereupon it will be "reflected" back into the bulk fluid in the cell or, under special circumstances, it will traverse the membrane. The translational diffusion of a reflected molecule will be seen to be *restricted* by the cell membrane if the diffusion is studied over times much longer than that required to traverse the cell cavity; for example, in the first high-resolution NMR studies of solute diffusion in erythrocytes, short diffusion times were used to obviate the effects of the restriction of motion by the membranes (Price et al., 1989). Hence, restricted diffusion has been observed in many cellular and geological samples (Callaghan,

1991). Expressions for spin-echo signal attenuation have been derived for molecules diffusing in different-shaped cavities constituting one compartment (Neuman, 1974; Kärger, 1985; Murday and Cotts, 1968), but no general analytical theory exists to describe rapid membrane transport in systems in which diffusion is restricted within boundaries of particular geometric form.

In the simple two-site theory of Kärger (1985) a useful simplification exists: for large Δ values, the quantity analogous to D_B in Equation (24) is much less than D_A and the expression for τ_B is written as

$$\tau_B = \{D_A - D_B(1 + 1/N) + D_B N\}/\{(KD_1(D_A + D_B - D_1) - KD_A D_B\} \tag{25}$$

where $N = P_A/P_B$. More rigorously, though, it is possible to conduct PFGSE experiments by varying the value of δ and regressing Equation (12) onto the data to obtain the relevant parameters including τ_A and τ_B (e.g., Waldeck *et al.*, 1993).

6.5. Slow Transmembrane Diffusion

Andrasko (1976a) studied the uptake of Li^+ by red blood cells by analyzing data from PFGSE experiments using theory that is essentially a simplified version of that in Section 6.4. If the observed nuclear magnetization undergoes slow exchange between two regions A and B, the attenuation of the signal amplitude in a PFGSE experiment with the possibility of $\delta \sim \Delta$ is given by a modification of Equation (24):

$$R = P_A \exp(-KD_A) + P_B \exp(-KD_B) \tag{26}$$

where, as distinct from Equation (24), $K = \gamma^2\delta^2 g^2 (\Delta - \delta/3)$. When the diffusion in region B is "restricted" the second term in this equation becomes $P_B R_r$, where R_r is the echo attenuation for the case of restricted diffusion. In this manner regions A and B may be identified with external and internal solutions, respectively. By choosing suitable experimental parameter values (δ, g, Δ), the first term in Equation (26), corresponding to the site with unrestricted diffusion, can be made negligible. Thus, the contribution of nuclei in the external medium to the spin-echo signal amplitude will be eliminated without disturbing the system, even when P_A is much greater than P_B. Under these conditions, the residual spin-echo signal arises from nuclei in molecules located inside the cells, and net transport of molecules or ions across the cell membranes may thus be measured.

As an aside it is worth noting that a further use of the PFGSE experiment is to measure the transverse relaxation rate constant, $1/T_2$, of nuclei in closed compartments. This may be achieved because it is possible to select values of δ, g, Δ which lead to the elimination of the contribution of molecules (nuclei) in the external solution to the echo amplitude, if the value of D_A and that of cell size are known; in other words the condition $(D_A\Delta)^{1/2} >>$ the largest cell dimension should be made to apply. Alternatively, the parameters can be chosen, after performing an experiment and graphing the result of R versus Δ, and ascertaining at what value of Δ the curve becomes effectively horizontal, thus implying the approach of D_B to its limiting value of zero.

In Andrasko's (1976a) experiments erythrocytes were suspended (haematocrit 0.5) in a solution of 155 mM LiCl. ^{7}Li NMR was used to obtain the spectra, and the diffusion coefficient for Li^+ in the extracellular medium was estimated to be 1.3×10^{-9} m^2 sec. The cellular uptake was described by a single-exponential rise, so, by fitting Equation (26) to data obtained from an experiment in which Δ was varied, the half time for lithium uptake was estimated to be 7.5 hr. This rate coresponds to an initial uptake velocity of 2.82×10^{-20} mol/cell per sec (or 1.2 mmol/liter RBC per hr) which, by way of comparison, is one third of the normal rate of Na^+ efflux from the cells via Na,K-ATPase (Grimes, 1980).

6.6. Rapid Transmembrane Diffusion

PFGSE experiments adapted to measure *rapid* membrane transport were first used to measure the rate of water exchange across the membranes of human erythrocytes (Andrasko, 1976b). The mean lifetime of water inside human erythrocytes at 24°C was estimated to be 17 msec; this estimate has subsequently been shown to be too high (Benga, 1989a). The water exchange was able to be inhibited by the sulfhydryl reagent, p-chloromercuribenzoate (pCMBS). More recently the PFGSE procedure has been used with ^{7}Li NMR to measure the rate of exchange of Li^+ across the membranes of synthetic phospholipid vesicles as mediated by the ionophore m139603 (Waldeck *et al.*, 1993). The estimates of the exchange rate constants were the same, within experimental error, as those determined using a magnetization-transfer procedure (1D EXSY, Section 5.8).

6.7. "Absorbing Wall" Experiment

In this theory it is supposed that molecules encountering the boundaries which restrict diffusion, in inidividual subregions of a heterogeneous system, effectively have their signal "annihilated" on encountering the boundary (Frey *et al.*, 1988). This situation can be realized in NMR experiments once it is recognized that with NMR one is concerned with the magnetization of nuclei rather than the molecules carrying the magnetization. Thus, no material (matter) is destroyed in this process. In NMR experiments the absorbing wall condition is fulfilled in a straightforward manner if the subregions are covered by a "layer of sites" which enhance the rate of relaxation, such as may occur with paramagnetic ions in solution.

The absorbing wall experiment has been modeled using ^{13}C-labeled bicarbonate diffusing inside human erythrocytes; the absorbing wall in this case resulting from the presence of Mn^{2+} in the extracellular fluid (Price and Kuchel, 1990). Thus, the signal arising from the ^{13}C-labeled bicarbonate ions that are transported through the membrane into the extracellular space is effectively "absorbed," or annihilated, while the signal from those molecules that remain inside the cells is readily detected. Furthermore, the rate of relaxation of the $H^{13}CO_3{}^-$ signal is dramatically decreased in the presence of the band 3 inhibitor DNDS, thus reducing the rate of relaxation by the "absorbing wall" and demonstrating transport modulation of the restricted diffusion of $H^{13}CO_3{}^-$ (Price and Kuchel, 1990).

7. CONCLUDING REMARKS

The study of membrane transport is a vast and rapidly expanding field. A wide variety of methods have been used to monitor membrane transport processes and in this chaper we have attempted to show how the technique of NMR spectroscopy may be used for this purpose.

NMR spectroscopy has enormous potential for application in the study of almost any membrane transport process whether fast (time scale $\sim$ 1 sec) or slow (time scale $\sim$ 1 min to hours). In particular, it offers a number of advantages over more conventional techniques. First, most solute molecules have at least one NMR-receptive nucleus which is naturally present, so there is generally no requirement to specifically synthesize expensive and potentially hazardous labeled compounds as is the case using conventional radioisotope techniques. On the other hand, some very subtle effects, such as kinetic isotope effects on the rates of reactions, can be studied using NMR spectroscopy if compounds are synthesized with isotopic atoms. Second, NMR magnetization-transfer procedures allow the measurement of very fast transport processes that are technically extremely difficult to study using other techniques. Rapid transport has been measured previously using a custom-built flow-tube aparatus (Brahm, 1977); however, this equipment is not commercially available and the technique entails the physical separation of the cells from the media, which therefore necessitates a significant amount of sample processing.

Finally, perhaps the most attractive feature of the NMR technique for measuring membrane transport lies in the fact that information about transport may be obtained simultaneously with information about the physiological and biochemical properties of the cells. NMR spectra of intact cells can be used to provide a wealth of information about parameters such as intracellular pH, viscosity, and metabolite levels while concomitantly allowing observation of transport processes. There is increasing interest among cell physiologists in the control of membrane tansport and in the coupling of processes at the cell membrane to intracellular phenomena. The potential of NMR spectroscopy to provide information about the mechanism of these coupling and control processes would appear to be enormous.

ACKNOWLEDGMENTS. The work was funded by grants from the Australian National Health and Medical Research Council (P.W.K), the Nuffield Foundation and the Lister Institute of Preventive Medicine (K.K.), and the University of Sydney Cancer Research Fund (G.F.K.). Dr. B. E. Chapman, Ms. A. J. Lennon, Dr. J. R. Potts, Dr. J. E. Raftos, Mr. A. R. Waldeck, and Dr. A. S.-L. Xu are thanked for valuable discussions and assistance with preparation of the manuscript. Mr. Bill Lowe is thanked for expert technical assistance, and Mrs. Merilyn Kuchel and Ms. Susan Rowland are thanked for expert proofreading of the manuscript.

8. REFERENCES

Abel, E. W., Coston, T.P.J., Orrell, K. G., Sik, V., and Stephenson, D., 1986, Two-dimensional NMR exchange spectroscopy. Quantitative treatment of multisite exchanging systems, *J. Magn. Reson.* **70**:34–53.

Abragam, A., 1978, *The Principles of Nuclear Magnetism,* Oxford University Press (Clarendon), London.

Akitt, J. W., 1983, *NMR and Chemistry: An Introduction to the Fourier Transform-Multinuclear Era,* 2nd ed., Chapman & Hall, London.

Alger, J. R., and Prestegard, J. H., 1979, Nuclear magnetic resonance study of acetic acid permeation of large unilamellar vesicle membranes, *Biophys. J.* **28**:1–14.

Alger, J. R., and Shulman, R. G., 1984, NMR studies of enzymatic rates *in vitro* and *in vivo* by magnetization transfer, *Q. Rev. Biophys.* **17**:83–124.

Allis, J. L., Dixon, R. M., Till, A. M., and Radda, G. K., 1989, ^{87}Rb NMR studies for the evaluation of K$^+$ fluxes in human erythrocytes, *J. Magn. Reson.* **85**:524–529.

Allis, J. L., Dixon, R. M., and Radda, G. K., 1990, A study of transverse relaxation of ^{87}Rb in agarose gels by triple-quantum filtration, *J. Magn. Reson.* **90**:141–147.

Anderson, S. E., Adorante, J. S., and Cala, P. M., 1988, Dynamic NMR measurement of volume regulatory changes in *Amphiuma* RBC Na$^+$ content, *Am. J. Physiol.* **254**:C466–C474.

Andrasko, J., 1976a, Measurement of membrane permeability to slowly perrmeating molecules by a pulse gradient NMR method, *J. Magn. Reson.* **21**:479–484.

Andrasko, J., 1976b, Water diffusion permeability of human erythrocytes studied by a pulsed field gradient NMR technique, *Biochim. Biophys. Acta* **428**:304–311.

Andrew, E. R., 1990, Magnetic resonance imaging, *Int. J. Mod. Phys. B* **4**:1269–1281.

Arnold, J. T., 1956, Magnetic resonance of protons in ethyl alcohol, *Phys. Rev.* **102**:136–150.

Ashley, D. L., and Goldstein, J. H., 1980, The application of dextran magnetite as a relaxation agent in the measurement of water exchange using pulsed nuclear magnetic resonance spectroscopy, *Biochem. Biophys. Res. Commun.* **97:**114–120.

Avison, M. J., Hetherington, H. P., and Shulman, R. G., 1986, Applications of NMR to studies of tissue metabolism, *Annu. Rev. Biophys. Biophys. Chem.* **15:**377–402.

Baine, P., 1986, Comparison of rate constants determined by two-dimensional NMR spectroscopy with rate constants determined by other NMR techniques, *Magn. Reson. Chem.* **24:**304–307.

Batley, M., and Redmond, J., 1982, ^{31}P NMR reference standards for aqueous samples, *J. Magn. Reson.* **49:**172–174.

Bax, A., 1984, *Two-Dimensional Nuclear Magnetic Resonance in liquids,* Delft University Press, Dordrecht.

Bendall, M. R., Pegg, D. T., and Dodrell, D. M., 1981, Polarization transfer pulse sequences for two-dimensional NMR by Heisenberg vector analysis, *J. Magn. Reson.* **45:**8–29.

Bendall, M. R., Pegg, D. T., Dodrell, D. M., and Field, J., 1983, Inverse DEPT sequence. Polarization transfer from a spin-$^{1}/_{2}$ nucleus to n spin-$^{1}/_{2}$ heteronuclei via correlated motion in the doubly rotating reference frame, *J. Magn. Reson.* **51:**520–526.

Benga, G., 1988, Water transport in red blood cell membranes, *Prog. Biophys. Mol. Biol.* **51:**193–245.

Benga, G., 1989a, Water exchange through the erythrocyte membrane, *Int. Rev. Cytol.* **114:**273–316.

Benga, G., ed., 1989b, *Water Transport in Biological Membranes,* Vol. 1, *From Model Membranes to Isolated Cells,* CRC Press, Boca Raton, FL.

Benga, G., Chapman, B. E., Gallagher, C. H., Cooper, D., and Kuchel, P. W., 1993, NMR studies of diffusional water permeability of red blood cells from macropodid marsupials (kangaroos and wallabies), *Comp. Biochem. Physiol.* **104A:**799–803.

Berg, O. G., and von Hippel, P. H., 1985, Diffusion-controlled macromolecular interactions, *Annu. Rev. Biophys. Biophys. Chem.* **14:**131–160.

Beutler, E., 1984, *Red Cell Metabolism: A Manual of Biochemical Methods,* 3rd ed., Grune & Stratton, New York.

Binsch, G. E., 1975, Band shape analysis, in: *Dynamic Nuclear Magnetic Resonance Spectroscopy* (L. M. Jackman and F. A. Cotton, eds.), pp. 45–81, Academic Press, New York.

Bleaney, B. I., and Bleaney, B., 1983, *Electricity and Magnetism,* 3rd ed., Oxford University Press, London.

Bloch, F., Hansen, W. W., and Packard, M., 1946, Nuclear induction, *Phys. Rev.* **69:**127.

Boulanger, Y., Vinay, P., and Desroches, M., 1985, Measurement of a wide range of intracellular sodium concentrations in erythrocytes by ^{23}Na nuclear magnetic resonance, *Biophys. J.* **47:**553–561.

Brahm, J., 1977, Temperature-dependent changes in chloride transport kinetics in human red blood cells, *J. Gen. Physiol.* **70:**283–306.

Brahm, J., 1983, Urea permeability of human red cells, *J. Gen. Physiol.* **82:**1–23.

Brauer, M., Spread, C. Y., Reithmeier, R.A.F., and Sykes, B. D., 1985, ^{31}P and ^{35}Cl nuclear magnetic resonance measurements of anion transport in human erythrocytes, *J. Biol. Chem.* **260:**11643–11650.

Bremer, J., Mendz, G. L., and Moore, W. J., 1984, Skewed exchange spectroscopy. Two-dimensional method for the measurement of cross relaxation in ^{1}H NMR spectroscopy, *J. Am. Chem. Soc.* **106:**4691–4696.

Brindle, K. M., and Campbell, I. D., 1987, NMR studies of kinetics in cells and tissues, *Q. Rev. Biophys.* **19:**159–182.

Brindle, K. M., Brown, F. F., Campbell, I. D., Grathwohl, C., and Kuchel, P. W., 1979, Application of spin-echo nuclear magnetic resonance to whole-cell systems. *Biochem. J.* **180:**37–44.

Brophy, P. J., Hayer, M. K., and Riddell, F. G., 1983, Measurement of intracellular potassium concentrations by NMR, *Biochem. J.* **210**:961–963.

Brown, F. F., 1983, The effect of compartmental location on the proton T_2^* of small molecules in cell suspensions: A cellular field gradient model, *J. Magn. Reson.* **54**:385–399.

Brown, F. F., Campbell, I. D., Kuchel, P. W., and Rabenstein, D. L., 1977, Human erythrocyte metabolism studies by ^1H spin echo NMR, *FEBS Lett.* **82**:12–16.

Brown, F. F., Sussman, I., Avron, M., and Degani, H., 1982, NMR studies of glycerol permeability in lipid vesicles, erythrocytes and the alga *Dunaliella*, *Biochim. Biophys. Acta* **690**:165–173.

Brown, F. F., Jaroszkiewicz, G., and Jaroszkiewicz, M., 1983, An NMR method for studying the intracellular distribution and transport properties of small molecules in cell suspensions: The chicken erythrocyte system, *J. Magn. Reson.* **54**:400–418.

Brown, R., 1828, in: *The Encyclopaedia Britannica*, 1988, Vol. 2, pp. 559–560, Encyclopaedia Britannica, Chicago.

Brown, T. R., 1980, Saturation transfer in living systems, *Philos. Trans. R. Soc. London Ser. B* **289**:441–444.

Brown, T. R., and Ogawa, S., 1977, ^{31}P nuclear magnetic resonance kinetic measurements on adenylate kinase, *Proc. Natl. Acad. Sci. USA* **74**:3627–3631.

Brown, T. R., Ugurbil, K., and Shulman, R. G., 1977, ^{31}P nuclear magnetic resonance measurements of ATPase kinetics in aerobic *Escherichia coli* cells, *Proc. Natl. Acad. Sci. USA* **74**:5551–5553.

Bruker Almanac, 1993, Bruker Analytische Messtechnik, Karlsruhe, Germany.

Bubb, W. A., Kirk, K., and Kuchel, P. W., 1988, Ethylene glycol as a thermometer for X-nucleus spectroscopy in biological samples, *J. Magn. Reson.* **77**:363–368.

Bulliman, B. T., and Kuchel, P. W., 1988, A series expression for the surface area of an ellipsoid and its application to the computation of the surface area of avian erythrocytes, *J. Theor. Biol.* **134**:113–123.

Bulliman, B. T., Kuchel, P. W., and Chapman, B. E., 1989, 'Overdetermined' one dimensional NMR exchange analysis: A 1-D counterpart of the 2-D EXSY experiment, *J. Magn. Reson.* **82**:131–138.

Bulsing, J. M., Brooks, W. M., Field, J., and Doddrell, D. M., 1984, Reverse polarization transfer through matrix order multiple-quantum coherence: A reverse POMMIE sequence, *Chem. Phys. Lett.* **104**:229–234.

Burum, D. P., and Ernst, R. R., 1980, Net polarization transfer via a J-ordered state for signal enhancement of low-sensitivity nuclei, *J. Magn. Reson.* **39**:163–168.

Bystrov, V. F., Dubrovina, N. I., Barsukov, L. I., and Bergelson, L. D., 1971, NMR differentiation of the internal and external phospholipid membrane surfaces using paramagnetic Mn^{2+} and Eu^{3+} ions, *Chem. Phys. Lipids* **6**:343–348.

Callaghan, P. T., 1984, Pulsed field gradient nuclear magnetic resonance as a probe of liquid state molecular organisation, *Aust. J. Phys.* **37**:359–387.

Callaghan, P. T., 1991, *Principles of Magnetic Resonance Microscopy*, Oxford University Press (Clarendon), London.

Campbell, I. D., Dobson, C. M., Jeminet, G., and Williams, R.J.P., 1974, Pulsed NMR methods for the observation and assignment of exchangeable hydrogens: Applications to bacitracin, *FEBS Lett.* **49**:115–119.

Campbell, I. D., Dobson, C. M., and Ratcliffe, R. G., 1977, Fourier transform NMR in H_2O. A method for measuring exchange and relaxation rates, *J. Magn. Reson.* **27**:455–463.

Campbell, I. D., Dobson, C. M., Ratcliffe, R. G., and Williams, R.J.P., 1978, Fourier transform NMR pulse methods for the measurement of slow-exchange rates, *J. Magn. Reson.* **29**:397–417.

Carr, H. Y., and Purcell, E. M., 1954, Effects of diffusion on free precession in nuclear magnetic resonance experiments, *Phys. Rev.* **94**:630–638.

Castle, A. M., Macnab, R. M., and Shulman, R. G., 1986, Measurement of intracellular sodium concentration and sodium transport in *Escherichia coli* by ^{23}Na nuclear magnetic resonance, *J. Biol. Chem.* **261**:3288–3294.

Cerdan, S., and Seelig, J., 1990, NMR studies of metabolism, *Annu. Rev. Biophys. Biophys. Chem.* **19**:43–67.

Cerdonio, M., Morantes, S., Torresani, D., Vitale, S., De Young, A., and Noble, R. W., 1985, Reexamination of the evidence for paramagnetism in oxy- and (carbonmonoxy) hemoglobins, *Proc. Natl. Acad. Sci. USA* **82**:102–103.

Chapman, B. E., and Kuchel, P. W., 1990, Fluoride transmembrane exchange in human erythrocytes measured with ^{19}F NMR magnetization transfer, *Eur. Biophys. J.* **19**:41–45.

Chapman, B. E., and Kuchel, P. W., 1993, Sensitivity in heteronuclear multiple quantum diffusion experiments, *J. Magn. Reson.* **102**:105–109.

Chapman, B. E., MacDermott, T. E., and O'Sullivan, W. J., 1973, Studies on manganese complexes of human serum albumin, *Bioinorg. Chem.* **3**:27–38.

Chapman, B. E., Kirk, K., and Kuchel, P. W., 1986, Bicarbonate exchange kinetics at equilibrium across the erythrocyte membrane by ^{13}C NMR, *Biochem. Biophys. Res. Commun.* **136**:266–272.

Chapman, B. E., Stewart, I. M., Bulliman, B. T., Mendz, G. L., and Kuchel, P. W., 1988, ^{31}P magnetization transfer in the phosphoglyceromutase–enolase coupled enzyme system, *Eur. Biophys. J.* **16**:187–191.

Cheshnovsky, D., and Navon, G., 1978, Nuclear magnetic resonance studies of carbonic anhydrase catalyzed reversible hydration of acetaldehyde by the saturation transfer method, *Biochemistry* **19**: 1866–1873.

Cheshnovsky, D., and Navon, G., 1980, NMR saturation transfer studies of the catalysis of the reversible hydration of acetaldehyde by carbonic anhydrase, in: *Nuclear Magnetic Resonance Spectroscopy in Molecular Biology* (B. Pullman, ed.), pp. 261-271, Reidel, Dordrecht.

Chu, C.-K., Xu, Y., Balschi, J. A., and Springer, C. S., 1990, Bulk magnetic susceptibility in NMR studies of compartmentalized samples: Use of paramagnetic reagents, *Magn. Reson. Med.* **13**:239–262.

Chu, S. C., Pike, M. M., Fossel, E. T., Smith, T. W., Balschi, J. A., and Springer, C. S., 1984, Aqueous shift reagents for high resolution cationic nuclear magnetic resonance. III. $Dy(TTHA)^{3-}$, $Tm(TTHA)^{3-}$ and $Tm(PPP)_2^{7-}$, *J. Magn. Reson.* **56**:33–47.

Clore, G. M., Kimber, B. J., and Gronenborn, A. M., 1983, The 1–1 hard pulse: A simple and effective method of water resonance suppression in FT ^{1}H NMR, *J. Magn. Reson.* **54**:170–173.

Conlon, T., and Outhred, R., 1972, Water diffusion permeability of erythrocytes using an NMR technique, *Biochim. Biophys. Acta* **288**:354–361.

Conlon, T., and Outhred, R., 1978, The temperature dependence of erythrocyte water diffusion permeability, *Biochim. Biophys, Acta* **511**:408–418.

Cramer, J. A., and Prestegard, J. H., 1977, NMR studies of pH-induced transport of carboxylic acids across phospholipid vesicle membranes, *Biochem. Biophys. Res. Commun.* **75**:295–301.

Crank, J., 1975, *The Mathematics of Diffusion,* 2nd ed. Oxford University Press (Clarendon), London.

Cusler, E. L. 1986, *Diffusion: Mass Transfer in Fluid Systems,* Cambridge University Press, London.

Dacie, J. V., and Lewis, S. M., 1975, *Practical Haematology,* Churchill Livingstone, Edinburgh.

Dadok, J., and Sprecher, R. F., 1974, Correlation NMR spectroscopy, *J. Magn. Reson.* **13**:243–248.

Davis, D. G., Murphy, E., and London, R. E., 1988, Uptake of cesium ions by human erythrocytes and perfused rate heart: A cesium-133 NMR study, *Biochemistry* **27**:3547–3551.

Degani, H., 1978, NMR kinetic studies of the ionophore X-537A-mediated transport of manganous ions across phospholipid bilayers, *Biochim. Biophys. Acta* **508**:364–369.

Degani, H., and Lenkinski, R., 1980, Ionophoric properties of angiotensin II peptides. Nuclear magnetic resonance kinetic studies of the hormone-mediated transport of manganese ions across phosphatidylcholine bilayers, *Biochemistry* **19**:3430–3434.

Degani, H., Simon, S., and McLaughlin, A. C., 1981, The kinetics of ionophore X-537-A-mediated transport of manganese through dipalmitoylphosphatidylcholine vesicles, *Biochim. Biophys. Acta* **646**:320–328.

Degani, H., Laughlin, M., Campbell, S., and Shulman, R. G., 1985, Kinetics of creatine kinase in heart: A ^{31}P nmr saturation- and inversion-transfer study, *Biochemistry* **24**:5510–5516.

Deutsch, C. J., and Taylor, J. S., 1987, ^{19}F NMR measurement of intracellular pH, in: *NMR Spectroscopy of Cells and Organisms* (R. Gupta, ed.), pp. 55–74, CRC Press, Boca Raton, FL.

Dobbs, E. R., 1984, *Electricity and Magnetism*, Routledge & Kegan Paul, London.

Duhm, J., and Behr, J., 1987, Role of exogenous factors in alterations of red cell Na^+–Li^+ exchange and Na^+–K^+ cotransport in essential hypertension, primary hyperaldosteronism, and hypokalaemia, *Scand. J. Clin. Lab. Invest.* **46**(Suppl. 180):82–95.

Dumoulin, C. L., and Williams, E. A., 1986, Suppression of uncoupled spins by single-quantum homonuclear polarization transfer, *J. Magn. Reson.* **66**:86–92.

Dwek, R. A., 1973, *Nuclear Magnetic Resonance (N.M.R.) in Biochemistry, Applications to Enzyme Systems*, Oxford University Press (Clarendon), London.

Eakin, R. T., Morgan, L. O., Gregg, C. T., and Matwiyoff, N. A., 1972, Carbon-13 nuclear magnetic resonance spectroscopy of living cells and their metabolism of a specifically labelled ^{13}C substrate, *FEBS Lett.* **28**:259–264.

Eigen, M., and De Maeyer, L., 1963, Relaxation methods, in: *Techniques of Organic Chemistry* (S. L. Freiss, E. S. Lewis, and A. Weissberger, eds.), Vol. VIII(II), pp. 895–1054, Wiley, New York.

Einstein, A., 1906, A new determination of molecular dimension, *Ann. Phys.* **19**:298–306.

Endre, Z. H., and Kuchel, P. W., 1986, Viscosity of concentrated solutions and of human erythrocyte cytoplasm determined from NMR measurement of molecular correlation times, *Biophys. Chem.* **24**:337–356.

Endre, Z. H., Chapman, B. E., and Kuchel, P. W., 1983a, Intra-erythrocyte microviscosity and diffusion of specifically labelled [glycyl-α-^{13}C]glutathione by using ^{13}C NMR, *Biochem. J.* **216**:655–660.

Endre, Z. H., Kuchel, P. W., and Chapman, B. E., 1983b, Cell volume dependence of ^{1}H spin-echo NMR signals in human erythrocyte suspensions: The influence of *in situ* field gradients, *Biochim. Biophys. Acta* **803**:137–144.

Endre, Z. H., Allis, J. L., Ratcliffe, P. J., and Radda, G. K., 1989, 87-Rubidium NMR: A novel method of measuring cation flux in intact kidney, *Kidney Int.* **35**:1249–1256.

Engler, R. E., Johnson, E. R., and Wade, C. G., 1988, Dynamic parameters from nonselectively generated 1D exchange spectra, *J. Magn. Reson.* **77**:377–381.

Espanol, M. C., and Mota De Freitas, D., 1987, ^{7}Li NMR studies of lithium transport in human erythrocytes, *Inorg. Chem.* **26**:4356–4359.

Espanol, M. C., Ramasamy, R., and Mota De Freitas, D., 1989, Measurement of lithium transport across human erythrocyte membranes by ^{7}Li NMR spectroscopy, in: *Biological and Synthetic Membranes* (D. A. Butterfield, ed.), pp. 33–43, Liss, New York.

Fabry, M. E., and Eisenstadt, M., 1975, Water exchange between red cells and plasma. Measurement by nuclear magnetic relaxation, *Biophys. J.* **15**:1101–1110.

Fabry, M. E., and San George, R. C., 1983, Effects of magnetic susceptibility on NMR signals arising from red cells: A warning, *Biochemistry* **22**:4119–4125.

Falke, J. J., Pace, R. J., and Chan, S. I., 1984a, Chloride binding to the anion binding site of band 3. A ^{35}Cl NMR study, *J. Biol. Chem.* **259**:6472–6480.

Falke, J. J., Pace, R. J., and Chan, S. I., 1984b, Direct observation of the transmembrane recruitment of band 3 transport sites by competitive inhibitors. A ^{35}Cl NMR study, *J. Biol. Chem.* **259**:6481–6491.

Fernandez, E., Grandjean, J., and Laszlo, P., 1987, Ion transport by lasalocid A across red-blood-cell membranes, *Eur. J. Biochem.* **167**:353–359.

Ferrige, A. G., Lindon, J. C., and Paterson, R. A., 1979. High resolution proton nuclear magnetic resonance studies of interaction between deoxyhaemoglobin and small molecules; dithionite and diphosphoglycerate, *J. Chem. Soc. Faraday Trans.* **75**:2851–2864.

Ford, W. T., Periyasamy, M., Spivey, H. O., and Chandler, J. P., 1985, Magnetization-transfer NMR determination of rates of exchange of solvent in and out of gel polymer beads, *J. Magn. Reson.* **63**:298–305.

Forsén, S., and Hoffman, R. A., 1963, A new method for the study of moderately rapid chemical exchange rates employing nuclear magnetic double resonance, *Acta Chem. Scand.* **17**:1787–1788.

Forsén, S., and Hoffman, R. A., 1964a, Study of moderately rapid chemical exchange reactions by means of nuclear magnetic double resonance, *J. Chem. Phys.* **39**:2892–2901.

Forsén, S., and Hoffman, R. A., 1964b, Exchange rates by nuclear magnetic multiple resonance. III. Exchange reactions in systems with several nonequivalent sites, *J. Chem. Phys.* **40**:1189–1196.

Forsén, S., and Lindman, B., 1981, Ion binding in biological systems, *Methods Biochem. Anal.* **27**:289–486.

Fourier, J.B.J., 1822, *Theorie Analytique de la Chaleur,* Paris.

Freeman, R., Mareci, R. H., and Morris, G. A., 1981, Weak satellite signals in high-resolution NMR spectra: Separating the wheat from the chaff, *J. Magn. Reson.* **42**:341–345.

Frei, K., and Bernstein, J., 1962, Method for determining magnetic susceptibilities by NMR, *J. Chem. Phys.* **37**:1891–1892.

Frey, S., Kärger, J., Pfeifer, H., and Walther, P., 1988, NMR self-diffusion measurements in regions confined by "absorbing" walls, *J. Magn. Reson.* **79**:336–342.

Fritz, O. G., and Swift, T. J., 1967, The state of water in polarised and depolarised frog nerves: A proton magnetic resonance study, *Biophys. J.* **7**:675–687.

Gadian, D. G., 1982, *Nuclear Magnetic Resonance and its Applications to Living Systems,* Oxford University Press, London.

Gadian, D. G., and Radda, G. K., 1981, NMR studies of tissue metabolism, *Annu. Rev. Biochem.* **50**:69–83.

Gary-Bobo, C., and Solomon, A. K., 1968, Properties of hemoglobin solutions in red cells, *J. Gen. Physiol.* **52**:825–853.

Geen, H., and Freeman, R., 1989, Band-selective excitation for multidimensional NMR spectroscopy, *J. Magn. Reson.* **87**:415–421.

Geen, H., Wimperis, S., and Freeman, R., 1989, Band-selective pulses without phase distortion. A simulated annealing approach. *J. Magn. Reson.* **85**:620–627.

Gesmar, H., and Led, J. J., 1986, Optimizing the multisite magnetization-transfer experiment, *J. Magn. Reson.* **68**:95–101.

Gibbs, S. J., and Johnson C. S., 1991, A PFG NMR experiment for accurate diffusion and flow studies in the presence of eddy currents, *J. Magn. Reson.* **93**:395–402.

Glasel, J. A., and Lee, K. H., 1973, On the interpretation of water nuclear magnetic resonance relaxation times in heterogeneous systems, *J. Am. Chem. Soc.* **96**:970–974.

Glickson, J. D., Dadok, J., and Marshall, G. R., 1974, Proton magnetic double resonance study of angiotensin II (Asn[1] Val[5]) in aqueous solution employing correlation spectroscopy. Assignment of peptide NH resonances and transfer of saturation from water, *Biochemistry* **13**:11–14.

Grandjean, J., and Laszlo, P., 1987, Cation transport across membranes: The NMR viewpoint, *Biochem. (Life Sci. Adv.)* **6**:1–7.

Grimes, A. J., 1980, *Human Red Cell Metabolism*, Blackwell Scientific, Oxford.

Günther, H., 1980, *NMR Spectroscopy: An Introduction*, Wiley, New York.

Gupta, R. K., and Gupta, P., 1982, Direct observation of resolved resonances from intra- and extracellular sodium-23 ions in NMR studies of intact cells and tissues using dysprosium (III) tripolyphosphate as paramagnetic shift reagent, *J. Magn. Reson.* **47:**344–350.

Gupta, R. K., and Redfield, A., 1970, Double nuclear magnetic resonance observation of electron exchange between ferri and ferrochrome c, *Science* **169:**1204–1206.

Gupta, R. K., Ferretti, J. A., and Becker, E. D., 1974, Rapid scan Fourier transform NMR spectroscopy, *J. Magn. Reson.* **13:**275–290.

Gupta, R. K., Benovic, J. L., and Rose, Z. B., 1978a, The determination of the free magnesium level in the human red blood cell by ^{31}P NMR, *J. Biol. Chem.* **253:**6172–6176.

Gupta, R. K., Benovic, J. L., and Rose, Z. B., 1978b, Magnetic resonance studies of the binding of ATP and cations to human haemoglobin, *J. Biol. Chem.* **253:**6165–6171.

Gupta, R. K., Gupta, P., and Moore, R. D., 1984, NMR studies of intracellular metal ions in intact cells and tissues, *Annu. Rev. Biophys. Bioeng.* **13:**221–246.

Gutowsky, H. S., 1975, Time dependent magnetic perturbations, in: *Dynamic Nuclear Magnetic Resonance Spectroscopy* (L. M. Jackman and F. A. Cotton, eds.), pp. 1–21, Academic Press, New York.

Gutowsky, H. S., and Holm, C. H., 1956, Rate processes and nuclear magnetic resonance spectra. II. Hindered internal rotations of amides, *J. Chem. Phys.* **25:**1228–1234.

Gutowsky, H. S., and Saika, A., 1953, Dissociation, chemical exchange, and the proton magnetic resonance in some aqueous electrolytes, *J. Chem. Phys.* **21:**1688–1694.

Hahn, E. L., 1950, Spin echoes, *Phys. Rev.* **80:**580–594.

Halliday, J. D., Richards, R. E., and Sharp, R. R., 1969, Chemical shifts in nuclear resonances of caesium ions in solution, *Proc. R. Soc. London Ser. A* **313:**45–69.

Hamasaki, N., Wyriwicz, A. M., Lubansky, H. J., and Omachi, A. A., 1981, A ^{31}P NMR study of phosphoenolpyruvate transport across the human erythrocyte membrane, *Biochem. Biophys. Res. Commun.* **100:**879–887.

Han, K. H., La Mar, G. N., and Nagai, K., 1989, Proton magnetic resonance study of the influence of chemical modification, mutation, quaternary state, and ligation state on the dynamic stability of the heme pocket in hemoglobin as reflected in the exchange of the proximal histidyl ring labile proton, *Biochemistry* **28:**2169–2178.

Harris, R., and Mann, B. E., 1978, *NMR and the Periodic Table*, Academic Press, New York.

Hele-Shaw, H. S., and Hay, A., 1901, XI. Lines of induction in a magnetic field, *Philos. Trans. R. Soc. London Ser. A.* **195:**303–327.

Helpern, J. A., Knight, R., Welch, K.M.A., and Smith, M. B., 1987, ^{87}Rb uptake in human erythrocytes: A potassium analogue for cation transport, Abstracts of the Sixth Annual Meeting of Soc. Magn. Reson. Med., p. 513.

Helpern, J. A., Welch, K.M.A., and Halvorson, H. R., 1989, Rubidium transport in human erythrocyte suspensions monitored by ^{87}Rb NMR with aqueous chemical shift reagents, *NMR Biomed.* **2:**47–54.

Hennig, J., and Limbach, H. H., 1982, Magnetization transfer in the rotating frame: A new simple kinetic tool for the determination of rate constants in the slow chemical exchange range, *J. Magn. Reson.* **49:**322–328.

Herbst, M. D., and Goldstein, J. H., 1984, Monitoring red cell aggregation with nuclear magnetic resonance, *Biochim. Biophys. Acta* **805:**123–126.

Hervé, M., Cybulska, B., and Gary-Bobo, C., 1985, Cation permeability induced by valinomycin, gramicidin D and amphotericin B in large lipidic unilamellar vesicles studied by ^{31}P-NMR, *Eur. Biophys. J.* **12:**121–128.

Hoffman, D. W., and Henkens, R. W., 1987, The rates of fast reactions of carbon dioxide and bicarbonate in human erythrocytes measured by carbon-13 NMR, *Biochem. Biophys. Res. Commun.* **143**:67–73.

Homans, S. W., 1989, *A Dictionary of Concepts in NMR*, Oxford University Press (Clarendon), London.

Hubbard, P. S., 1970, Non-exponential nuclear magnetic relaxation by quadrupole interactions, *J. Chem. Phys.* **53**:985–987.

Hughes, M. S., Flavell, K. J., and Birch, N. J., 1988a, Transport of lithium into human erythrocytes as studied by ^7Li nuclear magnetic resonance and atomic absorption spectroscopy, *Biochem. Soc. Trans.* **16**:827–828.

Hughes, M. S., Thomas, G.M.H., Partridge, S., and Birch, N. J., 1988b, An investigation into the use of a dysprosium shift reagent in the nuclear magnetic resonance spectroscopy of biological systems, *Biochem. Soc. Trans.* **16**:207–208.

Hunt, G.R.A., 1975, Kinetics of ionophore-mediated transport of Pr^{3+} ions through phospholipid membranes using ^1H NMR spectroscopy, *FEBS Lett.* **58**:194–196.

Hunt, G.R.A., and Jones, I. C., 1982, Lanthanide-ion transport across phospholipid vesicular membranes: A comparison of alamethicin 30 and A23187 using ^1H-NMR spectroscopy, *Biosci. Rep.* **2**:921–928.

Hunt, G.R.A., and Jones, I. C., 1983, A ^1H-NMR investigation of the effects of ethanol and general anesthetics on ion channels and membrane fusion using unilamellar phospholipid membranes, *Biochim. Biophys. Acta* **736**:1–10.

Hunt, G.R.A., Jones, I. C., and Veiro, J. A., 1984, Phosphatidic acid regulates the activity of the channel-forming ionophores alamethicin, melittin, and nystatin: A ^1H-NMR study using phospholipid membranes, *Biosci. Rep.* **4**:403–413.

Hunter, F. R., 1970, Facilitated diffusion in human erythrocytes, *Biochim. Biophys. Acta* **211**:216–222.

Ikehara, T., Yamaguchi, H., Hosokawa, K., Sakai, T., and Miyamoto, H., 1984, Rb^+ influx in response to changes in energy generation: Effect of the regulations of the ATP content of HeLa cells, *J. Cell. Physiol.* **119**:273–282.

Iles, R., 1981, Measurement of intracellular pH, *Biosci. Rep.* **1**:687–699.

Itada, N., and Forster, R. E., 1977, Carbonic anhydrase activity in intact red blood cells measured with ^{18}O exchange, *J. Biol. Chem.* **252**:3881–3890.

Jeener, J., Meier, G. H., Bachmann, P., and Ernst, R. R., 1979, Investigation of exchange processes by two-dimensional NMR spectroscopy, *J. Chem. Phys.* **71**:4546–4553.

Jelicks, L. A., and Gupta, R. K., 1989a, Double quantum NMR of sodium ions in cells and tissues. Paramagnetic quenching of extracellular coherence, *J. Magn. Reson.* **81**:586–592.

Jelicks, L. A., and Gupta, R. K., 1989b, Observation of intracellular sodium ions by double-quantum-filtered ^{23}Na NMR with paramagnetic quenching of extracellular coherence by gadolinium tripolyphosphate, *J. Magn. Reson.* **83**:146–151.

Jesson, J. P., Meakin, P., and Kniessel, J., 1973, Homonuclear decoupling and peak elimination in Fourier transform nuclear magnetic resonance, *J. Am. Chem. Soc.* **95**:618–620.

Jones, A. J., and Kuchel, P. W., 1980, Measurement of choline concentration and transport in human erythrocytes by ^1H NMR: Comparison of normal blood and that from lithium-treated psychiatric patients, *Clin. Chim. Acta* **104**:77–85.

Kaplan, J. I., and Fraenkel, G., 1980, *NMR of Chemically Exchanging Systems*, Academic Press, New York.

Kärger, J., 1971, Der einfluss der zweibereichdiffusion auf die spinechodampfung unter berucksichtigung der relaxation bei messungen mit der methode der gepulsten feldgradienten, *Ann. Phys.* **27**:107–109.

Kärger, J., 1985, NMR self-diffusion studies in heterogeneous systems, *Adv. Colloid Interface Sci.* **23**:129–148.

Kärger, J., Pfeifer, H., and Heink, W., 1988, Principles and applications of self-diffusion measurements by nuclear magnetic resonance, *Adv. Magn. Reson.* **12**:1–89.

Kendall, M., and Stuart, A., 1977, *The Advanced Theory of Statistics*, Vol. 1, Chapter 10, Charles Griffin, London.

King, G. F., and Boyd, C.A.R., 1991, Proton NMR studies of transmembrane solute transport, in: *Cell Membrane Transport: Experimental Approaches and Methodologies* (D. L. Yudilevich, R. Devés, S. Perán, and Z. I. Cabantchik, eds.), pp. 297–323, Plenum Press, New York.

King, G. F., and Kuchel, P. W., 1984, A proton NMR study of iminodipeptide transport and hydrolysis in the human erythrocytes, *Biochem. J.* **220**:553–560.

King, G. F., and Kuchel, P. W., 1985, Assimilation of α-glutamyl-peptides by human erythrocytes, *Biochem. J.* **227**:833–842.

King, G. F., York, M. J., Chapman, B. E., and Kuchel, P. W., 1983, Proton NMR spectroscopic studies of dipeptidase in human erythrocytes, *Biochem. Biophys. Res. Commun.* **110**:305–312.

King, G. F., Middlehurst, C. R., and Kuchel, P. W., 1986, Direct NMR evidence that prolidase is specific for the trans isomer of imidodipeptide substrates, *Biochemistry* **25**:1054–1062.

Kirk, K., 1990, NMR methods for measuring membrane transport rates, *NMR Biomed.* **3**:1–16.

Kirk, K., and Kuchel, P. W., 1985, Red cell volume changes monitored using a new NMR procedure, *J. Magn. Reson.* **62**:568–572.

Kirk, K., and Kuchel, P. W., 1986a, Equilibrium exchange of dimethyl methylphosphonate across the human red cell membrane measured using NMR spin transfer, *J. Magn. Reson.* **68**:311–318.

Kirk, K., and Kuchel, P. W., 1986b, Red cell volume changes monitored using ^{31}P NMR: A method and model, *Stud. Biophys.* **116**:139–140.

Kirk, K., and Kuchel, P. W., 1988a, The contribution of magnetic susceptibility effects to transmembrane chemical shift differences in the ^{31}P NMR spectra of oxygenated erythrocyte suspensions, *J. Biol. Chem.* **263**:130–134.

Kirk, K., and Kuchel, P. W., 1998b, Physical basis of the effect of haemoglobin on the ^{31}P NMR chemical shifts of some phosphoryl compounds, *Biochemistry* **27**:8803–8810.

Kirk, K., and Kuchel, P. W., 1988c, Characterisation of the transmembrane chemical shift differences in the ^{31}P NMR spectra of some phosphoryl compounds in erythrocyte suspensions, *Biochemistry* **27**:8795–8802.

Kirk, K., Kuchel, P. W., and Labotka, R. J., 1988, Hypophosphite ion as a ^{31}P nuclear magnetic resonance probe of membrane potential in erythrocyte suspensions, *Biophys. J.* **54**:241–247.

Kojima, S., Kanashiro, M., Hatashi, F., Yoshida, K., Abe, H., Imanishi, M., Kawamura, M., Kawano, Y., Ashida, T., Kimura, M., Kuramochi, M., Ito, K., and Omae, T., 1989, Clinical application of sodium-23 nuclear magnetic resonance for measurement of red cell sodium concentrations, *Scand. J. Clin. Lab. Invest.* **49**:489–495.

Kuchel, P. W., 1981, NMR of biological samples, *CRC Crit. Rev. Anal. Chem.* **12**:155–231.

Kuchel, P. W., 1987, Steady-state parameters of an enzyme from n.m.r. spin transfer with thermal variation, *Biochem. J.* **244**:247–248.

Kuchel, P. W., 1989, Biological applications of NMR, in: *Analytical NMR* (L. D. Field and S. Sternell, eds.), chapter 6, Wiley, New York.

Kuchel, P. W., 1990, Spin-exchange NMR spectroscopy in studies of the kinetics of enzymes and membrane transport, *NMR Biomed.* **3**:102–119.

Kuchel, P. W., and Bulliman, B. T., 1989, Perturbation of homogeneous magnetic fields by isolated single and confocal spheroids. Implicaitons for NMR spectroscopy of cells, *NMR Biomed.* **2**:151–160.

Kuchel, P. W., and Chapman, B. E., 1983, NMR spin exchange kinetics at equilibrium in membrane transport and enzyme systems, *J. Theor. Biol.* **105**:569–589.

Kuchel, P. W., and Chapman, B. E., 1993, Heteronuclear double-quantum coherence selection with magnetic gradients in diffusion experiments, *J. Magn. Reson.* **101**:53–59.

Kuchel, P. W., Chapman, B. E., Endre, Z. H., King, G. F., Thorburn, D. R., and York, M. J., 1984, Monitoring metabolic reactions in erythrocytes using NMR spectroscopy, *Biomed. Biochim. Acta* **43:**719–726.

Kuchel, P. W., King, G. F., and Chapman, B. E., 1987a, No evidence of high capacity α-glutamyldipeptide transport into human erythrocytes, *Biochem. J.* **242:**311–312.

Kuchel, P. W., Chapman, B. E., and Potts, J. R., 1987b, Glucose transport in human erythrocytes measured using ^{13}C NMR spin transfer, *FEBS Lett.* **219:**5–10.

Kuchel, P. W., Bulliman, B. T., Chapman, B. E., and Kirk, K., 1987c, The use of transmembrane differences in saturation transfer for measuring fast membrane transport: $H^{13}CO_3^-$ exchange across the human erythrocyte, *J. Magn. Reson.* **74:**1–11.

Kuchel, P. W., Bulliman, B. T., Chapman, B. E., Kirk, K., and Potts, J. R., 1987d, Fast transmembrane exchange in red cells studied with NMR, *Biomed. Biochim. Acta* **46:**S55–S59.

Kuchel, P. W., Bulliman, B. T., and Chapman, B. E., 1988a, Mutarotase equilibrium exchange kinetics studied by ^{13}C NMR, *Biophys. Chem.* **32:**89–95.

Kuchel, P. W., Bulliman, B. T., Chapman, B. E., and Mendz, G. L., 1988b, Variances of rate constants estimated from 2D NMR exchange spectra. *J. Magn. Reson.* **76:**136–142.

Kuchel, P. W., Chapman, B. E., and Xu, A. S.-L., 1992, Raes of anion transfer across erythrocyte membranes measured with NMR spectroscopy, in: *The Band 3 Proteins: Anion Transporters, Binding Proteins and Senescent Antigens* (E. Bamberg and H. Passow, eds.), pp. 105–119, Elsevier, Amsterdam.

Kuhn, W., Offermann, W., and Leibfritz, D., 1986, Influence of off-resonance irradiation upon T_1 in *in vivo* saturation transfer, *J. Magn. Reson.* **68:**193–197.

Labotka, R. J., 1984, Measurement of intracellular pH and deoxyhemoglobin concentration in deoxygenated erythrocytes by phosphorus-31 nuclear magnetic resonance, *Biochemistry* **23:**5549–5555.

Labotka, R. J., and Kleps, R. A., 1983, A phosphate-analogue probe of red cell pH using phosphorus-31 NMR, *Biochemistry* **22:**6089–6095.

Labotka, R. J., and Omachi, A., 1987a, Erythrocyte anion transport of phosphate analogues, *J. Biol. Chem.* **262:**305–311.

Labotka, R. J., and Omachi, A., 1987b, The pH dependence of red cell membrane transport of titratable anions. An NMR study, *Biomed. Biochim. Acta* **46:**S60–S64.

Labotka, R. J., and Schwab, C. M., 1990, A dialysis cell for nuclear magnetic resonance spectroscopic measurement of protein-small molecule binding, *Anal. Biochem.* **191:**376–383.

Led, J. J., and Gesmar, H., 1982, The applicability of the magnetization-transfer NMR technique to determine chemical exchange rates in extreme cases. The importance of complementary experiments, *J. Magn. Reson.* **49:**4444–4463.

London, R. E., and Gabel, S. A., 1989, Determination of membrane potential and cell volume by ^{19}F NMR using trifluoroacetate and trifluoroacetamide probes, *Biochemistry* **28:**2378–2382.

Lundberg, U. P., Harmsen, E., Ho, C., and Vogel, H., 1990, Nuclear magnetic resonance studies of cellular metabolism, *Anal. Biochem.* **191:**193–222.

Lunderg, U. P., Berners-Price, S. J., Roy, S., and Kuchel, P. W., 1992, NMR studies of erythrocytes immobilized in agarose and alginate gels, *Magn. Reson. Med.* **25:**273–288.

McCain, D. C., and Markley, J. L., 1985, Water permeability of chloroplast envelope membranes: In vivo measurement by saturation transfer, *FEBS Lett.* **183:**353–358.

McCall, D. W., Douglass, D. C., and Anderson, E. W., 1963, Self-diffusion studies by means of nuclear magnetic resonance spin-echo techniques, *Ber. Bunsenges. Phys. Chem.* **67:**366–340.

McConnell, H. M., 1958, Reaction rates by nuclear magnetic resonance, *J. Chem. Phys.* **28:**430–431.

McConnell, H. M., and Thompson, D. D., 1957, Molecular transfer of nonequilibrium nuclear spin magnetization, *J. Chem. Phys.* **26:**958–959.

Macey, R. I., 1984, Transport of water and urea in red blood cells, *Am. J. Physiol.* **246:**C195–C203.

Macey, R. I., and Farmer, R.E.L., 1970, Inhibition of water and solute permeability in human red cells, *Biochim. Biophys. Acta* **211:**104–106.

Macey, R. I., and Yousef, L. W., 1988, Osmotic stability of red cells in renal circulation requires rapid urea transport, *Am. J. Physiol.* **254:**C669–C674.

Maciel, G. E., and Natterstad, J. J., 1965, Carbon-13 chemical shifts of the carbonyl group. III. Solvent effects, *J. Chem. Phys.* **42:**2752–2759.

Maciel, G. E., and Ruben, G. C., 1963, Solvent effects on the ^{13}C chemical shift of the carbonyl group of acetone, *J. Am. Chem. Soc.* **85:**3903–3904.

Macura, S., and Ernst, R. R., 1980, Elucidation of cross relaxation in liquids by two-dimensional N.M.R. spectroscopy, *Mol. Phys.* **41:**95–117.

Macura, S., Huang, Y., Suter, D., and Ernst, R. R., 1981, Two-dimensional chemical exchange and cross-relaxation spectroscopy of coupled nuclear spins, *J. Magn. Reson.* **43:**259–281.

Mandelbrot, B. B., 1977, *Fractals: Form, Chance, and Dimension,* Freeman, San Francisco.

Marshall, W. E., Costello, A.J.R., Henderson, T. O., and Omachi, A., 1977, Organic phosphate binding to haemoglobin in intact erythrocytes determined by ^{31}P nuclear magnetic resonance spectroscopy, *Biochim. Biophys. Acta* **490:**290–300.

Mathur-De Vre, R., 1979, The NMR studies of water in biological systems, *Prog. Biophys. Mol. Biol.* **35:**103–134.

Maxwell, J. C., 1954, *A Treatise on Electricity and Magnetism,* 3rd ed., Vol. 2, Dover, New York.

Mayrand, R. R., and Levitt, D. G., 1983, Urea and ethylene glycol-facilitated transport in intact erythrocytes, *FEBS Lett.* **241:**188–190.

Mendz, G. L., Robinson, G., and Kuchel, P. W., 1986, Direct quantitative analysis of enzyme-catalyzed reactions by two-dimensional nuclear magnetic resonance spectroscopy: Adenylate kinase and phosphoglyceratemutase, *J. Am. Chem. Soc.* **108:**169–173.

Mendz, G. L., Bulliman, B. T., James, N. L., and Kuchel, P. W., 1989, Magnetic potential and field gradients of model cells, *J. Theor. Biol.* **137:**55–69.

Merck Index, 1989, 11th ed., p. 1523, Merck and Co., Rathway, NJ.

Messerle, B. A., Wider, G., Otting, G., Weber, C., and Süthrich, K., 1989, Solvent suppression using a spin lock in 2D and 3D NMR spectroscopy with H_2O solution, *J. Magn. Reson.* **85:**608–613.

Mills, R., 1973, Self-diffusion in normal and heavy water in the range 1–45°, *J. Phys. Chem.* **77:**685–688.

Moon, R. B., and Richards, J. H., 1973, Determination of intracellular pH by ^{31}P magnetic resonance, *J. Biol. Chem.* **248:**7276–7278.

Moore, W. J., 1981, *Physical Chemistry,* 5th ed., Longman, Harlow, Essex.

Morales, M. F., Horovitz, M., and Botts, J., 1962, The distribution of tracer substrate in an enzyme-substrate system at equilibrium. *Arch. Biochem. Biophys.* **99:**258–264.

Morariu, V. V., and Benga, G., 1977, Evaluation of a nuclear magnetic resonance technique for the study of water exchange through erythrocyte membranes in normal and pathological subjects, *Biochim. Biophys. Acta* **469:**301–310.

Morris, G. A., and Freeman, R., 1978, Selective excitation in Fourier transform nuclear magnetic resonance, *J. Magn. Reson.* **29:**433–462.

Murday, J. S., and Cotts, R. M., 1968, Self-diffusion coefficient of liquid lithium, *J. Chem. Phys.* **48:**4938–4945.

Naccache, P., and Sha'afi, R. I., 1974, Effect of pCMBS on water transfer across biological membranes, *J. Cell. Physiol.* **83,** 449–456.

Nakada, T., Kwee, I. L., Griffey, B. V., and Griffey, R. H., 1988, F-19 MR imaging of glucose metabolism in the rabbit, *Radiology* **168:**823–826.

Neuman, C. H., 1974, Spin echo of spins diffusing in a bounded medium, *J. Chem. Phys.* **11:**4508–4511.

Oderblad, E., Bhar, B. N., and Lindström, G., 1956, Proton magnetic resonance of human red blood cells in heavy-water exchange experiments, *Arch. Biochem. Biophys.* **63**:221–225.

Odoom, J. E., Campbell, I. D., Ellory, J. C., and King, G. F., 1990, Characterization of peptide fluxes into human erythrocytes: A proton-n.m.r. study, *Biochem. J.* **267**:141–147.

Ogino, T., Arata, Y., Fujiwara, S., Shaun, H., and Beppu, T., 1978, Use of proton correlation NMR spectroscopy in the study of living cells. Anaerobic metabolism of *Escherichia coli*, *J. Magn. Reson.* **31**:523–526.

Ogino, T., Den Hollander, J. A., and Shulman, R. G., 1983, ^{39}K, ^{23}Na and ^{31}P NMR studies of ion transport in *Saccharomyces cerevisiae*, *Proc. Natl. Acad. Sci. USA* **80**:5185–5189.

Ogino, T., Shulman, G. I., Avison, M. J., Gullans, S. R., Den Hollander, J. A., and Shulman, R. G., 1985, ^{23}Na and ^{39}K NMR studies of ion transport in human erythrocytes, *Proc. Natl. Acad. Sci. USA* **82**:1099–1103.

Ohgushi, M., Nagayama, K., and Wada, A., 1978, Dextran-magnetite: A new relaxation reagent and its application to T_2 measurements in gel systems, *J. Magn. Reson.* **29**:599–601.

Okerlund, L. S., and Gillies, R. J., 1988, Measurement of pH and Na$^+$ by nuclear magnetic resonance, in: *Na$^+$/H$^+$ Exchange* (S. Grinstein, ed.), pp. 21–43, CRC Press, Boca Raton, FL.

Oki, M., 1985, *Applications of Dynamic NMR Spectroscopy to Organic Chemistry*, pp. 1–37, VCH Publishers, New York.

Partridge, S., Hughes, M. S., Thomas, G.M.H., and Birch, N. J., 1988, Lithium transport in erythrocytes, *Biochem. Soc. Trans.* **16**:205–206.

Pekar, J., and Leigh, J. S., 1986, Detection of biexponential relaxation in sodium-23 facilitated by double-quantum filtering, *J. Magn. Reson.* **69**:582–584.

Pekar, J., Renshaw, P. F., and Leigh, J. S., 1987, Selective detection of intracellular sodium by coherence-transfer NMR, *J. Magn. Reson.* **72**:159–161.

Perrin, C. L., 1989, Optimum mixing time for chemical kinetics by 2D NMR, *J. Magn. Reson.* **82**:619–621.

Perrin, C. L., and Engler, R. E., 1990, Weighted linear-least-squares analysis of EXSY data from multiple 1D selective inversion experiments, *J. Magn. Reson.* **90**:363–369.

Perrin, C. L., and Gipe, R. K., 1984, Multisite kinetics by quantitative two-dimensional NMR, *J. Am. Chem. Soc.* **106**:4036–4038.

Pettegrew, J. W., Woessner, D. E., Minshew, N. J., and Glonek, T., 1984, Sodium-23 analysis of human whole blood, erythrocytes and plasma. Chemical shift, spin relaxation and intracellular sodium concentration studies, *J. Magn. Reson.* **57**:185-196.

Pettegrew, J. W., Post, J.F.M., Panchalingam, K., Withers, G., and Woessner, D. E., 1987, ^{7}Li NMR study of normal human erythrocytes, *J. Magn. Reson.* **1**:504–519.

Pike, M. M., Simon, S. R., Balschi, J. A., and Springer, C. S., 1982, High-resolution NMR studies of transmembrane cation transport: Use of an aqueous shift reagent for ^{23}Na, *Proc. Natl. Acad. Sci. USA* **79**:810–814.

Pirkle, J. L., Ashley, D. L., and Goldstein, J. H., 1979, Pulse nuclear magnetic resonance measurements of water exchange across the erythrocyte membrane employing a low Mn^{2+} concentration, *Biophys. J.* **25**:389–406.

Plateau, P., Dumas, C., and Gueron, M., 1983, Solvent-peak-suppressed NMR: Correction of baseline distortions and use of strong-pulse excitation, *J. Magn. Reson.* **54**:46–53.

Potts, J. R., and Kuchel, P. W., 1992, Anomeric preference of fluoro-glucose exchange across human red cell membranes: ^{19}F-n.m.r. studies, *Biochem. J.* **281**:753–759.

Potts, J. R., Kirk, K., and Kuchel, P. W., 1989, Characterisation of the transport of the non-electrolyte dimethyl methylphosphonate across the red cell membrane, *NMR Biomed.* **1**:198–204.

Potts, J. R., Hounslow, A. M., and Kuchel, P. W., 1990, Exchange of fluorinated glucose across the red cell membrane measured using ^{19}F NMR magnetisation transfer, *Biochem. J.* **266**:925–928.

Potts, J. R., Bulliman, B. T., and Kuchel, P. W., 1992, Urea exchange across the human erythrocyte membrane measured using ^{13}C NMR lineshape analysis, *Eur. Biophys. J.* **21**:207–216.

Prasad, K.V.S., Severini, A., and Kaplan, J. G., 1987, Sodium ion influx in proliferating lymphocytes: An early component of the mitogenic signal, *Arch. Biochem. Biophys.* **252**:515–525.

Prestegard, J. A., Cramer, J. A., and Viscio, D. B., 1979, Nuclear magnetic resonance determination of permeation coefficients for maleic acid in phospholipid vesicles, *Biophys. J.* **26**:575–584.

Price, W. S., and Kuchel, P. W., 1990, Restricted diffusion of bicarbonate and hypophosphite ions modulated by transport in suspensions of red blood cells, *J. Magn. Reson.* **90**:100–110.

Price, W. S., Chapman, B. E., Cornell, B. A., and Kuchel, P. W., 1989, Translational diffusion of glycine in erythrocytes measured at high resolution with pulsed field gradients, *J. Magn. Reson.* **83**:160–166.

Price, W. S., Kuchel, P. W., and Cornell, B. A., 1991, A ^{35}Cl and ^{37}Cl NMR study of chloride binding to the erythrocyte anion transport protein, *Biophys. Chem.* **40**:329–337.

Rabenstein, D. L., and Isab, A. A., 1982, Determination of the intracellular pH of intact erythrocytes by ^{1}H NMR spectroscopy, *Anal. Biochem.* **121**:423–432.

Raftos, J. E., Kirk, K., and Kuchel, P. W., 1988, Further investigation of the use of dimethylmethylphosphonate as a ^{31}P-NMR probe of red cell volume, *Biochim. Biophys. Acta* **968**:160–166.

Raftos, J. E., Bulliman, B. T., and Kuchel, P. W., 1990, Evaluation of an electrochemical model of erythrocyte pH buffering using ^{31}P NMR data, *J. Gen. Physiol.* **95**:1183–1204.

Redfield, A. G., 1978, Proton nuclear magnetic resonance in aqueous solutions, *Methods Enzymol.* **49**:253–270.

Redfield, A. G., 1985, Special problems of NMR in H_2O solution, in: *NMR in the Life Sciences* (E. M. Bradbury and C. Nicolini, eds.), pp. 1–10, Plenum Press, New York.

Renshaw, P. F., Blum, H., and Leigh, J. S., 1986, Applications of dextran magnetite as a sodium relaxation enhancer in biological systems, *J. Magn. Reson.* **69**:523–526.

Riddell, F. G., and Arumugam, S., 1988, Surface charge effects upon membrane transport processes: The effect of surface charge on the monensin-mediated transport of lithium ions through phospholipid bilayers studied by ^{7}Li NMR sepctroscopy, *Biochim. Biophys. Acta* **945**:65–72.

Riddell, F. G., and Arumugam, S., 1989, The transport of Li$^+$, Na$^+$ and K$^+$ ions through phospholipid bilayers mediated by the antibiotic M139603 studied by ^{7}Li, ^{23}Na and ^{39}K NMR, *Biochim. Biophys. Acta* **984**:6–10.

Riddell, F. G., and Hayer, M. K., 1985, The monensin-mediated transport of sodium ions through phospholipid bilayers studied by ^{23}Na NMR spectroscopy, *Biochim. Biophys. Acta* **817**:313–317.

Riddell, F. G., Arumugam, S., and Cox, B. G., 1988a, The monensin-mediated transport of Na$^+$ and K$^+$ through phospholipid bilayers studied by ^{23}Na and ^{39}K NMR, *Biochim. Biophys. Acta* **944**:279–284.

Riddell, F. G., Arumugam, S., Brophy, P. J., Cox, B. G., Payne, M.C.H., and Southon, T. E., 1988b, The nigericin-mediated transport of sodium and potassium ions through phospholipid bilayers studied by ^{23}Na and ^{39}K NMR spectroscopy, *J. Am. Chem. Soc.* **110**:734–738.

Robinson, G., Chapman, B. E., and Kuchel, P. W., 1984, ^{31}P NMR spin-transfer in the phosphoglyceromutase reaction, *Eur. J. Biochem.* **143**:643–649.

Robinson, G., Kuchel, P.W., Chapman, B. E., Doddrell, D. M., and Irving, M. G., 1985, A simple procedure for selective inversion of NMR resonances for spin transfer enzyme kinetic measurements, *J. Magn. Reson.* **63**:314–319.

Rogers, M. T., and Woodbrey, J. C., 1962, A proton magnetic resonance study of hindered internal rotation in some substitued N,N-dimethylamides, *J. Phys. Chem.* **66**:540–546.

Sandström, J., 1982, *Dynamic NMR Spectroscopy,* Academic Press, New York.

Savitz, D., Sidel, V. W., and Solomon, A. K., 1964, Osmotic properties of human red cells, *J. Gen. Physiol.* **48**:78–94.

Shami, Y., Carver, J., Ship, S., and Rothstein, A., 1977, Inhibition of Cl^- binding to anion transport protein of the red blood cell by DIDS (4,4′-diisothiocyano-2,2′-stilbene disulfonic acid) measured by [^{35}Cl] NMR, *Biochem. Biophys. Res. Commun.* **76**:429–436.

Shapiro, Y. E., Viktorov, A. V., Volkova, V. I., Barsukov, L. I., Bystrov, V. F., and Bergelson, L. D., 1975,^{13}C NMR investigation of phospholipid membranes with the aid of shift reagents, *Chem. Phys. Lipids* **14**:227–232.

Shaw, D., 1984, *Fourier Transform N.M.R. Spectroscopy*, 2nd ed., Elsevier, Amsterdam.

Shinar, H., and Navon, G., 1984, NMR relaxation studies of intracellular Na^+ in red blood cells, *Biophys. Chem.* **20**:275–283.

Shulman, R. G., 1979, *Biological Applications of Magnetic Resonance*, Academic Press, New York.

Shungu, D. C., and Briggs, R. W., 1988, Application of 1D and 2D ^{23}Na magnetization-transfer NMR to the study of ionophore-mediated transmembrane cation transport, *J. Magn. Reson.* **77**:491-503.

Sklenár, V., and Starcuk, Z., 1982, 1-2-1 pulse train: A new effective method of selective excitation for proton NMR in water, *J. Magn. Reson.* **50**:495–501.

Slonczewski, J. L., Rosen, B. P., Alger, J. R., and Macnab, R. M., 1981, pH homeostasis in *Escherichia coli:* Measurement by ^{31}P nuclear magnetic resonance of methylphosphonate and phosphate, *Proc. Natl. Acad. Sci. USA* **78**:6271–6275.

Springer, C. S., 1987, Measurement of metal cation compartmentation in tissue by high-resolution metal cation NMR, *Annu. Rev. Biophys. Biophys. Chem.* **16**:375-399.

Stein, W. D., 1986, *Transport and Diffusion Across Cell Membranes*, Academic Press, New York.

Stejskal, E. O., and Tanner, J. E., 1965, Spin diffusion measurements: Spin echoes in the presence of a time-dependent field gradient, *J. Chem. Phys.* **42**:288–292.

Stewart, I. M., Chapman, B. E., Kirk, K., Kuchel, P. W., Lovric, V. A., and Raftos, J. E., 1986, Intracellular pH in stored erythrocytes. Refinement and further characterisation of the ^{31}P NMR methylphosphonate procedure, *Biochim. Biophys. Acta* **885**:23–33.

Stilbs, P., 1987, Fourier transform pulsed-gradient spin-echo studies of molecular diffusion, *Prog. NMR Spectrosc.* **19**:1–45.

Tanford, C., 1966, *Physical Chemistry of Macromolecules*, Wiley, New York.

Taylor, J. S., and Deutsch, C., 1983, Fluorinated α-methylamino acids as ^{19}F NMR indicators of intracellular pH, *Biophys. J.* **43**:261–267.

Taylor, J. S., Deutsch, C., McDonald, G. G., and Wilson, D. F., 1981, Measurement of transmembrane pH gradients in human erythrocytes using ^{19}F NMR, *Anal. Biochem.* **114**:415–418.

Thomas, G.M.H., Hughes, M. S., Partridge, S., Olufunwa, R. I., Marr, G., and Birch, N. J., 1988, Nuclear magnetic resonance studies of lithium ion transport in isolated rat hepatocytes, *Biochem. Soc. Trans.* **16**:208.

Toon, M. R., and Solomon, A. K., 1991, Transport parameters in the human red cell membrane: Solute–membrane interactions of amides and ureas, *Biochim. Biophys. Acta* **1063**:179–190.

Tyrrell, H.J.V., and Harris, K. R., 1984, *Diffusion in Liquids: A Theoretical and Experimental Study*, Butterworth, London.

Ugurbil, K., 1985, Magnetization-transfer measurements of individual rate constants in the presence of multiple reactions, *J. Magn. Reson.* **64**:207–219.

Vandenberg, J. I., King, G. F., and Kuchel, P. W., 1985, The assimilation of tri- and tetrapeptides by human erythrocytes, *Biochim. Biophys. Acta* **846**:127–134.

Waldeck, A. R., and Kuchel, P. W., 1993, ^{23}Na-NMR study of ionophore-mediated cation exchange between two populations of liposomes, *Biophys. J.* **64**:1445–1455.

Waldeck, A. R., Lennon, A. J., Chapman, B. E., and Kuchel, P. W., 1993, Cation transport and diffusion in liposomes studied by ^{7}Li$^+$ and ^{23}Na$^+$ pulsed field gradient NMR, *Faraday Trans.*, in press.

Weith, J. O., Andersen, O. S., Brahm, J., Bjerrum, P. J., and Borders, C. L., Jr., 1982, Chloride–bicarbonate exchange in red blood cells: Physiology of transport and chemical modification of binding sites, *Philos. Trans. R. Soc. London Ser. B* **299**:383–399.

Williams, R.J.P., 1982, The chemistry of lanthanide ions in solution and in biological systems, *Struct Bonding (Berlin)* **50**:79–11.

Wittenkeller, L., Mota de Freitas, D., Geraldes, C.F.G.C., and Tomé, A.J.R., 1992, Physical basis for the resolution of intra- and extracellular ^{133}Cs NMR resonances in Cs$^+$-loaded human erythrocyte suspensions in the presence and absence of shift reagents, *Biochemistry* **31**:1135–1144.

Woessner, D. E., 1961, Nuclear transfer effects in nuclear magnetic resonance pulse experiments, *J. Chem. Phys.* **35**:41–48.

Xu, A. S.-L., and Kuchel, P. W., 1991, Difluorophosphate as a ^{19}F NMR probe of erythrocyte membrane potential, *Eur. Biophys. J.* **19**:327–334.

Xu, A. S.-L., Potts, J. R., and Kuchel, P. W., 1991, The phenomenon of separate intra- and extracellular resonances in ^{19}F NMR spectra used for measuring membrane potential, *Magn. Reson. Med.* **18**:193–198.

York, M. J., Kuchel, P. W., and Chapman, B. E., 1984, A proton nuclear magnetic resonance study of γ-glutamyl-amino acid cyclotransferase in human erythrocytes, *J. Biol. Chem.* **259**:15085–15088.

Young, J. D., and Ellory, J. C., 1977, Red cell amino acid transport, in: *Membrane Transport in Red Cells* (J. C. Ellory and V. L. Lew, eds.), pp. 301–325, Academic Press, New York.

Young, J. D., Wolowyk, M. W., Fincham, D. A., Cheeseman, C. L., Rabenstein, D. L., and Ellory, J. C., 1987, Conflicting evidence regarding the transport of α-glutamyl-dipeptides by human erythrocytes, *Biochem. J.* **242**:309–311.

Chapter 8

Determination of Soluble and Membrane Protein Structure by Fourier Transform Infrared Spectroscopy
I. Assignments and Model Compounds

Erik Goormaghtigh, Véronique Cabiaux,
and Jean-Marie Ruysschaert

1. INTRODUCTION

During the last five years, the use of infrared spectroscopy (IR)* to determine the structure of biological materials has dramatically expanded. However, IR's biggest advantage and highest potential over older techniques is in analyzing the components of biological membranes. IR is technically simple, requires little

---

*Abbreviations used in this chapter: ATR, attenuated total reflection spectroscopy; CD, circular dichroism; DMPC, dimyristoylphosphatidylcholine; DOPC, dioleoylphosphatidylcholine; FTIR, Fourier transform infrared spectroscopy; FWHH, full width at half-height; IR, infrared spectroscopy; NMR, nuclear magnetic resonance.

Erik Goormaghtigh, Véronique Cabiaux, and Jean-Marie Ruysschaert   Laboratoire de Chimie Physique des Macromolécules aux Interfaces, Université Libre de Bruxelles, B-1050 Brussels, Belgium.

Subcellular Biochemistry, Volume 23: Physicochemical Methods in the Study of Biomembranes, edited by Herwig J. Hilderson and Gregory B. Ralston. Plenum Press, New York, 1994.

material (less than 0.1 µg) when attenuated total reflection spectroscopy (ATR) is used. Spectra are recorded in a matter of minutes; the environment of the studied molecules can be modified so that their conformation can be studied as a function of temperature, pressure, and pH, as well as in the presence of specific ligands. Because of IR's long wavelength, light scattering problems are virtually nonexistent, and highly aggregated materials or large membrane fragments can be studied. Secondary structure evaluation is in most cases affected by neither amino acid side chains nor by the presence of disulfide bridges. In addition to the conformational parameters which can be deduced from the shape of the infrared spectra, the orientation of several molecular axes can be computed with polarized infrared spectroscopy. This allows more precise analysis of the general architecture of the membrane molecules within the biological membranes. The unique advantage of IR is that it allows simultaneous study of the structure of lipids and proteins in intact biological membranes without introduction of foreign probes.

The large number of established protein sequences, over 40,000 for only a few hundred known structures, has prompted investigators to resort to prediction methods to determine the localization of membrane embedded segments and the secondary structure of membrane proteins. The main problem in evaluating the potentialities of the predictive methods for membrane proteins is the scarcity of experimental data. Reasonably high resolution structures have been obtained for only three unrelated proteins: photosynthetic complexes (Michel, 1982; Deisenhofer and Michel 1989; Kühlbrandt *et al.*, 1994), bacteriorhodopsin (Henderson and Unwin, 1975; Henderson *et al.*, 1990), and bacterial porin (Wallian and Jap, 1990, Weiss *et al.*, 1990, 1991a,b; Cowan *et al.*, 1992). Recently, part of the transmembrane sequence of the nicotinic acetylcholine receptor was found to contain a β-sheet in a 9 Å resolution structure (Unwin *et al.*, 1993). This was confirmed by FTIR (Görne-Tshelnokow, 1994). IR now indicates that other proteins could contain lipid associated β-sheets (Sanders *et al.*, 1993; Goormaghtigh *et al.*, 1993; Ghadiri *et al.*, 1994; Li *et al.*, 1992). The widely believed 12 transmembrane-helix structure of the human glucose transporter (Mueckler *et al.*, 1985) has also been challenged recently (Fischbarg *et al.*, 1993). Localization of transmembrane segments is predicted very accurately for the α-helices but fails for porin unless specially designed algorthims are used (Schrimer and Cowan, 1993; Cowan and Rosenbusch, 1994). The large proportion of polar and charged amino acids and even of prolines in transmembrane helices, especially in transport proteins (Brandl and Deber, 1986), makes prediction a complex task. Prediction of secondary structures for soluble proteins is not yet very efficient (Nishikawa and Nogushi, 1991) and was found to be worse than random for membrane proteins by Wallace et al. (1986), probably because folding in the membrane environment can not be described by the rules elaborated for the soluble proteins.

X-ray crystallography provides more accurate structural data, but obtaining high quality crystals of membrane proteins is still extremely difficult. Moreover, the method is inherently static and the three-dimensional crystals necessary to obtain atomic resolution require the replacement of the lipid bilayer by detergents. As a complementary method, IR is useful in pointing out structure and orientation changes that occur in a protein on addition of ligands or modification of the environment. Such data are essential for the understanding of dynamic processes such as enzyme catalysis. IR also provides relevant information on the structure of rather short peptides, whose conformation changes from the solution state to the membrane-bound state.

Nuclear magnetic resonance spectroscopy remains limited to small proteins because the line-broadening effect associated with larger structures and non-isotropic motion of proteins within the lipid matrix. Large membrane vesicles containing both lipids and proteins or peptides are therefore still difficult to examine. Circular dichroism suffers from light scattering and is prone to errors when large membrane fragments or lipid vesicles have to be studied. Under such conditions, a major effort is needed to correct the CD spectrum for flattening effects. Fluorescence spectroscopy often requires the use of perturbating probes and is also sensitive to light scattering effects.

Application of Raman spectroscopy to the study of membrane proteins is promising, but is still limited by the low signal-to-noise ratio usually caused by interfering luminescence background with a strong contribution from chromophores. Raman spectroscopy does offer, however, a wealth of information on some amino acid side chains, which is complementary to the infrared data.

Infrared Spectroscopy has a characteristically short time scale ($\sim 10^{-13}$ sec). For the sake of the comparison, rapid motions occurs in a phospholipid bilayer such as trans-gauche isomerization ($5 \ 10^{-9}$ sec) or trans-to-gauche isomerization ($2.5 \ 10^{-8}$ sec) (Pastor and Venable 1988) and rotational diffusion (10^{-8} to 10^{-10} sec) (Peterson and Chan 1977; Speyer et al., 1989; Rommel et al., 1988); lateral diffusion constant is in the range of 10^{-8} cm^2/sec (Kimmich et al., 1983). For proteins, this motion is expected to be several orders of magnitude slower because of their mass. In turn, infrared spectroscopy yields an instantaneous picture of a molecule when other techniques yield averaged values. ESR and NMR, for example, are characterized by time scales in the range of $\sim 10^{-8}$ and $\sim 10^{-5}$ sec, respectively.

The advent of the new Fourier transform spectrophotometers, along with the possibilities of mathematical treatments of digitally stored spectra (subtraction, integration of surfaces, deconvolutions) has prompted new interest in this approach. During the last five years infrared spectroscopy has realized extraordinary achievements in the description of membrane proteins. For example, IR has provided a detailed picture of structural changes and proton movements from one amino acid to another during the photocycle of bacteriorhodopsin (part II), and has also delineated secondary structures of proteins in different environments, including the lipid bilayer (part III).

The purpose of this review is not only to illustrate the potentials of IR for investigating the structure of proteins but also to stress some of the difficulties and pitfalls of a technique whose use is rapidly expanding. The first part of the review summarizes the basic knowledge accumulated on the different vibrations of interest for the study of proteins. In a second part we examine the experimental problems related to the recording of protein spectra and the potentialities of infrared spectroscopy for the study of the structure of the amino acid side chains and for the recording of hydrogen/deuterium exchange kinetics, which are related to membrane insertion and protein folding. Chapter 10 deals more specifically with the application of IR for the purpose of extracting protein secondary structures from the complex infrared spectra.

2. BAND ASSIGNMENTS

The infrared spectrum of polypeptides contains contributions from the peptide amide group, called amide A, B, I, II, etc., and from relatively weaker contributions from the amino acid side chains. The former are of interest for the determination of the protein secondary structure, while the latter can be used to obtain information on the structure and the ionization state of the amino acid side chains. The first part of this review discusses these different aspects, but first, the assignments of the major band of absorption need to be briefly described.

2.1. Amide Vibrations

A typical IR spectrum of a protein is shown in Figure 1 before and after H/D exchange for 3 hours and 24 hours.

The peptide group, the structural repeat unit of proteins, gives rise to 9 characteristic bands named amide A, B, I, II, through VII. Amide A (~3300 cm^{-1}) and amide B (~3100 cm^{-1}) arise from a Fermi resonance between the N–H fundamental and the first overtone of amide II. Only Amide A is intense enough to be of any use. Amide I and II bands are the two major bands of the protein IR spectrum (Figure 1). Amide I (~1655 cm^{-1}) is associated with the peptide ν(C=O); the dipole of amide I is oriented in the CO–NH group plane close to a parallel to the C=O axis. It is conformational sensitive and shifts towards shorter wavenumbers by 2–10 cm^{-1} upon H/D exchange. Amide II (~1560 cm^{-1}) results from ν(C–N) and δ(N–H). Its transition dipole lays in the plane of the CO-NH group but roughly perpendicular to the transition dipole of amide I. This band is conformational sensitive and is shifted by ~100 cm^{-1} upon H/D exchange. Amide III and IV are very complex bands resulting from a mixture of several coordinate displacements. The out-of-plane motions are found in amide V (δ(N–H)), amide VI (δ(C=O), and amide VII (C–N torsion).

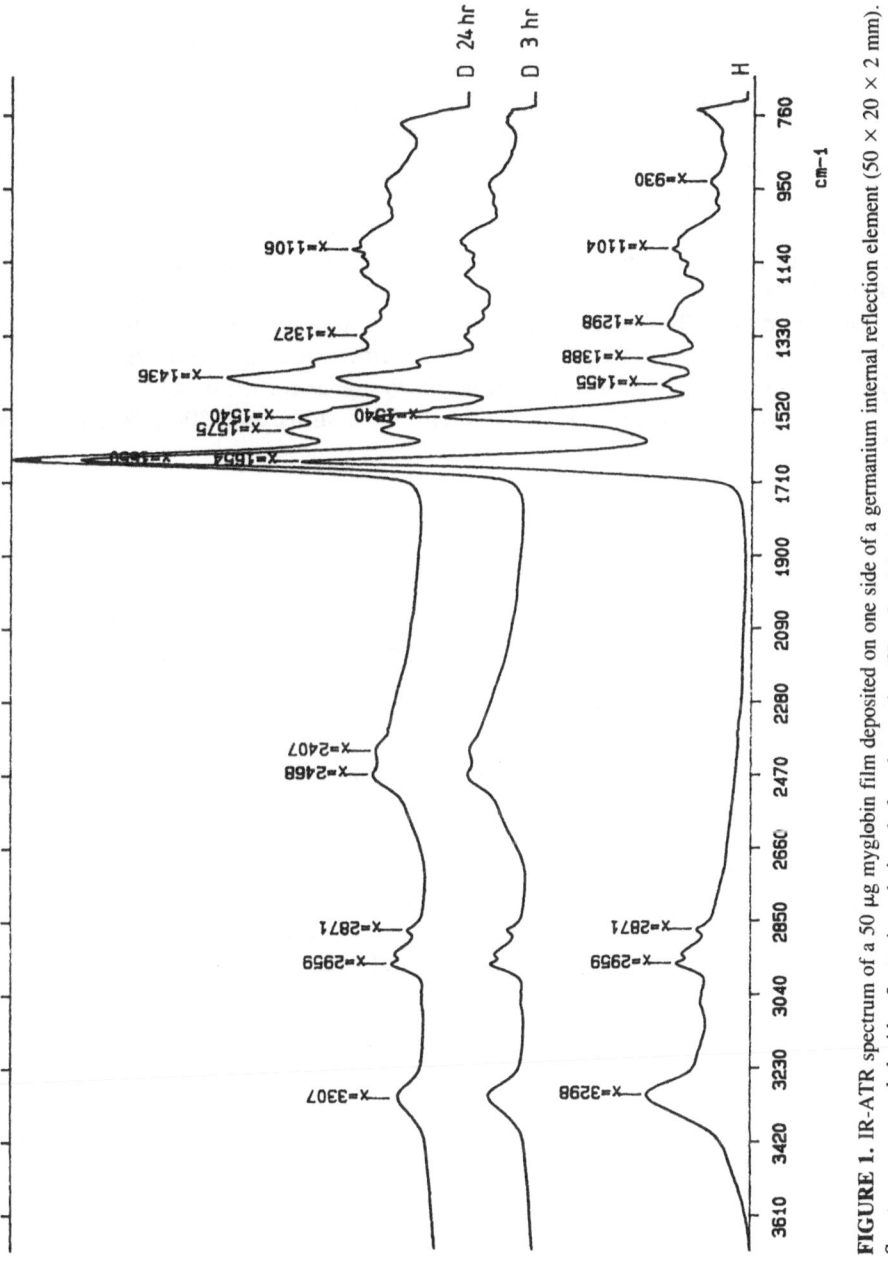

FIGURE 1. IR-ATR spectrum of a 50 μg myglobin film deposited on one side of a germanium internal reflection element (50 × 20 × 2 mm). Spectra were recorded with a 2 cm^{-1} resolution, before deuteration (H), after 3 hours deuteration (D 3 hrs), and after 24 hours deuteration (D 24 hrs). One hundred twenty-eight scans were averaged for each curve. The most intense amide vibrations are visible on the H spectrum at 3298 cm^{-1} (amide A), 1654 cm^{-1} (amide I), and 1540 cm^{-1} (amide II). Amide A is partially shifted on deuteration and amide II is (partially) shifted to 1436 cm^{-1}. Amide I remains almost unchanged at 1650 cm^{-1}.

2.1.1. Amide A

Amide A is essentially derived (more than 95%) from the amide ν(N–H). Highly localized, this mode of vibration does not depend on the chain conformation but is very sensitive to the strength of a hydrogen bond. The frequency of the unperturbed ν(N–H) was reported by Krimm and Bandekar (1986) to shift from 3225 cm^{-1} to 3280 cm^{-1} when the length of the hydrogen bond increases from ~2.69 to 2.85 Å, that is, a variation of *0.0022 Å/cm^{-1}*. This is in excellent agreement with the data collected on crystals for a wide variety of compounds for NH$\cdots$O, and also on hydrogen bonds (NH$\cdots$N or OH$\cdots$O) (Pimentel and Sederholm 1956).

Interaction of ν(N–H) with an overtone of an amide II mode (α helix) or with combination of amide II modes (β-sheet) further disturbs the frequency of amide A. The frequency of the ν(N–H) becomes higher for the α-helix (~3305 cm^{-1}) than for the β-sheet (~3275–3230 cm^{-1}). It shifts to 2380–2450 cm^{-1} region upon H/D exchange.

2.1.2. Amide I

Amide I is the most intense absorption band of the polypeptides. ν(C=O) has a predominant role in amide I: It accounts for 70–85% of the potential energy. ν(C–N) follows with 10–20% of the potential energy; The C–CN deformation may account for about 10%; Amide I also contains some in-plane NH bending, which is mainly responsible for the downshifts in the amide I frequency on N-deuteration (Krimm and Bandekar, 1986). Amide I is found between 1700 and 1600 cm^{-1}, but its exact frequency is determined by the geometry of the polypetide chain and hydrogen bonding. The energies involved per mole of H-bonded structure are small. The relation between the shift in ν(C=O) and the enthalpy of hydrogen bond formation have been reported by Tonge and Carey (1992). In the case of amide I, a 1 cm^{-1} shift corresponds to about 30 cal/mol. Such low energies could not be detected by other techniques, for instance, differential thermal analysis (George and Veis 1991). Amide I components do not shift by more than 20 cm^{-1} upon H/D exchange.

2.1.3. Amide II

Amide II occurs in the 1510–1580 cm^{-1} region. It is more complex than amide I, and, for this reason, few people have attempted to use its shape to quantify the secondary structure of proteins even though it is conformational sensitive. Amide II derives mainly from the in-plane N–H bending (40–60% of the potential energy). The rest of the potential energy arises from ν(C–N) (18–40%) and ν(C–C) (~10%). Upon H/D exchange, the in-plane N–D bending appears in the 940–1040 cm^{-1} region mixing with other modes and the ν(C–N) moves in the 1450–1490 cm^{-1} region where it mixes with other modes to yield a band called amide II'.

2.1.4. Amide III

Amide III occurs in the 1200–1400 cm^{-1} region. It is a very complex band, and although very conformational sensitive, it is very dependent on the details of the force field and on the nature of the side chains. The major contribution of amide III to the potential energy is the N–H in-plane bending, which is responsible for its sensitivity to H/D exchange. It accounts for 10–40% of the potential energy. Seventeen different contributions accounting for 0–30% (but not consistently found for 7 polypeptides) have been listed by Krimm and Bandekar (1986).

2.1.5. Amide V

Amide V occurs in the 610–710 cm^{-1} range and derives mainly from C–N torsion (40–75%) and from N–H out-of-plane bending (10–40%), which is responsible for its sensitivity to H/D exchange. Many other contributions to amide V have been described (Krimm and Bandekar, 1986). Amide V, also characterized by a strong dependency on the nature of the side chains, has been quoted to be located near 700 cm^{-1} for the β-sheet structure, near 620 cm^{-1} for the α-helix structure and near 650 cm^{-1} for disordered forms. The marked shift upon deuteration makes it possible to differentiate amide V from other vibrations. In practice, amide V is of little use for membrane proteins because of its superimposition with other bands arising from other membrane components and from buffering materials.

Before describing in more details the effects of the secondary structure on the frequency of the amide bands, it is useful to remember the general rules:

1. Hydrogen bonding induces a shift to lower frequencies for stretching modes, since it facilitates the stretching displacement. Conversely, it induces a shift to higher frequencies in the bending modes. In turn, amide A and amide V will move in opposite direction upon hydrogen bonding.
2. The shift of the hydrogen donor (e.g., N–H) is always larger than the shift of the hydrogen acceptor (e.g., C=O).
3. Coupling between the chemical repeat units within a unit cell results in splitting of the absorption bands. In general, a band will split into as many components as there are chemical repeat units. For example, the amide I band will split into four components in the antiparallel β-sheet

conformation, two in the parallel β-sheet, and three in the α-helix conformation.

We shall see that, for the study of membrane proteins in their natural hydrated environment, only amide I and amide II are of real use. The weak intensity and superimposition of the other amide vibrations with absorption bands of the membrane (phospholipid) and water (H_2O or D_2O) make detailed analysis difficult.

2.2. Amino Acid Side Chain Vibrations

The presence of nonamide bands must be recognized before attempting to extract more information from the shapes of amide I and II. The contribution of the side chain vibrations in the amide I and amide II regions has been thoroughly investigated by Venyaminov and Kalnin (1990a), and Chirgadze *et al.*, (1975). They used model compounds, such as N_{α}-acetylmethyl-esters of L-amino acids, and systematic comparison between different amino acids spectra by difference spectroscopy. Among the 20 amino acids, only nine (Asp, Glu, Asn, Gln, Lys, Arg, Tyr, Phe, and His) display a significant absorbance between 1800 and 1480 cm^{-1}. The contribution of the different amino acid side chains were fitted by a sum of Gaussian and Lorentzian contours whose parameters are reported in Table 1a for the deuterated species and in Table 1b for the protonated forms.

2.2.1. Glutamic and Aspartic Residues

Figure 2 shows the absorbance of glutamic acid and aspartic acid at pH 7.0 and pH 3.0, computed according to Venyaminov and Kalnin (1990a). Deuteration has little effect on the frequency of the COO^- and COOH(D) contribution. On the other hand, the protonation of the carboxyl group dramatically shifts the carboxyl contribution. It must be stressed that the pK used to define the proportion of ionized and protonated species in Figure 2 is an average value for proteins, which does not reflect individual large variations found in proteins. It appears that the side chains of Asp and Glu bring little contribution in the amide I region (located between ~1690 and 1613 cm^{-1}) but strongly overlap the amide II band (located between ~1580 and 1520 cm^{-1}). The carboxylate bonds are supposed to bear two equivalent oxygens in salts of amino acids. In proteins however, the position of the counterion is mainly determined by the protein structure. Therefore, it is possible that the two oxygens are no longer equivalent, causing a wide variability of band frequency (Engelhard *et al.*, 1985). The absorbances of the COO^- and COOH groups can be used to titrate a protein as a function of pH by measuring the intensity of the COO^- band in the 1500–1600

Table Ia
Absorptions Band Parameters of the Main Amino Acid Side Chains Between 1800 and 1400 cm^{-1} in D_2O^a

Vibration				ν_0 (cm^{-1})	Absorbance at max. (A_0) (l/mol/cm)	FWHH (cm^{-1})	f_G	Surface ($\times 10^{-4}$) l/mol/cm^2)
Asp	—COO$^-$	(ν_{as})	pH > pK	1584	820	34	0.4	8.8
	—COOH	(ν)	pH < pK	1713	290	45	0.8	3.5
Glu	—COO$^-$	(ν_{as})	pH > pK	1567	830	34	0.4	8.9
	—COOH	(ν)	pH < pK	1706	280	45	0.8	3.4
Arg	—CN$_3$H$_5^+$	(ν_{as})		1608	500	21	0.4	3.1
		(ν_s)		1586	460	22	0.4	3.4
Asn	—C=O	(ν)		1648	570	31	0.6	5.2
	—NH$_2$	(δ)						
Gln	—C=O	(ν)		1635	550	36	0.6	5.8
	—NH$_2$	(δ)						
Tyr	ring—OH		pH < pK	1615	160	9	0.0	0.5
				1515	500	7	0.4	1.1
	ring—O$^-$		pH > pK	1603	350	14	0.0	1.8
				1500	650	14	0.4	2.9
Terminal								
	—COO$^-$	(ν_{as})		1592	830	32	0.4	8.4
	—COOH	(ν)		1720	230	45	0.8	2.7

^{a}Bands are characterized by frequency at the maximum of absorption ν_0 (cm^{-1}), absorbance at the maximum A_0 (l/mol/cm), the full width at half height of the band (FWHH) (cm^{-1}), and the fraction of Gaussian component f_G. The integrated intensity of the peak (surface of the peak) is proportional to the product of the absorbance and of the FWHH (liter/mol/cm^2) and to the factor f_G. Estimated error is $\pm$ 3–5% on the absorbance, $\pm$ 5–10% on the surface, $\pm$ 1 cm^{-1} on ν_0, and $\pm$ 0.1 on f_G. The absorbance A_ν at any frequency ν can be computed by summation of the Lorentzian and Gaussian terms, respectively: $A\nu = (1 - f_G) \cdot A_0 \cdot$ FWHH^2 (FWHH$^2 + 4 \cdot (\nu - \nu_0)^2) + F_G \cdot A_0 \cdot e_{- \ln(2)} \cdot 2 \cdot (\nu - \nu_0)^2/$FWHH^2.
Band parameters are reported according to Chirgadze et al., (1975). Data reprinted with permission.

cm^{-1} region, which is deprived of lipid absorption. For membrane proteins, the COOH band is indeed superimposed to the phospholipid ν(C=O) band near 1735 cm^{-1} and can therefore not be used.

2.2.2. Asparagine, Glutamine, Arginine and Lysine Residues— N-glucosamine

Asparagine, glutamine, arginine (Figure 3) and lysine (Figure 4) absorb significantly in the amide I region. Since, as discussed later, the secondary structure of a protein is usually derived from an analysis of the shape of amide I, it might be important to subtract the side chain contributions from amide I prior to further analysis of its shape. For polyglutamic acid or polyarginine, the side-

Table Ib
Absorption Band Paramenters of the Amino Acid Side Chains
Between 1800 and 1400 cm^{-1} in H$_2$O^a

Vibration				ν_0 (cm^{-1})	A_0 (l/mol/cm)	FWHH (cm^{-1})	f_G	Surface ($\times 10^{-4}$) l/mol/cm²
Asp	—COO$^-$	(ν_{as})	pH > pK (~4.5)	1574 ± 2	380 ± 20	44 ± 2	0.3	5.5 ± 0.5
	—COOH	(ν)	pH < pK (~4.5)	1716 ± 2	280 ± 20	50 ± 2	0.6	4.1 ± 0.3
Glu	—COO$^-$	(ν_{as})	pH > pK (~4.4)	1560 ± 2	470 ± 30	48 ± 2	0.4	7.1 ± 0.7
	—COOH	(ν)	pH < pK (~4.4)	1712 ± 2	220 ± 10	56 ± 2	0.6	3.6 ± 0.4
Arg	—CN$_3$H$_5^+$	(ν_{as})		1673 ± 3	420 ± 40	40 ± 2	0.9	4.3 ± 0.6
		(ν_s)		1633 ± 3	300 ± 20	40 ± 4	0.5	3.6 ± 0.6
Lys	—NH$_3^+$	(δ_{as})		1629 ± 1	130 ± 10	46 ± 2	0.5	1.8 ± 0.2
		(δ_s)		1526 ± 3	100 ± 10	48 ± 2	0.7	1.3 ± 0.3
Asn	—C=O	(ν)		1678 ± 3	310 ± 20	32 ± 2	0.8	2.7 ± 0.3
	—NH$_2$	(δ)		1622 ± 2	160 ± 15	44 ± 2	0.0	2.5 ± 0.3
Gln	—C=O	(ν)		1670 ± 4	360 ± 20	32 ± 1	0.8	3.1 ± 0.3
	—NH$_2$	(δ)		1610 ± 4	220 ± 20	44 ± 2	0.0	3.5 ± 0.3
Try	ring—OH		pH < pK (~10)	1518 ± 1	430 ± 20	8 ± 1	0.5	1.0 ± 0.1
	ring—O$^-$		pH > pK (~10)	1602 ± 2	160 ± 20	14 ± 2	0.4	0.7 ± 0.1
				1498 ± 1	700 ± 10	10 ± 1	0.0	2.5 ± 0.2
His	ring			1596 ± 1	70 ± 10	14 ± 1	0.4	0.3 ± 0.1
Phe	ring			1494 ± 1	80 ± 10	6 ± 1	0.2	0.2 ± 0.1
Terminal								
	—COO$^-$	(ν_{as})		1598 ± 2	240 ± 15	47 ± 5	0.4	3.5 ± 0.4
	—COOH	(ν)		1740 ± 5	170 ± 15	50 ± 5	1.0	2.1 ± 0.4
	—NH$_3^+$	(δ_{as})		1631 ± 1	210 ± 10	54 ± 7	0.2	3.8 ± 0.7
		(δ_s)		1515 ± 5	200 ± 30	60 ± 3	0.0	4.3 ± 0.6
	—NH$_2$	(δ)		1560 ± 2	450 ± 30	46 ± 2	0.0	7.5 ± 0.9

^{a}Band parameters are reported by Venyaminov and Kalnin(1990a). Other parameters are explained in footnote of table Ia. Date reprinted by permission of ©John Wiley & Sons, Inc (1991); granted on January 29, 1993.

chain group absorption bands are comparable in intensities with the amide I band and can be confused with the amide I absorption resulting in incorrect interpretation of some particular conformation of the polypeptide chain.

N-glucosamine is present on several proteins as the result of posttranslational modifications and the importance of its contribution to the protein spectrum must be raised. The model compound N-acetyl-glucosamine presents two majors bands at 1638 and 1567 cm^{-1}, which are shifted to 1627 and 1479 cm^{-1} upon deuteration. Other characteristic peaks are located at 1442, 1380, and 1323 cm^{-1} (H form) and 1435 cm^{-1} (D form) (Caughey *et al.*, 1991). However, in the example of the scrapie-associated protein PrP-res 27–30, the contribution of N-glucosamine was found to be negligible (Caughey *et al.*, 1991).

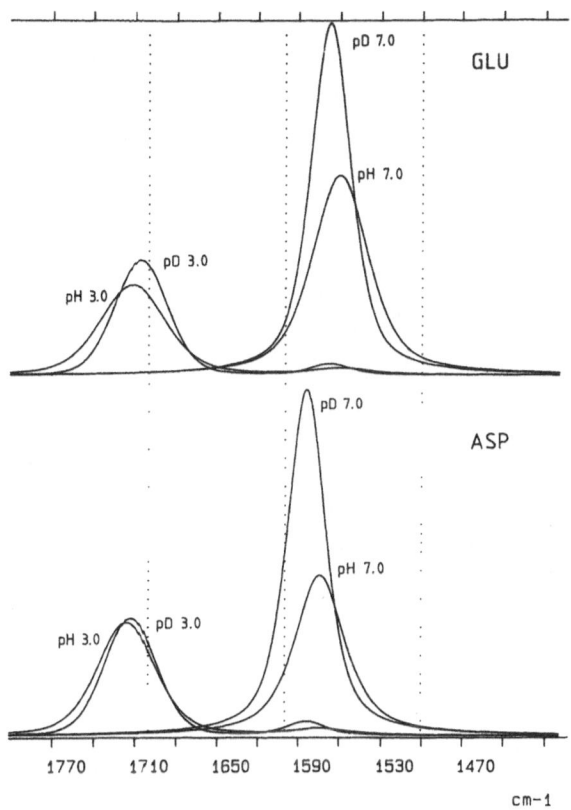

FIGURE 2. Amino acid side-chain absorption in the 1800–1480 cm⁻¹ domain according to Ven-yaminov and Kalnin (1990a), and Chirgadze *et al.*, (1974). Spectral parameters have been deter-mined for the amino acid side-chain absorption in water and in heavy water. The effect of protonation or deprotonation is indicated for Glu and Asp by reporting the spectra in H_2O at pH 3.0 and 7.0 and in D_2O at pD 3.0 and 7.0. The ordinate scale is the same for all the spectra: it is expressed in terms of molar extinction coefficient. The ordinate amplitude for the largest band of each panel is 1660 1 mol⁻¹ cm⁻¹. The shape of the different spectra presented is only a good approximation obtained from a curve fitting of more complex spectra.

2.2.3. Tyrosine, Histidine and Phenylalanine Residues

The other amino acids with significant absorption in the 1800–1480 cm⁻¹ range are tyrosine, histidine, and phenylalanine (Figure 4). Since these absorp-tion bands are weak (His) or are localized below the frequency range of amide I (Tyr, Phe) they are not expected to cause problems in determining the secondary structure of proteins.

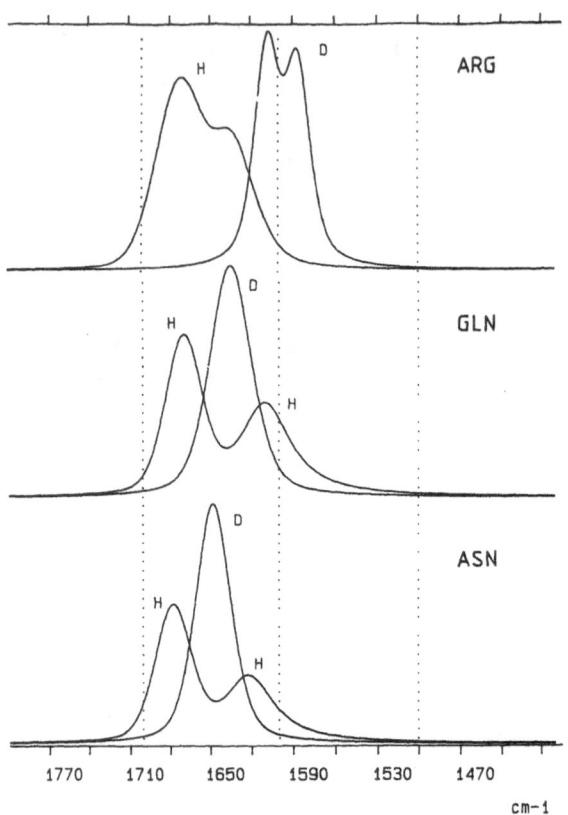

FIGURE 3. Amino acid side-chain absorption in the 1800–1480 cm⁻¹ domain according to Ven-yaminov and Kalnin (1990a), and Chirgadze *et al.*, (1974). The contribution of side chains of arginine, glutamine, and asparagine residues are reported for the deuterated form (D) and for the hydrogen form (H). The ordinate amplitude for the largest bands of each panel is 1660 1 mol⁻¹ cm⁻¹. Ionisation changes occur at too extreme pH to be biologically relevant and are not considered.

Tyrosine absorption near 1515 cm⁻¹ is easily recognized as a shoulder on amide II on the spectrum of most proteins. After H/D exchange by incubation of the protein in D_2O, Amide II is shifted towards lower frequencies and the sharp 1515 cm⁻¹ band now appears clearly distinct over the background. The intensity of the 1515 cm⁻¹ band can be used to scale the intensity of the overall side-chain group absorption with respect to the experimental protein spectrum when the knowledge of the protein concentration in the sample is lacking. This is always the case, for instance, in ATR spectroscopy.

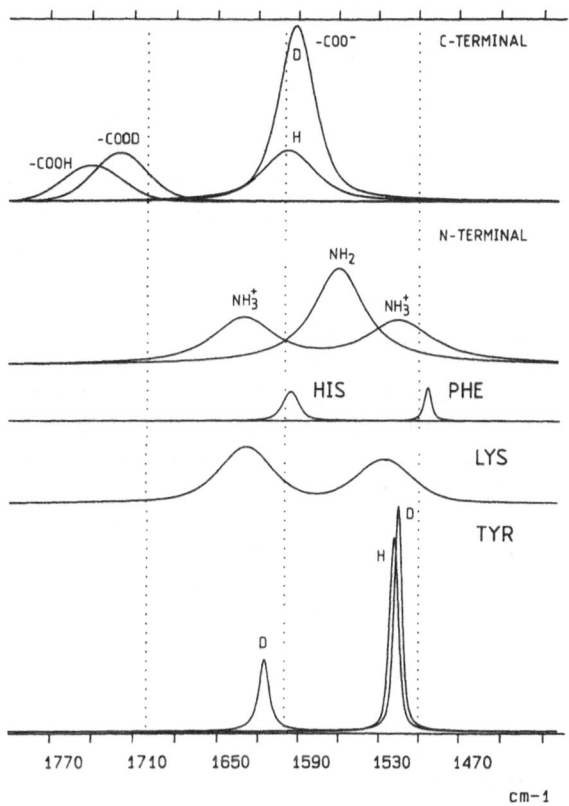

FIGURE 4. Amino acid side-chain absorption in the 1800–1480 cm⁻¹ domain, according to Ven-yaminov and Kalnin (1990a), and Chirgadze *et al.*, (1974). The contribution of side chains of tyrosine, lysine, histidine, phenylalanine residues, and of the carboxylic- and amino-terminal groups of a peptide are reported in the deuterated form (D) and in the hydrogen form (H). The total ordinate amplitude between the two horizontal abscissa axes is 1660 1 mol⁻¹ cm⁻¹ except for the terminal -COOH and -NH₂ spectra for which the same distance on the graph is worth 3320 1 mol⁻¹ cm⁻¹. For the clarity of the figure, the spectra are shifted along the ordinate scale. Ionisation changes occur at too extreme pH to be biologically relevant and are not considered.

2.2.4. Carboxyl and Amino Terminal Groups

For large proteins, the contribution of the terminal groups to the whole spectrum is weak and, therefore, could hardly interfere with secondary structure determination. However, synthetic short peptides are increasingly used to mimic particular regions of larger proteins, and the contribution of the terminal groups obviously grows as the length of the peptide decreases. Even though data reported in Table I may be valid in general, the study of free amino acids has shown that new association bands can occur. While at low concentration in free amino acid, the β-COO⁻ vibration is located at 1617 cm⁻¹ for proline, and at 1618 cm⁻¹ for glycine; increasing the concentration to 0.5 and 1 M yields a series of new sharp peaks between 1700 and 1590 cm⁻¹ due to intermolecular associations (Rudolph and Crowe, 1986). Such modifications of the band shape could occur in aggregated states (e.g., membranes) where local concentrations are high.

2.2.5. Trifluoroacetate Counterion

Trifluoroacetic acid, commonly used as a solvent in high pressure liquid chromatography (HPLC) purification of peptides, is also the most efficient agent in solubilizing hydrophobic peptides (Subrahmanyeswara *et al.*, 1988). In any case, it is difficult to remove trifluoroacetic acid completely for reasons which are not clear yet, and its presence brings an important contribution in the amide I range centered around 1673 cm⁻¹ (ν_{as} COO⁻) in D_2O (Laczko *et al.*, 1992). Clercx and Vandenbussche found two components for TFA, presumably assigned to the protonated and unprotonated forms at, respectively, 1783 cm⁻¹, and between 1700 and 1660 cm⁻¹, centered around 1686 cm⁻¹ (Clercx and Vandenbussche, personal communication).

3. SECONDARY STRUCTURES OF PEPTIDE MODEL COMPOUNDS

In the 1960s, a large number of synthetic polypeptides has been used to characterize the IR spectrum of the main secondary structures. The essential role of synthetic polypeptides in establishing the basis of secondary structure determination from IR spectra analysis is illustrated by polylysine. Polylysine adopts an α-helical, random, or β-sheet conformation according to the pH and temperature of the sample (Susi *et al.*, 1967). Polyglutamic acid is another much described example of polypeptide with different potential secondary structures (Zimmerman *et al.*, 1975; Zimmerman and Mandelkern, 1975). However, polypeptide structures do not always represent the secondary structures found in proteins, which are less regular than polypeptides that constitute a single amino acid species. Yet, experimental and theoretical work on a large number of synthetic polypeptides shed some light on the variability of the frequencies for each structure. Moreover, the large amount of data accumulated by normal mode analysis of numerous structures by Krimm and coworkers (Krimm and Bandekar, 1986) gave a deeper insight in the nature of the amide and other bands.

Recent progress in determining band intensities from *ab initio* calculations of dipole moment derivatives should make a powerful approach in the future to help assign the different IR absorption bands (Cheam and Krimm, 1985). Orientation of the transition dipoles also helps determine the orientation of secondary structures associated with oriented stacked membranes from polarized IR spectra. Cheam and Krimm (1985) provide examples of dipole intensities and orientations for N-methylacetamide and poly-(glycine I).

3.1. The Antiparallel-Chain Pleated Sheet

More than 20 synthetic polypeptides adopting an antiparallel-chain structure have been studied in the amide I region. Frequencies of the main absorption bands have been compiled by Chirgadze and Nevskaya (1976a). They conclude that the amide I frequencies are essentially determined by the properties of the backbone structure and are practically independent of the amino acid sequence, its hydrophilic or hydrophobic character, its size, and charge. The average frequency of the main component (B2) is 1629 cm^{-1} with a minimum of 1615 cm^{-1} for the β form of sodium polyglutamate and a maximum of 1637 cm^{-1} for poly(β-n-propylaspartate). The average B1 frequency is 1696 cm^{-1} with a lowest value of 1685 cm^{-1} for polyglycine. It is about ten fold weaker than the B2 component. The average splitting between B1 and B2 is 67 cm^{-1}.

The influence of the chain length and of the number of chains in the sheet has been investigated by Chirgadze and Nevskaya (1976a) by a modification of the perturbation theory of localized vibrations (Miyazawa, 1960; Krimm, 1962; and Hanlan, 1970). Interaction between nearest and next nearest neighbour peptide groups are associated with a transition dipole coupling. For two antiparallel chains, a gradual decrease of the main absorption band (B2) frequency for maximum value of ~8 cm^{-1} occurs when the chain length increases from 2 to 25 residues with no change of the $\nu_{B1}-\nu_{B2}$ splitting. Increasing the number of chains in the sheet dramatically decreases the frequency of the main absorption band by ~30 cm^{-1}, when the number of chain is increased from two to seven. For instance, the low frequency of the β-sheet form of polylysine (~1610 cm^{-1} in D$_2$O) (Carrier *et al.*, 1990a) can be explained by the formation of extensive intermolecular β-sheets as indicated by aggregation and gel formation at a concentration of 10%. Some proteins contain only short segments of β-sheet. Lysozyme, for instance, contains very short β-segments (3+3+6+4+3 residues) and, therefore, might yield atypical β-sheet bands. In interpretating polarization data obtained on membrane bound β-sheet, it is of interest to note that most finite β-sheets are twisted instead of planar, as they are in the extended structures of synthetic polypeptides. Deriving orientation of the β-sheet planes from dichroism measurements is complicated by this twisted geometry.

3.2. The Parallel-Chain Pleated Sheet

Very little information is available on parallel-chain pleated sheet structures in polypeptides, since this structure is not common in synthetic polypeptides. The effect of sheet length and number of chains in a sheet has been investigated for the antiparallel structure by Chirgadze and Nevskaya (1976b). The main band located near 1640 cm^{-1} is subject to frequency shifts very similar to those described for the antiparallel structure when the length of the sheet or the number of chains in the sheet is varied.

3.3. The α-Helical Structure

The mean frequency of a series of α-helical synthetic polypeptides compiled by Nevskaya and Chirgadze (1976) was found to be 1652 cm^{-1} for the main component of amide I and 1548 cm^{-1} for the main component of amide II. A few examples are reported in Table II. The effect of the number of residues involved in the α-helical structures was investigated for the other structures by Nevskaya and Chirgadze (1976). While for an infinite α-helix, the frequency of the amide I main component is 1653 cm^{-1} (in good agreement with the experimental data), it steadily increases when the length is decreased. Nevskaya and Chirgadze (1976) predicted a frequency of 1658 cm^{-1} for 14–16 residues, 1660 cm^{-1} for 11–13 residues, 1663 cm^{-1} for 8–10 residues, 1668 cm^{-1} for 5–7 residues and 1678 cm^{-1} for 1–4 residues. The frequency of the main component of amide II is not affected for a number of residues ≥6.

The half-width of the α-helix band depends on the stability of the helix. Most stable helices, such as polymethionine, poly(γ-ethylglutamate) and poly(γ-benzylglutamate), all in CHCl$_3$, display a half-width of ~15 cm^{-1} and a helix–coil transition free energy of ΔG > 300 cal/mol. Other polypeptides, such as polylysine (D$_2$O, 0.2M NaCl, pD 11) display half-widths of 38 cm^{-1} and helix–coil transition free energies ΔG ≃ 90 cal/mol. Polyglutamic acid displays intermediate half-widths and intermediate ΔG values according to the experimental conditions.

Polylysine has often been examined (Susi *et al.*, 1967; Nevskaya, 1967; Chirgadze *et al.*, 1973; Chirgadze and Brazhnikov, 1974) because it can adopt an α-helix, a β-sheet, or a random structure depending on the pH and temperature. α-helix formation can be induced in mild conditions after binding to a negatively charged membrane (Carrier *et al.*, 1990b). However, in the α-helix conformation, the amide I frequency is anomalously low, near 1633 cm^{-1} in D$_2$O (Carrier *et al.*, 1990a). This unusual low frequency α-helix component was explained by the fact that the α helical conformer consists of short helices, which have to be linked by nonhelical segments (Gill *et al.*, 1972). Torii and Tasumi (1992) explain the presence of α-helix component below 1640 cm^{-1} for myoglobin by a larger splitting for short helices.

Table II
IR Frequencies of the Amide Vibrations in Synthetic Polypeptides
Observed and Computed by Normal Mode Analysis[a]

	Amide A		
	Observed	Calculated	Reference
β rippled sheet	3272	3272	1
β pleated sheet	3242	3243	2
β pleated sheet	3230	3244	3
α_I helix	3279	3279	4
α_I helix	3301	3279	5
α_{II} helix	—	3288	6

	Amide I		
	Observed	Calculated	Reference
β rippled sheet	1636	1643	1
	1685	1689	
β pleated sheet	1632	1630	2
	1694	1695	
β pleated sheet	1693	1692	3
	1624	1630	
α_I helix	1658	1657	4
	1655		
α_I helix	1653	1657	5
	1655		
α_{II} helix	—	1667	6
		1660	
		1654	
3_{10} helix	1656	1665	7
		1661	
β turn I	1686	1689	8
	1655	1659	
β turn II	1680	1680	9
	1662	1664	
β turn III	1650	1651	10

	Amide II		
	Observed	Calculated	Reference
β rippled sheet	1517	1515	1
β pleated sheet	1524	1528	2
	1555	1562	
β pleated sheet	1560	1550	3

(*continued*)

Table II
(*Continued*)

	Amide II		
	Observed	Calculated	Reference
α_I helix	1545	1538	4
	1516	1519	
α_I helix	1550	1537	5
	1510	1517	
α_{II} helix	—	1540	6
	1515		
3_{10} helix	1545	1547	7
	1533		
β turn I	1568	1562	8
	1548	1544	
β turn II	1556	1545	9
β turn III	1543	1549	10
	1530	1537	

	Amide III		
	Observed	Calculated	Reference
β pleated sheet	1224	1231	2
β pleated sheet	1260	1248	3
	1223	1222	
α_I helix	1270	1278	4
	1265	1262	
α_I helix	1283	1287	5
	1263		
α_{II} helix	—	1272	6
3_{10} helix	1280	1287	7
β turn I	1294	1291	8
β turn II	1370	1375	9
	1336	1328	
	1241	1237	
β turn III	1272	—	10
	1241	1245	

	Amide V		
	Observed	Calculated	Reference
β rippled sheet	708	718	1
β pleated sheet	706	706	2

(*continued*)

Table II
(Continued)

	Amide V		
	Observed	Calculated	Reference
β pleated sheet	—	713	3
α_I helix	658	660	4
	618	608	
α_I helix	618	626	5
	618	678	
α_{II} helix	—	666	6
3_{10} helix	694	701	7
β turn I	645	657	9
β turn III	732	703	10
	682	676	

[a]Date reported in this table should be considered as examples and do not necessarily represent the structure. For instance, β turn type I here has frequencies significantly different from that computed for β turn type I in insulin (Bandekar and Krimm, 1980). The references reported here are: 1. Polyglycine I: anti//rippled sheet (Moore and Krimm, 1976b); 2. β-poly(L-alanine): anti//pleated sheet (Moore and Krimm, 1976b); 3. β-calcium-poly(L-glutamate): anti//pleated sheet (Sengupta et al., 1984); 4. α-poly(L-alanine): α_I helix (Dwivedi and Krimm, 1984);5. α-poly(L-glutamic acid): α_I helix (Sengupta and Krimm, 1985); 6. α_{II}-poly(L-alanine): α_{II} helix (Dwivedi and Krimm, 1984); 7. Poly(α-aminoisobutyric acid): 3_{10} helix (Krimm and Bandekar, 1986); 8. p-carbobenzoxyl-Gly-Pro-Leu-Gly-OH: β turn type I (Krimm and Bandekar, 1986); 9. Pro-Leu-Gly-NH$_2$: β turn type II (Krimm and Bandekar, 1986); 10. Cyclo(L-Ala-L-Ala-Aminocaproyl): β turn type III (Krimm and Bandekar, 1986).

Other parameters likely to modify the amide frequencies of α-helices are subtle variations of the α-helical structure. Solvent-induced curvature of α-helices are common in proteins (Blundell et al., 1983). Broadening of the amide A, I, and II bands of unstable α-helical polypeptides has been reported by Chirgadze et al., (1976). Polyglutamic acid forms a stable α-helix in water-dioxane and an unstable α-helix in water. The FWHH of the amide I band increases from 16 to 34 cm^{-1} from the stable to the unstable structure. In both states, the presence of helices was confirmed by CD.

3.4. The α_{II} Structure

For IR, it is important to differentiate the α_{II} helix from the usual α_I structure. Both helices have the same number of residues per turn (n = 3.6) and the same unit axial translation (h = 1.5Å) but different values of Φ and ψ angles. Moreover, the plane of the peptide group has a larger tilt with respect to the helix axis. The consequences of this tilt on the orientation of the transition dipole are important and should be evaluated before a quantitative interpretation of polarization spectra through orientation of α_{II} helices. The α_{II} helix has been suggested to be relevant to the structure of bacteriorhodopsin (Dwivedi and Krimm,

1984), which displays a very high frequency component (Rothschild and Clark, 1979). The problem of bacteriorhodopsin is further discussed in Chapter 10.

3.5. The 3_{10} Helical Structure

The 3_{10} helix differs from the α-helix in that the internal hydrogen bonding occurs between residues i and i+3 instead of i and i+4 in the α-helix. This structure is common in proteins and also occurs in poly(α-amino-isobutyric acid). Observed and calculated frequencies are reported in Table II.

3.6. Other Helices

Several other types of helices have been described for synthetic polypeptides and studied by IR, but they are not relevant for membrane proteins. The case of single-and double-stranded β-helices deserves a special mention because it is relevant for the structure of the pore-forming peptide gramicidin. The interested reader will find a normal mode analysis of the different β-helices in a paper by Naik and Krimm (1986a) and experimental spectra elsewhere (Urry *et al.*, 1983; Naik and Krimm, 1986b; Buchet *et al.*, 1985).

3.7. The β Turns

In contrast to the β-sheet and helices, turns are not as conspicuous because backbone torsion angles are nonrepeating in the polypeptide chain. Nonetheless turns are regular structures which are often exposed to the solvent and are thought to be important sites for enzymatic reactions (receptor binding, antibody recognition, glycosylation, phosphorylation, etc.). Their structure and function were reviewed by Rose *et al.*, (1985). Different types of β-hairpins in proteins have been compiled by Sibanda and Thornton (1991).

The β turn structure involves four amino acid residues, which form a loop so that the two chain segments separated by the turn adopt an antiparallel orientation and form a 4→1 (i to i+3) hydrogen bond. According to a search carried out on 30 nonhomologeous proteins of known crystal structure by Bandekar and Krimm (1979), 29% of the amino acids occur in β turns for 21% in β region, and 39% in α-helix structures. Numerous types of β turns have been identified; the main ones are type I (non helical), type II (nonhelical, requires Gly in position 3), and type III (corresponds to one turn of 3_{10} helix). Type I represents 42% of the observed turns, type II, 15%, and type III, 18% (Krimm and Bandekar, 1980). Despite the importance of the β-turn structure, only a few β-turn-forming model peptides exist because long polypeptide may not adopt an all-β turn structure. Some short peptides, however, have been shown to adopt a β-turn structure. The observed and computed frequencies of an example of β turn type I, II, and III are reported in Table II. These examples demonstrate that β-turn structures overlap the high frequency component of the β-sheet structure and the α-helix frequency domain of amide I. Moreover, in contrast to the α-helix and β-sheet structures, it appears that small variations of the β-turn conformation induce large frequency shifts in amide I. This effect also appears in the compilation of experimental and calculated frequencies reported by Lagant et al. (1984). The presence of bands between 1330 and 1290 cm^{-1} indicates specifically the presence of turns because amide III for both α-helices and β-sheets lie below 1300 cm^{-1}. The determination of the type of turn requires an analysis of the pattern of amide III bands between 1300 and 1235 cm^{-1} (Pande et al., 1986). Because of their weak intensity, assignment of these peaks to amide III is best based on their sensitivity to H/D exchange. Bandekar (1992) reported the observed and calculated frequencies for type I, II, and III in his review.

Assignment of β turns from normal mode analysis in insulin demonstrates the strong overlapping of the β turn (four different types) with the α-helix (Bandekar and Krimm, 1980), but an important contribution near 1680 cm^{-1} is now clearly assigned to β turns.

The IR spectrum of γ-turns (a reverse turn like the β turn, but with 3→1 hydrogen bonds, instead of 4→1 for the β turn) has been investigated by Bandekar and Krimm (1985) but will not be discussed here any further because of its infrequent occurrence in proteins.

4. DIPOLE ORIENTATION

Knowing the orientation of the dipole associated to the various amide transitions is required for computing from dichroism spectra obtained on oriented samples (stretched fibres, crystals, membrane multilayers, etc.) and the orientation of chemical groups (e.g. Michel-Villaz et al. 1979, Naberyk and Breton 1981, Breton and Nabedryk 1984, Goormaghtigh et al. 1987, 1989, 1990, 1991a, 1991b, 1993, Goormaghtigh and Ruysschaert 1990, Wald et al. 1990, Cabiaux et al., 1989). Recording linear dichroism spectra permits, in some instance, detection of several components underlying a broad amide band, confirming the results obtained by resolution enhancement techniques such as Fourier self-deconvolution. An example of agreement between dichroism data and Fourier self-deconvolution has been reported (Goormaghtigh et al., 1991a).

Amide A will not be discussed further because of it superimposes with lateral side chain contribution and with H$_2$O and HOD. Because the vibration is nearly confined to an oscillation of the hydrogen atom in the bond direction, the associated transition dipole direction will be almost parallel to hydrogen. However, this direction could be considerably displaced towards the direction of the hydrogen bond in cases where the angle between the vectors NH and CO is large

Table IIIA
Orientation of the Dipole Associated with Amide I and Amide II
for the α-helix with Respect to the C=O Bond.[a]

Origin of the Value	Amide I	Amide II	Reference
N-methylacetamide	15° to 25°	73° or −37°	[a]Krimm and Bandekar, 1986
Silk fibroin	19°		Suzuki, 1967
Poly(γ-benzyl-L-glutamate)	17°	77°	Tsuboi, 1972
N,N'-diacetlhexamethylene-diamine crystals	17°	77°	[a]Tsuboi, 1972
Acetanilide crystal	20°	72°	[a]Tsuboi, 1972
Ab initio methods of calculation	17°	69°	[a]Krimm and Bandekar, 1986
	9°	58°	[a]Krimm and Bandekar, 1986
	−19°	52°	[a]Krimm and Bandekar, 1986

[a]Dipole is in the plane of the peptide group and a positive value indicates a direction closer to a parallelism with the C–N bond.

(Fraser 1953). An angle of 8° with the C=O bond in poly(Glu(bzl)), which corresponds to an angle of 28° with the helix axis is reported by Tsuboi (1962). Angles of the dipole with the helix axis ranging between 17° and 28° have been reported from various sources by Fraser and MacRae (1973). Other examples are presented in Table IIIB.

For amide I and amide II, splitting into two or more components of uneven intensities makes the analysis more difficult. For the antiparallel β-sheet, the amide I component near 1630 cm^{-1} is polarized perpendicular to the sheet axis, while the weak component near 1680 cm^{-1} is polarized parallel to the sheet axis. The opposite occurs in amide II. The strongest component near 1530 cm^{-1} has its dipole oriented parallel to the sheet axis while the weaker components located near 1550 and 1510 cm^{-1} have a dipole perpendicular to the sheet axis. When a polypeptide contains only an antiparallel β-sheet, amide I and amide II appear with opposite polarization, and in theory the orientation of the sheet can be gained from the two amide dichroisms. Deviation of the expected orientation and twisting of the sheet structure, along with other experimental problems, such as disordering of the molecules and difficult evaluation of the dichroic ratios, makes the analysis frequently semiquantitative. The α-helix structure is even more difficult to discern because of the weak separation between components of different polarization in each amide band. The strongest component in amide I near 1655 cm^{-1} has a dipole orientation close to parallel to the helix axis; weaker components have opposite polarization. In amide II the polarization is opposite. The strongest component near 1548 cm^{-1} is polarized perpendicular to the helix axis. It must be stressed that in addition to the problem already mentioned for the β-sheet there is no consensus on exact orientation of the dipole to the helix axis.

Table IIIB
Orientation of the Dipole Associated with Amide A, Amide I and Amide II
for the α-helix with Respect to the Helix Axis[a]

Origin of the value	Amide A	Amide I	Amide II	Reference
Poly(pNO$_2$Benzylglutamate)	19°	31°		[b]Fraser and MacRae, 1973
Estimation used for bacteriorhodopsin	22°	27°	85°	Rothschild and Clark (1979)
	+7° or −8°	+5° or −7°	+5° or −9°	
Used for bacteriorhodopson				
αI structure assumed	17–25°	22–29°	82–88°	Draheim et al., 1991
αII structure assumed	40–51°	42–53°	86–91°	Draheim et al., 1991
Computed for bacteriorhodopsin		32°		[d]Thiaudière et al., 1993
		29°		[e]id.
Poly(benzyl-L-glutamate)	28°	39°	75°	observed by Tsuboi, 1962
	19°	28°	87°	calculated by Tsuboi, 1962
	29°	38°	83°	calculated by Tsuboi, 1962
	17°	27°	75	[a]Bazzi and Woody 1985
		29° to 34°		Miyazawa and Blout, 1961
		40°		[a]Bazzi and Woody, 1985
[c]Calculated	23°	35°	88°	[a]Fraser and MacRae, 1973
[d]Calculated	23°	35°	88°	[a]Fraser and MacRae, 1973
Poly(benzyl-L-aspartate) (amide I at 1664 cm^{-1})		40°		Bradbury et al., 1962
Used for bacteriorhodopsin		39°		Nabedryk and Breton 1981

[a]Dipole is in the plane of the peptide group.
[b]Reported from other sources, see references therein.
[c]Calculated for an α-helix with h = 1.5 Å and assuming transition dipole oriented at 20° for amide I and 73° for amide II relative to the C=O bond.
[d]Calculated for an α-helix with h = 1.525 Å and assuming transition dipole oriented at 20° for amide I and 73° for amide II relative to the C=O bond.
[d]Computed from the three dimensional structure of bacteriorhodopsin with the particular tilt angle of the 7 transmembranes helices and the ca. 10 helical residues lying flat on the membrane, assuming a mosiac spread order parameter = 1 or,
[e]= 0.9.

This orientation depends on the exact geometry of the helix, which is particular to each helix. Some proposed values appear in Table IIIA, expressed with respect to the C=O bond, and in table IIIB, expressed with respect to the helix axis. These values must be taken with caution, especially if the investigated helix is characterized by an unusual frequency. Values reported in Table IIIB range between 20° and 40° for α_I helices, but high values should be probably reserved for helices characterized by unusually high frequencies, such as bacteriorhodopsin or poly(benzyl-L-aspartate). In our experience with ATR, a value of 27° has been found to be preferable because, even in the case of bacteriorhodopsin, experimental dichroic ratios are too large to be compatible with a value as high as 40°. Indeed, dichroic ratios above 3.0 can be measured for the amide I of bacteriorhodopsin (Goormaghtigh *et al.*, 1990). Recent dichroic measurements obtained by transmission spectroscopy on bacteriorhodopsin indicate that angles corresponding to dihedral angles of a structure intermediate between α_I and α_{II} are consistent with CD data for a series of samples, including a random orientation obtained by ethanol treatment (Draheim *et al.*, 1991). The values of the dichroic ratio obtained from ATR experiments (Yang *et al.*, 1987, Marrero and Rothschild 1987, Goormaghtigh et al. 1990, Earnest *et al.*, 1990) is much larger than when measured by transmittance, even though the mosaic spread order parameter has been carefully controlled and demonstrates a high ordering of the membrane (Rothschild *et al.*, 1980) obtained by the isopotential spin drying method in the transmission experiments (Clark *et al.*, 1980; Clark and Rothschild, 1982). No explanation for this phenomenon has been put forward yet.

It must be stressed here that we have observed in many cases that the dichroism spectrum in the region of amide I for an α-helix structure presents a maximum on the high frequency side of the helix band. An example is reported in figure 5 showing the thermolytic fragment of colicin A reconstituted into DMPC vesicles (Goormaghtigh *et al.*, 1991b). Similarly, the dichroism spectrum of cytochrome c oxidase presents a maximum at 1662 cm^{-1}, while the amide I maximum related to the α-helices is present at 1655 cm^{-1} (Bazzi and Woody, 1985). The dichroism peak of the synthetic GALA peptide is shifted by only 2 cm^{-1} toward higher frequencies (Goormaghtigh *et al.*, 1991a). Two explanations can be put forward: Transmembrane helices might be characterized by a higher amide I frequency than usual helices. The most obvious example of this is found in bacteriorhodopsin (Rothschild and Clark, 1979). Such an effect could be explained if the transmembrane α-helix is slightly extended and the hydrogen bonds weaker. That the maximum of the dichroism spectrum reported in Figure 5 is on the high frequency side of the helix domain could indicate a better orientation of the helices characterized by a higher frequency, confirming their insertion into the oriented DMPC bilayer. Alternatively, splitting of the parallel and perpendicular components of amide I could explain this fact (Earnest *et al.*, 1990; Rath *et al.*, 1991). This splitting could reach 10 cm^{-1} in the special case of bacteriorhodopsin (Earnest *et al.*, 1990). Generally the main component of the α-helix contribution to amide I is located about 5 cm^{-1} higher than other components with the opposite polarization.

Values of the angle between the transition dipole and the helix axis have been discussed in the special case of the $\pi^{4.4}$LD helix: 22–23° for amide I, 46–81° for amide II, and 13–30° for amide A (Nabedryk *et al.*, 1982).

Beside the uncertainties about the orientation of the transition dipole moment pointed out here, knowing the sample refractive index is necessary to evaluate from dichroism the molecular orientations with respect to the ATR plate. The dependence of the refractive index on the wavenumber was introduced by Fringeli (Fringeli *et al.*, 1989), and used by Frey and Tamm (1991) for

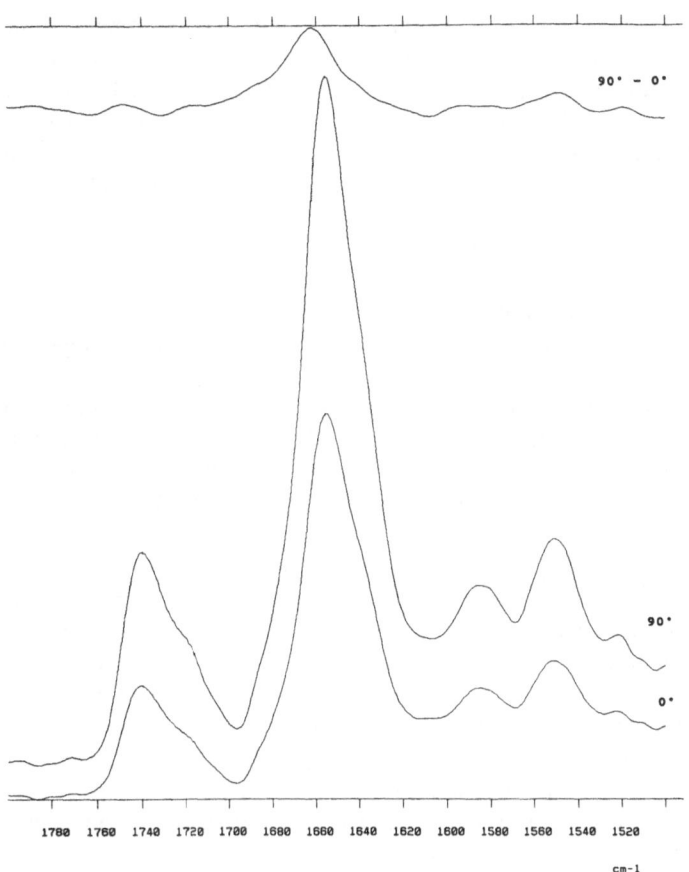

90° – 0°

90°

0°

1780 1760 1740 1720 1700 1680 1660 1640 1620 1600 1580 1560 1540 1520

cm-1

FIGURE 5. Infrared ATR spectra of the C-terminal thermolytic fragment of colicin A - DMPC (1/33 mole/mole) complex at pH 7.2 recorded with a 90° and 0° polarization in the plane of the germanium plate (Goormaghtigh and Ruysschaert, 1990). The dichroism spectrum, obtained by subtracting the 0° recorded spectrum from the 90° recorded spectrum, is plotted on top of the figure on a scale expanded 2 times.

melittin. Only recently, these values have been subject to discussion and investigated by experiments by comparing data obtained on two ATR plates of different refractive index (Buchet *et al.*, 1991). The influence of the numeric value of the real refractive index is far from negligible. For a 45° germanium ATR support, the values of the dichroic ratio R^{iso} for an isotropically oriented sample steadily increase as the refractive index of the thin film decreases. The values are worth 1.06, 1.14, 1.30 and 1.66 for film refractive index of 1.7, 1.5, 1.3 and 1.1 respectively as computed according to Goormaghtigh and Ruysschaert (1990).

To illustrate the dependence on the refractive index of the sample on the determination of a molecular axis orientation from a dichroic ratio, Figure 6 shows the dichroic ratios evaluated according to Goormaghtigh and Ruysschaert (1990) from the thin film hypothesis as a function of the orientation of the molecular axis. We suppose the transition dipole moment is parallel to the molecular axis in panel A, oriented at 27° to the molecular axis (the amide I dipole and the α-helix axis) in panel B, and is perpendicular to the molecular axis in panel C. The horizontal dotted lines indicate the error on the determined angle from a given dichroic ratio (R = 1.25, 1.5, and 1.9) if the refractive index of the sample is 1.5 ± 0.2. This error is limited to about 15° in panel A, but reaches 25–30° for a helix (panel B). Refractive indices of biomembranes have been reported to range from 1.35–1.70 (Buchet *et al.*, 1991).

Moreover, thick films yield higher values of the dichroic ratio for isotropically oriented samples. One report (Buchet *et al.*, 1991) describes an increase of R^{iso} from ~1.3 to ~2.0 when the amount of protein in sarcoplasmic reticulum membrane spread on a germanium ATR crystal is raised from ~7.5 to ~40 μg/cm² (proteins only, lipids not included). The same observation has been made in our laboratory (unpublished results). Limitations of the thin film hypothesis seem to be more stringent than expected, and the variations observed are probably to be related to Ez field component dependency on film thickness.

In conclusion, determination of secondary structure orientation is subject to several uncertainties including the orientation of the transition dipole with respect to the molecular structure and the refractive index. Other sources of uncertainty have been discussed earlier (Goormaghtigh and Ruysschaert, 1990). However, comparison of polarization measured on different absorption bands of the same sample, and on different, well-characterized samples, indicates that infrared dichroism remains a powerful tool to investigate the orientation of membrane-associated structures, as long as an accuracy of ±10° is acceptable.

5. EXTINCTION COEFFICIENTS

Polypeptides with well-defined structures provide us with potentially important information on the intensities of the amide components of each secondary structure. These intensities, called here integrated intensities because they represent the area under the IR spectrum obtained after numerical integration, should be known if a secondary structure of a protein is to be quantified after a curve fitting operation (Chapter 10). Indeed, the integrated intensities should be used to transform the intensities into an amount of protein folded in a given structure. However, lack of solid data has prompted all spectroscopists to assume that the integrated intensities of the components of amide I assigned to the different secondary structures are identical. Table IV reports data obtained on various

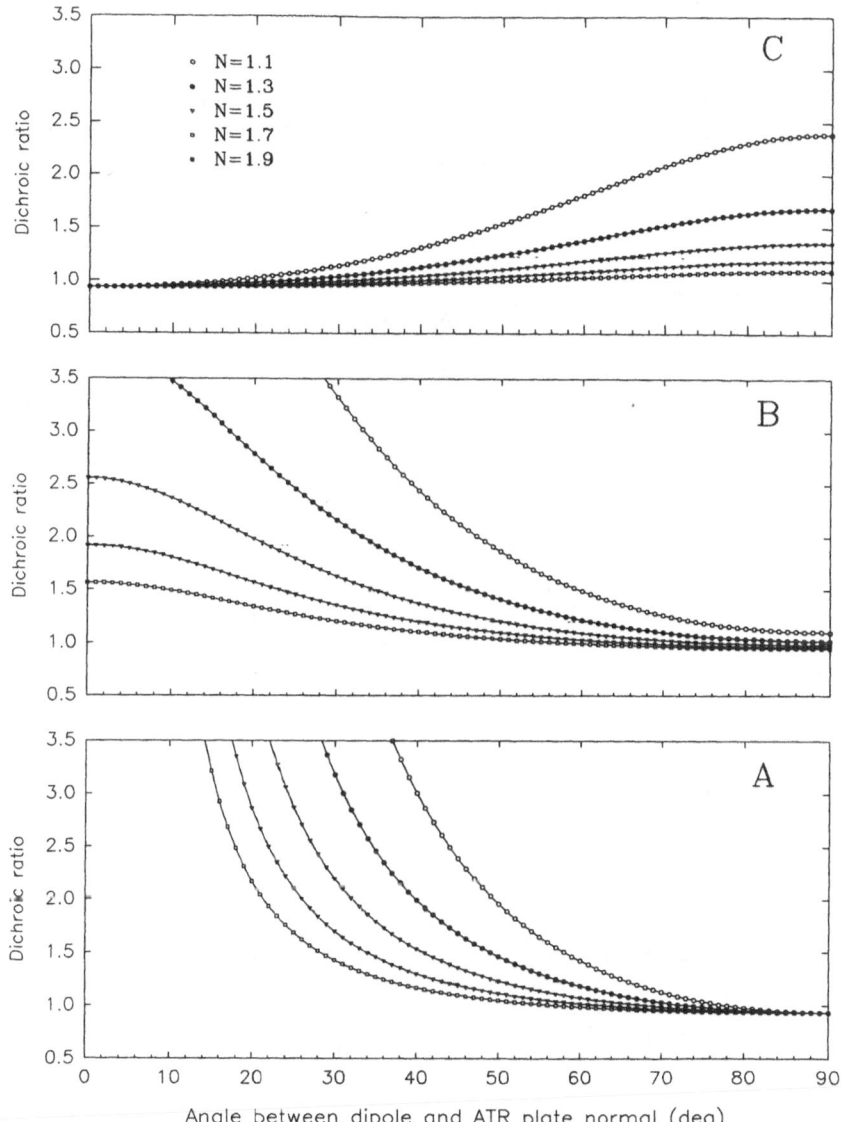

FIGURE 6. Dichroic ratio computed according to Goormaghtigh and Ruysschaert (1990) for the thin film hypothesis as function of the orientation of the molecular axis. The transition dipole moment is parallel to the molecular axis in panel A, oriented at 27° to the molecular axis in panel B, and perpendicular to the molecular axis in panel C. The refractive index of the germanium plate is 4.0, and the refractive index of the air surrounding the film is 1. Values of the electric field component Ex and Ey are 1.41 and 1.46 respectively. For a refractive index of the film of 1.1, 1.3, 1.5, 1.7, and 1.9, the value of Ez is 1.257, 0.89, 0.67, 0.52 and 0.42, respectively. The horizontal dotted lines are located between the refractive index lines of 1.3 and 1.7 at dichroic ratios of 1.25, 1.5, and 1.9.

Table IV
Frequency, Half Width, and Integrated Intensity of Amide I (Am I) and Amide II (Am II) for Different Secondary Structures before H/D Exchange (H-form) and after H/D Exchange (D-form)[a]

Structure		ν_0	A_0 (cm^{-1})	FWHH (l.mol^{-1} cm^{-1})	f_G (cm^{-1})	Surface ($\times 10^{-4}$ l/mol/cm^2)	Reference
α-helix	Am I	1647 ± 3	700	32 ± 1	0.2	7.6 ± 1.1	1
(H-form)							
	Am I	1651 ± 0	794	16 ± 1	0.3	4.2 ± 0.1	2
	Am II	1551 ± 3	310	30 ± 2	0.4	2.9 ± 0.6	1
		1520 ± 1	80	42 ± 4	0.6	1.1 ± 0.3	1
α-helix	Am I	1637 ± 3	544	32 ± 10	0.7	4.5 ± 0.1	2
(D-form)		1652					6
	Am II	1550 ± 1		14 ± 1		1.3 ± 0.1	2
		1717 ± 0		15 ± 1		~0.1	
β-sheet	Am I	1619 ± 1	980	22 ± 1	0.3	6.9 ± 1.2	1
(H-form)		1695 ± 6	180	10 ± 2	0.4	0.5	
	Am II	1533 ± 3	340	35 ± 3	0.7	3.3 ± 0.3	1
		1557 ± 6	145	41 ± 13	0.7	1.9 ± 1.3	
β-sheet	Am I	1620 ± 7	1017	17 ± 2	0.4	5.4 ± 0.6	3
(D-form)		1690 ± 8	198	9 ± 1	0.0	0.6 ± 0.2	
Ribonuclease			453				4
α-chymotrypsin			422				
Lysozyme			551				
		1634	720	45			6
	Am II	1520	290	30	0.6	2.5	3
		1555	70	25	0.6	0.5	
Unordered	Am I	1651 ± 2	320	55 ± 5	0.9	4.6 ± 0.4	1
(H-form)							
	Am II	1550 ± 3	210	59 ± 6	0.7	4.6 ± 0.4	1
Unordered	Am I	1650 ± 5	336	43 ± 13	0.8	3.8 ± 0.8	3
(D-form)		1672 ± 4	77	25 ± 2	0.7	0.5 ± 0.2	
Na$^+$K$^+$−ATPase		1647		50		3.8	5
	Am II	1533	225	62	1.0	3.4	3

[a]Data obtained for 1–5 different polypeptides with the same secondary structure have been averaged and are presented ± a standard deviation. Some data available for proteins are also presented. Bands are characterized by the frequency at the maximum of absorption ν_0 (cm^{-1}), the absorbance at the maximum A_0 (l/mol/cm), the full width at half height of the band FWHH (cm^{-1}), the integrated intensity of the peak (l/mol/cm^2) and the fraction of Gaussian component f_G. Curves can be constructed as explained on table I except for reference 6. Reference: 1. Venyaminov and Kalnin, 1990b; 2. Chirgadze and Brazhnikov, 1974; 3. Chirgadze *et al.*, 1973; 4. Rüegg *et al.*, 1975 (values of A_0 assuming a mean weight of 100 per amino acid residue); 5. Brazhnikov *et al.*, 1978; 6. Eckert *et al.*, 1977 (consensus data but not reported as a sum of Lorentzian and Gaussian). Part of the data reprinted with permission.

polypeptides for amides I and II before deuteration (H-form) and after deuteration (D-form). It must be stressed that these data could not be valid for proteins (as suggested by the very low frequencies of the α-helix), but they provide an idea of the difference of intensities that can be expected. The main difference of intensity in the H- and D-forms concerns the α-helix structure. Its intensity in H_2O solution is equal to that of the β-sheet but in D_2O is equal to that of the unordered structure. It has been suggested that these differences result from the different strength of the H and D bonds.

Polylysine has been investigated by Jackson et al. (1989). For a 0.5, 1, or 2% solution of polylysine, the random structure has the smallest amide I area; the amide I area for the α-helix is 13 $\pm$ 4% larger, and the amide I area of the β-sheet is 31 $\pm$ 11% larger.

Comparison of the amide I/amide II ratios for myoglobin, β-lactalbumin, and bovine serum albumin when the spectra are recorded by different methods (diffuse reflectance, KBr pellets, transmission (solution), ATR of the adsorbed proteins) shows differences up to a factor 2.5 (Ishida and Griffiths, 1993), mainly assigned to differences in the amide I absorptivity.

REFERENCES

Bandekar, J., 1992, Amide modes and protein conformation, Biochem. Biophys. Acta 1120: 123–143.

Bandekar, J., and Krimm, S., 1979, Vibrational analysis of peptides, polypeptides and proteins: characteristic amide bands of β-turns, Proc. Natl. Acad. Sci U.S.A. 76: 774–777.

Bandekar, J., and Krimm, S., 1980, Vibrational analysis of peptides, polypeptides and proteins. VI. Assignment of β-turns modes in insulin and other proteins, Biopolymers 19: 31–36.

Bandekar, J., and Krimm, S., 1985, Vibrational analysis of peptides, polypeptides and proteins. XXX. Normal mode analyses of γ-turns, Int. J. Peptide Protein Res. 26: 407–415.

Bazzi, M. D., and Woody, R. W., 1985, Oriented secondary structure in integral membrane proteins: circular dichroism and infrared spectroscopy of cytochrome c oxidase in multilamellar films, Biophys. J. 48: 957–966.

Blundell, T., Barlow, D., Borkako, N., and Thornton, J., 1983, Solvent-induced distorsions and the curvature of α-helices, Nature 306: 281–283.

Bradbury E. M., Brown L., Downie, A. R., Elliot, A., Fraser, R. D. B., and Handby, W. E., 1962, The structure of the β-form of poly-β-benzyl-L-aspartate, J. Mol. Biol. 5: 230–247.

Brandl, C. J., and Deber, C. M., 1986, Hypothesis about the function of membrane-buried proline residues in transport proteins, Proc. Natl. Acad. Sci. U.S.A. 83: 917–921.

Brazhnikov, E. V., Chetverin, A. B., and Chirgadze, Y. N., 1978, Secondary structure of Na^+, K^+ dependent adenosine triphosphatase, FEBS Lett. 93: 125–128.

Buchet R., Sandorty C., Trapane, T. L., and Urry, D. W., 1985, Infrared spectroscopic studies on gramicidin ion-channels: relation to the mechanism of anesthesia, Biochim. Biophys. Acta 821: 8–16.

Buchet, R., Varga, S., Seidler, N. W., Molnar, E., and Martonosi, A., 1991, Polarized infrared attenuated total reflectance spectroscopy of the Ca^{2+}−ATPase of sarcoplasmic reticulum, Biochim. Biophys. Acta 1068: 201–216.

Cabiaux, V., Goormaghtigh, E., Wattiez, R., Falmagne, P., and Ruysschaert, J. M., 1989, Secondary structure of diphtheria toxin interacting with asolectin liposomes: An infrared spectroscopy study, *Biochimie* **71:** 153–158.

Carrier, D., and Pézolet, M., 1984, Raman spectroscopy study of the interaction of poly-L-lysine with dipalmitoylphosphatidylglycerol bilayers, *Biophys. J.* **46:** 497–506.

Carrier, D., and Pézolet, M., 1986, Investigation of polylysine-dipalmitoylphosphatidylglycerol interactions in model membranes, *Biochemistry* **25:** 4167–4174.

Carrier, D., Mantsch, H. H., and Wong, P. T. T., 1990a, Pressure-induced reversible changes in secondary structure of Poly(L-lysine): an IR spectroscopic study, *Biopolymers* **29:** 837–849.

Carrier, D., Mantsch, H. H., and Wong, P. T. T., 1990b, Protective effect of lipidic surfaces against pressure-induced conformational changes of poly(L-lysine), *Biochemistry* **29:** 254–258.

Caughey, B. W., Dong, A., Bhat, K. S., Ernst, D., Hayes, S. F., and Caughey, W. S., 1991, Secondary structure analysis of the scrapie-associated protein Prp 27–30 in water by infrared spectroscopy, *Biochemistry* **30:** 7672–7680.

Cheam, T. C., and Krimm, S., 1985, Infrared intensities of amide modes in N-methylacetamide and poly-L-glycine I from ab initio calculations of dipole moment derivatives of N-methylacetamide, *J. Chem. Phys.* **82:** 1631–1641.

Chirgadze, Y. N., and Brazhnikov, E. V., 1974, Intensities and other spectral parameters of infrared amide bands of polypeptides in the α-helical form, *Biopolymers* **13:** 1/01–1712.

Chirgadze, Y. N., and Nevskaya, N. A., 1976a, Infrared spectra and resonance interaction of amide-I vibration of the antiparallel-chain pleated sheet, *Biopolymers* **15:** 609–625.

Chirgadze, Y. N., and Nevskaya, N. A., 1976b, Infrared spectra and resonance interaction of amide I vibration of the parallel-chain pleated sheet, *Biopolymers* **15:** 627–636.

Chirgadze, Y. N., Brazhnikov, E. V., and Nevskaya, N. A., 1976, Intramolecular distorsion of the α-helical structural of polypeptides, *J. Mol. Biol.* **102:** 781–792.

Chirgadze, Y. N., Fedorov, O. V., and Trushina, N. P., 1975, Estimation of amino acid residue side-chain absorption in the infrared spectra of protein solution in heavy water, *Biopolymers* **14:** 679–694.

Chirgadze, Y. N., Shestopalov, B. V., and Venyaminov, S. Y., 1973, Intensities and other spectral parameters of infrared amide bands of polypeptides in the β- and random forms. *Biopolymers* **12:** 1337–1351.

Clark, N. A., and Rothschild, K., 1982, Preparation of oriented multilamellar arrays of natural and artificial biological membranes, *Methods Enzymol.* **88:** 326–333.

Clark, N. A., Rothschild, K., Luippold, D. A., and Simon, B. A., 1980, Surface-induced lamellar orientation of multilayer membrane arrays: theoretical analysis and a new method with application to purple membrane fragments, *Biophys. J.*, **31:** 65–96.

Cowan, S. W., and Rosenbusch, J., 1994, Folding pattern diversity of integral membrane proteins, *Science* **264:** 914–916.

Cowan, S. W., Schirmer, T., Rummel, G., Steiert, M., Ghosh, R., Pauptit, R. A., Jansonius, J. N., and Rosenbusch, J. P., 1992, crystal structures explain functional properties of two E. coli porins, *Nature* **358:** 727–733.

Deisenhofer, J., and Michel, H., 1989, The photosynthetic reaction center from the purple bacterium *Rhodopseudomonas viridis*, EMBO J. **8:** 2149–2170.

Draheim, J. E., Gibson, N. J., and Cassim, J. Y., 1991, Dramatic in situ conformational dynamics of the membrane protein bacteriorhodopsin, *Biophys. J.* **60:** 89–100.

Dwivedi, A. M., and Krimm, S., 1984, Vibrational analysis of peptides, polypeptides and proteins. XVIII Conformational sensitivity of the α-helix spectrum: α_I and α_{II}-Poly(L-alamine), *Biopolymers* **23:** 923–943.

Earnest, T. N., Herzfeld, J., and Rothschild, K. J., 1990, Polarized FTIR of bacteriohodopsin: transmembrane α-helices are resistant to hydrogen-deuterium exchange, *Biophys. J.* **58:** 1539–1546.

Eckert, K., Grosse, R., Malur J., and Repke, K. R. H., 1977, Calculation and use of protein-derived conformation-related spectra for the estimate of the secondary structure of proteins from their infrared spectra, *Biopolymers* **16**, 2549–2563.

Engelhard, M., Gerwert, K., Hess, B., Kreutz, W. and Siebert, F., 1985, Light-driven protonation changes of internal aspartic acids of bacteriorhodopsin: An investigation by static and time-resolved infrared difference spectroscopy using (4-^{13}C)Aspartic acid labeled purple membrane, *Biochemistry* **24**: 400–407.

Fischbarg, J., Cheung, M., Czegledy, F., Li, J., Iserovich, P., Kuang, K., Hubbard, J., Garner, M., Rosen, O. M., Golde, D. W., and Vera, J. C., 1993, Evidence that facilitative glucose transporters may fold as β-barrels, *Proc. Natl. Acad. Sci. U.S.A.* **90**: 11658–11662.

Fraser, R. D. B., 1953, The interpretation of infrared dichroism in fibrous protein structures, *J. Chem. Physics* **21**: 1511–1515.

Fraser, R. D. B., and MacRae, T. P., 1973, *Conformation in fibrous proteins and related polypeptides* Academic press, New York.

Frey, S., and Tamm, L. K., 1991, Orientation of melittin in phospholipid bilayers. A polarized attenuated total reflection infrared spectroscopy study, *Biophys. J.* **60**: 922–930.

Fringeli, U. P., Apell, H. J., Fringeli, M., Lauger, P., 1989, Polarized infrared absorption of Na$^+$/K$^+$-ATPase studied by attenuated total reflection spectroscopy, *Biochem Biophys Acta*, **984**: 301–312.

George, A., and Veis, A., 1991, FTIRS in H$_2$O demonstrates that collagen monomers undergo a conformational transition prior to thermal self-assembly in vitro, *Biochemistry* **30**: 2372–2377.

Gill, T. J., Ladaulis, C. T., Kunz, H. W., and King, M. F., 1972, Studies of intramolecular transitions and intermolecular interactions of polypeptides by fluorescence techniques, *Biochemistry* **11**: 2644–2653.

Goormaghtigh, E., and Ruysschaert, J. M., 1990, Polarized attenuated total reflection spectroscopy as a tool to investigate the conformation and orientation of membrane components, in: *Molecular Description of Biological Membranes by Computer-Aided Conformational Analysis* (R. Brasseur, ed.) CRC Press Inc., Boca Raton, Florida.

Goormaghtigh, E., Brasseur, R., Huart, P., and Ruysschaert, J. M., 1987, Study of the adriamycin-cardiolipin complex structure using Attenuated Total Reflection I.R. Spectroscopy, *Biochemistry* **26**: 1789–1794.

Goormaghtigh, E., Cabiaux, V., and Ruysschaert, J. M., 1990, Secondary structure and dosage of soluble and membrane proteins by attenuated total reflection Fourier-transform infrared spectroscopy on hydrated films, *Eur. J. Biochem.* **193**: 409–420.

Goormaghtigh, E., Cabiaux, V., De Meutter, J., Rosseneu, M., and Ruysschaert, J. M., 1993, Secondary structure of the particle associating domain of apolipoprotein B-100 in low-density lipoprotein by attenuated total reflection infrared spectroscopy, *Biochemistry* **32**: 6104–6110.

Goormaghtigh, E., De Meutter, J., Cabiaux, V., Szoka, F., and Ruysschaert, J. M., 1991a, Secondary structure and orientation of the amphipatic peptide GALA in lipid structures: An infrared spectroscopy approach, *Eur. J. Biochem.* **195**: 421–429.

Goormaghtigh, E., Martin, I., Vandenbranden, M., Brasseur, R., and Ruysschaert, J. M., 1989a, Secondary structure and orientation of a chemically synthesized mitochondrial signal sequence in phospholipids bilayers, *Biochem. Biophys. Res. Commun.* **158**: 610–616.

Goormaghtigh, E., Vigneron, L., Knibiehler, M., Lazdunski, C. and Ruysschaert, J. M. 1991b, Secondary structure of the membrane-bound form of the pore-forming domain of colicin A: a FTIR-ATR study, *Eur. J. Biochem.*, **202**: 1299–1305.

Görne-Tschelnokow, U., Stecker, A., Kaduk, C., Naumann, D., and Hucho, F. 1994, The transmembrane domains of the nicotinic acetylcholine receptor contain α-helical and β structures. *EMBO. J.* **13**: 338–341.

Ghadiri, M. R., Granja, J. R., and Buehler, L. K., 1994, Artificial transmembrane ion channels from self-assembling peptide nanotubes. *Nature* **368**: 301–304.

Hanlon, S., 1970, Infrared studies on biopolymers and related models, in: *Spectroscopic Approaches to Biomolecular Conformation*, pp. 161–215, D. W. Urry, ed. American Medical Association, Chicago.

Hefele-Wald, J., Goormaghtigh, E., De Meutter, J., Ruysschaert, J. M., and A. Jonas 1990, Investigation of the lipid domains and apoliporotein orientation in reconstituted high density lipoproteins by fluorescence and IR methods, *J. Biol. Chem.* **275**: 20044–20050.

Henderson, R., and Unwin, P. N. T., 1975, Three-dimensional model of purple membrane obtained by electron microscopy, *Nature* **257**: 28–32.

Henderson, R., Baldwin, J. M., Ceska, T. A., Zemlin, F., Beckmann, E., and Downing, K. H., 1990, Model for the structure of bacteriorhodopsin based on high-resolution electron cryo-microscopy, *J. Mol. Biol.* **213**: 899–929.

Ishida, K. P., and Griffiths, P. R., 1993, Comparison of the amide I/II intensity ratio of solution and solid-stateproteins sampled by transmission, attenuated total reflectance, and diffuse reflectance spectroscopy, *Appl. Spectrosc.* **47**: 584–589.

Jackson, M., Haris, P. I., and Chapman, D., 1989, Conformational transitions in poly(L-lysine): studies using Fourier transform infrared spectroscopy. *Biochem. Biophys. Acta* **998**: 75–79.

Kimmich, R., Schnur, G., and Scheuermann, A., 1983, Spin-lattice relaxation and line shape parameters in nuclear magnetic resonance of lamellar lipid systems: Fluctuation and spectroscopy of disordering mechanisms, *Chem. Phys. Lipids* **32**: 271–322.

Krimm, 1962, Infrared spectra and chain conformational of proteins, J. Mol. Biol. **4**: 528–540.

Krimm, S., and Bandekar, J., 1980, Vibrational analysis of peptides, polypeptides and proteins. V. Normal vibrations of β-turns, *Biopolymers* **19**: 1–29.

Krimm, S., and Bandekar, J., 1986, Vibrational spectroscopy and conformational of peptides, polypeptides and proteins, *Advances* Prot Chem **38**: 181–364.

Kühlbrandt, W., Wang, D. A., and Fujiyoshi, Y., 1994, Atomic model of plant light-harvesting complex by electron crystallography, *Nature* **367**: 614–621.

Lacsko, I., Hollosi, M., Ürge, L., Ugen, K. E., Weiner, D. B., Mantsch, H. H., Thurin, J., and Ötvös, Jr., 1992, Synthesis and conformational studies of N-glycosylated analogues of the HIV-1 principal neutralizing determinant, *Biochemistry* **31**: 4282–4288.

Lagant, P., Vergoten, G., Fleury, G., and Loucheux-Lefebvre, A.-H., 1984, Vibrational normal modes of folded prolyl-containing peptides: application to β turns, *Eur. J. Biochem.* **139**: 149–154.

Laroche, G., Carrier, D., and Pézolet, M., 1988, Study of the effect of poly(L-lysine) on phosphatidic acid and phosphatidylcholine/ phosphatidic acid bilayers by Raman spectroscopy, *Biochemistry* **27**: 6220–6228.

Li M., Smith, J. L., Clark, D. C., Wilson, R., and Murphy D. J., 1992, Secondary structures of a new class of lipid body proteins from oilseeds, *J. Biol. Chem.* **267**: 8245–8253.

Marrero, H. and Rothschild, K. J. 1987, Conformational changes in bacteriorhodopsin studied by infrared attenuated total reflection, *Biophys. J.* **5**: 629–635.

Michel, H., 1982, Three-dimensional crystals of a membrane protein complex. The photosynthetic reaction centre from Rhodopseudomonas viridis, *J. Mol. Biol.* **158**: 567–572.

Miyazawa, T., 1960, Perturbation treatment of the characteristic vibrations of polypeptide chains in various configurations, *J. Chem. Phys.* **32**: 1647–1652.

Miyazawa, T., and Blout, E. R., 1961, The infrared spectra of polypetides in different conformations: amide I and amide II bands, *J. Am. Chem. Soc.* **83**: 712–719.

Moore, W. H., and Krimm, S., 1976*a*, Vibrational analysis of peptides, polypeptides and proteins. III. β-poly (L-alanine) and β-poly (L-alanylglycine), *Biopolymers* **15**: 2465–2483.

Moore, W. H., and Krimm, S., 1976*b*, Vibrational analysis of peptides, polypeptides and proteins. I. Polyglycine I, *Biopolymers* **15**: 2439–2464.

Mueckler, M., Caruso, C., Baldwin, S. A., Panico, M., Blench, I., Morris, H. R., Allard, W. J., Lienhard, G., and Lodish, H. F., 1985, Sequence and structure of a human glucose transporter, *Science* **229**: 941–945.

Nabedryk, E., Gingold, M. P., and Breton, J., 1982, Orientation of gramicidin A transmembrane channel. Infrared dichroism study of grammicidin in vesicles, *Biophys. J.* **38**: 243–249.

Nabedryk, E., and Breton, J., 1981, Orientation of intrinsic proteins in photosynthetic membranes. Polarized infrared spectroscopy of chloroplasts and chromatophores, Biochem. *Biophys. Acta* **635**: 515–524.

Naik, V. M., and Krimm, S., 1986*a*, Vibrational analysis in the structure of gramicidin A. I. Normal mode analysis, *Biophys. J.* **49**: 1131–1145.

Naik, V. M., and Krimm, S., 1986*b*, Vibrational analysis of the structures of gramicidin A. II. Vibrational spectra, *Biophys. J.* **49**: 1147–1154.

Nevskaya, N. A., and Chirgadze, Y. N., 1976, Infrared spectra and resonance interactions of amide I and II vibrations of α-helix, *Biopolymers* **15**: 637–648.

Nishikawa, K., and Nogushi, T., 1991, Predicting protein secondary structure based on amino acid sequence, *Methods Enzymol.* **202**: 31–44.

Pande, J., Pande, C., Glig, D., Vasak, M., Callender, R. and Kagi, J. H. R., 1986, Raman, infrared, and circular dichroism spectroscopic studies on metallothionein: a predominantly "turn"-containing protein, *Biochemistry* **25**: 5526–5532.

Pastor, R. W., and Venable, R. M., 1988, Brownian dynamics simulation of a lipid chain in a membrane bilayer, *J. Chem. Phys.* **89**: 1112–1127.

Peterson, N. O., and Chan, S. I., 1977, More on the motional state of lipid bilayer membranes: interpretation of order parameters obtained from nuclear magnetic resonance experiments, *Biochemistry* **16**: 2657–2667.

Pimentel, G. C., and Sederholm, C. H., 1956, Correlation of infrared frequencies and hydrogen bond distances in crystals, *J. Chem. Phys.* **24**: 639–641.

Rath, P., Bousché, O., Merill, A. R., Cramer, W. A., and Rothschild, K. J., 1991, FTIR evidence for a predominantly alpha-helical structure of the membrane bound channel forming C-terminal peptide of colicin E1, *Biophys. J.* **59**: 516–522.

Romel, E., Noack, F., Meier, P., and Kothe, G., 1988, Proton spin relaxation dispersion studies of phospholipid membranes, *J. Phys. Chem.* **92**: 2981–2987.

Rose, G. D., Gierasch, L. M., and Smith, J. A., 1985, Turns in peptides and proteins, *Adv. Prot. Chem.* **37**: 1–101.

Rothschild, K. J., and Clark, N. A., 1979*a*, Anomalous amide I infrared absorption of purple membrane, *Science* **204**: 311–312.

Rothschild, K. J., and Clark, N. A., 1979*b*, Polarized infrared spectroscopy of oriented purple membrane, *Biophys. J.* **25**: 473–488.

Rothschild, K. J., Sanches, R., Hsiao, T. L., and Clark, N. A., 1980, A spectroscopic study of rhodopsin alpha-helix orientation, *Biophys. J.* **31**: 53–64.

Rudolph, A. S., and Crowe, J. H., 1986, Biophys. J. A calorimetric and infrared spectroscopic study of the stabilizing solute proline, *Biophys. J.* **50**: 423–430.

Rüegg, M., Metzeger, V., and Susi, H., 1975, Computer analyses of characteristic infrared band of globular proteins, *Biopolymers* **14**: 1465–1471.

Sanders, J. C., Haris, P. I., Chapman, D., and Hemminga, M. A., 1993, Secondary structure of M13 coat protein in phospholipids studied by circular dichroism, Raman, and Fourier transform infrared spectroscopy, *Biochemistry* **32**: 12446–12454.

Schrimer, T., and Cowan, S. W., 1993, Prediction of membrane-spanning β-strands and its application to maltoporin, *Prot. Sci.* **2**: 1361–1363.

Sengupta, P. K., and Krimm, S., 1985, Vibrational analysis of peptides, polypeptides and proteins. XXXII. α-poly(L-glutamic acid), *Biopolymers* **24:** 1479–1491.

Sengupta, P. K., Krimm, S., and Hsu, S. L., 1984, Vibrational analysis of peptides, polypeptides and proteins. XXI. β-calcium-poly(L-glutamic acid), *Biopolymers* **23:** 1565–1594.

Sibanda, B. L., and Thornton, J. M., 1991, Conformation of β hairpins in protein structures: classification and diversity in homologous structures, *Methods Enzymol.* **202:** 59–82.

Speyer, J., Weber, R., Das Gupta, S., and Griffin, R., 1989, Anisotropic ^{2}H NMR spin-lattice relaxation in L-(alpha) phase cerebroside bilayers, *Biochemistry* **28:** 9569–9574.

Subrahmanyeswara, U., Hennessey, J. P., and Scarborough, G. A., 1988, Protein chemistry of the Neurospora crassa plasma membrane H$^+$-ATPase, *Anal. Biochem.* **173:** 251–264.

Susi, H., Timaseff, S. N., and Stevens, L., 1967, Infrared spectra and protein conformations in aqueous solutions: I. The amide I band in H$_2$O and D$_2$O solutions, *J. Biol. Chem.* **242:** 5460–5466.

Suzuki, E., 1967, A quantitative study of the amide vibrations in the infrared spectrum of silk fibroin, *Spectrochimica Acta* **23A:** 2303–2308.

Thiaudière, E., Soekarjo, M., Kuchinka, E., Kuhn, A., and Vogel, H., 1993, Strustural characterization of membrane insertion of M13 procoat, M13 coat, and Pf3 coat proteins, *Biochemistry* **32:** 12186–12196.

Tonge, P. J., and Carey, P. R., 1992, Forces, bond length, and reactivity: fundamental insight into the mechanism of enzyme catalysis, *Biochemistry* **31:** 9122–9125.

Torii, H., and Tasumi, M., 1992, Model calculations on the amide I infrared bands of globular proteins, *J. Chem. Phys.* **96:** 3379–3387.

Tsuboi, M., 1962, Infrared dichroism and molecular conformation of α-form poly-γ-benzyl-L-glutamate, *J. Polymer Sci.* **59:** 139–153.

Unwin, N., 1993, Nicotinic acetylcholine receptor at 9 Å resolution, *J. Mol. Biol.* **229:** 1101–1124.

Urry, D. W., Shaw, R. G., Trapane, T. L., and Prasad, K. U., 1983, Infrared spectra of the gramicidin A transmembrane channel: the single-stranded-beta 6-helix, *Biochem. Biophys. Res. Commun.* **114:** 373–379.

Venyaminov, S. Y., and Kalnin, N. N., 1990a, Quantitative IR spectrophotometry of peptides compounds in water (H$_2$O) solutions. I. Spectral parameters of amino acid residue absorption band, *Biopolymers* **30:** 1243–1257.

Venyaminov, S. Y., and Kalnin, N. N., 1990b, Quantitative IR spectrophotometry of peptides compounds in water (H$_2$O) solutions. II. Amide absorption bands of polypeptides and fibrous proteins in α-, β- and random conformations, *Biopolymers* **30:** 1259–1271.

Walian, P. J., and Jap, B. K., 1990, Three-dimensional diffraction of PhoE porin to 2.8 Å resolution, *J. Mol. Biol.* **215:** 429–438.

Wallace, B. A., Cascio, M., and Mielke, D. L., 1986, Evaluation of methods for the prediction of membrane protein secondary structures, *Proc. Natl. Acad. Sci. USA* **83:** 9423–9427.

Weiss, M. S., Kreusch, A., Schiltz, E., Nestel, U., Welte, W., Weckesser, J., and Schulz, G. E., 1991b, The structure of porin from Rhodobacter capsulatus at 1.8 Å resolution, *FEBS Lett.* **280:** 379–382.

Weiss, M. S., Wacker, T., Weckesser, J., Welte, W., and Schulz, G. E., 1990, The three-dimensional structure of porin from Rhodobacter capsulatus at 3 Å resolution, *FEBS Lett.* **267:** 268–272.

Weiss, M. S., Abele, U., Weckesser, W., Schiltz, E., and Schulz, G. E., 1991a, Molecular architecture and electrostatic properties of a bacterial porin, *Science* **254:** 1627–1630.

Yang, P. W., Stewart, L. C. and Mantsch, H. H., 1987, Polarized attenuated total reflectance spectra of oriented purple membranes, *Biochem. Biophys. Res. Commun.* **145:** 298–302.

Zimmerman, S. S., and Mandelkern, L., 1975, The precipitation of poly-L-glutamic acid. I. α-precipitation, *Biopolymers* **14:**567–584.

Zimmerman, S. S., Clark, J. C., and Mandelkern, L., 1975 The precipitation of poy-L-glutamic acid. II. β-precipitation, *Biopolymers* **14:** 585–596.

Chapter 9

Determination of Soluble and Membrane Protein Structure by Fourier Transform Infrared Spectroscopy
II. Experimental Aspects, Side Chain Structure, and H/D Exchange

Erik Goormaghtigh, Véronique Cabiaux,
and Jean-Marie Ruysschaert

1. INTRODUCTION

Increasingly, infrared spectra discussed in the literature are modified by mathematical treatments such as smoothing or deconvolution which respectively destroy or enhance defined original features of the raw spectra. In some instances, features that do not belong to the protein spectrum are modified in such a way that they now seem to belong, leading to flawed interpretation of the results. Here we summarize the main causes of interference in protein spectra in relation to the experimental set up used for the recording of the spectra.

Erik Goormaghtigh, Véronique Cabiaux, and Jean-Marie Ruysschaert   Laboratoire de Chimie Physique des Macromolécules aux Interfaces, Université Libre de Bruxelles, B-1050 Brussels, Belgium.

Subcellular Biochemistry, Volume 23: Physicochemical Methods in the Study of Biomembranes, edited by Herwig J. Hilderson and Gregory B. Ralston. Plenum Press, New York, 1994.

Determination of protein structure has been a major challenge during the last decade. (The determination of the secondary structure will be reviewed in Chapter 10.). The study of the structure and ionisation of the amino acid side chains has developed simultaneously and has brought a wealth of information on proteins whose biological mechanism of action involves some specific amino acid ionisation or structural changes, particularly for membrane proteins involved in transport processes. Another contribution to our knowledge of membrane protein assemblies comes from amide hydrogen/deuterium exchange kinetics. Even though this approach is not yet fully developed, we believe that with the help of the new infrared technology, it will become part of standard analyses in the near future.

2. EXPERIMENTAL ASPECTS

Before analyzing in more detail the shape of the infrared spectra of proteins, the sources of possible spectral interference should be discussed.

2.1. Amino Acid Side Chain Vibrations

The contribution of amino acid side chain absorption of a protein to its infrared spectrum can be evaluated from the parameters given in Chapter 8 and must be considered for potential interference before amide vibrations can be studied in detail. Figure 1 shows the infrared spectrum of the plasma membrane H^+-ATPase (spectrum a), the contribution of the amino acid side chains (spectrum c) and the result of the subtraction from the experimental spectrum and the side chain contribution (spectrum b). The contribution of the amino acid side chains (spectrum c) has been obtained by addition of their individual contributions in proportion to their abundance in the protein obtained from the primary sequence. Scaling of the intensity is based on the tyrosine band at 1515 cm^{-1}, which is clearly resolved after hydrogen/deuterium exchange. In the present example, the side-chain contribution represents only ~8% of the integrated intensity of amide I and its overall rather broad shape makes the shape of the difference spectrum (spectrum b) similar to the shape of the experimental spectrum (spectrum a). In the present case, secondary structure estimation is affected less than 1% by the side chain absorbance. In the general, the effect of the side chain contribution on the secondary structure is expected to amount to less than 8%. Fabian *et al.* (1992) arrived at a similar conclusion (10%) for streptokinase in D_2O. Dong et al. (1992) found that absorption of lysine to presented two maxima, at 1624, and 1521 cm^{-1} in H_2O, but with little influence on the shape of cytochrome c, whose sequence contains about 20% lysine. However, in other proteins the contribution of amino acid side-chain group absorption may account for ~20% of the integrated intensities of amide I and II (Venyaminov and

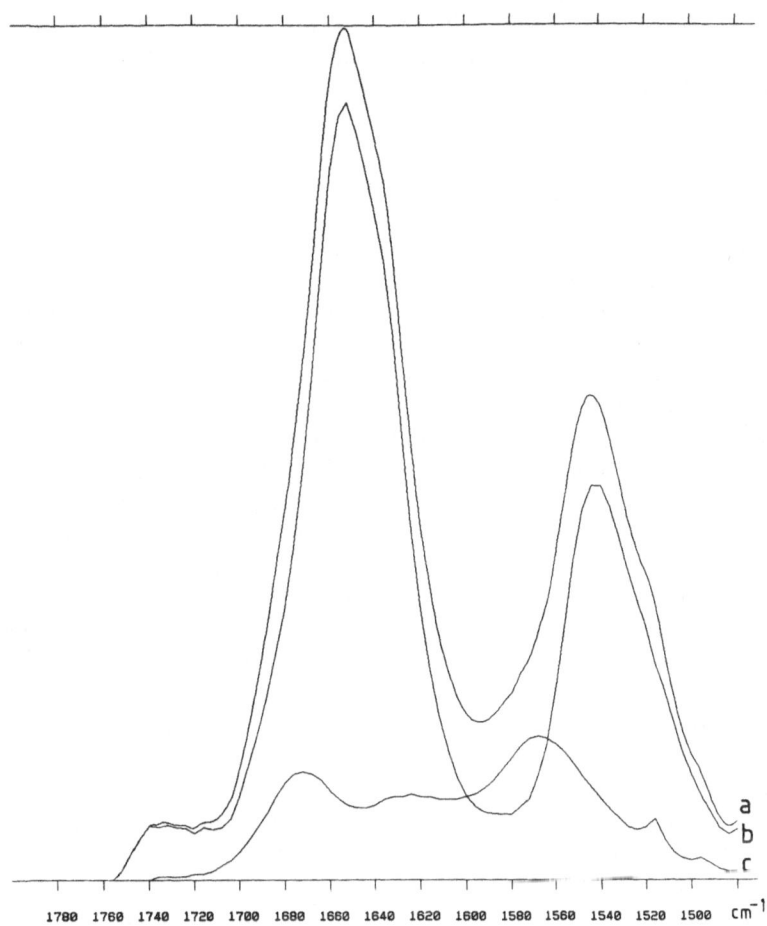

1780 1760 1740 1720 1700 1680 1660 1640 1620 1600 1580 1560 1540 1520 1500   cm⁻¹

FIGURE 1. Subtraction of amino acid side chain contributions from the ATR spectrum of 10 μg of Neurospora crassa plasma membrane H⁺-ATPase film deposited on a germanium internal reflection element (50 × 20 × 2 mm). Spectrum a. FTIR-ATR spectrum of 10 μg of Neurospora crassa plasma membrane ATPase in the presense of lysophosphatidylglycerol. The contribution of the amino acid side-chain absorption in the region of amide I-amide II, spectrum c, has been scaled on the intensity of the tyrosine contribution at 1515 cm⁻¹ after complete H/D exchange performed on the sample used for recording spectrum a (not shown). Subtraction of the side-chain contribution from spectrum a yields spectrum b. Spectra a, b, and c are presented on the same absorbance scale.

Kalnin, 1991; Chirgadze *et al.*, 1975). For short peptides or homo-polypeptides the side-chain group absorption may be much more important. Note that when the contribution of amino acid side chains is subtracted from the infrared spectrum of a protein, it is assumed that the spectral parameters of the side-chain bands in the model compounds are only slightly influenced by the molecular environment. This assumption is based on the high localization of the vibrational modes within the considered groups. In practice, we have observed in proteins variations of only a few wavenumbers with respect to the frequencies reported in Chapter 8.

2.2. Atmospheric Water and CO_2 Absorption

The atmosphere contains CO_2 and H_2O vapor, both of which absorb strongly in the infrared spectral region. The asymmetric stretch absorption of CO_2 at 2350 cm^{-1} falls outside the region of interest for the study of biological membranes. The absorption of H_2O vapor is so strong in the amide I region (usually > 1.0 absorbance units) that careful subtraction of a baseline is not always able to deal satisfactorily with this problem. Extensive purging of the spectrophotometer with dry air or nitrogen is a prerequisite for obtaining spectra of high quality. The influence of traces of atmospheric water absorption on the study of amide I has been pointed out previously (Lee *et al.*, 1985; Dong *et al.*, 1990) and is illustrated in Figure 2. To the spectrum of myoglobin (curve b), we have numerically added 2.5% of the spectrum of atmospheric water (curve a) to obtain curve c. Curve c is then deconvoluted with a Lorentzian lineshape (FWHH = 18 cm^{-1}) and apodized by a Gaussian lineshape to obtain a K factor 2 (Kauppinen *et al.* 1981a) (curve d). It can be seen that while the contribution of the atmospheric water to spectrum c was almost not detectable, after Fourier self deconvolution (curve d) all the resolved features coincide with the atmospheric water bands whose positions are marked by the vertical dotted lines on figure 2. The presence of a sharp shoulder at 1685 cm^{-1} can be considered as the signature of atmospheric water absorption. For a β-sheet rich protein, the shoulder can be confused with high frequency component of the β-sheet, and other signatures of the atmospheric water bands must be sought after. Exact numerical subtraction of the atmospheric water bands is always difficult, since the shape and frequency of the OH bending bands change with the temperature of the spectrophotometer chamber. Subtraction coefficients can be evaluated by subtracting a second derivative spectrum of water vapor from a second derivative of the sample spectrum so that the 1800–1700 cm^{-1} region is featureless (Petrelski *et al.*, 1991).

In our experience, we have found it useful to take advantage of the intrinsic bandwidth difference existing between the absorption bands of the water vapor and these of the liquid or solid sample. When a nominal resolution of 4 cm^{-1} is chosen, atmospheric water bands are broad and rather featureless. Their subtraction from a protein spectrum is difficult (Figure 3), and the coefficient applied for the subtraction may vary by about 50% depending on the operator. In the example used in figure 3, some oversubtraction has occurred. Conversely, when the spectrum of the same sample in the same conditions is recorded with a nominal resolution of 0.5 cm^{-1}, the sharp atmospheric water bands appear clearly re-

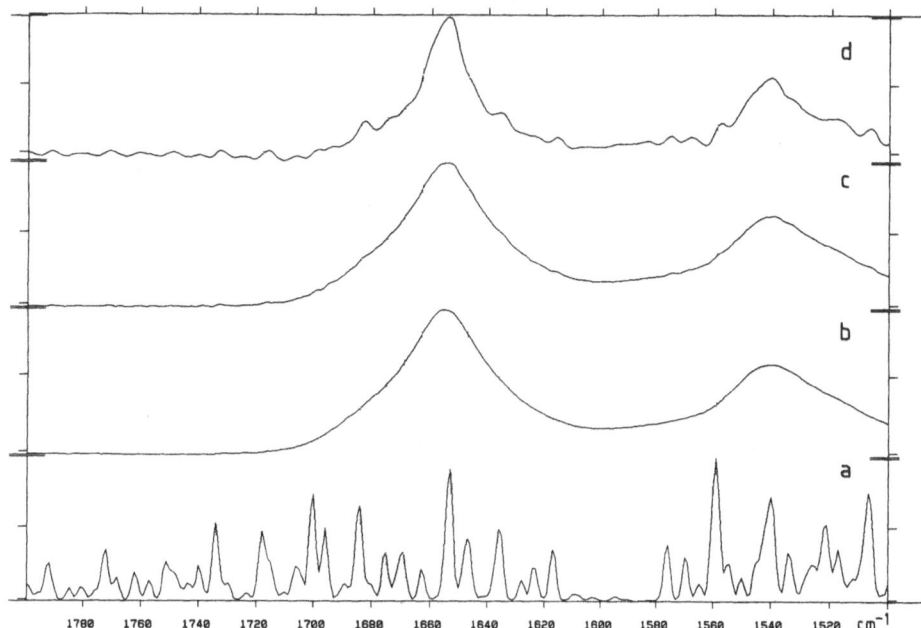

FIGURE 2. Influence of atmospheric water absorption on the analysis of protein spectra by Fourier self-deconvolution in the 1800—1500 cm^{-1} range. Spectra are obtained by ATR with a resolution of 2 cm^{-1} on a germanium crystal and are the average of 256 scans. The spectrum of the atmospheric water, spectrum a, was recorded after opening the lid of the spectrophotometer against a background obtained after purging the spectrophotometer with dry air. Spectrum b is the spectrum of a 20 μg film of myoglobin recorded after purging the spectrophotometer with dry air. The absorbance amplitude is 0.29 for spectrum a and 0.32 for spectrum b. Spectrum c was computed by adding 2.5% of the intensity of spectrum a to spectrum b. Spectrum d shows the results of a Fourier self-deconvolution applied on spectrum c: FWHH deconvoluting Lorentzian: 18 cm^{-1}, Gaussian apodization K = 2. The vertical dotted lines indicate the main atmospheric water bands in the amide I region. Their corresponding frequencies are 1734, 1718, 1684, 1675, 1669, 1663, 1653, 1647, 1636, 1617, and 1559 cm^{-1}.

solved from the broad amide bands (compare Figures 3a and Figure 4a). The very sharp features arising from imperfect subtraction completely disappear if the difference spectrum is now convoluted with a 4 cm^{-1} (FWHH) Gaussian lineshape (figure 4d). Fourier self-deconvolution of spectrum 3c and spectrum 4d yields quite different results (not shown). Spectrum 3c yields a spectrum with a maximum at 1650 cm^{-1} and a shoulder at 1656 cm^{-1}, but spectrum 4d yields a maximum at 1656 cm^{-1} and a shoulder at 1649 cm^{-1}. It is our view that correct subtraction of the atmospheric water absorption can be achieved only with the 0.5 cm^{-1} resolution. Improving the resolution by almost one order of magnitude (from 4 cm^{-1} to 0.5 cm^{-1}) takes full advantage of the width difference existing

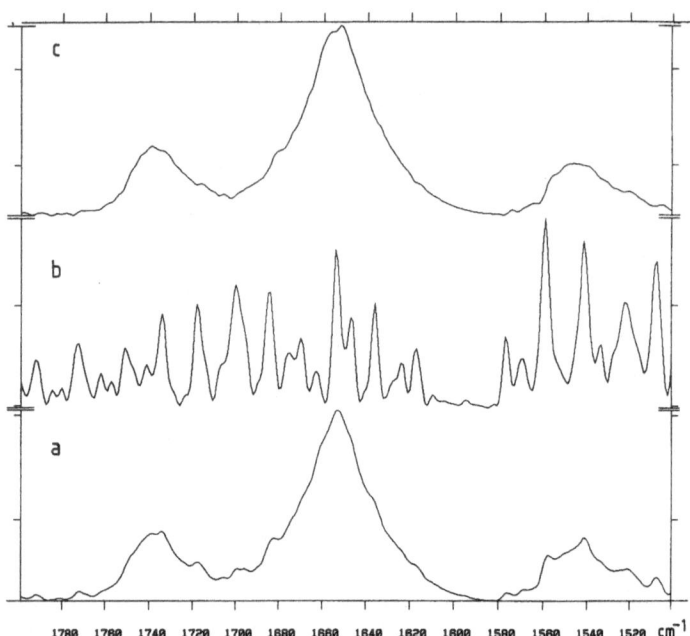

FIGURE 3. Subtraction of H_2O vapour from a protein spectrum recorded by ATR at a resolution of 4 cm^{-1} on a germanium crystal with a sampling every 1 cm^{-1}. 256 scans were accumulated for each experience. Spectrum a: 20 μg of protein, spectrum b: H_2O vapour recorded as described on Figure 2, spectrum c: subtraction spectrum a − spectrum b*0.08. The strongest water bands near 1559 and 1541 cm^{-1} have been cancelled.

between the absorption bands of the vapours and liquids or solids. At 0.5 cm^{-1} the very sharp features resulting from imperfect subtraction (Figure 4c) have no chance to reappear during the Fourier self-deconvolution analyses usually applied.

In the case of attenuated total reflection spectroscopy (ATR), water vapor bands are distorted with respect to transmission spectra. This makes it even more difficult to correctly subtract the atmospheric water contribution because the reference spectrum to be subtracted should have a similar balance between the water vapor spectrum recorded through the ATR plate by reflection and outside the ATR plate by transmission. In practice, this is never the case, as the mere presence of a sample film reduces the proportion of atmospheric water sampled by reflection. In turn, derivative-like lines appear after subtraction of the atmospheric water contribution. Purging the spectrophotometer with dry air and using high resolution spectra for the subtraction as described above yields satisfactory results.

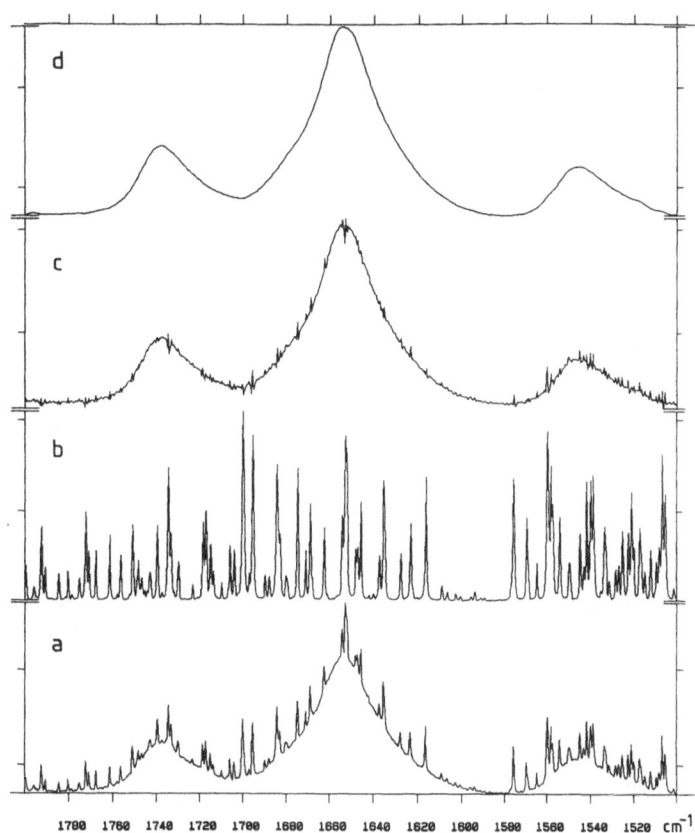

1780 1760 1740 1720 1700 1680 1660 1640 1620 1600 1580 1560 1540 1520 cm⁻¹

FIGURE 4. Subtraction of H_2O vapor from a protein spectrum recorded by ATR at a resolution of 0.5 cm⁻¹ on a germanium crystal with a sampling every 0.25 cm⁻¹. 256 scans are accumulated for each experience. Spectrum a: 20 μg of protein; spectrum b: H_2O vapor recorded as described in Figure 2; spectrum c: subtraction spectrum a − spectrum b*0.2; spectrum d: smoothing of spectrum c to a resolution of 4 cm⁻¹ by apodization of the Fourier transformed spectrum c by a Gaussian lineshape with a full width at half height of 4 cm⁻¹.

2.3. H_2O Solutions

Subtraction of the water contribution from spectra of aqueous solutions began with the availability of computerized manipulation of the spectra (Chapman *et al.*, 1980). The spectrum of liquid H_2O and D_2O appears in Figure 5, curve b and d respectively. Clearly, H_2O overlaps the amide I region, delimited by the vertical dotted lines, and makes it difficult to obtain the spectrum of a protein in water. In pure water, the area of the bending band near 1650 cm⁻¹ is

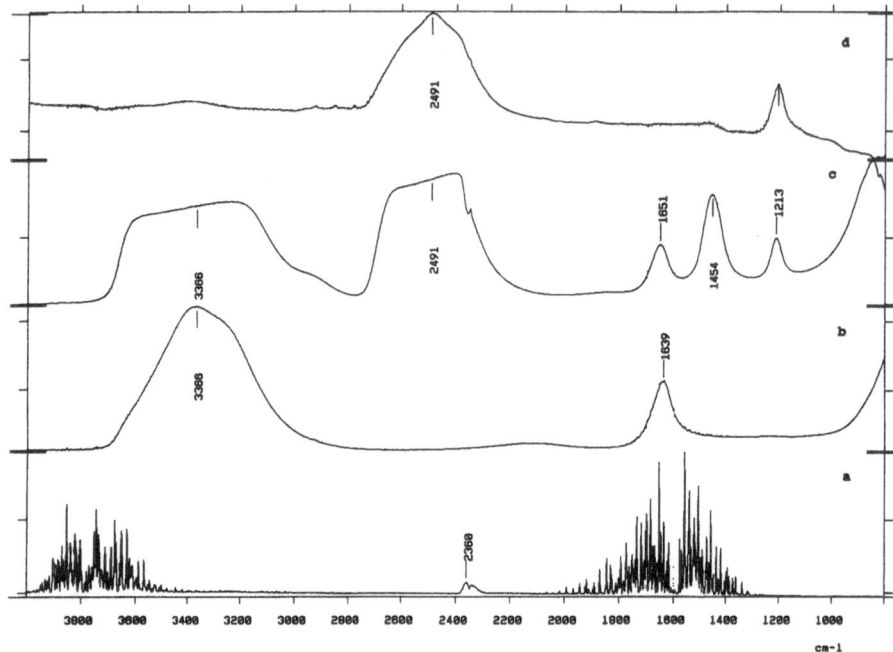

FIGURE 5. Atmospheric H_2O (spectrum a), liquid H_2O (spectrum b) and D_2O (spectrum d) reported between 4000 and 800 cm^{-1}. Spectrum a has been obtained as explained for spectrum a of figure 2. Spectra b and d are the spectra of respectively pure H_2O and pure D_2O recorded by ATR. Spectrum c has been recorded on a 25 μm thick liquid cell filled with a mixture of 50% H_2O and 50% D_2O (v/v). The vertical dotted lines delimit the amide I absorption region (1700–1600 cm^{-1}) for proteins.

about 20% of the area of the stretching band. This problem is overcome by replacing H_2O with D_2O, however, the H/D exchange then affects the amide vibrations as discussed below.

Several problems arise when the contribution of water must be subtracted from the spectrum of a solution or suspension of a protein or membrane.

First, the subtraction of a very strong band (extinction coefficient = 21 10³ M^{-1} cm^{-2} at ~1650 cm^{-1} and 63 10³ M^{-1} cm^{-1} at 3300 cm^{-1} to be compared with ~500 M^{-1} cm^{-1} for the amide I (Braiman and Rothschild, 1988, see also Arrondo *et al.*, 1993) (i.e. an absorbance about 200 fold higher in the amide I region for a same weight of water and protein) leads to substantial decrease of the signal-to-noise ratio.

Second, there are no objective scaling factors to be applied to the water spectrum, which is subtracted from the solution spectrum. Personal bias and uncertainty can result in significant differences in the spectrum of the same protein corrected for H_2O absorption by different researchers. Some authors consider the absorbance at 2130 cm^{-1} to quantify the intensity of the water spectrum to be subtracted (Vincent *et al.*, 1984). A resulting straight base line between 2000 and 1750 cm^{-1} (Haris *et al.*, 1986; Dong *et al.*, 1990) or between 1900 and 1750 cm^{-1} and from 3800 cm^{-1} (Sarver and Krueger, 1991) has also been quoted. Others use computer algorithms to improve the choice of the subtraction coefficients to subtract the H_2O spectrum (Powell *et al.*, 1986; Dousseau *et al.*, 1989).

Third, the interaction of a protein with water may considerably change the shape of the H_2O band and thus the subtraction of the reference spectrum of pure water (or buffer) may be not fully adequate (Cortijo *et al.*, 1982). Deformation of the H_2O band upon interaction with sucrose has been reported by Rothschild and Clark (1979). Pure water absorbs at 1645 cm^{-1} and water with sucrose yields an additional band near 1650 cm^{-1}. Similarly, water with salts brings a new maximum near 1650 cm^{-1} as reported by Venyaminov and Kalnin (1991b) (Figure 6). These numbers have to be kept in mind when comparing spectra of a protein obtained in the presence or in the absence of ligands such as Ca^{++}, Mg^{++}, ATP, and others. In the case of lipid-protein complexes, the water trapped in the lipid vesicles or bound to the lipid layers is expected to display a quite particular IR spectrum.

Because there is no way to estimate the impact of the protein on the H_2O

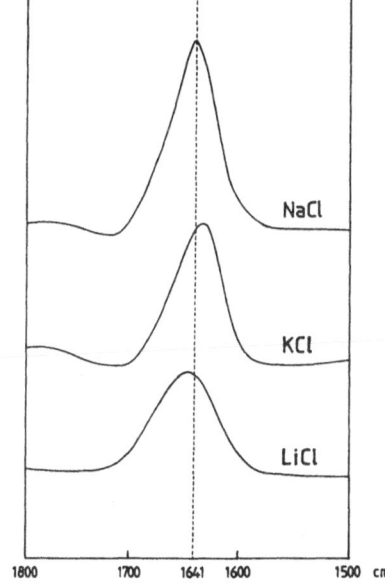

FIGURE 6. Infrared spectrum of bound H_2O in the presence of different salts according to Venyaminov and Kalnin (1990). The ordinate scale is proportional to the extinction coefficient which is worth ~100 liter/mol^{-1}/cm^{-1} for NaCl, 50 liter/mol^{-1}/cm^{-1} for LiCl and 72 liter/mol^{-1}/cm^{-1} for KCl. Data reprinted by permission of © John Wiley & Sons, Inc (1990); granted on January 29, 1993.

spectrum, one can only hope it is small or avoid the use of water. Most authors describing the use of infrared spectroscopy to study H_2O solution of proteins use thin-cell path length (typically 6 μm) and high protein concentration (typically 20–50 mg/ml). Such experimental conditions can help resolve the two first problems mentioned but it does not address the third one. Proteins not soluble in such high concentrations can only be studied in D_2O.

2.4. D_2O Solutions

An obvious way to overcome the problems related to the high liquid H_2O absorbance in the amide I region is to replace H_2O by D_2O. The main drawback of this approach is that the H/D exchange of the amide protons can be very slow for some membrane proteins, such as bacteriorhodopsin (70% unexchanged after 10 hr (Downer *et al.*, 1986)) or be extremely fast for others, such as the erythrocyte glucose transporter (more than 80% exchanged after 1 hr (Alvarez *et al.*, 1987)). Such differences are interesting in their own right, since they are relevant of the accessibility of the polypeptide backbone to the solvent (discussed below), but they make it difficult to compare the spectra of different proteins, as the extent of deuteration rules the shape of the amide bands. However it appears that rather short deuteration times are usually sufficient to shift the disordered components of amide I, whereas further deuteration which affects the ordered structures, does not significantly modify the assignments of their resolved components (Goormaghtigh *et al.*, 1993). When studying hydrogen/deuterium exchange, the presence of the mixed molecule HOD has to be taken into account, because it has a large absorbance in the amide II' region (Figure 5). The relative contribution of H_2O, D_2O, and HOD to the spectrum as a function of the H_2O/D_2O ratio can be estimated (Figure 7).

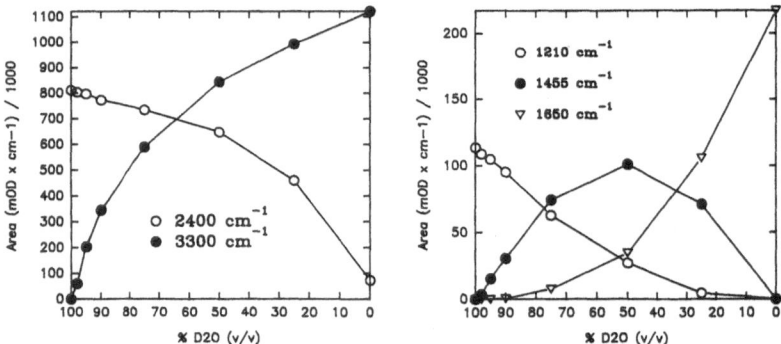

FIGURE 7. Evolution of the area under the stretching and bending bands of H_2O/D_2O mixtures as a function of the proportion of D_2O. Spectra were recorded by transmission through a 25μm demountable cell.

Another often neglected potential problem of the use of D_2O is its effect on chemical equilibrium for enzymes such as the maize plasmalemma proton AT-Pase, (Döring and Böttger, 1992), and on the secondary structure itself. Examples of such perturbations are summarized by Vasilescu and Katona (1986). An NMR study of the staphylococcal nuclease equilibrated in H_2O/D_2O mixtures indicates that the H/D fraction at a particular site varies from 0.3 to 1.5 for an average value of 0.85 (Loh and Markley, 1994). Values lower than unity indicate an equilibrium constant in favor of the H form. The low values are associated with strongly hydrogen bonded amide (e.g., the helices). The higher values are associated with exposed groups. Loh and Markley's study (1995) shows the different behavior of H and D in secondary structure stability. On the other hand, Reilly *et al.*, (1992) indicate that repressors and regulators as well as complexes have the same secondary structures and thermostabilities in H_2O and in D_2O. In spite of the two problems mentioned here, D_2O is widely used as a solvent for the study of biological molecules. Its weak absorption in the amide I region allows the use of lower protein concentrations (typically 1–10 mg/ml) and larger spacers (typically 25 μm but possibly up to 100 μm).

2.5. Spectra of Thin Films by Attenuated Total Reflection

Various forms of attenuated total reflection (ATR) spectroscopy of aqueous solutions have been developed, including the use of optical fibers (Simhony *et al.*, 1986, 1988) for the study of proteins (Simhony *et al.*, 1987). In the silver halide infrared optical fibers, the absorbance is linear with the film thickness up to 0.55 μm (Simhony and Katzir, 1988). The most usual design is the trapezoidal plate, which allows molecular orientation to be determined through polarization spectroscopy for oriented membrane systems. Preparation of oriented membranes and determination of orientations has been reviewed recently (Goormaghtigh and Ruysschaert, 1990) and will not be discussed here. The preference for ATR-IR over classical transmission spectroscopy stems from its high sensitivity. ATR-IR also provides much more structural information when working with membrane samples, because it gives access to the orientation of several chemical bonds with respect to the internal reflection element plane. However, this procedure raises a number of questions concerning the validity of the obtained information for biological components such as proteins and proteins embedded in lipid membranes.

2.5.1. Interaction Between the Film and ATR Plate

A major question when working with ATR-IR is the extent of protein secondary structure modification on interaction with the ATR internal reflection element. Such modifications appear, for example as a loss of helix and a gain of β-structures for bovine serumalbumin (BSA) adsorbed from a solution on a germanium plate (Lenk *et al.*, 1989). Adsorption is linear with the bulk concentration up to a concentration of 60 mg/ml, and intensity of the major bands of proteins is proportional to the amount adsorbed up to 0.2 μg/cm^2 (Fink and Gendreau, 1984; Fink *et al.*, 1987). The adsorption-denaturation problem can be solved as follows: first, when preparing a film of 100 μg of protein as described by Goormaghtigh *et al.*, (1990) on about 4 cm^2 of the ATR plate, the mean thickness of the film is 190 nm, considering an average density of 1.3 for the protein in the film. The thickness of the film is such that only a small fraction of its molecules are in contact with the plate. The film remains thin with respect to the IR wavelength ($\sim$5 μm) and, according to the ATR theory (Harrick 1967), all its molecules will contribute about evenly to the spectrum even though only a small fraction of them are in direct contact with the ATR support. Second, if there were secondary structure changes for the fraction of protein in contact with the ATR support, these changes should be more visible when working with small amounts of materials. However, in experiments in which α-chymotrypsin, myoglobin, and lysozyme were used, the spectra of films made up of only 1 μg of protein were identical to those obtained with 100 μg of protein, indicating that the hydrophilic surface of the germanium is not denaturing for the protein tested (unpublished results). Since in the usual experimental conditions only a small fraction of the protein molecules is in contact with the germanium surface, an effect of an interaction between the sample and the germanium plate would not threaten the validity of the analysis. It is important to realize that thicker films are not correctly described by the theory used for thin films (Harrick, 1987). Moreover, ordering was shown to decrease as the film thickness increased (Rothschild *et al.*, 1980). Experimentally, thick films yield higher values of the dichroic ratio for isotropically oriented samples. One report (Buchet *et al.*, 1991) describes an increase of R^{iso} from $\sim$1.3 to $\sim$2.0 when the amount of protein in sarcoplasmic reticulum membrane spread on a germanium ATR crystal is raised from $\sim$7.5 to $\sim$40 μg/cm^2 (proteins only, lipids not included).

A strong adhesion of the membrane film on the ATR crystals is, however, an advantage when an aqueous medium is to be flowed onto the film to control pH or solute concentration. In our experience, flowing an aqueous solution onto a membrane film deposited on germanium or KRS-5 ATR plates results in a rapid removal of the film. Desorption also occurs on the mere incubation of the film in the presence of an excess of buffer. Three different approaches have been designed to overcome this problem.

1. *Modification of the ATR plate surface.* The hydrophilic surface of germanium, similar to that of glass, enables chemical modification by silanization. Uncharged and negatively charged model membrane adhesion is achieved when germanium is coated with a monomolecular layer of aminopropylsilane, but erythrocytes remain more stable when adsorbed onto polymerized aminosilane coating. Furthermore, enzymatic active

acetylcholinesterase can be covalently bound to aminosilane coating by means of carbodiimide (Hofer and Fringeli, 1979).

2. *Coating of the germanium surface with a lipid bilayer.* First a DPPA monolayer is transferred at 40 mN/m from the air–water interface in the presence of 0.1 mM $CaCl_2$. The bilayer is then completed by incubation with POPC liposomes. This yields a stable bilayer (Fringeli *et al.*, 1989). The experiment has been repeated in our laboratory (unpublished data) and was shown to quantitatively yield a monolayer of DPPA and a bilayer after incubation with POPC. The final assembly was stable over a period of several hours. Proteoliposomes are then adsorbed or fused with the bilayer. A similar approach was used by Frey and Tamm (1991), but the first monolayer was made out of POPC.

3. *Cleaning by immersion in concentrated chromic acid and presence of adequate ions.* 30 min of immersion in concentrated chromic acid of germanium crystals yields a surface on which purple membranes tightly stick for hours in the presence of an aqueous environment, provided that a high salt concentration is maintained (Marrero and Rothschild, 1987). Similar results were obtained for the nicotinic acetylcholine receptor (Baenziger *et al.*, 1992a,b; 1993)

2.5.2. Hydration of the Films

The localization of the water stretching modes, and especially of its isotopically labeled analogs D_2O and $H_2^{18}O$ in spectral regions not overlapped by other vibrations, makes infrared spectroscopy a useful tool for quantification of the amount of water present. The sensitivity of the stretching frequency on the strength of the hydrogen bond is much higher than for any other technique, allowing accurate information on the molecular structure to be gained. D_2O content in biological samples can be accurately determined even for low isotopic enrichment (from 0.05 to 1.00%) in the sample, with no loss of precision for the lowest enrichment or in the presence of the second isotope $H_2^{18}O$ (Karasov *et al.*, 1989).

In ATR spectroscopy, it is of great interest to be able to evaluate the amount of water present in the film. Fringeli (1981) did this in a study on an L-α-dipalmitoylphsophatidylcholine (DPPC) layer using an absorption coefficient (3400 cm^{-1}, 22°C) = 1.3 ± 0.1 10^{-18} /reflection/molecule/cm^2 for H_2O and an absorption coefficient (1200 cm^{-1}, 22°C) = 1.8 ± 0.1 10^{-19} /reflection/molecule/cm^2 for the CH_2 wagging representing the all-trans chains. Another study (Okamura *et al.*, 1990) reports the almost linear relation between the intensity ratio (Y) $\nu(OH)(3380\ cm^{-1})/\nu(CH_2)(2918\ cm^{-1})$ as a function of the H_2O content (X) by weight of dry DPPC expressed in %: Y ≈ 0.053 X. Orientation of the water was also evaluated in this study. Dosage and water behavior in the phase transition of DPPC is reported by Okamura *et al.*, 1990 and Mellier and Diaf, 1988. Poole and Finney (1984) have reported the effect of sequential hydration of lysozyme and α-lactalbumin by water containing 60% D_2O in order to remove the H_2O bending band near 1645 cm^{-1}. Hydration was carried out step by step from the dry state ($\sim$8 mol HOD/mol protein) to the fully hydrated protein ($\sim$250 mol HOD/mol protein) by incubation in vessels in which the vapor pressure is controlled by saturated salt solutions in 60% D_2O (Hodgmann 1952). Hydration allows new hydrogen bonds to appear and induces a low frequency shift of the amide I as well as an increased intensity from the increased transition dipole moment. For both proteins, most of the hydration effects on amide I are over when the hydration reaches $\sim$100 mol HOD/mol protein—in rather good agreement with the value of 180 mol of "bound" water/mol lysozyme as determined by NMR (Kakalis and Baianu, 1988). Lysozyme activity begins at about 0.2 g water/g protein (Frauenfelder and Gratton, 1986). The analysis of different lysozyme properties as a function of the hydration level indicates that the carboxylate ionization (like the solution state) measured at 1580 cm^{-1} appears at 0.05 g water/g protein and is completely hydrated at 0.2 g water/g protein (Careri *et al.*, 1980). At 0.2–0.25 g water/g protein, all hydrogen bonding sites are covered, coinciding with the recovery of the enzymatic activity. The coverage of nonpolar elements continues up to 0.4 g water/g protein. Hydration does not seem to alter the secondary structure of the protein, as has been observed on model systems (Combelas *et al.*, 1974). A systematic study of the dehydration by lyphilization and rehydration was carried out by Petrelski *et al.* (1993). They quantified the similarity between second-derivatives spectra through a correlation coefficient $r = \Sigma x_i y_i / (\Sigma x_i^2 \Sigma y_i^2)^{1/2}$, where x_i and y_i represent the spectral absorbance values of the reference and sample spectra at the ith frequency position. Spectra were compared in the amide I region, between 1720 and 1610 cm^{-1} for aqueous solutions, dry powder (about 0.04 g water/g protein, mainly compressed in KBr pellets) and the rehydrated dry protein. Clearly, in some of the proteins tested secondary structure changes occur upon dehydration. These are not reversible upon rehydration. Such conditions correspond to a loss of activity for the lactate dehydrogenase. Importantly, the presence of sucrose prevents secondary structure changes and enzymatic activity loss. Estimation of the amount of residual (tritiated) water indicates that water is displaced from the dried protein by direct interaction between the sugar and the protein, supporting the view that interactions between the carbohydrate hydroxyl groups and the protein are responsible for the observed structure preservation upon dehydration.

2.5.3. Buffers and Salts

Buffers and salts can be used in ATR, provided that no absorption bands show up in the spectral region to be studied. However, the weight of the buffering molecules and salt ions per volume unit should not exceed the weight of the membrane molecules in the same volume unit if the multilayers are prepared by the solvent evaporation technique (Goormaghtigh and Ruysschaert, 1990). Indeed, excess of salt mechanically disrupts the flat orientation of the multilayers and prevents good contact between the IRE and the sample. Dramatically weak signal and poor orientation result from excess of buffer and salt. Best results are obtained when the membranes are dispersed in pure water. In addition to absorbance and mechanical effects, salts tend to accumulate hydration water, as indicated by a broad absorbance in the 3600–2400 cm^{-1} region. Near 1650 cm^{-1}, the contribution of the water is superimposed with the amide I contribution, but its presence is revealed by an anomalously low amide II/amide I ratio (normal value around 0.4–0.5). In experiments (unpublished results), we made films with 50 μg of bovine trypsin inhibitor in the presence or in the absence of 180 μg of a Mes-KOH-KCl buffer. In the presence of the buffer, more than 70% (10% in the absence of buffer) of the amide I integrated intensity was removed upon deuteration, partly because of H_2O removal and partly because of additional swelling of the film upon further hydration by D_2O. In conclusion, the shape and intensity of amide I can be affected by the presence of hydration water, especially when a rather large amount of buffer is present.

2.5.4. Study of the Water Molecules

Film containing limited amounts of water, opens the possibility to investigate water's structure through its IR spectrum by ATR as well as by transmission. Isolated water molecules were shown to absorb at 3725, 3627, and 1600 cm^{-1} for the monomer, at 3691, 3546, and 1620 cm^{-1} for the dimer and at 3510, 3355, 3318, 3322, and 1633 cm^{-1} for trimers and tetramers. Intensities increase by about tenfold on H bond formation (Van Thiel et al., 1957). Experimental studies of small clusters of water molecules indicate a very complex behavior in view of the theoretical models available (Strauss 1986). Liquid water spectrum is fitted by four Gaussian components whose frequencies are 3240, 3435, 3540, and 3620 cm^{-1} for respective half-widths of 310, 260, 150 and 140 cm^{-1} and respective contributions of 42%, 43%, 7% and 8% (Walrafen 1967). Very few studies have been done on the water $\nu(OH)$ in protein or membrane samples. Yet, the sensitivity of the absorption frequency of the $\nu(OH)$ vibration in H_2O is so large that infrared spectra should be able to detect tiny structural modifications in H_2O. Difference spectra obtained before and after illumination of a film of bacteriorhodopsin containing about 50% H_2O by weight at different temperatures, distinctly display a sharp peak at 3486 cm^{-1} for the L form on a broader band in the 3560–3450 cm^{-1} region, among other features. Replacement of H_2O by $H_2^{18}O$ or by D_2O shows which spectral changes are really due to water (Maeda et al., 1992). Frequencies found are not similar to any cluster of H_2O (dimer, trimer, etc.) that may be present on the surface of the protein. Tentatively, the water molecules observed are assigned to a narrow channel between

Asp-96 and the protonated Schiff base (Maeda *et al.*, 1992). Similar results were found for rhodopsin (Maeda *et al.*, 1993). Based on the extinction coefficient of H_2O, the authors estimate that only one or a few molecules at most are responsible for the changes observed.

2.5.5. Experimental Evidences for the Validity of ATR Studies

Experimental evidence indicates that studying the structure of proteins and lipids on thin film by ATR is a valid approach:

- For enzyme activity, the changes in absorbance in the visible light region of the spectrum of suspensions and films of purple membrane on illumination are indistinguishable (Korenstein and Hess, 1977; Gerwert *et al.*, 1990), demonstrating that the conformational changes related to the enzyme catalytic cycle are not affected in the film sample. In our experience, the activity of Neurospora plasma membrane proton ATPase is fully recovered after resuspension of the film.
- IR spectra are well related to the structure determined by X-ray fraction data. The high concentration of protein in the hydrated film is probably similar to that found in protein crystals and may favor the comparison between the crystal structures obtained from X-ray diffraction and the structure of the protein in the hydrated films. It must be noted that hydration of films by equilibration with H_2O- or D_2O-saturated nitrogen gas results in fully hydrated proteins as judged from the intensity of the O–H and O–D stretching bands. This might explain the similarity between solution and film data.
- There is no apparent difference between spectra recorded by the present method and by transmission in solution (Goormaghtigh *et al.*, 1989). Even small peptides, such as the 21 amino acid synthetic signal peptide of PhoE, retains on the Germanium plate the secondary structure they had in the solvent used for the preparation of the film (Demel *et al.* 1990). The most critical step in the film preparation is encountered when the sample is dried before rehydration with D_2O. This step is essential for the formation of oriented lipid multilayers (Fringeli and Günthard, 1981). It must be noted that in the "dry" film the hydration water molecules are not removed, which explains why the secondary structure is usually not modified. Should a problem occur at this stage, the addition of a protectant such as trehalose (Goormaghtigh *et al.*, 1988; Buchet *et al.*, 1992; Crowe *et al.*, 1984, 1985, 1987; Crowe and Crowe, 1986; Carpenter and Crowe, 1989) would probably take care of it. Furthermore, the shape of the spectrum of protein films deposited on polyethylene cards is not modified by incubation at 75% relative humidity for 24 h, indicating that enough

hydration water remains associated under usual room conditions (Sarver and Krueger, 1993). Alzheimer amyloid peptides were found to have the same spectra in the air-dried sample used for x-ray diffraction and in the solution, with the exception of a shift of the β-maximum from about 1618 to about 1625 cm^{-1}; indicating a weaker hydrogen bonding (Fraser et al., 1991). On the other hand, lyophilization may alter more drastically the shape of the amide I band (Petrelski et al., 1993, see above).

- pH structure dependency is similar in the solution and "dry" state, demonstrating that the protein retains both its ionization state and secondary structure on drying (Fraser et al., 1991).
- The similar hydrogen/deuterium exchange behaviors of proteins in solution, films, and crystals is discussed in section 4.2.1.
- Concerning membrane organization, it must be noted that long-range lateral diffusion coefficients of phospholipid analogs occurs in both leaflets of an oriented bilayer (Tamm and McConnel, 1985; Tamm, 1988), demonstrating that the behavior of bilayers adsorbed on the ATR crystal is similar to the behavior of vesicle bilayers. Similar lateral diffusion rates and transition temperatures have also been reported by Tiede (1985) for films and vesicle suspensions.

3.   AMINO ACID SIDE CHAIN STRUCTURE

Raman spectroscopy is specially suited for studying the amino acid side chain (essentially the aromatic ones) in proteins, but some interesting data can also be gathered from the infrared spectra. One problem with infrared spectroscopy is the weak intensity of side chain absorption bands compared with the amide bands. We briefly describe a few examples.

3.1.   Ionization of Glutamic and Aspartic Residues

Monitoring the protonation/deprotonation of carboxylic acids can be achieved by recording spectra at different pH and comparison with a reference pH spectrum by subtraction. Titration curves are obtained and mean pK determined. An example of pK determination is provided in the work of Tong et al., (1989) on cytochrome c and N-trifluoroacetyllysine ferricytochrome c.

The molar extinction coefficient of the $\nu_{as}(COO^-)$ is about 800 liter mol^{-1}/cm^{-1} for Asp and Glu in D$_2$O (Chirgadze et al., 1975), that is, in the range of the values between 300 and 1000 liter mol^{-1}/cm^{-1} found for amide I in D$_2$O (Chirgadze and Brazhnikov, 1974; Chirgadze et al., 1973). In a protein such as diphtheria toxin, which exists in a water soluble form at neutral pH and as a membrane form at acidic pH (Cabiaux *et al.* 1989a,b), the 12% content of Asp and Glu implies a contribution of the COO^- at neutral pH of about 12% of the amide I intensity. However, each single glutamic or aspartic acid accounts for less than 0.5% of the amide I intensity. Monitoring single residue ionization changes is therefore on the edge of feasibility, but sensitivity increases as the molecular weight of the protein fragment decreases. Using various well characterized fragments of diphtheria toxin might therefore be necessary.

Monitoring of protonation/deprotonation of glutamic and aspartic residues in bacteriorhodopsin and rhodopsin is more sensitive because of the low noise difference spectra recorded in the presence or in the absence of light. Protonation/deprotonation is of interest because Asp and Glu residues are common in the predicted transmembrane helices of many integral membrane proteins, including the proton transporting bacteriorhodopsin and several ATPases. For bacteriorhodopsin, the catalytic cycle can be initiated by switching on the light. The sample is unmodified except for the specific reactions following illumination; difference spectroscopy can pick up single residue protonation. Addition of chemical ligands disturbs the sample more. For this reason, release of free ATP or Ca^{++} induced by light from the so-called caged-ATP or Ca^{++} already in place in the sample could be a way to overcome this difficulty. But in practice, the newly formed chemicals and the fact that the reaction can be initiated only once, limits the study of the catalytic cycle of ATPases. Sensitivity of absorbance change was claimed to reach the absorbance of a single $C=O$ group in the Ca^{++}-ATPase (0.1% of the total protein absorbance or $5 \ 10^{-3}$ AU out of 0.5 AU). Carboxylic acid protonation seems to occur in the E_2P state (Barth *et al.*, 1991). For bacteriorhodopsin, unambiguous assignment of an IR band to Asp and Glu was made possible by growing *H. halobium* on isotopically labeled amino acids. The aspartic 1765 and 1755 cm^{-1} bands are shifted to 1725 and 1715 cm^{-1} in ^{13}C-Asp-bacteriorhodopsin in H_2O. The glutamic band at 1740 cm^{-1} is shifted to 1730 cm^{-1} in D_2O (Engelhard *et al.*, 1985). Measurements of the COO^- band is made difficult by the superimposition with other intense bands. The absence of significant H/D exchange is a criteria which should be met to assign a feature to a COO^- group. More details about assignments to Asp and Glu are reported by Dollinger *et al.*, (1986). Introduction of site-directed mutagenesis allows definition of the position of the amino acids involved in the protonation/deprotonation on the primary structure of the protein (Rothschild *et al.*, 1990). Clear separation of the Br, L, M, and N species of the photocycle (Ormos *et al.*, 1992) combinated with fast recording with a resolution in the range of 1 msec (which describes changes in bacteriorhodopsin, including sequential determination of side-group protonation, chromophore isomerization, and backbone motion (Gerwert *et al.*, 1990)) allows a complete picture of the light-induced pumping, while polarization measurements (Braiman and Rothschild, 1988) define the reorientation of the different molecular groups of interest. Similar approaches have been used for rhodopsin (Jansen *et al.*, 1990). A limited secondary structure change, which could play a role in proton pumping, has also been discovered (Ormos 1991).

Details of the formation of hydrogen bonds between phosphate and glutamic acid ($COH \cdots ^-OP \rightleftharpoons CO^- \cdots HOP$) have been described by Burget and Zundel (1987). When poly-L-glutamic acid interacts with phosphate, a continuum at 1900 cm^{-1} indicates that dihydrogen phosphates form complexes with the Glu residues in which the charge fluctuates and shows large proton polarizabilities due to correlated proton motion (Fritsch *et al.*, 1984). The intensity of these continua decreases in the series K$^+$, Na$^+$, and Li$^+$ systems because local fields caused by cations polarize H-bonds and decrease the proton polarizability. Hence, increasing cation fields hinders the charge fluctuation within the chain of hydrogen bonds (Burget and Zundel 1987). Increasing the hydration tends to decrease the continuum. We suggest that similar proton binding with large polarizibility could play a role in the transport of protons in membrane protein pumps (Zundel, 1986; Mertz and Zundel, 1986).

3.2. Proline Isomerization

Integral membrane proteins involved in transport or serving as receptors often contain proline residues in hydrophobic transmembrane segments (Brandl and Deber, 1986). It has been suggested that isomerization about the peptide bond (Brandl and Deber, 1986) or protonation at the amino nitrogen (Dunker, 1982) could play a role in transport mechanism by allowing structural modifications. The incorporation of ^{15}N-Pro and D$_7$-Pro in bacteriorhodopsin has identified one band near 1424 cm^{-1} that involves the proline ring and another band near 1456 cm^{-1} that does involve the ring hydrogen. A shift from 1423 to 1429 cm^{-1} occurs in the br$\rightarrow$K reaction of bacteriorhodopsin, which could be due to isomerization induced by the trans$\rightarrow$13-cis isomerization of the retinylidene chromophore. This shift can also be monitored by difference infrared spectroscopy (Rothschild *et al.*, 1984, 1988, 1989 (for halorhodopsin)).

3.3. -S-H in Cysteine

Cysteine presents no significant absorption in the amide I–amide II region, but the S-H vibration can provide information on the cysteine state. For instance, the high frequency at 2592 cm^{-1} for the ν(S-H) vibration in hemoglobin indicates that there is no hydrogen bonding for the thiol group (Moh *et al.*, 1987). Moreover, the effects of ligands can be followed qualitatively by the intensity of their absorption. The formation of $SH \cdots N \rightleftharpoons S^- \cdots H^+N$ was studied in detail by Kristof and Zundel (1982). A shift from 2552 to 2524 cm^{-1} with propylamine indicates weak hydrogen bonds. Strong hydrogen bonding shifts this band to about 2400 cm^{-1}.

3.4. Tyrosine

The 1516 cm^{-1} sharp peak arising from tyrosine residues is easily recognizable, especially when H/D exchange has shifted the amide II band to lower wavelengths. The intensity of this band was reported to be sensitive to the environment of the tyrosine side chain. Insertion of this residue into a lipid environment (DPPC) was reported to increase the I_{1545}/I_{1515} ratio from 1.4 to 2.6 for the myelin protein also called lipophilin (Nedelec *et al.*, 1989). On the other hand, the polarization of this band can yield useful information on the orientation of the tyrosine cycles with respect to a perpendicular to the cell window or to the ATR plate. Such measurements have, for instance, indicated that the tyrosine cycles are preferentially oriented along the normal to the membrane in porin (Nabedryk *et al.*, 1988).

Because of the presence of Tyr in the transmembrane segments of several integral proteins and because of their proton transfer capabilities, knowledge of the state of protonation of tyrosine residues in proton transport membrane proteins is essential for understanding the mechanism of proton transport. The use of bacteriorhodopsin containing a variety of labeled tyrosine (Dollinger *et al.*, 1986; Lin *et al.*, 1987; Roepe *et al.*, 1988) showed that the tyrosinate band near 1277 cm^{-1} is for diagnosing deprotonation. Chang *et al.*, (1991) arrived at the same conclusion by monitoring the ν(O-H) at 3671 and 3641 cm^{-1}.

Difference spectroscopy applied at the Desulfovibrio cytochrome c_3 among different redox states shows difference signals at 1512 cm^{-1} assigned to tyrosines and at 1614 or 1618 cm^{-1} assigned to phenylalanine (Schlereth *et al.*, 1993), demonstrating the high sensibility of difference infrared spectroscopy.

3.5. Histidine

The contribution of the imidazole ring of histidine to the photosystem II spectrum upon illumination has been evaluated by MacDonald and Barry (1992). Histidine protonation induces the appearance of bands at 1653, 1626, and 1540 cm^{-1} and the disappearance of bands at 1587 and 1494 cm^{-1}.

3.6. Phosphorylated Residues

Because phosphorylation is a posttranslational modification of proteins involved in numerous biochemical processes, the presence and ionization state of the phosphate monoester group is of major interest. $\nu_{as}(PO_2^{2-})$ is expected near 1090–1100 cm^{-1} and $\nu_s(PO_2^{2-})$ near 980 cm^{-1} while $\nu_{as}(PO_2^-)$ is expected near 1230–1190 cm^{-1} and $\nu_s(PO_2^-)$ near 1085–1080 cm^{-1}. These bands can be unambiguously identified by their frequency and pH titration behavior, and by comparison with the dephosphorylated protein. Sensitivity is sufficient to appreciate that the 979 cm^{-1} band (ν_s(PO$_2$$^{2-}$) of diphosphorylated ovalbumin is twice as intense as the band of the monophosphorylated ovalbumin (Sanchez-Ruiz and Martinez-Carrion 1988), that is, an intensity of 0.005 AU for a 73 mg/ml solution in a 25 μm thick cell. The pK of the phosphoserine residues was found to be close to six.

4. HYDROGEN/DEUTERIUM EXCHANGE

Crystallography has created rigid models of proteins, which hopefully represent the most abundant structure. However, proteins are not rigid and their functioning often needs other structural forms. Understanding the functions of a protein at the molecular level requires a description of the protein structure fluctuations in time. The rate of exchange of protein hydrogens is one way to approach a description of these fluctuations. Indeed, H/D exchange rate measurements yield information not available from equilibrium methods because the contributions to rates from each state are weighed by both the probability of the state and the state intrinsic rate constant. In addition to the basic understanding of protein structures, hydrogen exchange measurements yield information on the effects of the environment, such as the presence of specific ligands on the structural characteristics or on the shielding of the amide bonds from the solvent by the hydrophobic lipid bilayer of the membrane. In most previous studies, measurements of all the exchangeable protons were carried out by tritium/hydrogen exchange kinetics. More recently, IR and NMR offer the possibility of recording the exchange by deuterium of only the amide protons and of obtaining data on a number of peptide bonds but remains limited to a few proteins.

4.1. Factors Determining the Exchange Rate Constants

Interpretation of exchange data requires knowledge of the physico-chemical factors affecting the exchange rates. The two main factors are: For freely accessible hydrogens, rates of exchange depend only on hydrogen exchange chemistry which is catalysed by OH$^-$ and H$^+$: $k = k_0 + k_{OH}$ [OH$^-$] $+ k_H$ [H$^+$], where k_{OH} and k_H depend on both the diffusion rate constant ($\sim$10^{-10} sec^{-1}), which determines the fastest possible exchange, and on the difference between the pK of the proton donor and acceptor (significant only if the pK difference is small (<1)). Approximate peptide group rate constants are $k_{OH}\approx$10^7 M^{-1}/sec^{-1} and $k_H\approx$10^{-1} M^{-1}/sec^{-1}, that is, the exchange at neutral pH is dominated by OH$^-$ catalysis. The previous equation shows that the exchange rates strongly depend on pH. For the random chain of poly-DL-alanine the slowest amide proton exchange rate is found near pH 3.0, where the half-time reaches a value of about

100 min at 0°C. It decreases to less than one second near pH 7.0 (Englander and Kallenbach, 1984). A shift of one pH unit is in principle able to modify the rate by a factor of 10. pH control is therefore a crucial point for all exchange measurements. Peptide H-exchange also depends on temperature. An increase in the rate by about three-fold per 10°C increment necessitates temperature control. As reported by Rosenberg (1986), a simplified expression of k has been derived for poly-DL-alanine: $k = 50 \ (10^{-pH} + 10^{pH-6}) \times 10^{0.05(t°-20)}$. It must be noted that when individual peptide groups are being studied, the effect of the environment created by the side chains must be considered, because induction effects (Molday *et al.*, 1972) and charge effects modify local pK and pH.

The second factor affecting exchange rates is that in large proteins, a broad distribution of exchange rate constants indicates that usually more than half of the protons have exchange rates much slower than expected for freely exchangeable groups. Secondary structures stabilized by hydrogen bonds are resistant to exchange to an extent that depends on their stability. Burial of peptide groups in the hydrophobic region of the folded protein prevents the access of the solvent and catalysts required for the exchange or stabilizes the secondary structure. The models of mechanisms that can explain the experimental data have been reviewed by Englander and Kallenbach (1984) and will not be discussed here. It is enough to say that H-bonding is the major determinant of slow exchange, but burial of amino acid residues into the protein hydrophobic core or in the hydrophobic part of the membrane slows down the exchange rates whether or not the hydrogen is involved in a H-bond, even though buried water exchanges rapidly with the solvent (Tüchsen *et al.*, 1987).

4.2. Experimental Data

4.2.1. Data Recording

Infrared spectroscopy is a convenient way to monitor amide H/D exchange in protein in various conditions; a time resolution of about 10 sec is easily reached by most commercial FTIR spectrophotometers. The area of amide II permits quantification of the percentage of unexchanged amide groups, and the area of amide II′ measures the percentage of exchanged amides. Amide I and the lipid $\nu(C=O)$ band are usually used for calibration, and ratios amide II/amide I or amide II′/amide I are reported as a function of the time. The exchange can be followed by solubilizing a lyophilized protein in D_2O or by ATR techniques. In ATR, a chamber surrounding the film is flushed with D_2O-saturated nitrogen. ATR spectra recorded on polycrystalline silver halide fibers demonstrated on μg quantities that the exchange of trypsin mixed with its inhibitor [soybean trypsin inhibitor (SBTI)] is significantly slower than expected for the sum of the parameters for the two individual proteins (Chiacchiera and Kosower, 1992). An example of an exchange experiment carried out by ATR-FTIR spectroscopy on the gastric H^+, K^+-ATPase is shown in Figure 8 (Goormaghtigh and Soumarmon, unpublished data).

Just as it is legitimate to question the validity of protein structure determination on thin film by ATR spectroscopy, it is necessary to question the validity of exchange measurements carried out on thin films. In favor of ATR-FTIR spectroscopy experiments, note that exchange in lysozyme in the dry, powder condition as a function of the hydration degree indicates that the exchange rate is slow but becomes equal to the rate found in solution when the hydration level exceeds 0.4 g water per gram protein, that is, when enough water is present to produce a monolayer of water around the protein (Englander and Kallenbach, 1984). Schinkel *et al.*, (1985) found that the rate of exchange reaches a stable value when 0.15 g of H_2O/g lysozyme is present (120 mol H_2O/mol protein), corresponding to a coverage of less than half of the protein surface with a monolayer of water. In conclusion, film samples are expected to display essentially the same exchange behavior as solution samples provided that a conservative value of ~0.5 g H_2O/g protein is present.

Our own experiments (unpublished data) indicate that the rate of exchange is not affected by the amount of materials deposited on the internal reflection plate up to the highest amount tested (250 μg spread on ~4 cm²), indicating that the successive layers do not hinder the penetration of the D_2O vapor. These experiments were realized with membranes containing different ATPases and are in agreement with the diffusion coefficient of water across the lipid bilayers. This diffusion reaches tens of μm/sec (Graziani and Livne, 1972; Fettiplace and Haydon, 1980; Finkelstein, 1987). In the absence of a potential lipid barrier, these results are also in agreement with the similar exchange rates of individual amides in solution and in crystals (Rashin, 1987; Bentley *et al.*, 1983; Wlodawer *et al.*, 1984). One recent report, however, indicates that in the crystal form some specific sites could be stabilized, slowing down the exchange (Gallagher *et al.*, 1992). One peculiarity of ATR is its sensitivity to the distance between the sample and the internal reflection element surface (Goormaghtigh and Ruysschaert, 1990). In turn, there is no use placing more than 100 μg on the plate, as further addition does not result in a significant increase of the spectral intensity. Another implication is that swelling of the layers, which occurs in the presence of D_2O-saturated atmosphere (nitrogen), increases the average distance between the sample and the internal reflection element surface. This results in an intensity loss across all the absorption bands. The intensity of the "dry" film is recovered by flushing the sample chamber with dry nitrogen. It is therefore necessary for ATR experiments to report the amide II and II' area to another band to cancel this effect, which is unrelated to the exchange process. Amide I or the lipid $v(C=O)$ are used for this purpose in our laboratory. The kinetic exchange curves measured for amide II and amide II' for the H^+, K^+-ATPase are shown in

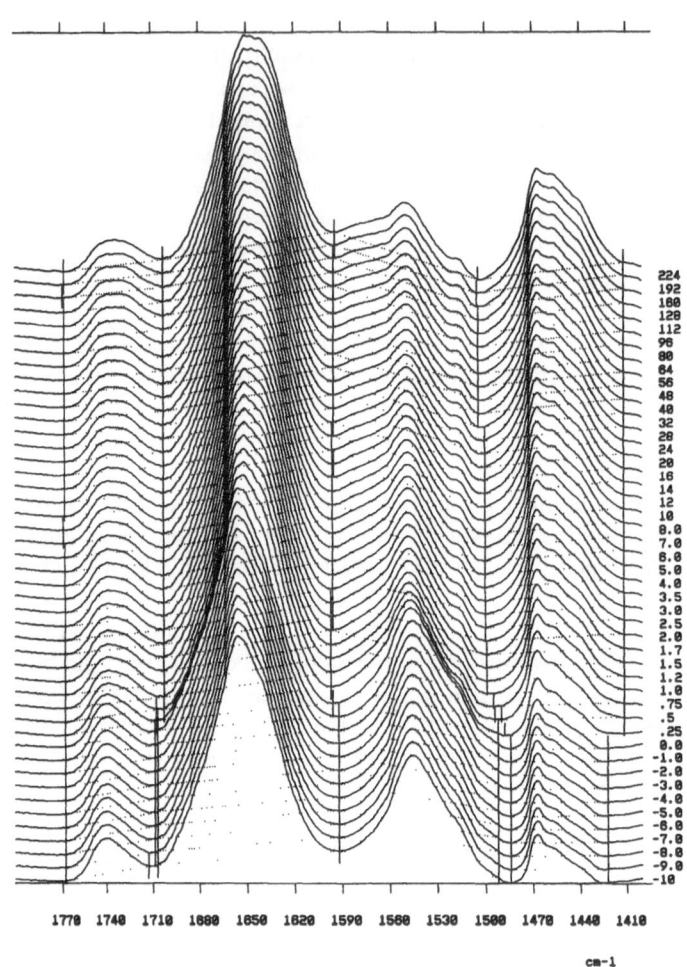

FIGURE 8. Evolution of the infrared spectrum of a 50 μg gastric H^+, K^+-ATPase film recorded by ATR on a germanium ATR plate as a function of the deuteration time obtained by exposure of the film to D_2O-saturated N_2. Deuteration time appears in min on the right margin of the figure. Deuteration starts at time = 0 min. Integration of the four bands; lipid $\nu(C=O)$ between 1765 and 1717 cm^{-1}, amide I between 1700 and 1595 cm^{-1}, amide II between 1593 and 1507 cm^{-1} and amide II' between 1493 and 1417 cm^{-1} is automatically performed by our program after adjustment within 5 or 15 cm^{-1} of these limits in order to find the curve minimum. These limits are indicated by the vertical lines for each spectrum.

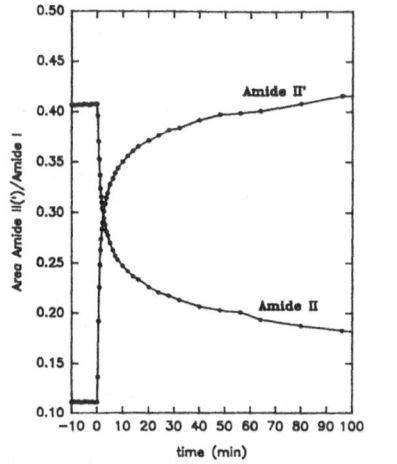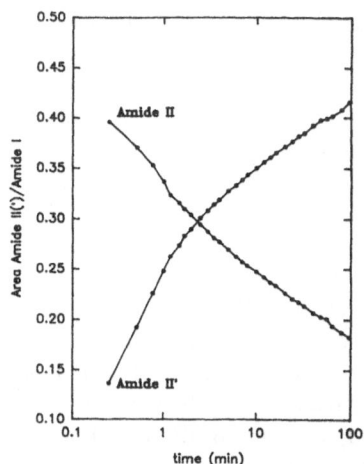

FIGURE 9. Area of amide II and amide II′ divided by the area of amide I of the corresponding spectrum for the spectra of Figure 8 reported as a function of the deuteration time (left panel) and as a function of log10 of the deuteration time (right panel).

Figure 9 as a function of the time(t) or of log(t). It must be noted that the relatively faster increase of amide II′ at the beginning of kinesis might be related to the fast appearance of small amounts of HOD, which absorbs in this region (see Figures 5 and 7).

The data described above demonstrate that both ATR-FTIR spectroscopy and transmission spectroscopy are able to produce high resolution data for hydrogen exchange studies. However, to the best of our knowledge, no systematic analysis of such data has yet been reported. We therefore report below some results obtained so far by means of IR, though no high time resolution has been sought. Rather than focusing on the mechanism of exchange of the different secondary structures, which is of significance for understanding structure stability and transient unfolding (Rath *et al.*, 1992), we have chosen here to consider variations observed in the exchange rates that can be correlated to accessibility changes of the amide sites to the solvent. Enzyme catalysis is thought to proceed through protein cleft closing and opening. Locking one of these states with appropriate ligands could result in solvent accessibility changes to a significant surface of the enzyme. For membrane proteins, burial into the hydrophobic lipid domain is also expected to modify the solvent accessibility to the embedded protein domains.

4.2.2. Exchange of Membrane Embedded Proteins

On one hand is bacteriorhodopsin, whose exchange is extremely slow. Monitoring the exchange between 1.5 to 25 hr after addition of D_2O was carried out by Downer *et al.*, (1986) at different pH. Almost independent of the pH (between pH 5 and pH 7), the authors found that a plateau was reached after about 10 h, where some 70% of the amide groups remain unexchanged (60% for rhodopsin). These values roughly correspond to the embedded helical fraction of the protein. They also agree with ATR data obtained from a film that showed 80% of unexchanged amides after 2 h exposure to D_2O and 73% unexchanged amides after 48 hr exposure to D_2O (Earnest *et al.*, 1990). Estimation of H/D exchange at different pH values for rhodopsin (Haris *et al.*, 1989) indicates that the amount of unexchanged amide groups after 2 hr falls to 48% at pH 5 and to 28% at pH 2. On the other hand is the human erythrocyte glucose exchanger, whose exchange rate is fast (Alvarez *et al.*, 1987): at pD 7, 81% of the amide groups are exchanged after 1 hr in the presence of D_2O. Very little exchange takes place during the following three days. A model of this protein proposes 12 α-helices crossing the membrane accounting for about 50% of the 492 amino acids of the protein. This exchange behavior contrasts with the behavior of bacteriorhodopsin (or rhodopsin) described above and provides an indication that there is a solvent accessibility for the transmembrane helices in the latter protein but not in the former. Since H/D exchange is very slow in the phospholipid membrane (Holloway and Buchheit, 1990), this approach can be a useful tool to diagnose the presence of aqueous channels or aqueous pockets in membrane proteins. The channel forming peptide of colicin E1 also shows a surprisingly rapid H/D exchange: 80% of the amide groups are exchanged after 5 min of exposure of a film to D_2O vapor and only 5% is left after 72 hr. At least 40% of the amino acids were inserted in the lipid bilayer in these experiments (Rath *et al.*, 1991). Exchange kinetics for individual backbone amides have been found by NMR to present a periodicity in melittin, which is reconstituted in bilayers (PC/PS 88/12 mol/mol) with an exchange rate twenty-fold slower for the amino acids located on the hydrophobic face of the helix (Dempsey and Butler, 1992). Insertion of the model peptide Lys_2-Gly-Leu_{24}-Lys_2-Ala results in the presence of an amide population, which is virtually unexchangeable (Zhang *et al.*, 1992). The same observation is made for glycophorin reconstituted in egg-phosphatidylcholine (Sami and Dempsey, 1988, Challou *et al.* 1994). However, the mere presence of a large aqueous channel is not sufficient to ensure fast exchange. The β-sheet protein porin from *E. coli* outer membrane, for instance, possesses a large channel, but its exchange kinetics are similar to those of rhodopsin (Kleffel *et al.*, 1985). Obviously, the intrinsic stability of the secondary structure can dominate the process. Similarly, stable α-helices from pulmonary surfactant SP-C (Vandenbussche *et al.*, 1992a; Pastrana *et al.*, 1991) and SP-B (Vandenbussche *et al.*, 1992b) exchange very slowly, while binding of cytochrome c to the negatively charged surface of DMPG or DOPG vesicles results in destabilization of the α-helices as indicated by differential scanning calorimetry and by the immediate and complete H/D exchange recorded by FTIR

(Muga *et al.*, 1991) in agreement with high resolution ^{1}H NMR studies (De Jongh *et al.*, 1992). The temperature dependency of the exchange of free and lipid-bound cytochrome c was carefully investigated with FTIR by Heimburg and Marsh (1993). In conclusion, useful classification of the proteins seems possible from the previous examples, but it is too soon to retrieve general rules relating H/D exchange kinetics, protein structure, and their location in the membrane.

4.2.3. Structure Effects on the Exchange

While several infrared studies indicate that E1E2 ATPases do not experience large secondary structure changes in their catalytic cycle (with the notable exception of Chevterin and Brazhnikov (1985)), one paper reports an increased H/D exchange rate for the H^+/K^+-ATPase in the presence of the inhibitor omeprazole (Mitchell *et al.*, 1987). Analysis of the spectra indicates that the change does not occur at the level of an α-helix. Obviously, hydrogen exchange measurements offer promising potentialities to quantify structural changes in terms of surface of the protein accessible to the solvent in the catalytic cycle, which up to now is only qualitatively characterized by difference to proteolytic susceptibility or by fluorescence assays. The effects of ligand binding on other classes of proteins have been reported. For instance, a clear effect of Ca^{++} on the rate of exchange has been presented by Rainteau *et al.*, (1989) for calmodulin and by Jackson *et al.*, (1991) for paravalbumin. Whatever the ligand, we think that it is important to report effects of the ligands on the exchange rate of an unrelated protein, because it is hard to predict what would be its role on the exchange mechanism in the absence of any specific structural changes. For instance, Ca^{++} effectively collapses surface potential related to negatively charged groups and therefore could increase the pH in the vicinity of these charges. None of the previous reports take these effects into consideration.

Obviously, the stability of the protein is an important determinant for the exchange by an unfolding mechanism in which exchange takes place from a denatured state. Destabilizing mutations and urea (Kim et al., 1993; Kim and Woodward, 1993) increase the rate of exchange of the unfolding mechanism. In the case of enzyme specific ligand, it is difficult to appreciate if reduced exchange in the presence of the ligand arise from reduced protein surface accessible to the solvent or from an increase in protein stability. Both effects seem related, as a more compact structure with a small surface in contact with the buffer is expected to be more stable.

4.3. Data Description

Careful data description and treatment has been elaborated for tritium/hydrogen exchange and has not been extended to data obtained by infrared spectroscopy until now. However, some data interpretations previously designed are expected to improve their significance and interpretation. We outline here one possible approach, which highlights the distribution of the rate constants in the protein.

Each individual site is supposed to exchange via first-order kinetics. For N amino acids in a protein, the number of sites occupied by a hydrogen H(t) as a function of the time t is given by:

$$H(t) = \sum_{j=1}^{N} \exp(-k_j t)$$

The large number of sites, even in a small protein, makes it impossible to obtain the individual rate constant kj. One approach to this problem is to fit the curve (H(t)) function of t by a small number (M) of exponents, each representing a class (A_i) of amide groups (Benson *et al.*, 1964; El Antri *et al.*, 1990).

$$H(t) = \sum_{i=1}^{M} A_i \exp(-k_i t)$$

This approach was found to be too simplistic in several instances (Woodward and Rosenberg, 1970; Laiken and Printz, 1970). Another approach, which does not make any assumption, is to replace the previous sum expression by an integral

$$H(t) = \int_0^\infty f(k) \exp(-kt)\, dk = \mathscr{L}\{f(k)\}$$

The sum expression can be regained if $f(k) = \delta(k-kj)$ where the Kronecker delta indicates a discrete distribution of the rate constants. The advantage of the integral formulation is that without further assumption on the distribution of the rate constants f(k), the inverse Laplace transform $\mathscr{L}^{-1}$ immediately yields the distribution shape $f(k) = \mathscr{L}^{-1}\{H(t)\}$. Knox and Rosenberg (1980) suggested a dimensionless presentation of the distribution function obtained after rewriting of the integral expression H(t)

$$H(t) = \int_0^\infty k\, f(k) \exp(-kt)\, d(\ln(k)).$$

Solving the Laplace transform can be approached analytically after fitting H(t) to a suitable function, or numerically. For reasons detailed by Gregory and

Lumry (1985), the numerical approach is subject to several artifacts; if not carefully treated (Provencher, 1976; Provencher and Dovi, 1979; Provencher, 1982a,b; Halvorson, 1992; Ameloot, 1992).

4.4. Amide I Shape in the Course of the Exchange

In Chapter 10, on protein secondary structure determination, we maintain that a few minutes of H/D exchange is sufficient to shift the random structure from about 1650 cm^{-1} to about 1640 cm^{-1}. This shift distinguishes an α-helix from random conformation, because the α-helix component experiences only a moderate low-frequency shift upon deuteration. We question this generally accepted assertion, based on polypeptides, which is the basis of several secondary structure determination methods. For this purpose, in Figure 10 we show the difference spectra obtained by subtracting a less deuterated spectrum from a more deuterated spectrum. The amide deuteration increment per protein molecule is constant for each difference ($\sim$7% of the total deuteration). Analysis of Figure 10 indicates three things. First, the problem is more complex than a long range shift from 1650 to 1640 cm^{-1} and limited shifts for the other structures. The larger shifts are observed between the 1690–1660 cm^{-1} region to the 1630–1570 cm^{-1} region at the beginning of kinesis (spectrum 1, Figure 10) and from $\sim$1655 cm^{-1}–$\sim$1608 cm^{-1}, when more than half of the exchange is over (spectrum 9, Figure 10). It must be noted that superimposition of positive and negative lobes can cancel larger effects. However, the effects observed at the beginning of kinesis concern about 10% (roughly 5% positive and 5% negative) of the intensity of the original spectra; this is far from negligible. Second, the pattern of the shifts in the amide I is roughly constant as the deuteration proceeds; and third, the intensity of the difference spectra decreases as the deuteration proceeds. In conclusion, the data reported in Figure 10 demonstrate that the arguments used for secondary structure determination are not as well established as generally assumed. Similar data have been observed in our laboratory for the H^+-ATPase from Neurospora crassa.

5. LIPID/PROTEIN RATIOS

The lipid/protein ratio can be evaluated from the ratio of the lipid band around 1735 cm^{-1} (ester ν(C=O)) and the protein band near 1650 cm^{-1} for a wide range of ratios, as illustrated in Figure 11. Figure 11 reports the spectra of 100 μg DMPC mixed with 1.78 μg lysozyme (curve a), 17.8 μg lysozyme (curve b) and 100 μg lysozyme (curve c). The ν(C=O) lipid band at 1736 cm^{-1} is virtually absent for the pure protein and no absorption occurs in the amide I region between 1700 and 1600 cm^{-1} for the pure lipid (not shown). The ratio of the integrated intensities of the amide I band to the lipid ν(C=O) can therefore be

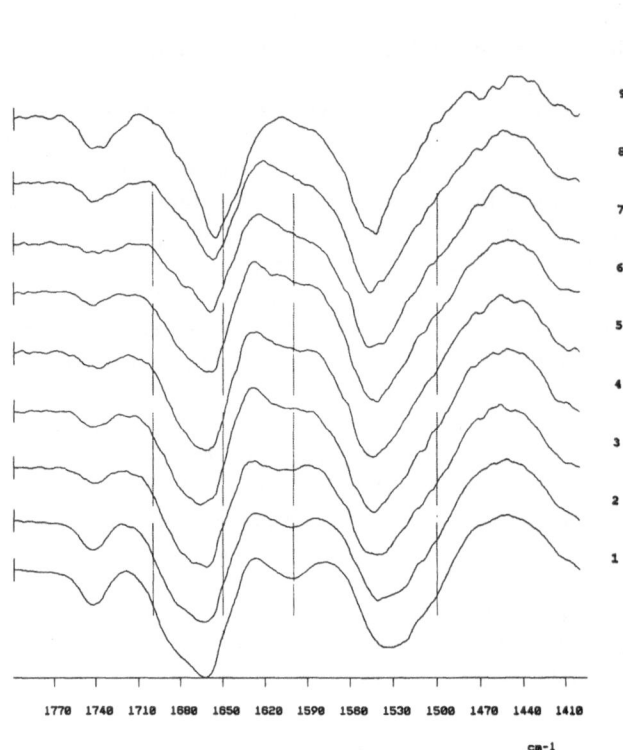

1770 1740 1710 1680 1650 1620 1590 1580 1530 1500 1470 1440 1410

cm-1

FIGURE 10. Difference spectra obtained by subtraction of a less deuterated spectrum from a more deuterated spectrum (see Figure 8 for conditions). Deuteration time intervals have been chosen in such a way that a similar amount of amide protons ($\pm 7\%$) has been exchanged between the two spectra used for the subtraction. Deuteration times and amplitude in mOD units are as follows: 1. 0.25–0.00 min, ΔmOD = 66; 2. 0.50–0.25 min, ΔmOD = 35; 3. 0.75–0.50 min, ΔmOD = 14; 4. 1.25–0.75 min, ΔmOD = 15; 5. 2.00–1.25 min, ΔmOD = 11; 6. 5.0–2.0 min, ΔmOD = 11; 7. 12.0–5.0, ΔmOD = 13; 8. 28.0–12.0 min, ΔmOD = 10; 9. 64.0–28.0 min, ΔmOD = 8. For the original spectra the ΔmOD = 260. The vertical dotted lines are positioned at 1700, 1650, 1600, and 1500 cm^{-1}.

taken as a measure of the protein/lipid ratio. The relation between these two quantities is shown in Figure 12 for lysozyme. Clearly, an accurate determination of the lipid/protein ratio can be obtained for a wide range of values. When the lipid concentration is known, the protein concentration is easily obtained. When an α-helix rich protein such as myoglobin or a β-sheet rich protein such as

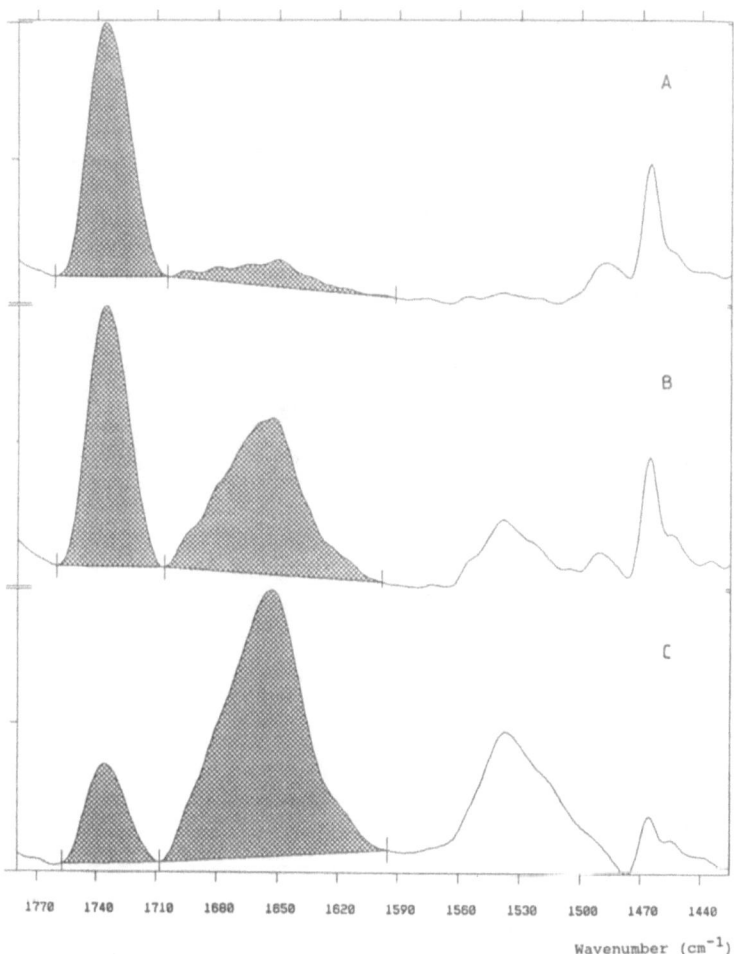

FIGURE 11. IR-ATR spectra of 100μg DMPC in the presence of 1.78μg lysozyme (A), 17.8 μg lysozyme (B) and 100μg lysozyme (C). The shaded area represents the lipid ν(C=O) integrated intensity (left-hand side) and the protein amide I integrated intensity (right-hand side) that are used to evaluate the lipid/protein ratios.

α-chymotrypsin is used, similar relations are obtained (not shown). The determination of protein concentration by IR-ATR spectroscopy is very accurate when a calibration curve is available (correlation coefficient for the linear regression =0.9992 for the example in Figure 12). The protein concentration assay described is most useful for membrane associating proteins and especially short peptides with atypical amino acid composition, whose concentration is difficult

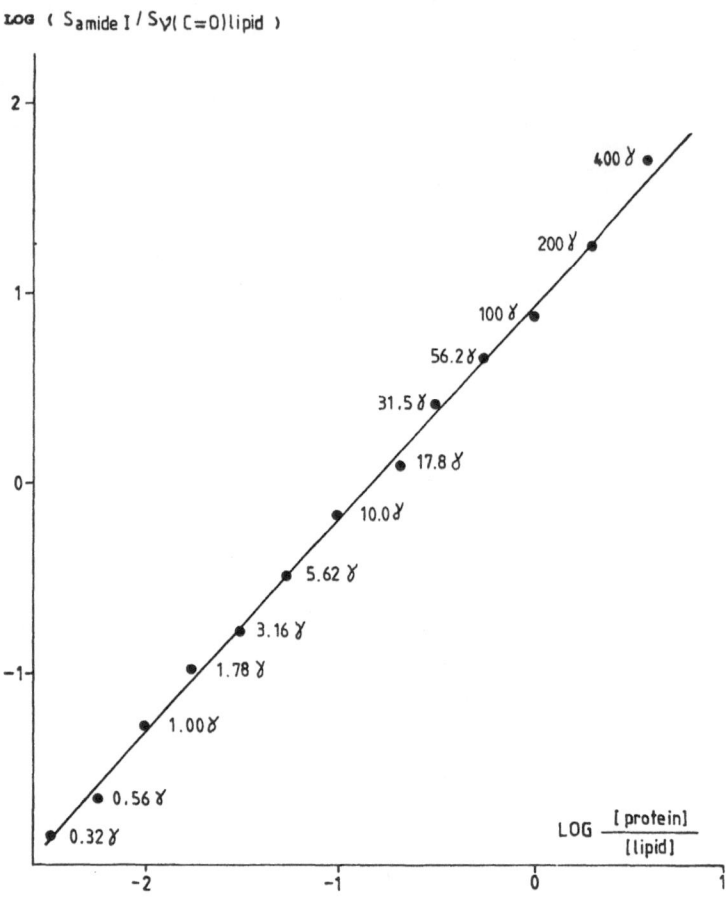

FIGURE 12. Calibration curve for lysozyme/DMPC ratio assay. 100 μg of DMPC were mixed with the indicated amount of lysozyme. Log of the amide I area/lipid ν(C=O) area ratio (see Figure 11) is reported as a function of Log [protein]/[lipid] (w/w). The linear regression correlation coefficient is 0.9992.

to evaluate by the usual colorimetric methods. Applications to membrane research are numerous. For example, once the reconstitution of a protein in a lipid bilayer is achieved, the fraction of the protein not absorbed to the bilayer is usually eliminated by a separation technique such as chromatography or centrifugation on a density gradient. The IR-ATR technique provides a fast procedure for accurately determining the efficiency of the reconstitution (lipid/protein ratio), a step which otherwise requires an assay for both the lipid and the protein. Moreover, the lipids present are likely to interfere with the colorimetric assay of the protein. Asolectin for instance yields a strong Lowry response but is quite suitable for the IR-ATR technique (Cabiaux et al., 1989b). In the absence of a calibration curve, it is tempting to use an average calibration curve. Using the relation $S_{amideI}/S\nu_{(C=O)lipid} = 5*[Protein]/[DMPC]$, where S_{amideI} is the integrated area of amide I and $S\nu_{(C=O)lipid}$ is the integrated area of the band at 1736 cm^{-1}. This yields a correct estimate of the [Protein]/[DMPC] ratio (w/w) with a maximum deviation of 50% for all the proteins tested so far. A secondary structure analysis of the proteins and a knowledge of the specific extinction coefficients for the different secondary structures should help improve the latter application in the future.

It must be stressed that the approach described above does not take into account polarization effects. Tamm and Tatulian (1993) presented a more rigorous approach for ATR spectra and, for a 45° beveled germanium ATR element, arrived at the following expression for the lipid/protein molar ratio L/P: $L/P = 0.208 (n_{res}-1) (\sigma_P/\sigma_L) (A^{o}{}_{L}/A^{o}{}_{P})$ using the integrated molar absorption coefficient of the methylen stretching vibrations (integrated from 2980 to 2800 cm^{-1}) of $1.32\ 10^8$ cm/mol of lipid and of the amide I vibration (integrated from 1690 to 1600 cm^{-1}) of $2.74\ 10^7$ cm/mol of peptide bond (Fringeli et al., 1989) and refractive indices of 4.0, 1.43, and 1.33 for the germanium, the sample, and D_2O respectively. n_{res} is the number of residues in the protein and $\sigma_i = (S_i \sin^2\alpha_i)/2 + (1-S_i)/3$ where S_i is the order parameter and α the angle of the transition moment to the molecular axis.

An absolute dosage of the lipids can easily be obtained by introducing a reference amount of predeuterated lipid in the sample and by measuring the $\nu(CH_2)/\nu(CD_2)$ ratios (Pidgeon et al., 1989).

REFERENCES

Alvarez, J., Lee, D. C., Baldwin, S. A., and Chapman, D., 1987a, Fourier-transform infrared spectroscopy study of the structure and conformational changes of the human erythrocyte glucose transporter, J. Biol. Chem. 262: 3502–3509.

Ameloot, M., 1992, Laplace deconvolution of fluorescence decay surfaces, Methods Enzymol. 210: 279–304.

Arrondo, J. L. R., Muga, A., Castresana, J. and Goñi, F. M., 1993, Quantitative studies of the structure of proteins in solution by Fourier-transform infrared spectroscopy, Prog. Biophys. Molec. Biol. 59: 23–56.

Baenziger, J. E., Miller, K. W., and Rothschild, K. J., 1992a, Incorporation of the nicotinic acetylcholine receptor into planar multilamellar films: Characterization by fluorescence and Fourier transform infrared difference spectroscopy, Biophys. J. 61: 983–992.

Baenziger, J. E., Miller, K. W., McCarthy, M. P., and Rothschild, K. J., 1992b, Probing conformational changes in the nicotinic acetylcholine receptor by Fourier-transform infrared difference spectroscopy, Biophys. J. 62: 64–66.

Baenzinger, J. E., Miller, K. W., and Rothschild, K. J., 1993, Fourier-transform infrared difference spectroscopy of the nicotinic acetylcholine receptor: Evidence for specific protein structural changes upon desensitization, *Biochemistry* **32**: 5448–5454.

Barth, A., Kreutz, W., and Mäntele, W., 1990, Molecular changes in the sarcoplasmic reticulum calcium ATPase during catalytic activity. A Fourier transform infrared (FTIR) study using photolysis of caged ATP to trigger the reaction cycle, *FEBS Lett.* **277**: 147–150.

Barth, A., Mäntele, W., and Kreutz, W., 1991, Infrared spectroscopy signals arising from ligand binding and conformational changes in the catalytic cycle of sarcoplasmic reticulum calcium ATPase, *Biochim. Biophys. Acta* **1057**: 115–123.

Benson, E. S., Hallaway, B. E., and Lumry, R. W., 1964, Deuterium-hydrogen exchange analaysis of pH-dependent transitions in bovine plasma albumin, *J. Biol. Chem.* **239**: 122–129.

Bentley, G. A., Delepierre, M., Dobson, C. M., Wedin, R. E., Mason, S. A. and Poulsen, F. M., 1983, Exchange of individual hydrogens for a protein in a crystal and in solution, *J. Mol. Biol.* **170**: 243–247.

Braiman, M. S., and Rothschild, K. J., 1988, Fourier transform infrared techniques for probing membrane protein structure, *Ann. Rev. Biophys. Chem.* **17**: 541–570.

Brandl, C. J., and Deber, C. M., 1986, Hypothesis about the membrane-buried proline residues in transport proteins, *Proc. Natl. Acad. Sci. U.S.A.* **83**: 917–921.

Breton, J., and Nabedryk, E., 1984, Transmembrane orientation for alpha-helices and the organization of chlorophylls in photosynthetic pigment-protein complexes, *FEBS Lett.* **176**: 355–359.

Buchet, R., Varga, S., Seilder, N. W., Molnar, E., and Martonosi, A., 1991, Polarized infrared attenuated total reflectance spectroscopy of the Ca^{2+}-ATPase of the sarcoplasmic reticulum, *Biochim. Biophys. Acta* **1068**: 201–216.

Burget, U., and Zundel, G., 1987, Glutamic acid-dihydrogen phosphate hydrogen-bonded networks: Their proton polarizability as a function of cations present, *Biophys. J.* **52**: 1065–1070.

Cabiaux, V., Brasseur, R., Wattiez, R., Falmagne, P., Ruysschaert, J. M., and Goormaghtigh, E., 1989a, Secondary structure of diphtheria toxin and its fragments interacting with acidic liposomes studied by polarized infrared spectroscopy, *J. Biol. Chem.*, **264**: 4928–4938.

Cabiaux, V., Goormaghtigh, E., Wattiez, R., Falmagne, P., and Ruysschaert, J. M., 1989b, Secondary structure of diphtheria toxin interacting with asolectin liposomes: An infrared spectroscopy study, *Biochimie* **71**: 153–158.

Careri, G., Gratton, E., Yang, P. H., and Rupley, J. A., 1980, Correlation of IR spectroscopic, heat capacity, diamagnetic susceptibility and enzymatic measurements on lysozyme powder, *Nature* **284**: 572–573.

Carpenter, J. F., and Crowe, J. H., 1989, An infrared spectroscopic study of the interactions of carbohydrates with dried proteins, *Biochemistry* **28**: 3916–3922.

Challou, N., Goormaghtigh, E., Cabiaux, V., Conrath, K., and Ruysschaert, J. M., 1994, Sequence and structure of the membrane-associated peptide of glycophorin A. *Biochemistry* **33**: 6902–6910.

Chang, C-W., Sekiya, N., and Yoshihara, K., 1991, O-H stretching vibration in Fourier-transform difference infrared spectra of bacteriorhodopsin, *F.E.B.S. Lett.* **287**: 157–159.

Chapman, D., Gomez-Fernandez, J. C., Goñi, F. M., and Barnard, M., 1980, Difference infrared spectroscopy of aqueous model and biological membranes using an infrared data station, *J. Biochem. Biophys. Methods* **2**: 315–323.

Chetverin, A. B., and Brazhikov, E. V., 1985, Do sodium and potassium forms of Na, K-ATPase differ in their secondary structure, *J. Biol. Chem.* **260**: 7817–7819.

Chiacchiera, S. M., and Kosower, E. M., 1992, Deuterium exchange on micrograms of proteins by attenuated total reflection Fourier-transform infrared spectroscopy on silver halide fiber, *Anal. Biochem.* **201**: 43–47.

Combelas, P., Garrigou-Lagrange, C., and Lascombe, J., 1974, Etude par spectroscopie infrarouge de la transformation hélice-chaîne statistique des polypetides. II. Etude de l'interaction de quelques polypetides à chaînes latérales polaires avec l'acide trifluoroacétique, *Biopolymers* **13:** 577–589.

Cortijo, M., Alonso, A., Gomez-Fernandez, J. C., and Chapman, D., 1982, Instrinsic protein-lipid interactions: Infrared spectroscopic studies of Gramicidin A, bacteriorhodopsin and Ca^{++}AT-Pase in Biomembranes and reconstituted systems, *J. Mol. Biol.* **157:** 597–618.

Crowe, J. H., and Crowe, L. M., 1986, Water and carbohydrate interactions with membranes: studies with infrared spectroscopy and differential scanning calorimetry methods, *Methods. Enzymol.* **127:** 693–703.

Crowe, J. H., Crowe, L. M., and Chapman, D., 1984, Infrared spectroscopy studies on the interactions of water and carbohydrates with a biological membrane, *Arch. Biochem. Biophys.* **232:** 400–407.

Crowe, J. H., Crowe, L. M., and Chapman, D., 1985, Interaction of carbohydrates with dry dipalmitoylphosphatidylcholine, *Arch. Biochem. Biophys.* **236:** 289–296.

Crowe, J. H., Spargo, B. J., and Crowe, L. M., 1987, Preservation of dry liposomes does not require retention of residual water, *Proc. Natl. Acas. Sci. U.S.A.* **84:** 1537–1540.

De Jongh, H. H. J., Killian, A., and de Kruijff, B., 1992, A water-lipid interface induces a highly dynamic folded state in apocytochrome c and cytochrome c which may represent a common folding intermediate, *Biochemistry* **31:** 1636–1643.

Demel, R., Goormaghtigh, E., and de Kruijff, B., 1990, Lipid and peptide specificities in signal peptide–lipid interactions in model membranes, *Biochim. Biophys. Acta*, **1027:** 155–162.

Dempsey, C. E., and Butler, G. S., 1992, Helical structure and orientation of melittin in dispersed phospholipid membranes from amide exchange analysis in situ, *Biochemistry* **31:** 11973–11977.

Dollinger, G., Eisenstein, L., Lin, S-L., Nakanishi, K., and Termini, J., 1986, Fourier transform infrared difference spectroscopy of bacteriorhodopsin and its photoproducts regenerated with deuterated tyrosine, *Biochemistry* **25:** 6524–6533.

Dollinger, G., Eisenstein, L., Lin, S-L., Nakanishi, K., Odashima, K. and Termini, J., 1986, bacteriorhodopsin: Fourier-transform infrared difference methods for studies of protonation of carboxyl groups, *Methods Enzymol.* **127:** 649–662.

Dong, A., Huang, P., and Caughey, W. S., 1990, Protein secondary structure in water from second-derivative amide I infrared spectra, *Biochemistry* **29:** 3303–3308.

Dong, A., Huang, P., and Caughey, W. S., 1992, Redox-dependent changes in β-extended chain and turn structures of cytochrome c in water solution determined by second derivative amide I infrared spectra, *Biochemistry* **31:** 182–189.

Döring, O., and Böttger, M., 1992, Effect of D$_2$O on maize plasmalemma ATPase and electron transport coupled proton pumping, *Biochem. Biophys. Res. Commun.* **182:** 870–876.

Dousseau, F., Therrien, M., and Pézolet, M., 1989, On hte spectral subtraction of water from the FT–IR spectra of aqueous solutions of proteins, *Appl. Spectroscopy* **40:** 538–542.

Downer, N. W., Bruckman, T. J., and Hazzard, J., 1986, Infrared spectroscopic study of photorecep-tor membrane and purple membrane: Protein secondary structure and hydrogen deuterium exchange, *J. Biol. Chem.* **261:** 3640–3647.

Dunker, A. K., 1982, Proton motive force tranducer and its role in proton pumps, proton engine, tabacco mosaic virus assembly and hemoglobin allosterism, *J. Theor. Biol.* **97:** 95–127.

Earnest, T. N., Herzfeld, J., and Rothschild, K. J., 1990, Polarized FTIR of bacteriorhodopsin: transmembrane α-helices are resistant to hydrogen-deuterium exchange, *Biophys. J.* **58:** 1539–1546.

El Antri, S., Sire, O., and Alpert, B., 1990, Relationship between protein/solvent proton exchange and progressive conformation and fluctuation changes in hemoglobin, *Eur. J. Biochem.* **191:** 163–168.

Engelhard, M., Gerwert, K., Hess, B., Kreutz, W., and Siebert, F., 1985, Light-driven protonation changes of internal aspartic acids of bacteriorhodopsin: An investigation by static and time-resolved infrared difference spectroscopy using (4-^{13}C)Aspartic acid labeled purple membrane, *Biochemistry* **24:** 400–407.

Englander, S. W., and Kallenbach, N. R., 1984, Hydrogen exchange and structral dynamics of protein and nucleic acids, *Quarterly Review Biophysics* **16:** 521–655.

Fabian, H., Nauman, D., Misselwitz, R., Ristau, O., Gerlach, D., and Welfe, H., 1992, Secondary structure of streptokinase in aqueous solution: A Fourier transform infrared spectroscopic study, *Biochemistry* **31:** 3532–3538.

Fettiplace, R., and Haydon, D. A., 1980, Water permeability of lipid membranes, *Physiol. Rev.* **60:** 510–550.

Fink, D. J., and Gendreau, R. M., 1984, Quantitative surface studies of protein adsorption by infrared spectroscopy: corrections for bulk concentrations, *Anal. Biochem.* **139:** 140–148.

Fink, D. J., Hutson, T. B., Chittur, K. K., and Genderau, R. M., 1987, Quantitative surface studies of protein adsorption by infrared spectroscopy. II. Quantification of adsorbed and bulk proteins, *Anal. Biochem.* **165:** 147–154.

Finkelstein, A., 1987, *Water Movement trough Lipid Bilayers, Pores, and Plasma Membranes. Theory and reality,* John Wiley & Sons, New York.

Fraser, P. E., Nguyen, J. T., Surewicz, W. K., and Kirschner, D. A., 1991, pH-dependent structural transitions of Alzheimer amyloid peptides, *Biophys. J.* **60:** 1190–1201.

Fraser, P. E., Nguyen, J. T., Surewicz, W. K., and Mantsch, H. H., 1991, pH-dependent structural transitions of Alzheimer amyloid peptides, *Biophys. J.* **60:** 1190–1201.

Frauenfelder, H., and Gratton, E., 1986, Protein dynamics and hydration, *Methods Enzymol.* **127:** 207–216.

Frey, S., and Tamm, L. K., 1991, Orientation of melittin in phospholipid bilayers. A polarized attenuated total reflection infrared spectroscopy study, *Biophys. J.* **60:** 922–930.

Fringeli, U. P., 1981, A new crystalline phase of L-α-dipalmitoylphosphatidylcholine monohydrate, *Biophys. J.* **34:** 173–187.

Fringeli, U. P., and Günthard, H. H., 1981, Infrared membrane spectroscopy, in: *Membrane Spectroscopy,* (E. Grell ed.) pp 270–332, Springer-Verlag, Berlin.

Fringeli, U. P., Apell, H. J., Fringeli, M., Lauger, P., 1989, Polarized infrared absorption of Na$^+$/K$^+$-ATPase studied by attenuated total reflection spectroscopy, *Biochim Biophys Acta.* **984:** 301–312.

Fritsch, J., Zundel, G., Hayd, A., and Maurer, M., 1984, Proton polarizability of hydrogen-bonded chains: an ab initio SCF calculation with a model related to the conducting system in bacteriorhodopsin, *Chem. Phys. Lett.* **107:** 65–69.

Gallagher, W., Tao, F., and Woodward, C., 1992, Comparison of hydrogen exchange rates for bovine pancreatic trypsin inhibitor in crystals and in solution, *Biochemistry* **31:** 4673–4680.

Gerwert, K., Souvigner, G., and Hess, B., 1990, Simultaneous monitoring of light-induced changes in protein side-group protonation, chromophore isomerization, and backbone motion of bacteriorhodopsin by time-resolved Fourier-transform infrared spectroscopy, *Proc. Natl. Acad. Sci. U.S.A.* **87:** 9774–9778.

Goormaghtigh, E., and Ruysschaert, J. M., 1990, Polarized attenuated total reflection spectroscopy as a tool to investigate the conformation and orientation of membrane components, in: *Molecular Description of Biological Membranes by Computer-Aided Confromationazl Analysis* (R. Brasseur, ed.), Vol. I, p. 285–329, CRC Press Inc., Boca Raton, Florida.

Goormaghtigh, E., Cabiaux, V., and Ruysschaert, J. M., 1990, Secondary structure and dosage of soluble and membrane proteins by attenuated total reflection Fourier-transform infrared spectroscopy on hydrated films, *Eur. J. Biochem.* **193:** 409–420.

Goormaghtigh, E., Cabiaux, V., De Meutter, J., Rosseneu, M., and Ruysschaert, J. M., 1993, Secondary structure of the particle associating domain of apolipoprotein B-100 in low-density lipoprotein by attenuated total reflection infrared spectroscopy, *Biochemistry* **32:** 6104–6110.

Goormaghtigh, E., De Meutter, J., Vanloo, B., Brasseur, R., Rosseneu, M., and Ruysschaert, J. M., 1989, Evaluation of the secondary structure of apo B100 in low density lipoprotein (LDL) by infrared spectroscopy, *Biochim. Biophys. Acta* **1006:** 147–150.

Goormaghtigh, E., Ruysschaert, J. M., and Scarborough, G. A., 1988, High-yield incorporation of the Neurospora plasma membrane H+ ATPase into proteoliposomes: Lipid requirement and secondary structure of the enzyme by IR spectroscopy, *Prog. Clin. Biol. Res.* **273:** 51–56.

Graziani, Y., and Livne, A., 1972, Water permeability of bilayer lipid membranes: Sterol-lipid interactions, *J. Membr. Biol.* **7:** 275–284.

Gregory, R. B., and Lumry, R., 1985, Hydrogen-exchange evidence for distinct structural classes in globular proteins, *Biopolymers* **24:** 301–326.

Halvorson, H. R., 1992, Padé–Laplace algorithm for sum of exponentials: selecting appropriate exponential model and initial estimates for exponential fitting, *Methods Enzymol.* **210:** 54–67.

Haris, P. I., Coke, M., and Chapman, D., 1989, Fourier transform infrared spectroscopic investigation of rhodopsin structure and its comparison with bacteriorhodopsin, *Biochim. Biophys. Acta* **995:** 160–167.

Haris, P. I., Lee, D. C. and Chapman D. 1986, A Fourier transform infrared investigation of the structural differences between ribonuclease A and ribonuclease S, *Biochim. Biophys. Acta* **874:** 255–265.

Harrick, N. J., 1967, *Internal Reflection Spectroscopy,* John Wiley & Sons, New York.

Heimburg, T., and Marsh, D., 1993, Investigation of secondary and tertiary structural changes of cytochrome c in complex with anionic lipids using amide hydrogen exchange measurements: An FTIR study, *Biophys. J.* **65:** 2408–2417.

Hodgmann, C. D., (ed.), 1952, *Hanbook of Chemistry and Physics,* pp. 2147–2148, Chemical Rubber, Cleveland, Ohio.

Hofer, P., and Fringeli, U. P., 1979, Structural investigation of biological material in aqueous environment by means of infrared-ATR spectroscopy, *Biophys. Struct. Mech.* **6:** 67–80.

Holloway, P. W., and Buchheit, C., 1990, Topography of the membrane-binding domain of cytochrome b5 in lipids by Fourier-transform infrared spectroscopy, *Biochemistry* **29:** 9631–9637.

Kackson, M., Haris, P. I., and Chapman, D., 1991, Fourier-transform infrared spectroscopic studies of Ca2+-binding proteins, *Biochemistry* **30:** 9681–9686.

Kakalis, L. T., and Baianu, I. C., 1988, Oxygen-17 and deuterium nuclear magnetic relaxation studies of lysozyme hydratation in solution: Field dispersion, concentration, pH/pD and protein activity dependences, *Arch. Biochem. Biophys.* **267:** 829–841.

Kauppinen, J. K., Moffat, D. J., Mantsch, H. H., and Cameron, D. G., 1981a, Fourier self-deconvolution: A method for resolving intrinsically overlapped bands, *Appl. Spectrosc.* **35:** 271–276.

Kim, K. S., and Woodward, C. K., 1993, Protein internal flexibility and global stability: Effect of urea on hydrogen exchange rates of bovine pancreatic trypsin inhibitor, *Biochemistry* **32:** 9609–9613.

Kim, K. S., Fuchs, J. A., and Woodward, C. K., 1993, Hydrogen exchange identifies native-state motional domains important in protein folding, *Biochemistry* **32:** 9600–9608.

Kleffel, B., Garavito, R. M., Baumeister, W., and Rosenbusch, J. P., 1985, Secondary structure of a channel-forming protein: Porin from E. coli outer membranes, *EMBO J.* **4:** 1589–1592.

Knox, D. G., and Rosenberg, A., 1980, Fluctuations of protein structure as expressed in the distribution of hydrogen exchange rate constants, *Biopolymers* **19:** 1049–1068.

Korenstein, R., and Hess, B., 1977, Hydration effects on the photocycle of bacteriorhodopsin in thin layers of purple membrane, *Nature* **270:** 184–186.

Krasov, W. H., Han, L. R., and Munger, J. C., 1988, Measurements of ^2H_2O by IR absorbance in doubly labeled H_2O studies of energy expenditure, *Am. J. Physiol.* **253:** R174–R177.

Kristof, W., and Zundel, G., 1982, Proton transfer and polarizability of hydrogen bonds formed between cysteine and lysine residues, *Biopolymers* **21:** 25–42.

Laiken, S., and Printz, M. P., 1970, Kinetic class analysis of hydrogen-exchange data, *Biochemistry* **9:** 1547–1553.

Lenk, T. J., Ratner, B. D., Gendreau, R. M., and Chittur, K. K., 1989, IR spectral changes of bovine serum albumin upon surface adsorption, *J. Biomed. Mater. Res.* **23:** 549–569.

Lin, S-L., Eisenstein, L., Govindjee, R., Konno, K., and Nakanishi, K., 1987, Deprotonation of tyrosines in bacteriorhodopsin as studied by Fourier transform infrared spectroscopy with deuterium and nitrate labeling, *Biochemistry* **26:** 8327–8331.

Loh, S. N., and Markley, J. L., 1994, Hydrogen bonding in proteins as studied by amide hydrogen D/H fractionation factors: Application to staphylococcal nuclease, *Biochemistry* **33:** 1029–1036.

MacDonald, G. M., and Barry, B. A., Difference FT–IR study of a novel biochemical preparation of photosystem II. *Biochemistry* **31:** 9848–9856.

Maeda, A., Ohkita, Y. J., Sasaki, J., Shichida, Y., and Yoshizawa, T., 1993, Water structural changes in lumirhodopsin, metarhodopsin I, and metarhodopsin II upon photolysis of bovin rhodopsin: Analysis by Fourier-transform infrared spectroscopy, *Biochemistry* **32:** 12033–12038.

Maeda, A., Sasaki, J., Shichida, Y., and Yoshizawa, T., 1992, Water structural changes in the bacteriorhodopsin photocycle: Analysis by Fourier transform infrared spectroscopy, *Biochemistry* **31:** 462–467.

Marrero, H., and Rothschild, K. J., 1987, Conformational changes in bacteriorhodopsin studied by infrared attenuated total reflection, *Biophys. J.* **5:** 629–635.

Mellier, A., and Diaf, A., 1988, Infrared study of phospholipid hydration. Main phase transition of saturated phosphatidylcholine/ water multilamellar samples, *Chem. Phys. Lipids* **46:** 51–56.

Merz, H., and Zundel, G., 1986, Thermodynamics of proton transfer in carboxylic acid-retinal Schiff base hydrogen bonds with large proton polarizability, *Biochem. Biophys. Res. Commun.* **138:** 819–825.

Michel-Villaz, M., Saibil, H. R., and Chabre, M., 1979, Orientation of rhodopsin α-helices in retinal rod outer segment membranes studied by infrared linear dichroïsm, *Proc. Natl. Acad. Sci. U.S.A.* **76:** 4405–4408.

Mitchell, R. C., Haris, P. I., Fallowfield, C., Keeling, D. J., and Chapman, D., 1988, Fourier-transform infrared spectroscopy studies on gastric H^+/K^+-ATPase, *Biochim. Biophys. Acta* **947:** 31–38.

Moh, P. P., Fiamingo, F. G., and Alben, J., 1987, Conformational sensitivity of beta-93 cystein SH to ligation of hemoglobin observed by FT–IR spectroscopy, *Biochemistry* **26:** 6243–6249.

Molday, R. S., Englander, S. W., and Kallen, R. G., 1972, Primary structure effects on peptide group hydrogen exchange, *Biochemistry* **11:** 150–158.

Muga, A., Mantsch, H. H., and Surewitcz, W. K., 1991, Membrane binding induces destabilization of cytochrome c structure, *Biochemistry* **30:** 7219–7224.

Nabedryk, E., and Breton, J., 1981, Orientation of intrinsic proteins in photosynthetic membranes: Polarized infrared spectroscopy of chloroplasts and chromatophores, *Biochim. Biophys. Acta* **635:** 515–524.

Nabedryk, E., Garavito, R. M., and Breton, J., 1988, The orientation of β-sheets in porin. A polarized Fourier-transform infrared spectroscopic investigation, *Biophys. J.* **53:** 671–676.

Nedelec, J-F., Alfsen, A., and Lavialle, F., 1989, Comparative study of myelin proteolipid apoprotein solvation by multilayer membranes of synthetic DPPC and biological lipid extract from bovine brain. An FT–IR investigation, *Biochimie* **71:** 145–171.

Okamura, E., Umemura, J., and Takenaka, T., 1990, Orientation studies of hydrated dipalmitoylphosphatidylcholine multibilayers by polarized FTIR–ATR spectroscopy, *Biochim. Biophys. Acta* **1025**: 94–98.

Ormos, P., 1991, Infrared spectroscopic demonstration of a conformational change in bacteriorhodopsin involved in proton pumping, *Proc. Natl. Acad. Sci. U.S.A.* **88**: 473–477.

Ormos, P., Chu, K., and Mourant, J., 1992, Infrared study of the L, M, and N intermediates of bacteriorhodopsin using the photoreaction of M, *Biochemistry* **31**: 6933–6937.

Pastrana, B., Mautone, A. J., and Mendelsohn, R., 1991, Fourier-transform infrared studies of secondary structure and orientation of pulmonary surfactant SP–C and its effects on the dynamic surface properties of phospholipid? *Biochemistry* **30**: 10058–10064.

Petrelski, S. J., Byler, D. M., and Liebman, M. N., 1991, Comparison of various molecular forms of bovine trypsin: correlation of infrared spectra with X-ray crystal structures, *Biochemistry* **30**: 133–143.

Petrelski, S. J., Tedeschi, N., Arakawa, T., and Carpenter, J. F., 1993, Dehydration-induced conformational transitions in protein and their inhibition by stabilizers, *Biophys. J.* **65**: 661–671.

Pidgeon, C., Apostol, G., and Markovich, R., 1989, Fourier-transform infrared assay of liposomal lipids, *Anal. Biochem.* **181**: 28–32.

Poole, P. L., and Finney, J. L., 1984, Sequential hydration of dry proteins: a direct difference IR investigation of sequence homologs lysozyme and α-lactalbumin, *Biopolymers* **23**: 1647–1666.

Powell, J. R., Wasacz, F. M., and Jakobsen, R. J., 1986, An algorithm for the reproducible spectral subtraction of water from the FTIR spectra of proteins in dilute solutions and adsorbed monolayers, *Appl. Spectrosc.* **40**: 339–344.

Provencher, S. W., 1976, A Fourier method for the analysis of exponential decay curves, *Biophys. J.* **16**: 27–41.

Provencher, S. W., 1982, A constrained regularization method for inverting data represented by linar algebraic or integral equations, *Computer Phys. Commun.* **27**: 213–227.

Provencher, S. W., 1982, CONTIN: A general purpose constrained regularization program for inverting noisy linear algebraic and integral equations, *Computer Phys. Commun.* **27**: 229–242.

Provencher, S. W., and Dovi, V. G., 1979, Direct analysis of continuous relaxation spectra, *J. Biochem. Biophys. Methods* **1**: 313–318.

Rashin, A. A., 1987, Correlation between the calculated local stability and hydrogen exchange rates in proteins, *J. Mol. Biol.* **198**: 339–349.

Rath, P., Bouché, O., Merril, A. R., Cramer, W. A., and Rotschild, K. J., 1991, Fourier-transform infrared evidence for a predominantly α-helical structure of the membrane bound channel forming COOH-terminal peptide of colicin E1, *Biophys. J.* **59**: 516–522.

Reilly, K. E., Becka, R., Thomas, G. J., 1992, Deuterium exchange of operator 8CH groups as a Raman probe of repressor recognition: interactions of wild-type and mutant G repressors with operator $O_L 1$, *Biochemistry* **31**: 3118–3125.

Roepe, P. D., Ahl, P. L., Herzfeld, J., Lugtenburg, J., and Rothschild, K. J., 1988, Tyrosine protonation changes in bacteriorhodopsin: A Fourier transform infrared study of BR548 and its primary photoproduct, *J. Biol. Chem.* **263**: 5110–5117.

Rosenberg, A., 1986, Use of hydrogen exchange kinetics in the study of the dynamic properties of biological membranes, *Methods Enzymol.* **127**: 630–648.

Rothschil, K. J., Sanches, R., Hsiao, T. L., and Clark, N. A., 1980, A spectroscopic study of rhodopsin alpha-helix orientation, *Biophys. J.* **31**: 53–64.

Rothschild, K. J., and Clark, N. A., 1979, Polarized infrared spectroscopy of oriented purple membrane, *Biophys. J.* **5.25**: 473–488.

Rothschild, K. J., Bousché, O., Braiman, M. S., Hasselbacher, C. A., and Spudich, J. L., 1988, Fourier-transform infrared study of the halorhodopsin chloride pump, *Biochemistry* **27**: 2420–2424.

Rothschild, K. J., Braiman, M. S., He, Y-W., Marti, T., and Khorana, H. G., 1990, Vibrational spectroscopy of bacteriorhodopsin mutants: Evidence for the interaction of aspartic acid 212 with tyrosine 185 and possible role in the proton pump mechanism, *J. Biol. Chem.* **265:** 16985–16991.

Rothschild, K. J., He, Y-W., Gray, D., Roepe, P. D., Pelletier, S. L., Brown, R. S., and Herzfeld, J., 1989, Fourier transform infrared evidence for proline structural changes during the bacteriorhodopsin photocycle, *Proc. Natl. Acad. Sci. U.S.A.* **86:** 9832–9835.

Rothschild, K. J., Roepe, P., Lugtenburg, J., and Pardoen, J. A., 1984, Fourier transform infrared evidence for Schiff base alteration in the first step of the bacteriorhodopsin photocycle, *Biochemistry* **23:** 6103–6109.

Sami, M., and Dempsey, C., 1988, Hydrogen exchange from the transbilayer hydrophobic peptide of glycophorin reconstituted in lipid bilayers, *F.E.B.S. Lett.* **240:** 211–215.

Sanchez-Ruiz, J. M., and Martinez-Carrion, M., 1988, A Fourier-transform infrared spectroscopic study of the phosphoserine residues in hen egg phosvitin and ovalbumin, *Biochemistry* **27:** 3338–3342.

Sarver, R. W., and Krueger, W. C., 1993, Infrared investigation of proteins deposited on polyethylene films, *Anal. Biochem.* 519–525.

Sarver, R. W., and Kruerger, W. C., 1991, Protein secondary structure from Fourier-transform infrared spectroscopy: A database analysis, *Anal. Biochem.* **194:** 89–100.

Schinkel, J. E., Downer, N. W., and Rupley, J. A., 1985, Hydrogen exchange of lysozyme powders. Hydration dependence of internal motions, *Biochemistry* **24:** 352–366.

Schlereth, D., Fernandez, V. M., and Mäntele, W., 1993, Protein conformational changes in tetraheme cytochromes detected by FTIR spectroelectrochemistry: Desulfovibrio desulfuricans Norway 4 and Desulfovibrio gigas cytochromes c₃, *Biochemistry* **32:** 9199–9208.

Simhony, S., Katzir, A., and Kosower, E. M., 1988, Fourier-transform infrared spectra of organic compounds in solution and as thin films obtained by using an attenuated total internal reflectance fiber-optic cell, *Anal. Chem.* **60:** 1908–1910.

Simhony, S., Kosower, E. M., and Katzir, A., 1986, Novel attenuated total internal reflectance spectroscopic cell using infrared fibers for aqueous solutions, *Appl. Phys. Lett.* **49:** 253–254.

Simhony, S., Kosower, E. M., and Katzir, A., 1987, Fourier-transform infrared spectra of aqueous protein mixtures using a novel attenuated total internal reflectance cell with infrared fibers, *Biochem. Biophys. Res. Commun.* **142:** 1059–1063.

Simhony, S., Schnitzer, I., Katzir, A., and Kosower, E. M., 1988, Evanescent wave infrared spectroscopy of liquids using silver halide optical fibers, *J. Appl. Phys.* **64:** 3732–3734.

Strauss, H. L., 1986, Vibrational methods for water structure in non polar media, *Methods Enzymol.* **127:** 106–113.

Swedberg, S. A., Pesek, J. J., and Fink, A. L., 1990, Attenuated total reflection Fourier-transform infrared analysis of an acyl-enzyme intermediate of α-chymotrypsin, *Anal. Biochem.* **186:** 153–158.

Tamm, L. K., 1988, Lateral diffusion and microscope studies on a monoclonal antibody specifically bound to supported phospholipid bilayers, *Biochemistry* **27:** 1450–1457.

Tamm, L. K., and McConnel, H. M., 1985, Supported phospholipid bilayers, *Biophys. J.* **47:** 105–113.

Tamm, L. K., and Tatulian, S. A., 1993, Orientation of functional and nonfunctional PTS permease signal sequences in lipid bilayers. A polarized attenuated total reflection infrared study, *Biochemistry* **32:** 7720–7726.

Tiede, D. M., 1985, Incorporation of membrane proteins into interfacial films: Model membranes for electrical and structural characterisation, *Biochim. Biophys. Acta* **811:** 357–359.

Tonge, P., Moore, R., and Wharton, C. W., 1989, Fourier-transform infra-red studies of the alkaline isomerization of mitochondrial cytochrome c and the ionization of carboxylic acids, *Biochem. J.* **258:** 599–605.

Tüchsen, E., Hayes, J. M., Ramaprased S., Copie, V., and Woodward, C., 1987, Solvent exchange of buried water and hydrogen exchange of peptide NH group hydrogen bonded to buried waters in bovine pancreatic trypsin inhibitor, *Biochemistry* **26:** 5163–5172.

Van Thiel, M., Backer, E. D., and Pimentel, G. C., 1957, Infrared studies of hydrogen bonding of water by the matrix isolation technique, *J. Chem. Phys.* **27:** 486–490.

Vandenbussche, G., Clercx, A., Clercx, M., Curstedt, T., Johansson, J., Jörnvall, H., and Ruysschaert, J. M., 1992b, Secondary structure and orientation of the surfactant protein SP-B in a lipid environment. A Fourier-transform infrared spectroscopy study, *Biochemistry* **31:** 9169–9176.

Vandenbussche, G., Clercx, A., Curstedt, T., Johansson, J., Jörnvall, H., and Ruysschaert, J. M., 1992a, Structure and orientation of the surfactant-associated protein C in a lipid bilayer, *Eur. J. Biochem.* **203:** 201–209.

Vasilescu, V., and Katona, E., 1986, Deuteration as a tool in investigating the role of water in the structure and function of excitable membrane, *Methods Enzymol.* **127:** 662–678.

Venyaminov, S. Y., and Kalnin, N. N., 1990a, Quantitative IR spectrophotometry of peptides compounds in water (H_2O) solutions. I. Spectral parameters of amino acid residue absorption band, *Biopolymers* **30:** 1243–1257.

Venyaminov, S. Y., and Kalnin, N. N., 1990b, Quantitative IR spectrophotometry of peptides compounds in water (H_2O) solutions. II. Amide absorption bands of polypeptides and fibrous proteins in α-, β- and random conformations, *Biopolymers* **30:** 1259–1271.

Vincent, S. S., Star, C. J., and Levin, I. W., 1984, Infrared spectroscopic study of the pH-dependent secondary structure of brain clathrin, *Biochemistry* **23:** 625–631.

Walrafen, G. E., 1967, Raman spectral studies of the effects of temperature on water structure, *J. Chem. Phys.* **47:** 114–126.

Wlodawer, A., Walter, J., Huber, K., and Sjölin, L., 1984, Structure of bovine pancreatic trypsin inhibitor. Results of joint neutron and X-ray refinement of crystal form II, *J. Mol. Biol.* **180:** 301–329.

Woodward, C. K., and Rosenberg, A., 1970, Oxidized Rvase as a protein model having no contribution to the hydrogen exchange rate from conformational restriction, *Proc. Natl. Acad. Sci. U.S.A.* **66:** 1067–1074.

Zhang, Y-P., Lewin, R. N., Hodges, R. S., and McElhaney, R. N., 1992, FTIR spectroscopic studies of the conformation and amide hydrogen exchange of a peptide model of the hydrophobic transmembrane α-helices of membrane proteins, *Biochemistry* **31:** 11572–11578.

Zundel, G., 1986, Proton polarizability of hydrogen bonds: Infrared methods, relevance to electrochemical and biological systems, *Methods Enzymol.* **127:** 439–455.

Chapter 10

Determination of Soluble and Membrane Protein Structure by Fourier Transform Infrared Spectroscopy

III. Secondary Structures

Erik Goormaghtigh, Véronique Cabiaux,
and Jean-Marie Ruysschaert

1. INTRODUCTION

The basic knowledge accumulated over the last twenty years on the different vibrations of polypeptides were reviewed in Chapter 8. Because of the complexity of naturally occurring proteins, most of these data have been obtained from the study of model compounds, from simple amino acid derivatives to large synthetic polypeptides, which can be crystallized in a single secondary structure. This chapter covers biologically synthesized proteins. Data on this subject are much more recent because the advent of the new generation of Fourier transform spectrophotometers only now provides high quality spectra. Simultaneously, manipulations of the spectra have been made possible by the concomitant digi-

Erik Goormaghtigh, Véronique Cabiaux, and Jean-Marie Ruysschaert   Laboratoire de Chimie Physique des Macromolécules aux Interfaces, Université Libre de Bruxelles, B-1050 Brussels, Belgium.

Subcellular Biochemistry, Volume 23: Physicochemical Methods in the Study of Biomembranes, edited by Herwig J. Hilderson and Gregory B. Ralston. Plenum Press, New York, 1994.

talization of the spectra and the availability of low cost computers in laboratories. It was only in 1986 that the race for determination of secondary structure from manipulated IR spectra started with a paper by Byler and Susi (1986), although it is only fair to say that the results of several attempts to obtain secondary structures had been published before. The number of papers using infrared spectroscopy (IR) to obtain secondary structures has been growing exponentially ever since. One purpose of the present review is to point out, through the description and the comparison of the different methods, that interpretation of the results still needs caution. Indeed, while some spectral features of the main secondary structures are well established, others are not. Moreover, no agreement exists on a "correct" mathematical treatment of the spectra. Both the intrinsic uncertainties in the assignments and the methodological diversity open the door to flawed conclusions if the user is not properly aware of these problems. Such warnings have been issued previously (Haris and Chapman, 1992; Surewicz *et al.*, 1993; Haris and Chapman, 1992).

Proteins are a major component of biological membranes and, in most instances, no other techniques but IR are available for their study. As far as their IR spectra are concerned no difference has been found between soluble and membrane proteins. We now discuss the data obtained for both soluble and membrane proteins.

2. DETERMINATION OF PROTEIN SECONDARY STRUCTURES BY FOURIER SELF-DECONVOLUTION TECHNIQUES

It has long been known that the shape of the Amide I band of globular proteins is characteristic of their secondary structure (Susi *et al.*, 1967; Timaseff *et al.*, 1967). A paper published in 1977 (Eckert *et al.*, 1977) demonstrated the suitability of a method to establish the β-conformation percentage in proteins. Yet, only in 1986, with a publication by Byler and Susi (1986), was the determination of secondary structures from IR spectra really started. This was boosted by the availability of high signal-to-noise ratio digital spectra obtained by the new class of FTIR spectrophotometers and by easy access of spectroscopists to computers and software able to perform many mathematical operations on the digitalised spectra in a short period of time.

Experimental limitations of the technique are few: the buffer should not contain carboxylic acids such as acetate or EDTA and no carbonate that absorb in the amide I region. The most commonly used buffers, such as Mes, Tris, Hepes, phosphate, are quite acceptable. Among the phospholipids, the spectrum of phosphatidylserine and of sphingomyelin interferes to some extent with secondary structure determination, but complex mixtures such as asolectin do not (Cabiaux *et al.*, 1989). The structure of the phospholipids (Fringeli and Günthard, 1981; Casal and Mantsch, 1984) and even of DNA (Taillandier *et al.* 1984) can be studied simultaneously.

Before briefly describing the methods currently used to determine the secondary structure of a protein from an analysis of its IR spectrum, it is useful to remember that these methods were assessed by comparison with X-ray structures. A problem here is that once an X-ray structure is determined, the way the three-dimensional structure is analysed to yield a percentage of an α-helix, or β-sheet, depends, sometimes strongly, on how the different secondary structures are defined. Among the various algorithms written so far, the approach developed by Levitt and Greer (1977) was so comprehensive that it was used by most authors who developed the IR approach. The analysis of Levitt and Greer uses dihedral angles and bond distances associated with the α-carbons in conjunction with hydrogen bonding. The importance of consistently using the same algorithm to evaluate the secondary structure has been illustrated by Byler and Susi (1986). They studied papain, whose β-structure content varies between 14% and 29% and α-structure content between 19 and 29%, according to the criteria retained to define these structures. Several examples of such discordance are illustrated on Figure 6 of the paper by Yang *et al.* (1986). The diagram indicates a difference of 30 to 50% for the β-sheet content of lysozyme, RNAase A, papain, and lactate dehydrogenase, when estimated from the X-ray data according to different criteria. As mentioned by H. Mantsch during the 10th International Biophysics Congress (1990), the way the different authors retrieve the secondary structure from the Levitt and Greer paper may also differ. Myoglobin was found to contain 77% α-helix by Dousseau and Pézolet (1990), 84% by Kalnin *et al.*, (1990), 87% by Byler and Susi (1986) and Goormaghtigh *et al.* (1990), and 88% by Lee *et al.*, (1990). All of these authors used Levitt and Greer's method.

The criteria developed by Levitt and Greer (1977) were first chosen because their study covered a large number of proteins. Yet there is no guarantee that the dependency of the shape of the IR spectrum on the structure adheres to the same criteria even though a satisfactory match between the two structure determinations has been generally recognized. Some authors, however, found it useful to define two kinds of α-helices, assigned to the core and to the ends of the helix, respectively.

2.1. Fourier Self-Deconvolution-Curve Fitting Methods

The concept of Fourier self-deconvolution is based on the assumption that a spectrum of narrow bands, each characteristic of a secondary structure, is broadened in the liquid or solid state so much that the bands overlap and cannot be distinguished in the amide envelope. Deconvolution by the broadening function is then intended to return to a spectrum of narrow bands well resolved from each other. A curve fitting can then be applied to estimate quantitatively that area of each component. The shape of the bands fitted is usually Gaussian or Lorentzian or a mixture of both, as already described in 1967 by E. Suzuki for silk fibroin (Suzuki, 1967; Fraser and Suzuki, 1966). In a previous review we discussed the procedure used to obtain the self-deconvoluted spectra (Goormaghtigh and Ruysschaert, 1990). Here we will focus on their application to protein research and on a critical appraisal of the results obtained so far.

The pioneer work by Susi and Byler (1986) deserves special mention. They recorded the spectra of 20 proteins of well characterized secondary structure equilibrated in D_2O at a concentration of approximately 50 mg/ml in a 75 μm path length cell. Amide I was deconvoluted with a Lorentzian line-shape function (FWHH = 13 cm^{-1}) and a "resolution enhancement" factor of 2.4 was applied (Kauppinen *et al.*, 1981a). The deconvoluted spectrum was fitted with Gaussian band shapes by an iterative curve fitting procedure; which allows modification of the frequency, width, and intensity of the components, the other parameters being fixed at the will of the operator until a good fit is obtained. The proportion of a given secondary structure is obtained by dividing the area of the component assigned to this secondary structure by the total area of amide I. The results prove that it is possible to achieve a fitting that yields a secondary structure in close agreement (within 4% at most) with the X-ray structure (analysed according to Levitt and Greer (1977)) for most proteins, while maintaining the assignments reported in Table I. Interestingly, the same authors have compared H_2O and D_2O solutions. In H_2O solutions, frequency and width change of the solvent band is too important to allow an accurate measurement of the protein amide I band shape. They concluded that subtraction procedures are prone to serious errors for aqueous solutions in the region of 1640 cm^{-1} (Susi and Byler, 1986). For amide II, even though less overlapped by the H_2O spectrum, frequency shifts occur after addition of H_2O and make it less useful as a conformational indicator in H_2O solution than in the solid state. Dry samples have been successfully investigated using the same method by Yang *et al.* (1985).

The group of H. Mantsch has played an important role in the extension of the use of Fourier self-deconvolution-curve fitting techniques to assess the secondary structures of the proteins in solution or suspension. Their investigations confirm the location of the main components of amide I. Their work has been reviewed elsewhere (Surewicz and Mantsch 1988a) and is not further discussed here. Arrondo and Goñi developed a related approach, which has recently been reviewed and compared with other methods (Arrondo *et al.*, 1993).

More recently we have developed a similar approach directed towards the elucidation of membrane protein structure. Proteins, whether embedded in a lipid bilayer or not, are deposited as a thin film on a germanium internal reflection plate. After hydration of the film with D_2O (by flushing the sample chamber with

D_2O-saturated N_2), spectra are recorded by attenuated total reflection. The spectral analysis methodology was designed to avoid the intervention of a skilled operator. A program generates Lorentzian components of amide I at predetermined frequencies (Goormaghtigh *et al.*, 1990). These frequencies, chosen because they often appear in deconvoluted spectra, are retained only if the intensity of the spectrum is sufficient at that frequency. A fitting is first realized on a spectrum after its Fourier self-deconvolution by a Lorentzian band shape (FWHH = 30 cm^{-1}) and apodization by a Gaussian line shape chosen to yield "resolution enhancement" factors between 1.8 and 2.2. The resulting parameters are then used as input parameters for a new curve fitting carried out on a nondeconvoluted spectrum to avoid introducing artifacts related to the deconvolution procedure. Figure 1 illustrates this procedure for papain, which contains both α-helix and β structures. The final result of a similar analysis appears in Figure 2 for the α-helix rich protein myoglobin and for the β-sheet rich protein α-chymotrypsinogen. Since the first fitting stabilizes the frequency of the components, the second fitting generally does not modify the frequency of the components by more than 1 cm^{-1}. After fitting, the components are assigned to a secondary structure according to their frequency, as described in Table I. Because this method was entirely automated, the mean deviation between the X-ray data analyzed by Levitt and Greer (1977) and the IR results is higher than for the previous approach and amounts to ~9% for 14 proteins tested (Goormaghtigh *et al.*, 1990). Interestingly, membrane protein spectra and their dependency on the secondary structure are similar to that found for globular proteins. Figures 3 and 4 illustrate this point by comparing the IR spectra of deuterated and non-deuterated membrane and globular proteins. The membrane protein bacteriorhodopsin and the globular protein myoglobin are both rich in α-helices and present a maximum of amide I in the expected range of frequency (see Table I). The same observation is made for the β-sheet rich membrane protein porin and globular protein α-chymotrypsinogen.

Many researchers have now reported their assignments for the different secondary structures, most of them with the help of Fourier self-deconvolution or second derivatives, which are two equivalent techniques. Most of these assignments have been arrived at based on a reasonable discussion aimed at reconciling the known data about the protein structure and the most likely assignments of the IR frequencies. A noncomprehensive sampling of assignments reported in the literature is presented in Table II. Table II is only intended to give the reader a broad overview on the present consensus on this matter. Data available on structures such as the 3_{10} helix or the different types of turns are scarce and much work remains to be done before such structures can be correctly identified and quantified. The lack of experimental investigation is particularly problematic in the case of a structure such as the 3_{10} helix (- C=O. . .HN- bonding of the i+4 type instead of i+3 in the α-helix), which is now reported to be the third

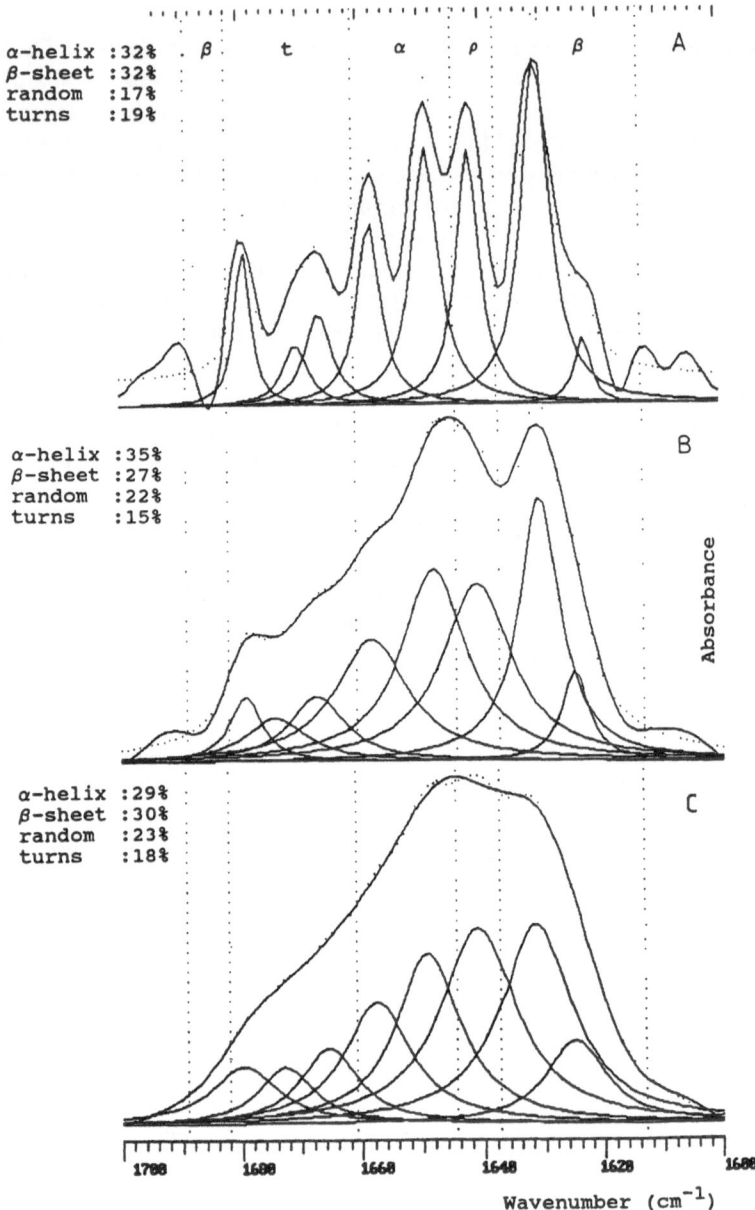

FIGURE 1. Secondary structure determination of papain deposited as a film on a germanium crystal and deuterated by flushing with D$_2$O-saturated N$_2$ for 4 hours. The figure represents the IR-ATR spectrum of the film in the amide I region at 3 stages of the deconvolution process: K = 1 (C), K = 1.8 (B) and K = 2.4 (A). The intensity, width and frequency of the Lorentzian band used for the curve fitting were chosen by the program on the basis of the shape of the most deconvoluted spectrum (A) (see Goormaghtigh *et al.*, 1990). An iterative least square curve fitting program adjusted then

principal structure occurring in globular proteins. Twenty-four percent of those proteins occur as an amino or carboxy terminal extension of an α-helix (Toniolo and Benedetti 1991). Data reported in Table II display frequencies scattered between 1639 and 1660 cm^{-1} for the deuterated form. The characteristic bands of some structures have been reported in only one or two papers and do not appear in the table. Examples of these include the loop structure whose deuterated form is reported at 1655 cm^{-1} in a human growth factor (Petrelski et al., 1991c) and trypsin (Petrelski et al., 1991a), and at 1656 cm^{-1} in interleukins (Wilder et al., 1992); the parallel β-sheet in the reticulum membrane at 1643 cm^{-1} (H-form, Arrondo et al., 1988a) or the parallel β-sheet in β-barrels in the triose phosphate isomerase (Castersana et al., 1988) at 1638 cm^{-1} (D-form). Susi and Byler (1987) found the parallel β-components in flavodoxin and triose phosphate isomerase in D_2O between 1626 and 1639 cm^{-1} and between 1673 and 1675 cm^{-1} for a high frequency weak shoulder. Many other structures such as β-ribbon (Kennedy et al., 1991), β-edges (Arrondo et al., 1989b, Castersana et al., 1988) have been reported too little for any consensus on these structures. Some structures appear only after denaturation (e.g., thermal denaturation) of the protein and yield frequencies in the 1614–1620 cm^{-1} region of amide I as reported for cholera toxin (Surewicz et al., 1990), cytochrome c (Muga et al., 1991b), acetylcholine receptor (Naumann et al., 1993) or acetylcholinesterase (Görne-Tschelnokow et al., 1993). Such low frequencies are associated with aggregated structures, with a hydrogen bonding pattern similar to that found in intermolecular β-strands (Muga et al., 1991a; Purcell and Susi, 1984; Toniolo et al., 1987). The low frequencies also are often found in amyloid peptides (Gasset et al., 1992, 1993). In agreement with theoretical works (see Chapter 8) a good correlation seems to exist between the frequency of the β-sheet component and the length of the sheet. A calibration curve reporting the frequency (from 1645 to 1628 cm^{-1}) as a function of the length of the sheet (from 6 to 12 amino acid residues) has been reported by Kleffel et al., (1985). On the high frequency side of the β-sheet domain, bands at 1647–1649 cm^{-1} have been assigned to strands bound to D_2O rather than to NH groups with the support of CD and NMR data (Wilder et al., 1991). Among the α-helix components, some unusual low frequency components in amide I (especially after deuteration) have been gathered at the end of the list in Table II. Explanations put forward for this anomalous behavior vary from author to author (shorter helices, longer helices,

---

these input parameters and a baseline to produce fitted bands which appear under spectrum A. These adjusted parameters were then reported as input parameters to fit spectrum B and spectrum C after modification of their width as described by Goormaghtigh et al., (1990). The result of the fitting appears under curve B and C. The dotted lines limit the region assigned to the different secondary structures: β = β sheet, α = α helix, t = β turns and ρ = random. The sum of the components is represented by the dotted spectrum. The secondary structure computed at each stage of the deconvolution appears on the left side of the figure.

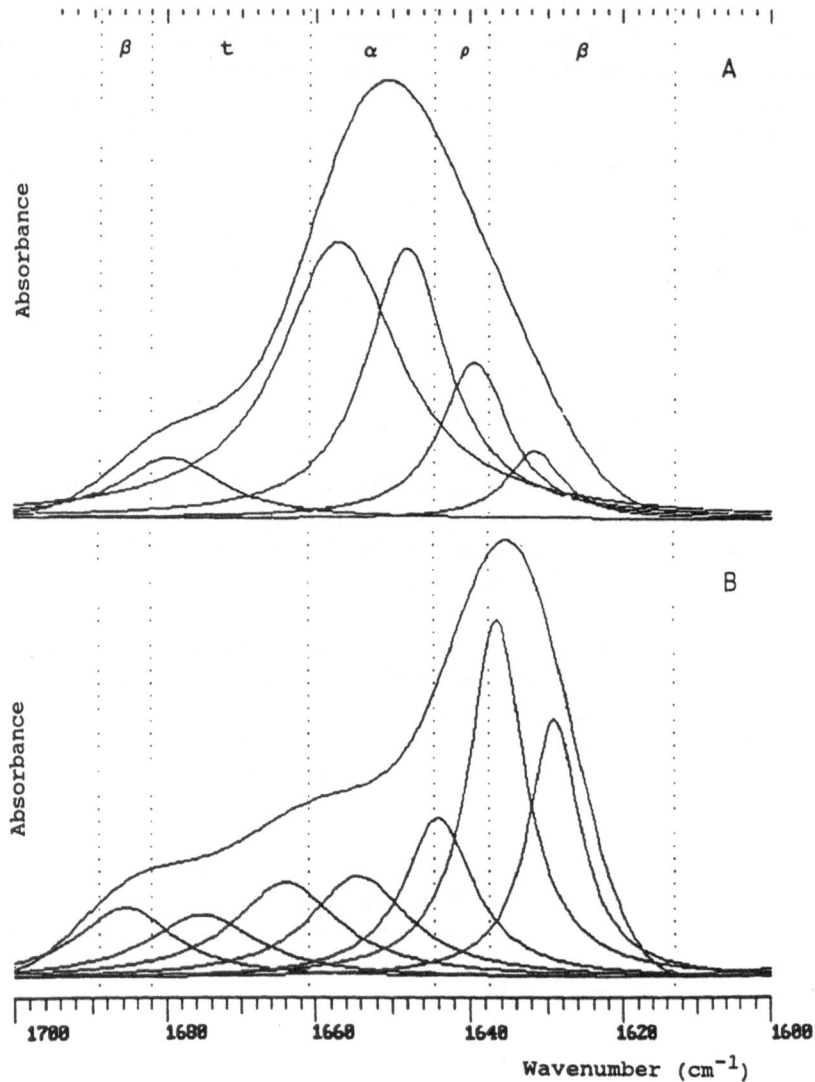

FIGURE 2. Results of the deconvolution-curve fitting procedure described in figure 1 for an α-helix rich protein myoglobin (A) and for a β-sheet rich protein α-chymotrypsin (B). The vertical dotted lines have the same meaning as in figure 1.

Table I

Assignment of the Different Components of Amide I Obtained after Fourier Self-Deconvolution and Curve Fitting of Amide I[a]

	Byler and Susi, 1986			Gormaghtigh et al., 1990			
	Mean (cm^{-1})	RMS (cm^{-1})	Max (cm^{-1})	Mean (cm^{-1})	RMS (cm^{-1})	Max (cm^{-1})	Limits of the structure (cm^{-1})
Turns	1694	1.7	2	—	—	—	
	1688	1.1	2	—	—	—	
	1683	1.5	2	1678	2.1	5	1682–1662
	1670	1.4	2	1670	2.9	5	
	1663	2.2	4	1664	1.0	3	
α-helix	1654	1.5	3	1656	1.5	3	
				1648	1.6	3	1662–1645
Unordered	1645	1.6	4	1641	2.0	3	1645–1637
β- or	1624	2.4	4	1624	2.5	5	
extended	1631	2.5	3	1633	2.1	4	1637–1613
structures	1637	1.4	3	—	—	—	
	1675	2.6	4	1685	2.1	4	1689–1682

[a]Proteins were deuterated in solution (Byler and Susi, 1986) or deposited on a hydrated film on a Germanium ATR plate (Goormaghtigh et al., 1990). The mean frequency of each component is reported with the root mean square (RMS) deviation and the maximum deviation. The limits of each structure reported by Goormaghtigh et al., (1990) are the frequency limits used by the program to assign the components to a secondary structure.

bifurcated hydrogen bonds,) but could include a hydration factor responsible for the low frequency down-shift observed in the presence of D_2O. The common feature of these helices is that, in the present state of our knowledge, the anomalous frequency of amide I can not be predicted. On the other hand, rhodopsin and bacteriorhodopsin helices were separated from the rest of the α-helix containing proteins on the basis of their anomalously high amide I frequency. This high frequency of bacteriorhodopsin has been attributed to helix–helix coupling (Hunt et al., 1988; Hunt 1988) or alternatively to an $α_{II}$ conformation (Krimm and Dwivedi, 1982; Dwivedi and Krimm, 1984) or to a 3_{10} helix (Chapman et al., 1988; Haris and Chapman, 1988). Theoretical works such as this reported by Torii and Tasumi (1992) should help refine the assignments in the future.

Besides structural and hydrogen bonding effects, the nature of the "amino acids" may occasionally affect the frequency of the amide I. The imino group of proline, for instance, is known to absorb near 1630 cm^{-1} (extinction coefficient

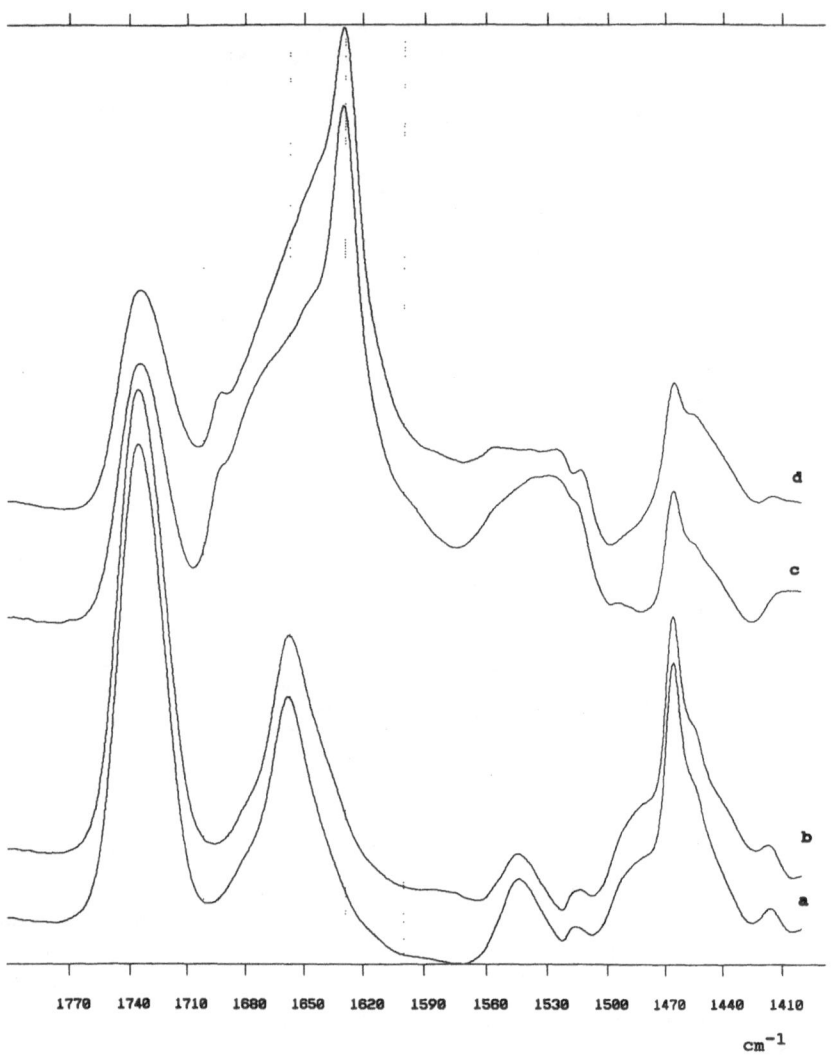

FIGURE 3. Comparison of IR spectra of deuterated and undeuterated proteins in the amide I-amide II region. Spectra are recorded by ATR of a thin membrane-protein film. Deuteration was obtained by flushing the chamber containing the film with D_2O-saturated N_2 for 4 hours. a. bacteriorhodopsin - DMPC H-form; b. bacteriorhodopsin - DMPC deuterated form; c. porin - DMPC H-form; d. porin - DMPC deuterated form. Reconstitution of these two membrane proteins in DMPC vesicles was described elsewhere (Goormaghtigh *et al.*, 1990). The vertical dotted lines are located at 1700, 1657, 1628.5 and 1600 cm^{-1}. The maximum of amide II is located at 1544 cm^{-1} for bacteriorhodopsin (curve a) and at 1529.5 cm^{-1} for porin (curve c).

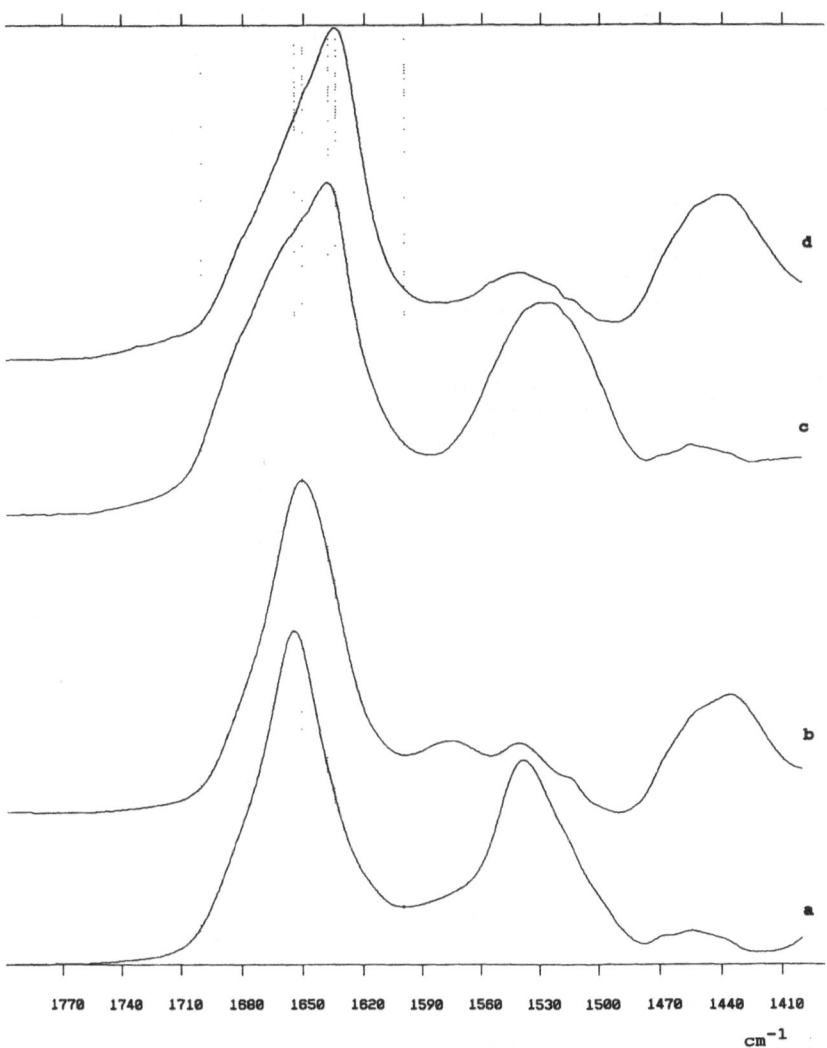

1770 1740 1710 1680 1650 1620 1590 1560 1530 1500 1470 1440 1410

cm^{-1}

FIGURE 4. Comparison of IR spectra of deuterated and undeuterated proteins in the amide I-amide II region. Spectra are recorded by ATR of a thin protein film. Deuteration was obtained by flushing the chamber containing the film with D_2O-saturated N_2 for 4 hours. a. myoglobin H-form; b. myoglobin deuterated form; c. α-chymotrypsinogen H-form; d. α-chymotrypsinogen deuterated form. The vertical dotted lines are located at 1700, 1654.5, 1650.5, 1638, 1634.5 and 1600 cm^{-1}. The maximum of amide II is located at 1538.5 cm^{-1} for myoglobin (curve a) and at 1526.7 cm^{-1} for α-chymotrypsinogen (curve c).

Table II
Sampling of Assignments for the Different Secondary Structures Reported for the Different Amide Bands[a]

α-Helix

Amide I

Frequency (cm^{-1})			
H-form	D-form	Protein	Reference
1655	1650	Cytochrome b$_5$	Holloway and Mantsch, 1989
			Holloway and Buchheit, 1990
1657	1655	Cytochrome b$_5$	Holloway and Mantsch, 1989
1656	—	Adenylate kinase	Arrondo *et al.*, 1989a
—	1653	Methionine apoprepressor	Yang *et al.*, 1989
1654	1654	Hydrophobic myelin protein (lipophilin)	Surewicz *et al.*, 1987a
—	1653	Myelin basic protein	Surewicz *et al.*, 1987b
—	1653	Enkephalin	Surewicz and Mantsch, 1988b
—	1652	Cholera toxin fragment B	Surewicz and Mantsch, 1990
—	1652	Cholera toxin fragment A	Surewicz and Mantsch, 1990
—	1655	Mitochondrial cytochrome oxidase	Bazzi and Woody, 1985
1651	—	Mitochondrial complex III	Valpuesta *et al.*, 1986
1652	1642	Myelin membrane	Ayala *et al.*, 1987
1656	1654	Mitochondrial uncoupling protein	Rial *et al.*, 1990
1653	—	Melittin	Lavialle *et al.*, 1982
1655–1651	—	Phosvitin	Renugopalakrishnan *et al.*, 1985
1655	1651	Ribonuclease	Ollinger *et al.*, 1986
1658	1653	Glucose transporter	Alvarez *et al.*, 1987a
1656	1656	H$^+$/K$^+$ ATPase	Mitchell *et al.*, 1988
	1656	Na$^+$K$^+$ − ATPase	Brazhnikov *et al.*, 1978
—	1655	Ca^{++} − ATPase	Buchet *et al.*, 1989
—	1657	Ca^{++} ATPase	Arrondo *et al.*, 1987
—	1650		
—	1657	Ca^{++} − ATPase	Teruel *et al.*, 1990
1656	—	Ca^{++} − ATPase	Lee *et al.*, 1985
1656	1653	β-galactosidase	Arrondo *et al.*, 1989b
—	1653	Triose phosphate isomerase	Castersana *et al.*, 1988
—	1654	Calmodulin, troponin C	Trewhella *et al.*, 1989
—	1660	Glucose transporter	Chin *et al.*, 1986
1657	1657	Glucose transporter	Cairns *et al.*, 1987
1656	1654	Colicin El channel peptide	Rath *et al.*, 1991
—	1655	Trypsin	Petrelski *et al.*, 1991a
—	1651	α-lactalbumin	Petrelski *et al.*, 1990b

(continued)

Table II (*Continued*)

Amide I

Frequency (cm^{-1})			
H-form	D-form	Protein	Reference
	1659		
—	1650	Lysozyme	Petrelski *et al.*, 1990b
	1657		
—	1650	Cytochrome oxidase signal sequence	Goormaghtigh *et al.*, 1989a
—	1648	Cytochrome c (SDS)	Muga *et al.*, 1991b
—	1649	Cytochrome c(lipid bound)	Muga *et al.*, 1991a
	1654	Cytochrome c(unbound)	
—	1655	Colicin E$_1$ channel peptide	Suga *et al.*, 1991
1657	1654	Photosystem II	He *et al.*, 1991
1657	1647	Phospholipase A$_2$	Kennedy *et al.*, 1990
1657	1650	Azurin	Surewicz *et al.*, 1987d
1648	—	Clathrin	Vincent *et al.*, 1984
1653	1649	Calmodulin $\pm$ Ca^{2+}	Rainteau *et al.*, 1989
1657	1657	Photosystem II	He *et al.*, 1991

"Unusual" frequencies for α-helices after deuteration

1648	—	Adenylate kinase	Arrondo *et al.*, 1989a
—	1644	Calmodulin	Trewhella *et al.*, 1989
	1644	Troponin c	
1654	1647	Paravalbumin	Jackson *et al.*, 1991
1652	1642	Aib containing peptide	Kennedy *et al.*, 1991
1654	1645	Melittin in POPC-POPG	Frey and Tamm, 1991
1654	1644	Lys2-Gly-Leu24-Lys2-Ala in DPPC	Zhang *et al.*, 1992a
1655	1648	Acetylcholinesterase	Görne-Tschelnokow *et al.*, 1993

H-form average frequency: 1654 $\pm$ SD = 2.6 cm^{-1} (extremes: 1657–1648 cm^{-1})
D-form average frequency: 1652.0 $\pm$ SD = 4.5 cm^{-1} (extremes: 1660–1642 cm^{-1})

Amide II

1546		Clatherin	Vincent *et al.*, 1984
1547		Mitochondrial cytochrome c oxidase	Bazzi and Woody, 1985
1545		Melittin	Lavialle *et al.*, 1982
1545		Ca^{++} ATPase	Lee *et al.*, 1985
1548		Glucose transporter	Alvarez *et al.*, 1987a
1547		Glucose transporter	Cairns *et al.*, 1987
1545		Colicin E$_1$ channel peptide	Rath *et al.*, 1991
1548		Photosystem II	He *et al.*, 1991

(*continued*)

Table II   (*Continued*)

Amide II

Frequency (cm⁻¹)			
H-form	D-form	Protein	Reference
1545		Calmodulin $\pm$ Ca^{2+}	Rainteau *et al.*, 1989
1550		Melittin in POPC-POPG	Frey and Tamm, 1991

H-form average frequency: $1546 \pm SD = 1.7$ cm⁻¹ (extremes: 1545–1550 cm⁻¹)

Amide III

1288		Phosvitin	Renugopalakrishnan *et al.*, 1985
1300		Albumine	Wasacz *et al.*, 1987
1288		γ-globuline	Wasacz *et al.*, 1987

Amide A

3290	—	Na$^+$, K$^+$ −ATPase	Brazhnikov *et al.*, 1986
3305	—	Glucose transporter	Chin *et al.*, 1986
3297	—	Colicin E$_1$ channel peptide	Rath *et al.*, 1991

Rhodopsin and Bacteriorhodopsin Helices

Amide I

Frequency (cm⁻¹)			
H-form	D-form	Protein	Reference
1665	—	Rhodopsin	Rothschild *et al.*, 1980
1657	1656	Rhodopsin	Haris *et al.*, 1989
1659	—	Bacteriorhodopsin	Rothschild and Clarck, 1979a,b
			Earnest *et al.*, 1990
1662		Alamethicin	Haris and Chapman, 1988

Amide II

1545	—	Rhodopsin	Rothschild *et al.*, 1980a
			Rothschild *et al.*, 1980b
1548		Rhodopsin	Haris *et al.*, 1989
1545	—	Bacteriorhodopsin	Rothschild and Clarck, 1979a,b
			Earnest *et al.*, 1990
1545		Alamethicin	Haris and Chapman, 1988

Amide A

3295	—	Bacteriorhodopsin	Earnest *et al.*, 1990
3310	—	Rhodopsin	Rothschild *et al.*, 1980a,b

(*continued*)

Table II (*Continued*)

3_{10} **Helix**

Amide I

Frequency (cm^{-1})

H-form	D-form	Protein	Reference
1643	1639	Cytochrome b$_5$	Holloway and Mantsch, 1989
1662	—	Alamethicin	Haris and Chapman, 1988
1640(sh)			
1665	—	Hemoglobin ($-65°$)	Casal et al., 1988
1664	—	Triose phosphate isomerase	Castersana et al., 1988
1650—1659	1639	α-lactalbumin	Petrelski et al., 1991b
1650—1659	1642	Lysozyme	Petrelski et al., 1991b
1660		Cytochrome c	Dong et al., 1992
1666—1662	1661—1659	Difference Aib-containing peptides	Kennedy et al., 1991
	1660	Insulin, despentapeptide Insulin, desoctapeptide Insulin	Wei et al., 1991

Amide A

3300		(Leu-Leu-D-Phe-Pro)	Kamegai et al., 1986

Disordered Structures

Amide I

Frequency (cm^{-1})

H-form	D-form	Protein	Reference
1657	1645	Cytochrome b$_5$	Holloway and Mantsch, 1989
1656	—	Adenylate kinase	Arrondo et al., 1989
—	1639	Methionine aporepressor	Yanng et al., 1987
—	1646	Myelin basic protein	Surewicz et al., 1987a
—	1646	Natriuretic peptide	Surewicz et al., 1987c
1656	1642	Mitochondrial uncoupling protein	Rial et al., 1990
—	1648	α$_1$-antitrypsin	Haris et al., 1990
1646	1643	Melittin	Lavialle et al., 1982
1658	1645	Glucose transporter	Alvarez et al., 1987a
—	1647	Na$^+$ K$^+$ ATPase	Brazhnikov et al., 1978
—	1643	Ca^{++} ATPase	Arrondo et al., 1987
—	1645	Ca^{++} ATPase	Teruel et al., 1990

(*continued*)

Table II   (*Continued*)

Amide I

Frequency (cm⁻¹)			

H-form	D-form	Protein	Reference
1642		Ca⁺⁺ ATPase	Lee *et al.*, 1985
—	1654	Ca⁺⁺ ATPase	Buchet *et al.*, 1990
1656	1639	β galactosidase	Arrondo *et al.*, 1990b
—	1645	Trypsin	Petrelski *et al.*, 1991a
1657	1647	Phospholipase A₂	Kennedy *et al.*, 1990
1657	1650	Azurin	Surewicz *et al.*, 1987d
1654	—	Clathrin	Vincent *et al.*, 1984
—	1641	Insulin, despentapeptide	Wei *et al.*, 1991
		Insulin, desoctapeptide	
		Insulin	

H-form average frequency: $1654.1 \pm SD = 4.8$ cm⁻¹ (extremes: 1657–1642 cm⁻¹)
D-form average frequency: $1644.8 \pm SD = 3.6$ cm⁻¹ (extremes: 1654–1639 cm⁻¹)

Amide II

1545	1458	Melittin	Lavialle *et al.*, 1982
1515	—	Ca⁺⁺ ATPase	Lee *et al.*, 1985

Amide III

1260		γ-globuline	Wasacz *et al.*, 1987

Amide A

3255—3260		glucose transporter	Chin *et al.*, 1986

β-Sheet Structure

Amide I

Frequency (cm⁻¹)			

H-form		D-form		Protein	Reference
—	1630	—	1625	Cytochrome b₅	Holloway and Mantsch, 1989
—	1641	—	1636	Cytochrome b₅	Holloway and Mantsch, 1989
—	1638	—	—	Adenylate kinase	Arrondo *et al.*, 1989a

(*continued*)

Table II (*Continued*)

Amide I

Frequency (cm^{-1})					
H-form		D-form		Protein	Reference
—	1627	—	—	Adenylate kinase	Arrondo *et al.*, 1989a
—	—	1675	1625	Methionine apoprepressor	Yang *et al.*, 1989
—	1634	—	1635	Hydrophobic myelin protein (lipophilin)	Surewicz *et al.*, 1987a
	1622		1621		
—	—	1675	1635	Myelin basic protein	Surewicz *et al.*,1987b
			1625		
—	—	1680	1637	Natriuretic peptide	Surewicz *et al.*, 1987c
			1626		
			1618		
			1615		
—	—	—	1626	Enkephalin	Surewicz and Mantsch, 1988b
—	—	1694	1636	Concanavalin A	Arrondo *et al.*, 1988b
1695	1634	—	—	Concanavalin A	Alvarez *et al.*, 1987b
—	—	—	1632	Cholera toxin fragment B	Surewicz and Mantsch, 1990
—	—	—	1634	Cholera toxin fragment A	Surewicz and Mantsch, 1990
—	—	1675	1632	Na$^+$, K$^+$−ATPase	Brazhnikov *et al.*, 1978
—	1638	—	—	α-chymotrypsin	Swedberg *et al.*, 1990
1686	1632	—	—	Phosvitin	Ranugopalakrishnan *et al.*,1989
—	1631	—	—	Ca^{++}-ATpase	Lee *et al.*, 1985
—	—	1677	1630	Ca^{++}-ATPase	Arrondo *et al.*, 1987
—	—	—	1634	Ca^{++}-ATPase	Teruel *et al.*, 1990
			1623		
—	—	1678	1638	Ca^{++}-ATPase	Buchet *et al.*, 1989
1679	1634	1679	1634	Ca^{++}-ATPase	Villalain *et al.*, 1989
—	1623	—	1624		
1689	1639	1685	1631	Ribonuclease	Haris *et al.*, 1989
1687	1641	1676	1637	Ribonuclease	Ollinger *et al.*, 1986
—	1639	—	1634	H$^+$/K$^+$ ATPase	Mitchell *et al.*, 1988
	1635		1630	Myelin	Ayala *et al.*, 1987
1687	—	1672	1635	Mitochondrial uncoupling protein	Rial *et al.*, 1990
		1694	1634	α_1-antitrypsin	Haris *et al.*, 1990
—	1627	—	—	β-galactosidase	Arrondo *et al.*, 1989b
	1338		1634		
—	—	1673	1635	Calmodulin	Trewhella *et al.*, 1989
			1628		
—	1636	—	1636	Scrapie protein	Caughey *et al.*, 1991

(*continued*)

Table II   (*Continued*)

Amide I

\multicolumn{2}{l}{Frequency (cm^{-1})}			
H-form	D-form	Protein	Reference

H-form		D-form		Protein	Reference
	1627		1627		
1693	1638	1678	—	Photosystem II	He *et al.*, 1991
	1628				
1674	1636	1674	1637	Azurin	Surewicz *et al.*, 1987d
—	1630	—	—	Clathrin	Vincent *et al.*, 1984
—	—	—	1620	Apo B-100	Gotto *et al.*, 1968
—	—	—	1622	Apo B-100	Goormaghtigh *et al.*, 1989b
			1614		
—	—	—	1620	Apo B-100	Herzyk *et al.*, 1987
1675	1630	—	—	Porin	Kleffel *et al.*, 1985
1679	1631	—	—	Porin	Nabedryk *et al.*, 1988
		—	1618	Alzheimer's peptide	Fraser *et al.*, 1991

H-form average frequency: 1633.0 ± SD = 5.2 cm^{-1} (extremes: 1641–1623 cm^{-1})
1684.4 ± SD = 6.9 cm^{-1} (extremes: 1695–1674 cm^{-1})
D-form average frequency: 1630.2 ± SD = 6.2 cm^{-1} (extremes: 1638–1615 cm^{-1})
1679.0 ± SD = 6.6 cm^{-1} (extremes: 1694–1672 cm^{-1})

Amide II

1524	—	Phosvitin	Renugopalakrishnan *et al.*, 1985
1530	—	Glucose transporter	Alvarez *et al.*, 1987a
1526	—	α-chymotrypsin	Swedberg *et al.*, 1990
1532	—	Concanavalin A	Alvarez *et al.*, 1987b
1525	—	Porin	Kleffel *et al.*, 1985
1530	—	Porin	Nabedryk *et al.*, 1988

H-form average frequency: 1527.8 ± SD = 3.0 cm^{-1} (extremes: 1525–1532 cm^{-1})

Amide III

1237	—	Concanavalin A	Alvarez *et al.*, 1987b
1261	—	Phosvitin	Renugopalakrishnan *et al.*, 1985
1236	—	α-chymotrypsin	Swedberg *et al.*, 1990

Amide A

3820	—	Calmodulin binding peptide	Vorherr *et al.*, 1990

(*continued*)

Table II *(Continued)*

β-Sheets Stabilized by Strong, Possibly Bifurcated Hydrogen Bonds or Additional Bonding
to Other Sheets or Amino Acids

Amide I

Frequency (cm⁻¹)			
H-form	D-form	Protein	Reference
—	1625	Concanavalin A	Arrondo *et al.*, 1988b
1624	1624	Mitochondrial uncoupling protein	Rial *et al.*, 1990

Free Solvated (C═O) Groups

Amide I

Frequency (cm⁻¹)			
H-form	D-form	Protein	Reference
1684—1671		Robinia lectin	Wantyghem *et al.*, 1990
1682		Aib containing peptide	Kennedy *et al.*, 1991

Turns

Amide I

Frequency (cm⁻¹)			
H-form	D-form	Protein	Reference
1678	1653	Cytochrome b₅	Holloway and Mantsch, 1989
1667	1662		
1681	1674	Adenylate kinase	Arrondo *et al.*, 1989a
1671	1661		
—	1665	Methionine aporepressor	Yang *et al.*, 1987
—	1663	Myelin basic protein	Surewicz *et al.*, 1987b
—	1666	Natriuretic peptide	Surewicz *et al.*, 1987c
	1663		
—	1680	Enkephalin	Surewicz and Mantsch, 1988b
—	1665	Myelin	Ayata *et al.*, 1987
1662	—	Phosvitin	Renugopalakrishnan *et al.*, 1985
1667	1664	Ribonuclease	Haris *et al.*, 1986
1667	1662	Ribonuclease	Ollinger *et al.*, 1986

(continued)

Table II   (*Continued*)

Amide I

Frequency (cm⁻¹)			
H-form	D-form	Protein	Reference
—	1692	Ca⁺⁺-ATPase	Arrondo *et al.*, 1987
	1683		
	1677		
	1688		
—	1691	Ca⁺⁺-ATPase	Teruel *et al.*, 1990
	1667		
1686	1673	β-galactosidase	Arrondo *et al.*, 1989b
1678	1685		
1667	1663		
	1675	Triose phosphate	Castersana *et al.*, 1988
	1670	Isomerase	
	1663	Calmoduline, troponine c	Trewehellaa *et al.*, 1989
1672	1685	Trypsin	Petrelski *et al.*, 1991a
	1686	α-lactalbumine	Petrelski *et al.*, 1991b
	1668		
	1684	Lysozyme	Petrelski *et al.*, 1991b
	1666		
	1676	Human properdin	Perkins *et al.*, 1989
	1663		
	1637	HIV gp120 peptide in TFE	Lacsko *et al.*, 1992

H-form average frequency: $1672.4 \pm SD = 7.0$ cm⁻¹ (extremes: 1686–1662 cm⁻¹)
D-form average frequency: $1671.4 \pm SD = 10.0$ cm⁻¹ (extremes: 1691–1653 cm⁻¹)

Amide II

1563	—	Concanavaline A	Alvarez *et al.*, 1987b
	1532	HIV gp120 peptide in TFE	Lacsko *et al.*, 1992

Amide III

1318	—	Concanavaline A	Alvarez *et al.*, 1987b
1343			

Amide A

3330	—	Leu-Pro-Try-Ala	Tinker *et al.*, 1988

(*continued*)

Table II (*Continued*)

Proline Rich Peptides, Left Helices and Collagen Triple Helices

Amide I

Frequency (cm^{-1})			
H-form	D-form	Protein	Reference
1660		Type I collagen	George and Veis, 1991
1645		-1660:Pro internal to triple helix (increases in fibrils)	
1631		-1631: Pro facing aqueous phase	
1623		Polyproline solution	Lazarev *et al.*, 1985
1640		Polyproline aggregated or crystal	Lazarev *et al.*, 1985
1650			
	1626	(Gly Pro Pro)$_8$ hydrated film	Lazarev *et al.*, 1985
	1643		
	1640	(Gly Pro Pro)$_8$ dehydrated	
	1665	film	
1655		Native collagen (monomers)	Jakobson *et al.*, 1983
1650		Fibrils	
1640			

Amide II

H-form	D-form	Protein	Reference
1560		Type I collagen native fibrils	George and Veis, 1991
1558			
1559		Native collagen	Lazarev *et al.*, 1985
1555		Fibrils	

Amide A

H-form	D-form	Protein	Reference
3330		Collagen	Doyle *et al.*, 1975
3300–3335		Various Pro, Ala, Gly poly-peptides (frequency is function of the rise/residue (2.85–3.10 Å)	Doyle *et al.*, 1975

[a]Most of the data were determined with the help of Fourier self-deconvolution or second derivatives, two equivalent techniques used to narrow the components associated with the different secondary structures. Data already reported in Table I are not considered here. It must be emphasized that assignments are often speculative and data are presented here only to illustrate the present consensus on the assignments of IR bands of proteins. The H/D exchanged frequencies reported here were observed after a deuteration procedure, which varies from author to author. Should the deuteration procedure be identical, the extent of deuteration varies anyway from protein to protein. Complete deuteration is rarely achieved. Even if the deuteration extent is small, frequencies are reported here under the "D-form" header. In some instances, no exchange has taken place for a given secondary structure justifying in the view of the authors the lack of frequency shift (e.g. the α-helix of the H$^+$/K$^+$−ATPase (Mitchell *et al.*, 1988)).

~3 10^4 liter/mol^{-1} cm^{-1}) in an unordered conformation instead of ~1645 cm^{-1} for the amides (Lazarev *et al.*, 1985). Polypeptides such as (Gly-Pro-Ala)$_n$ or (Gly-Pro-Pro)$_n$ in the unordered conformation therefore display two peaks. The low frequency one is assigned to prolines and its intensity is proportional to the amount of proline in the polypeptide (Doyle *et al.*, 1975; Lazarev *et al.*, 1985).

2.2. Effect of H/D Exchange Extent on Structure Determination

The secondary structure evaluation supposes the random structure to have completed the H/D exchange. Under such conditions its contribution to amide I shifts from ~1655 cm^{-1} where it is superimposed on the α-helix component to ~1642 cm^{-1}. Because only the random structure must be completely exchanged prior to secondary structure evaluation, we expect the necessary deuteration time to lie in the range of a few minutes. Our method of secondary structure evaluation was successfully tested without taking into account the degree of deuteration reached (Goormaghtigh *et al.*, 1990). However, it is worth establishing the dependency of the secondary structure evaluation on the deuteration time. Apolipoprotein B-100 in native human low density lipoproteins was tested with respect to this question (Goormaghtigh *et al.*, 1993). In order to closely monitor the effects of the exchange, spectra were recorded at logarithmically spaced intervals starting with 15 second intervals at the beginning of the deuteration. The areas of amide I and amide II are reported in Figure 5, panels 1 and 3, as a function of the deuteration time. Equilibrium is definitely not reached after the three hours of deuteration (more than 25% of the amide groups had not deuterated). The decrease of the surface of amide II can be described as the sum of the individual exponential decays for each peptide group. Some reach a 50% decay point in seconds, others in weeks, depending on the secondary structure in which the amide group is involved, accessibility to solvent (tertiary structure folding), and pH. An "equilibrium" is therefore not expected after only a few hours. This is particularly evident in panel 2 of Figure 5, where the same data are reported on a logarithm time scale. In order to evaluate the importance of the deuteration extent on the secondary structure evaluation, a secondary structure evaluation was carried out by the automatic procedure described in details in (Goormaghtigh *et al.*, 1990) for each individual spectrum. The results are reported in panel 4 of Figure 5 as a function of the deuteration time. Clearly, the secondary structure evaluation remains unchanged (within 1%) after only a few minutes, and the method developed for the determination of the secondary structure yields reproducible results when applied to different data.

2.3. Critical Evaluation of the Deconvolution Results

The mathematical technique of spectrum self-deconvolution—useful for biological samples—has been developed essentially by Kauppinen *et al.*,

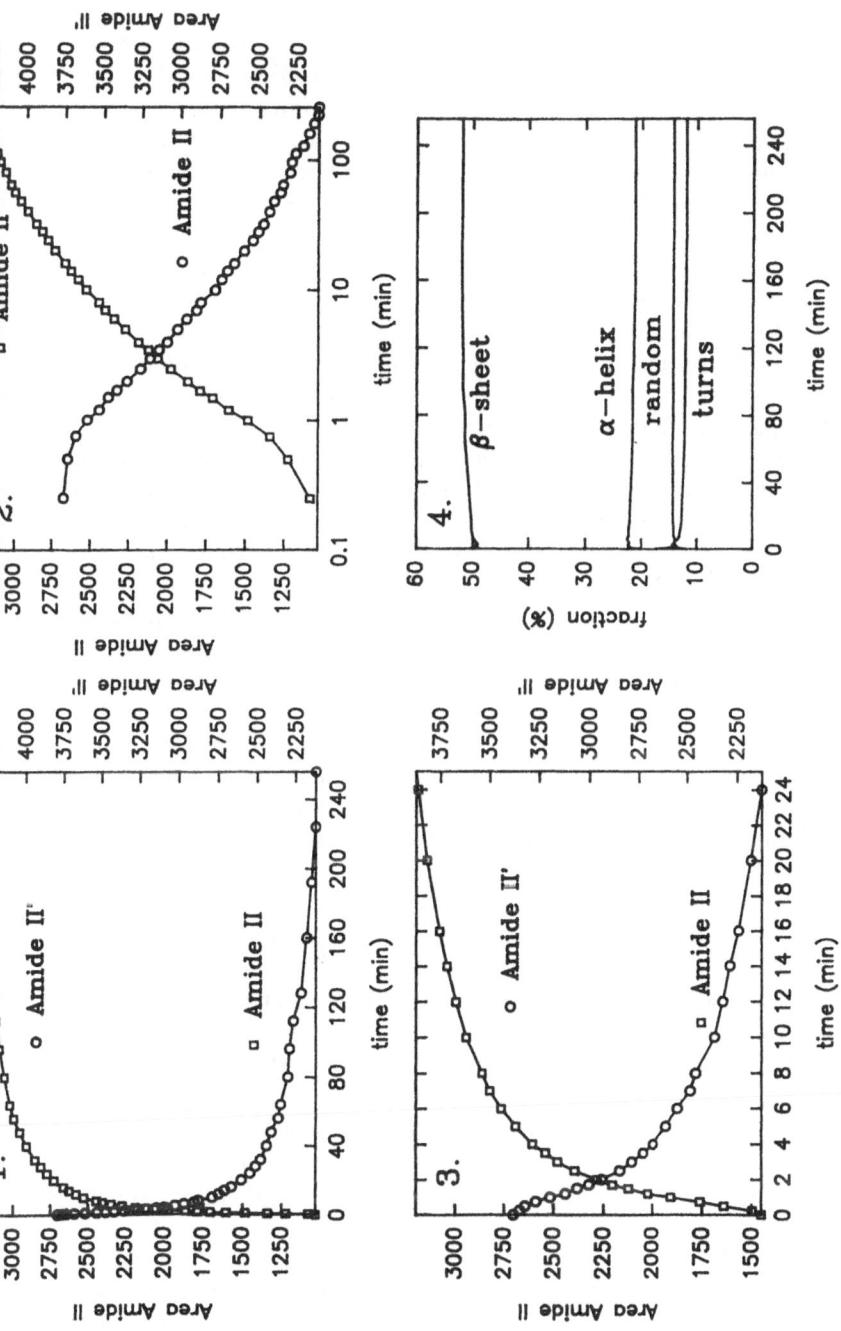

FIGURE 5. Deuteration rate of apo B-100 and its effect on the secondary structure determination. The deuteration process starts at time = 0. From time = 0 min to time = 256 min, spectra were recorded at logarithm spaced time intervals starting by 0.25 min intervals. The area of amide II (mOD × cm⁻¹) integrated between 1593 and 1507 ± 5 cm⁻¹ and amide II' integrated between 1493 and 1417 ± 5 cm⁻¹ is reported in panel 1 between 0 and 256 min and in panel 3 between 0 and 25 min. Curves of panel 1 have been replotted on panel 2 on a logarithm time scale. Evaluation of the secondary structure

(1981a,c, 1982) and by Yang and Griffiths (Yang and Griffiths, 1983, Yang and Griffiths, 1984, Griffiths 1983). Fourier transform is now easily and quickly computed thanks to the development of fast Fourier transform algorithms and is used in all domains of science (Bracewell 1990) and in infrared spectroscopy in particular (Bates, 1976; Griffith, 1978; Ferraro and Basile, 1987). The deconvolutions mentioned here are all realized by means of Fourier transform methods rather than by direct methods (Blass and Halsey, 1981, Nikolov and Kantchev, 1987). The goal of self-deconvolution is to enhance the spectral "resolution" so that the different components of a complex band can be resolved and eventually submitted to a curve-fitting procedure for quantitative analysis. The principle of self-deconvolution consists of considering that, in the liquid or solid state, the vibrational absorption bands are intrinsically broadened (or convoluted) by a function. The self-deconvolution uses this function to "unbroaden" (or deconvolute) the spectrum. The procedure encounters a major theoretical problem, which is the knowledge of the shape of the convoluting function. This function is considered to be essentially Lorentzian (Oxtoby 1981) but its exact shape is still to be defined. However, a pure Lorentzian shape has been assumed by many investigators for self-deconvolution of proteins. The FWHH (full width at half height) is chosen by trial and error. If the FWHH is too small, little narrowing is obtained. If it is too large, negative side-lobes appear. In practice, an apodization function prevents the high frequency components of the spectrum to become predominant. One defines a "resolution enhancement factor" K, which is a function of the ratio of the width of the deconvoluting function on the width of the apodization function. We have already mentioned that the physical meaning for the "resolution enhancement factor" is lost as soon as the knowledge of the exact shape of the deconvolution function is missing, which is virtually always the case (Goormaghtigh and Ruysschaert, 1990). The effect of the deconvolution parameters on the shape of the deconvoluted spectrum is illustrated in Figure 6, where the spectrum of papain has been analyzed as described in Figure 1, except for a deconvolution performed with a FWHH of the deconvoluting Lorentzian of 20 and 40 cm^{-1} (instead of 30 cm^{-1} as in Figure 1). The reasons for the differences obtained for the different deconvolution parameters appear most clearly when the deconvolution–apodization function shape in the transformed domain, $DA(x)$, is compared for the various FWHH; used (Figure 7). This shape is defined by (Goormaghtigh *et al.* 1990) $DA(x) = \sqrt{\ln 2} \exp (2\pi\sigma_1 |x| - \pi^2 \sigma_g^2 x^2 / \ln 2)$ where $\sigma_1 = \frac{1}{2}$ FWHH of the deconvoluting Lorentzian and $\sigma_g = \frac{1}{2}$ FWHH of the apodization Gaussian. (x is expressed in cm.) The maximum of the function occurs at (Goormaghtigh and Ruysschaert 1990) $x_{max} = \ln 2 \, \sigma_1 / \pi \sigma_g^2$.

Figure 7 shows that the most enhanced frequencies are different for given values of K as the FWHH varies. It appears that different combinations of Lorentzian deconvolution and "resolution enhancement" factors K can enhance

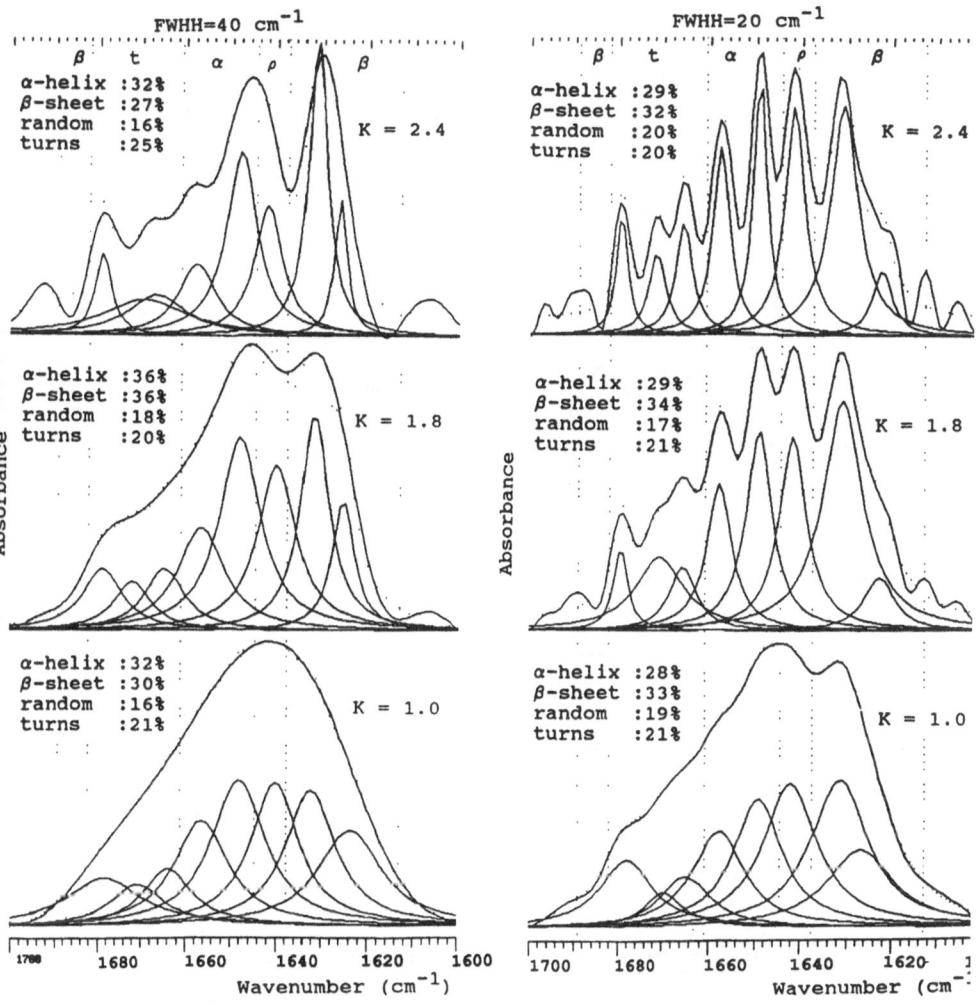

FIGURE 6. Secondary structure determination of papain performed on the sample described in figure 1 according to the same procedure except that the FWHH of the deconvoluting Lorentzian was 40 cm⁻¹ (left panel) and 20 cm⁻¹ (right panel) instead of 30 cm⁻¹ used in figure 1.

the same frequencies, but only the width of the enhanced frequency range is modified. Figure 7 shows why the spectrum deconvoluted with FWHH = 20 cm⁻¹ (K = 1.0) resembles the spectrum deconvoluted with FWHH = 40 cm⁻¹ (K = 1.8) in Figure 6. Since there is until now no satisfactory description of the shape of the function that should be used to specifically remove the broadening of the components of amide I' brought from different origins in the liquid or solid

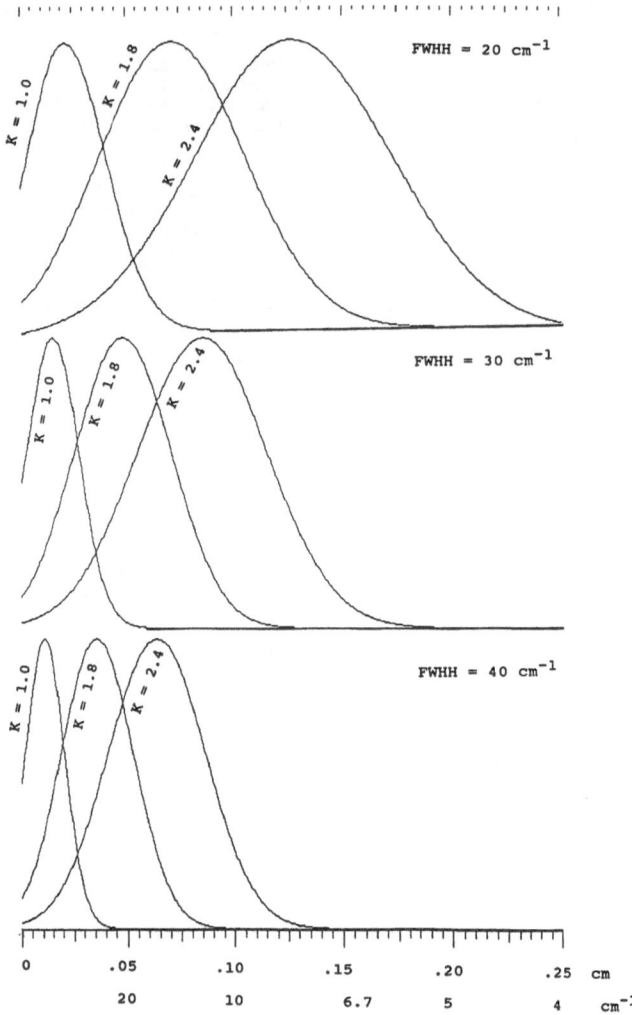

FIGURE 7. Shape of the deconvolution-apodization function in the transformed domain DA(x) (see text) for the three deconvolutions tested on the spectrum of papain and three values of the constant K used. The value of 20 cm^{-1} (top) for the FWHH of the deconvoluting Lorentzian was used in the left panel of figure 6. A value of 40 cm^{-1} (bottom) was used for the deconvolution reported in the right panel of figure 6 and a value of 30 cm^{-1} (middle) was used in the deconvolution reported in figure 1. Each curve has been rescaled to the same amplitude. The maximum amplitude of the curves is identical for a given value of K whatever the value of FWHH (see text). It is 45 for K = 2.4, 7.9 for K = 1.8 and 1.7 for K = 1.0. The first abscissa scale is the transformed domain variable x while the second abscissa scale reports $\nu = 1/x$ in the frequency domain.

state, Fourier self-deconvolution must be considered as a trick that allows a unique curve-fitting to be performed for a given spectrum and for given self-deconvolution parameters. Yet, as demonstrated in Figures 1 and 6, the procedure described here ultimately yields similar secondary structure estimations for the different deconvolutions. These considerations are important for understanding that the exact number and frequency of the components revealed by Fourier self-deconvolution depends on arbitrarily chosen parameters. The result can therefore only be arbitrary.

Fourier deconvolution selectively enhances specific frequencies, which depend on both the width of the deconvolution function and on the value of the factor K. It is important to realize that a noise is no longer random after Fourier self-deconvolution, because some frequencies are specifically enhanced (Surewicz et al., 1993). The noise appears then as the usual IR absorption bands and is easily confused with real signals. Only a preliminary check of the noise level or a careful analysis of the region of the spectrum free of any real signal can prevent confusion of IR signals with noise. Upon scan accumulation, the noise should decrease as the square root of the number of averaged scans. The same reasoning applies to the sharp peaks of atmospheric water that overlap amides I and II.

In view of the above comments, it would be futile to attempt to assign a single secondary structure to a single feature of a Fourier self-deconvoluted spectrum because the resolved features have little meaning for secondary structures.

2.4. Critical Evaluation of the Decomposition of Amide Bands into Primary Components

The above discussion indicates that no physical meaning should be given to features appearing only after Fourier self-deconvolution. Structural reasons also tend to indicate that it is not reasonable to expect a single component for each secondary structure. We shall consider the case of the α-helix for the sake of the discussion, but the same arguments stand for the other structures.

Theoretical work indicates that the frequency of amide I for an α-helix depends on the length of the helix (Nevskaya and Chirgadze, 1976) as discussed in Chapter 8. In proteins, this effect is probably illustrated by the difference between the spectra of the helical proteins (80 and 90% of helical content, respectively) myoglobin and tropomyosin (Dousseau and Pézolet, 1990). While the former is composed of small regions of α-helices separated by turns or coil conformation and has an amide I localized near 1655 cm^{-1} ($\sim$1548 cm^{-1} for amide II), the latter contains long strands of extended helices and its amide I is located near 1646 cm^{-1} ($\sim$1548 cm^{-1} for amide II). Another example is the Mg–ATP binding peptide of adenylate kinase. Two helical regions (24%), 3 stretches of β-strand (38%), and 38% of aperiodic structures are determined by

2-D NMR. Analysis of its infrared spectrum detects less than 10% of α-helix and 60% of turns and aperiodic structures. This disagreement seems to be related to the short length of the helices or/and to dynamic disorder (Fry *et al.*, 1988). A look at table II indicates that the helix component in amide I can occur anywhere between 1664 and 1642 cm^{-1}.

The central and end regions of the α-helix could yield different spectra (Van Wart and Scheraga 1978) and several users of multicomponent analysis by factor analysis have been compelled to account separately for the core region of the helices and for the two first and two terminal amino acid residues of the helix (Dousseau and Pézolet, 1990; Kalnin *et al.*, 1990). Obviously the example mentioned above about the effects of the length of the helices could be discussed here as well, since it is not possible to distinguish the end of helix effects and length effects.

Any distortion of an α-helix would also shift the frequency of the amide I associated band. Such distortions have been described as common in protein crystals (Blundell *et al.*, 1983) and have been shown to affect the IR spectrum of the helix (Chirgadze *et al.* 1976). As already mentioned for synthetic polypeptides (Chapter 8), the width of the helical contribution to amide I varies from 15 cm^{-1} to 40 cm^{-1} according to the stability of the helix measured as the free energy of the helix-coil transition. These effects are likely derived from subtle geometric features of the helices. The presence of no less than six amide III bands between 1320 and 1260 cm^{-1} assigned to α-helices in bovine serum albumin also suggests slight variations in the helical geometry, symmetry, or interactions (Anderle and Mendelsohn, 1987). An example of frequency shifts in the amide I region ($\pm$ 3 cm^{-1}) due to stretching of the helical structure of the petide Lys_2-Gly-Leu_{24}-Lys_2-Ala by varying the thickness of the lipid bilayer in which the peptide is inserted is provided by Zhang *et al.*, (1992b). Barlow and Thornton (1988) pointed out that only 15% of the α-helices investigated in crystals are linear, the rest of the helices being characterized by bends and kinks. Blundel *et al.*, (1983) describe how the solvent induces bending of helices.

Beside the structural effects, the width of the amide I band may change with the degree of hydration of the helix (Jakobson *et al.*, 1986).

Incomplete deuteration is also a source of multiplicity for the helix-associated frequencies. It seems impossible to obtain a complete deuteration for many proteins without denaturation of the original structure.

Three components are theoretically expected in the amide I range for the α-helix structure. Polarization experiments could allow the resolution of two of these components (see Chapter 8).

In conclusion, it appears wiser to consider the individual components revealed by Fourier self-deconvolution as a mixture of many subcomponents with slightly different frequencies and to study the spectrum by domains rather than by individual components. As discussed before (Goormaghtigh *et al.*, 1990), dichroism measurements carried out on oriented membrane proteins are one way to determine whether two helical components result from an arbitrary splitting or have a real origin.

2.5. General Discussion of the Method

The decomposition of the amide I band into components related to a secondary structure by Fourier self-deconvolution followed by a curve fitting involves both technical and methodological problems, some of which have been detailed above. Because this approach is increasingly popular among researchers, it is useful to point out here the main traps and limitations which should be considered before the interpretation of the results.

1. Parameters used for self-deconvolution are not unique and are just a little bit better than a guess. Different laboratories use different parameters, which yield different results (Figures 6, 7). Moreover these parameters are often different from one protein to the other within the same research group. They are chosen in order to obtain a good-looking resolved spectrum, that is, displaying between three and nine resolved features, including shoulders. None of these criteria have, however, any reasonable basis. The study by Lee *et al.*, (1990) (see next point) indicates that deconvolution does not improve the information contained in the amide I band.

2. A number of resolved features can be related to nonamide bands. Take, for example, noise, side chain vibrations, atmospheric water, and liquid water, including new water bands of associated-water. The relative importance of these contributions depends on the sequence and pH (amino side chain vibration) and on unpredictable effects of the protein and salts on the associated-water bands. The problem of the noise and of the atmospheric water bands can be managed satisfactorily (see Chapter 9).

3. Least squares methods in curve fitting have long been used for decomposing amide I and II into its components (Fraser and Suzuki, 1973). Yet, it is recognized that the result of the fitting depends to some extent on the input parameters because of the multiplicity of the solutions. The limitations of the curve-fitting techniques have been discussed elsewhere (Maddams, 1980). The curve-fitting procedure is a further subject of disagreement among various laboratories. First, the fitting lineshape used is sometimes Lorentzian (Goormaghtigh *et al.*, 1990), sometimes Gaussian (Byler and Susi, 1986). While the latter group deconvolutes with a Lorentzian function and justifies the choice of a Gaussian lineshape for the fitting because the Lorentzian character of the bands has been removed by the deconvolution process, the former fits ultimately undeconvoluted spectra with Lorentzian lineshapes. The way the fitting is conducted also differs from lab to lab, some preferring to fit all the parameters in a single step, while others allow the least square procedure to fit only one parameter at a time.

Finally, as already discussed, it is well known that fitting procedures do not yield a unique answer (Goormaghtigh and Ruysschaert, 1990). In our hands, modifying in "reasonable" limits the approximate frequency and width of the bands to be fitted can change the secondary structure by about 10–15%. In order to avoid tendentious manipulation of the approximate parameters to be fitted, we use an automatic procedure that chooses these parameters (Goormaghtigh *et al.*, 1990).

4. The assignment of the fitted bands to a secondary structure is also a matter of debate. It is impossible to assign all the bands unambiguously in all instances because overlapping between the maximum of different structures can occur. Again, it is then difficult for the researcher to realize the assignment without preconception of the structure to be obtained. In our automatic procedure we have clearly defined the frequency ranges assigned to each structure (Goormaghtigh *et al.*, 1990) but this process was not adequate for all the proteins. An investigated example of incorrect interpretation is the diagnosis of five α-helices and four strands of β-sheet in bacteriorhodopsin (in agreement with CD data) while it is now known that seven helices and no β-sheets cross the membrane. The shoulder previously assigned to β-sheet near 1640 cm^{-1} (Jap *et al.*, 1983) is now assigned (in part) to the $\nu(C{=}N)$ of the protonated Schiff base of the retinylidene chromophore (Earnest *et al.*, 1990). It is now suggested that details of the helix geometry could be gained more certainly from the analysis of the infrared spectra, once a high resolution structure is available (Glaeser *et al.*, 1991, Rothschild *et al.*, 1992). It has also been suggested that both loops and turns absorb in the helix region (see pg. 411).

5. It is implicitly assumed that the integrated extinction coefficients are similar for all the structures. This assumption is, at best, a rough approximation. The reader is referred to Chapter 8 to appreciate how risky this assumption is. Variations up to 30% have been reported for polylysine and other proteins on conformational change. Moreover, we have little evidence that the extinction coefficient is the same for the same structure in different proteins.

6. The mere idea of assigning one single peak (or two in some instances) to each described secondary structure is conceptually wrong since it has been shown that many parameters including the length, stability, precise geometry of a structure, temperature, hydration, can be the source of variation in the frequency observed.

In conclusion, these techniques work rather satisfactorily as far as the evaluation of the secondary structure is concerned, but a gap exists between the explanations most often put forward to explain the success of the method and its true basis. In the present state of our knowledge we are entitled to say that a curve fitting of amide I with five to nine bell-shaped functions effectively results in components whose integrated surface is in rather good correlation with the proportion of each secondary structure after assignments according to criteria discussed above and in Table I. However, we are not yet able to accurately link a secondary structure characterized by Φ and Ψ angles and hydrogen bonding to an amide I component of well-defined shape including its frequency, width, and intensity.

3. DETERMINATION OF PROTEIN SECONDARY STRUCTURES BY PATTERN RECOGNITION METHODS

Presented as a quite different approach to obtain the secondary structure of the protein are pattern recognition methods that have long been used for analysis of circular dichroism and Raman spectra. Usually in a first (calibration) step, spectra of proteins of known structures are recorded and data points are arranged to form a matrix. Diagonalization provides eigenvalues and eigenspectra. Combination of the eigenspectra can reconstruct any of the original spectra. Only the most significant eigenspectra are necessary for the reconstruction, while a number of them account for insignificant features (noise).

Up to this point, the analysis proceeds without reference to the protein structure. The problem of choosing the criteria used to analyse the X-ray data can therefore be tackled, from this point on, more efficiently than by the Fourier transform method. It must be stressed that the eigenspectra do not necessarily represent a single secondary structure. In the next step, the contribution of each eigenspectrum to the experimental spectra is compared with the secondary structures assigned to the corresponding protein, and a linear regression is used to establish a correlation between the weight of a eigenspectrum and a secondary structure. The eigenspectra with the best correlation to a structure are assigned to this structure. Finally, when a protein of unknown structure is analyzed, a linear regression determines the weight of each eigenspectrum to apply in order to reconstruct the experimental spectrum of the protein of unknown structure. Similar approaches are currently used in circular dichroism (CD) spectroscopy. The review by Yang and colleagues (1986) describes the listing of the program used by Provencher and Glöckner and by Hennessey and Johnson in CD spectroscopy (see also Hennessey and Johnson, 1981; Crompton and Johnson, 1986). In IR, quantitative determination of components of complex mixtures such as protein, glucose, triglycerides, cholesterol, urea, and uric acid in human blood plasma was obtained from ATR spectra by a partial least-squares method (Janatsch *et al.*, 1989; Heise *et al.*, 1989). The application of this approach (with several variations) to the determination of protein secondary structures from infrared spectra was delayed until 1990 when four papers were published almost simultaneously by Lee *et al.* (1990), Dousseau and Pézolet (1990), Kalnin *et al.* (1990), and Sarver and Krueger (1991). Spectra of aqueous solution of protein, 20–50 mg/ml, were recorded in 6 μm pathlength cells. Water contribution was subtracted in each case, but amino acid side-chain contribution was subtracted only by Kalnin *et al.*, (1990), according to Venyaminov and Kalnin (1990). In order to validate the technique, each protein, in turn, is eliminated from the calibration set and its secondary structure predicted from the generated eigenspectra. One of the main advantages of the method is that it allows one to investigate which parameters—frequency range, spectral manipulation (baseline subtraction or subtraction of nonpeptide contribution, deconvolution), sample preparation (deuteration), choice of calibration set or choice of criteria used to define the secondary structure—improve the correlation between the data and the secondary structure. Some of these parameters have already been tested in the above-mentioned papers. A synthesis of the most significant conditions tested and of some of the results is reported in Table III. It appears from Table III that no real consensus exists on the procedure to follow. Amazingly, H/D exchange and Fourier self-deconvolution, even with very conservative parameters, as well as second derivatives do not improve the correlation between the spectral shape and the secondary structures. These remarks question the theoretical ground of the Fourier self-deconvolution–based methods described in section 2. Though not reported in Table III it appears from various studies that it is not possible to determine the difference between parallel or antiparallel β-sheet or between different types of β-turns. Difficulties in the definitions of these structures might be partly responsible for this lack of success. The approach was later assessed by Thiaudière *et al.* (1993) using a set of 12 reference proteins. Similar results were obtained on thin films deposited on polyethylene cards (Sarver and Krueger, 1993) or, by the same method, on films of myoglobin dried on ZnSe windows (Van Hoek *et al.*, 1993). The structure of the water channel CHIP28 on ZnSe windows was also evaluated by the same authors and was found to be in good agreement with CD determinations.

Interestingly, the factor analysis method allows the reconstruction of eigenspectra of the pure secondary structures by combining the correct proportion the eigenspectra obtained by the calibration step. This offers a unique opportunity to compare their shape, which here is generated without preconceived idea about the assignment generally accepted (see Table I). The main frequencies of such reconstructed spectra are reported in Table IV. They roughly confirm the frequency limits assigned to each structure but some shoulder overlap frequency domains are assigned to other structures.

In conclusion, the advantages of the pattern recognition methods lie in the fact that none of the tedious technical and theoretical assumptions necessary for the Fourier self-deconvolution based methods are necessary (see section 3). Moreover, it opens the possibility of resolving structures which can be totally overlapped in some frequency regions but not in others. In our view, it is also a great advantage that it can be used without personal intervention of the user. Improvement of the method can be obtained by seeking more experimental or

Table III
Compilation of the Most Significant Conditions Tested and Results for the Determination of the Secondary Structure of Proteins by Multicomponent Analysis as Reported by Lee et al. (1990), Dousseau and Pézolet (1990), Kalnin et al. (1990) and Sarver and Krueger (1991)[a]

	Lee	Dousseau	Kalnin	Sarver
Number of proteins in calibration set	18	13	13 +6 poly- peptides	17
Normalization	Yes (improves results)	Yes (improves results)	No	Yes
Subtraction of side chain contribution	No	No	Yes (improves results)	No
Number of significant eigenspectra	11 or 9		11	4
Frequency ranges tested (cm^{-1})	1700–1600 1800–1500 1800–1600 1700–1500	1700–1600 1720–1480	1800–1480	1700–1600
Best frequency range (cm^{-1})	1700–1600	1720–1480	1800–1480	1700–1600
Types of α-helices	1 type	ordered unordered	ordered unordered	1 type
Types of β-sheet	1 type	1 type	ordered unordered	1 type
Other structures defined	turns	remain	turns	—
Effect of deuteration	worsens results	worsens results	improves results	—
Effect of Fourier self-deconvolution	worsens results	—	—	—
Effect of second derivatives	worsens results	—	—	—
Best average deviation[a]				
α-helix	3.9%	4.9%	7%	10%
β-sheet	8.3%	3.7%	4%	6%

[a] Values obtained with the best methodology, sometimes also by removing one protein from the protein calibration set.

analytic conditions to find better correlation coefficients. In particular, the search for new algorithms used to define the secondary structures from the X-ray data could result in new analyses better related to the spectral features. Results presented in Table III indicate that prediction of secondary structures with the pattern recognition method is at least as good as with the previously described technique. The work of Sarver and Krueger (1991), which closely parallels the CD paper written by Hennessey and Johnson (1981), compares CD and IR data.

Table IV
Spectral Features (Frequency) of Reconstructed Spectra of Pure Secondary Structures Reconstructed from Eigenspectra after Multicomponent Analysis as Reported by Lee *et al.* (1990), Dousseau and Pézolet (1990), Kalnin *et al.* (1990) and Sarver and Krueger (1991).

	Frequencies[a]			
	Lee	Dousseau[b]	Kalnin[c]	Sarver[d]
α-helix structure	1695			
	1659 (major)	1657 (major)	1651 (major)	1652
	1641		1634	
			1550 (AmII)	
β-sheet structure	1682	1685	1695	
		1665	1661—1648 broad	
			1526 (Am II)	
	1634 (major)	1637 (major)	1635 (major)	1638
Turns	1684 (major)	1682—1658 (broad)	1672—1651 broad	1660
	1654			
	1633			
			1544 (Am II)	

[a]Approximate values.
[b]For classical least square method for Dousseau and Pézolet.
[c]For ordered structures.
[d]All bands very broad (FWHH $\sim$ 30–50 cm^{-1}).

As expected, CD does slightly better for α-helices and IR does better for β-sheets. Both techniques are quoted as equally poor for β-turns. The presence of the tested protein in the database considerably improves the prediction in the work of Sarver and Krueger (1991). It must be noted here that the method for extracting secondary structures from X-ray data should be specifically chosen either for CD or for IR, but Sarver and Krueger did not make use of the Levitt and Greet (1977) analysis commonly used in IR. Bandekar (1992) took advantage of the IR and CD spectra complementarity in creating a mixed spectrum made out of the amide I region and of the CD spectrum recorded between 178 and 240 nm. Analysis by a factor analysis approach yields better results than either the IR or the CD spectrum alone.

One potential problem of the pattern recognition method lies in the crucial choice of the calibration protein set, which should adequately cover all the structures that may be encountered. Moreover, as already indicated by the differences in the best conditions of its application, it appears that this approach is very sensitive to parameters and computing methods that are not yet clearly defined. More systematic work needs to be published before assessing its validity in a wide range of circumstances. It must be pointed out for the theoretical evaluation of the method that formal equivalence between factor analysis and curve fitting (expression of the unknown spectrum as a linear combination of reference spectra) has been recognized by Rao and Zerbi (1984). One problem pointed out for the factor analysis approach is that the presence of spurious components in a single spectrum of the database will influence the analysis of all the components.

4. DERIVATIVE SPECTROSCOPY

4.1. Qualitative Aspects

Derivative spectroscopy (essentially second and fourth derivatives) has been widely used as a tool to narrow the band width of infrared spectra. Classically realized by the method of Savitzki and Golay (1964), it is advantageously performed in the Fourier domain by multiplying the interferogram by $(i2\pi x)^n$ where n is the order of the derivative and x is the variable in the transformed domain associated with the frequency in cm^{-1}. The interferogram is therefore multiplied by $-(2\pi x)^2$ for the second derivative or by $(2\pi x)^4$ for the fourth derivative (Kauppinen et al., 1981b). This operation is similar to the deconvolution. Indeed, in the case of a deconvolution by a Lorentzian line shape, the interferogram is multiplied by a e^x type function. In turn, an apodization is also requested for derivation.

For a spectrum made out of Lorentzian bands, the second derivative is 2.7 times narrower than the original line and 3.9 times narrower for the fourth derivative. As for self-deconvolution with K = 2.7 or K = 3.9, the second and fourth derivatives require a signal-to-noise ratio better than 200 and 2000 respectively (Kauppinen et al., 1981b). Yet, the Fourier self-deconvolution method is always advantageous, because for a same-line narrowing it does not yield the additional lobes found in derivative spectra. Similarly, smoothing of spectra in the Fourier domain is advantageous over smoothing in the frequency domain by the Savitzky-Golay algorithm so far as spectral distortions are considered (Kauppinen et al., 1982). One argument developed in the literature to justify the use of derivative spectra obtained by direct methods is that the artifacts related to the manipulation of Fourier transform would not appear. This argument does not stand in view of the above discussion since both approaches are essentially identical and prone to the same artifacts. Assignment of spectral features resolved by derivative spectroscopy are reported in Table II along with data obtained by Fourier self-deconvolution.

Beside its application in IR (e.g. Susi and Byler, 1983; Byler and Susi, 1986; Lee et al. 1985), second and fourth derivatives have been used to study absorbance spectra in the visible domain (Butler and Hopkins 1970a,b, Butler 1979, Talsky et al., 1978). In particular, this technique was used to analyse the

UV-visible spectra of aromatic amino acids (Padros *et al.*, 1964), the environment of tryptophan in melittin, cytochrome c, and bacteriorhodopsin (Dunach *et al.*, 1983), the tryptophan fluorescence spectrum of Ca^{++}-ATPase in its catalytic cycle (Restall *et al.*, 1986) and to estimate the amount of phenylalanine in proteins (Ichikawa and Terada, 1979) to quote only a few applications of this band narrowing technique.

4.2. Quantitative Determination of Secondary Structures

Determination of secondary structures from second-derivative spectra of amide I of proteins in H_2O solution has been attempted by Dong *et al.*, (1990). The authors tested 12 proteins of known structure and found a correlation between the secondary structure and the main resolved peaks, in agreement with the other works quoted in the present paper (see Table II): α-helix near 1656–1654 cm^{-1}, β-sheet between 1642 and 1624 cm^{-1} with a small component near 1675 cm^{-1}, turns at 1688, 1680, 1672, and 1666 cm^{-1}, and random structures at 1650 cm^{-1}. The authors considered the area under each resolved peak (usually more than ten) as a measure of the amount of the related secondary structure. This approach shares the main problems discussed for the Fourier self-deconvolution–curve-fitting methods but raises additional questions. First, the choice of the baseline is complicated by the presence of the positive lobes expected in derivative spectroscopy. Different baselines must be used on a same spectrum when a single component is very intense, but the position of the real baseline remains unknown. Second, upon second derivatization the bandwidth is inversely proportional to the square of the original one and the height of the peak is proportional to the square of the original height (with an opposite sign), assuming a Lorentzian lineshape. The authors do not discuss the relationship between the original surface of a band and the surface of its derivative. A reasonably good correlation is, however, found between the area of the peaks on the second-derivative of amide I and the secondary structure of the protein. This method was used to evaluate the structure of human class I and class II major histocompatibility complexes in conjunction with CD (Gorga *et al.*, 1989), the scrapie-associated protein (Caughey *et al.*, 1991) and pentraxin female protein (Dong *et al.*, 1992).

5. ISOTOPIC LABELLING OF PROTEINS

The availability of ^{13}C and ^{15}N uniformly labeled proteins creates a new breakthrough in the study of the interaction between proteins. Such a study has been reported by Haris *et al.*, (1992) for the interaction between HPr and IIA proteins from *E. coli* in the phosphoenolpyruvate-dependent phosphotransferase system responsible for the concomitant transport and phosphorylation of many hexoses and hexitols in bacteria. Upon $^{13}C/^{15}N$ labelling, the shift in the main spectral features for the protein dissolved in D_2O are as follows: $1684 \rightarrow 1640$, $1655 \rightarrow 1610$, and $1633 \rightarrow 1591$ cm and $1565 \rightarrow 1524$ cm^{-1} for the carboxylate; that is, a shift of 42–45 cm^{-1}. Labelling with ^{15}N alone induces a small shift (~ 2 cm^{-1}) in the amide I but of ~ 25 cm^{-1} in the amide II. Even though some overlap occurs in the amide I region for the $^{13}C/^{15}N$ labelled protein with the unlabelled protein, it becomes much easier to resolve the individual contributions in a protein mixture and should make it possible to study the effects of one protein on the structure of the other. Similarly, the shift in the amide II region could allow the H/D exchange rates to be measured individually in a mixture of two proteins. We anticipate, however, that problems in obtaining a close-to-100% labelled protein and the overlap of amide I in labeled and unlabeled proteins will prevent this approach from being used to study all protein–protein interactions. It should, rather, bring all its potentialities when the structure of the two proteins is rather different so that the labeling can further enhance the frequency split.

6. REFERENCES

Alvarez, J., Haris, P. I., Lee, D. C., and Chapman, D., 1987b, Conformational changes in concanavalin A associated with demetallization and α-methylmannose binding studied by Fourier-transform infrared spectroscopy, *Biochim. Biophys. Acta* **916**: 5–12.

Alvarez, J., Lee, D. C., Baldwin, S. A., and Chapman, D., 1987a, Fourier-transform infrared spectroscopy study of the structure and conformational changes of the human erythrocyte glucose transporter, *J. Biol. Chem.* **262**: 3502–3509.

Anderle, G., and Mendelsohn, R., 1987, Thermal denaturation of globular proteins. Fourier-infrared studies of the amide III spectra region, *Biophys. J.* **52**: 69–74.

Arrondo, J. L. R., Mantsch, H. H., Mullner, N., Pikula, S., and Martonosl, A., 1987, Infrared spectroscopy characterization of the structural changes connected with the E_1–E_2 transition in the Ca^{2+}-ATPase of sarcoplasmic reticulum, *J. Biol. Chem.* **262**: 9037–9043.

Arrondo, J. L. R., Muga, A., Castresana, J., and Goñi, F. M., 1993, Quantitative studies of the structure of proteins in solution by Fourier-transform infrared spectroscopy, *Prog. Biophys. Molec. Biol.* **59**: 23–56.

Arrondo, J. L. R., Muga, A., Castresana, J., Bernabeu, C., and Goñi, F. M., 1989b, An infrared spectroscopic study of β-galactosidase structure in aqueous solutions, *FEBS Lett.* **252**: 118–120.

Arrondo, J. L. R., Muga, A., Pardo, A., and Goñi, F. M., 1988a, A FT-IR study of sarcoplasmic reticulum solubilization by triton X-100, *Mikrochim. Acta* **1**: 385–388.

Arrondo, J. L. R., Young, N. M., and Mantsch, H. H., 1988b, The solution structure of convanavalin A probed by FT–IR spectroscopy, *Biochim. Biophys. Acta* **952**: 261–268.

Arrondo, S. L. R., Gilles, A. M., Bârzu, O., Fernandjian, S., Yang, P. W., and Mantsch, H. H., 1989a, Investigation of adenylate kinase from *Escherichia coli* and its interaction with nucleotides by Fourier-transform infrared spectroscopy, *Biochem. Cell. Biol.* **67**: 327–331.

Ayala, G., Carmona, P., de Cozar, M., and Monreal, J., 1987, Vibrational spectra and structure of myelin membranes, *Eur. Biophys. J.* **14**: 219–225.

Bandekar, J., 1992, Amide modes and protein conformation, *Biochim. Biophys. Acta* **1120:** 123–143.

Barlow, D. J., and Thornton, J. M., 1988, Helix geometry in proteins, *J. Mol. Biol.* **201:** 601–619.

Bates, J. B., 1976, Fourier transform infrared spectroscopy, *Science* **191:** 31–57.

Bazzi, M. D., and Woody, R. W., 1985, Oriented secondary structure in integral membrane proteins: Circular dichroism and infrared spectroscopy of cytochrome c oxidase in multilamellar films, *Biophys. J.* **48:** 957–966.

Blass, W. E., and Halsey, G. W., 1981, in: *Deconvolution of Absorption Spectra*, Academic Press, New York.

Blundell, T., Barlow, D., Borkako, N., and Thornton, J., 1983, Solvent-induced distorsions and the curvature of α-helices, *Nature* **306:** 281–283.

Bracewell, R. N., 1990, Numerical transform, *Science* **248:** 697–704.

Brazhnikov, E. V., Chetverin, A. B., and Chirgadze, Y. N., 1978, Secondary structure of Na$^+$, K$^+$ dependent adenosine triphosphatase, *FEBS Lett.* **93:** 125–128.

Buchet, R., Carrier, D., Wong, P. T. T., Jona, I., and Martonosi, A., 1990, Pressure effects on sarcoplasmic reticulum: A Fourier transform infrared spectroscopic study, *Biochim. Biophys. Acta* **1023:** 107–118.

Buchet, R., Jona, I., and Martonosi, A., 1989, Correlation of structure and function in the Ca^{++}AT-Pase of sarcoplasmic reticulum: A Fourier transform infrared spectroscopy (FTIR) study on the effects of dimethyl sulfoxide and urea, *Biochim. Biophys. Acta* **983:** 167–178.

Butler, W. L., 1979, Fourth-derivative spectra, *Methods Enzymol.* **56:** 501–515.

Butler, W. L., and Hopkins, D. W., 1970a, Higher derivative analysis of complex absorption spectra, *Photochem. Photobiol.* **12:** 439–450.

Butler, W. L., and Hopkins, D. W., 1970b, An analysis of fourth-derivative spectra, *Photochem. Photobiol.* **12:** 451–456.

Byler, D. M., and Susi, H., 1986, Examination of the secondary structure of proteins by deconvolved FTIR spectra, *Biopolymers* **25:** 469–487.

Cabiaux, V., Goormaghtigh, E., Wattiez, R., Falmagne, P., and Ruysschaert, J. M., 1989, Secondary structure of diphtheria toxin interacting with asolectin liposomes: An infrared spectroscopy study, *Biochimie* **71:** 153–158.

Cairns, M. T., Alvarez, J., Panico, M., Gibbs, A. F., Morris, H. R., Chapman, D., and Baldwin, S. A., 1987, Investigation of the structure and function of the human erythrocyte glucose transporter by proteolytic dissection, *Biochim. Biophys. Acta* **905:** 295–310.

Casal, H. L., and Mantsch, H. H., 1984, Polymorphic phase behavior of phospholipid membranes studied by infrared spectroscopy, *Biochim. Biophys. Acta* **779:** 381–401.

Casal, H. L., Köhler, U., Mantsch, H. H., and Arrondo, J. L. R., 1988, FT–IR spectra of proteins at low temperatures, *Mikrochim. Acta* **1:** 195–197.

Castresana, J., Muga, A., and Arrondo, J. L. R., 1988, The structure of proteins in aqueous solutions: an assessment of triose phosphate isomerase structure by Fourier-transform infrared spectroscopy, *Biochem. Biophys. Res. Commun.* **152:** 69–75.

Caughey, B. W., Dong, A., Bhat, K. S., Ernst, D., Hayes, S. F., and Caughey, W. S., 1991, Secondary structure analysis of the scrapie-associated protein Prp 27–30 in water by infrared spectroscopy, *Biochemistry* **30:** 7672–7680.

Chapman, D., Jackson, M., and Haris, P. I., 1988, Investigation of membrane protein structure using Fourier-transform infrared spectroscopy, *Biochem. Soc. Trans.* **17:** 617–619.

Chin, J. J., Jung, E. K. Y., and Jung, C. Y., 1986, Structural basis of human erythrocyte glucose transporter function in reconstituted vesicles: α helix orientation, *J. Biol. Chem.* **261:** 7101–7104.

Chirgadze, Y. N., Brazhnikov, E. V., and Nevskaya, N. A., 1976, Intramolecular distrosion of the α-helical structural of polypeptides, *J. Mol. Biol.* **102:** 781–792.

Chirgadze, Y. N., Fedorov, O. V., and Trushina, N. P., 1975, Estimation of amino acid residue side-chain absorption in the infrared spectra of protein solution in heavy water, *Biopolymers* **14:** 679–694.

Compton, L. A., and Johnson, W. C., 1986, Analysis of protein circular dichroism spectra for secondary structure using a simple matrix multiplication, *Anal. Biochem.* **155:** 155–167.

Dong, A., Caughey, B., Caughey, W. S., Bhat, K., and Coe, J. E., 1992, Secondary structure of the pentraxin female protein in water determined by infrared spectroscopy: Effects of calcium and phosphorylcholine, *Biochemistry* **31:** 9364–9370.

Dong, A., Huang, P., and Caughey, W. S., 1990, Protein secondary structure in water from second-derivative amide I infrared spectra, *Biochemistry* **29:** 3303–3308.

Dong, A., Huang, P., and Caughey, W. S., 1992, Redox-dependent changes in β-extended chain and turn structures of cytochrome c in water solution determined by second-derivative amide I infrared spectra, *Biochemistry* **31:** 182–189.

Dousseau, F., and Pézolet, M., 1990, Determination of the secondary structure contents of proteins in aqueous solutions from their amide I and amide II infrared bands. Comparison between classical and partial least-squares methods, *Biochemistry* **29:** 8771–8779.

Doyle, B. B., Bendit, E. G., and Blout, E. R., 1975, Infrared spectroscopy of collagen and collagenlike polypeptides, *Biopolymers* **14:** 937–957.

Dunach, M., Sabes, M., and Padros, E., 1983, Fourth-derivative analysis of tryptophan environment in proteins. Application to melittin, cytochrome c and bacteriorhodopsin, *Eur. J. Biochem.* **134:** 123–128.

Dwivedi, A. M., and Krimm, S., 1984, Vibrational analysis of peptides, polypeptides, and proteins. XVIII Conformational sensitivity of the α-helix spectrum: α_I and α_{II}-Poly(L-alamine), *Biopolymers* **23:** 923–943.

Earnest, T. N., Herzfeld, J., and Rothschild, K. J., 1990, Polarized FTIR of bacteriorhodopsin: Transmembrane α-helices are resistant to hydrogen–deuterium exchange, *Biophys. J.* **58:** 1539–1546.

Eckert, K., Grosse, R., Malur, J. and Repke, K. R. H., 1977, calculation and use of protein-derived conformation-related spectra for the estimate of the secondary structure of proteins from their infrared spectra, *Biopolymers* **16:** 2549–2563.

Ferraro, J. R., and Basile, L. S., (eds.) 1987, in: *Fourier Transform Infrared Spectroscopy,* Vols 1–3, Academic Press, New York.

Fraser, P. E., Nguyen, J. T., Surewicz, W. K., and Kirschner, D. A., 1991, pH-dependent structural transitions of Alzheimer amyloid peptides, *Biophys. J.* **60:** 1190–1201.

Fraser, R. D. B., and Suzuki, E., 1966, Resolution of overlapping absorption bands by least square procedures, *Anal. Chem.* **38:** 1770–1773.

Fraser, R. D. B., and Suzuki, E., 1973, The use of least squares in data analysis in: *Molecular Biology Part C: Physical Principles and Techniques of Protein Chemistry* (Sydney J. Leach, ed.), Academic Press, New York.

Frey, S., and Tamm, L. K., 1991, Orientation of melittin in phospholipid bilayers. A polarized attenuated total reflection infrared spectroscopy study, *Biophys. J.* **60:** 922–930.

Fringeli, U. P., and Günthard, H. H., 1981, Infrared membrane spectroscopy, in: *Membrane Spectroscopy* (E. Grell ed.), pp. 270–332, Springer-Verlag, Berlin.

Fry, D. C., Byler, D. M., Susi, H., Brown, Z. M., Kuby, S. A., and Mildvan, A. S., 1988, Solution structure of the 45-residue MgATP-binding peptide of adenylate cyclase as examined by 2-D NMR, FTIR and CD spectroscopy, *Biochemistry* **27:** 3588–3598.

Gasset, M., Baldwin, M. A., Fletterick, R., and Prusiner, S. B., 1993, Perturbation of the secondary structure of the scrapie prion protein under conditions that alter inffectivity, *Proc. Natl. Acad. Sci. USA* **90:** 1–5.

Gasset, M., Baldwin, M. A., Lloyd, D. H., Gabriel, J. M., Holtzman, D. M., Cohen, F., Fletterick,

R., and Prusiner, S. B., 1992, Predicted α-helical regions of the prion protein when synthesized as peptides form amyloid, *Proc. Natl. Acad. Sci. USA* **89**: 10940–10944.

George, A., and Veis, A., 1991, FTIRS in H_2O demonstrates that collagen monomers undergo a conformational transition prior to thermal self-assembly *in vitro, Biochemistry* **30**: 2372–2377.

Glaeser, R. M., Downing, K. H., and Jap, B. K., 1991, What spectroscopy can still tell us about the secondary structure of bacteriorhodopsin, *Trends Biochem. Sci.* **59**: 934–938.

Goormaghtigh, E., and Ruysschaert, J. M., 1990, Polarized attenuated total reflection spectroscopy as a tool to investigate the conformation and orientation of membrane components, in: *Molecular Description of Biological Membranes by Computer-Aided Conformational Analysis,* (R. Brasseur, ed.) Vol. 1, p. 285–329, CRC Press Inc., Boca Raton, Florida.

Goormaghtigh, E., Cabiaux, V., and Ruysschaert, J. M., 1990, Secondary structure and dosage of soluble and membrane proteins by attenuated total reflection Fourier-transform infrared spectroscopy on hydrated films, *Eur. J. Biochem.* **193**: 409–420.

Goormaghtigh, E., De Meutter, J., Vanloo, B., Brasseur, R., Rosseneu, M., and Ruysschaert, J. M., 1989b, Evaluation of the secondary structure of apo B100 in low density lipoprotein (LDL) by infrared spectroscopy, *Biochim. Biophys. Acta* **1006**: 147–150.

Goormaghtigh, E., Martin, I., Vandenbranden, M., Brasseur, R., and Ruysschaert, J. M., 1989a, Secondary structure and orientation of a chemically synthesized mitochondrial signal sequence in phospholipids bilayers, *Biochem. Biophys. Res. Commun.* **158**: 610–616.

Goormaghtigh, E., Ruysschaert, J. M., and Scarborough, G. A., 1988, High yield incorporation of the neurospora plasma membrane H^+ ATPase into proteoliposomes: Lipid requirement and secondary structure of the enzyme by IR spectroscopy, *Prog. Clin. Biol. Res.* **273**: 51–56.

Gorga, J. C., Dong, A., Manning, M. C., Woody, R. W., Conghey, W. S. and Strominger, J. L., 1989, Comparison of the secondary structures of human class I and class II major histocompatibility complex antigens by Fourier-transform infrared and circular dichroism spectroscopy, *Proc. Natl. Acad. Sci. U.S.A.* **86**: 2321–2325.

Görne-Tschelnokow, U., Naumann, D., Weise, C., and Hucho, F., 1993, Secondary structure and temperature behavior of acetylcholinesterase. Studies by Fourier-transform infrared spectroscopy, *Eur. J. Biochem.* **213**: 1235–1242.

Gotto, A. M., Levy, R. I., and Frederickson, 1968, Observation on the conformation of human beta lipoprotein: Evidence for the occurrence of beta structure, *Proc. Natl. Acad. Sci. USA* **60**: 1436–1441.

Griffith, P. R. (ed.), 1978, in: *Transform Techniques in Chemistry,* Plenum Press, New York.

Griffiths, P. R., 1983, Fourier-transform infrared spectroscopy, *Science* **222**: 297–302.

Haris, P. I., and Chapman, D., 1988, Fourier-transform infrared spectra of the polypeptide alamethicin and a possible structural similarity with bacteriorhodopsin, *Biochim. Biophys. Acta* **943**: 375–380.

Haris, P. I., and Chapman, D., 1992, Does Fourier-transform infrared spectroscopy provide useful information on protein structure? *Trends Biochem. Sci.* **17**: 328–333.

Haris, P. I., Chapman, D., Harrison, R. A., Smith, K. F., and Perkins, S. J., 1990, Conformational transition between native and reactive center cleaved forms of α_1-antitrypsin by Fourier-transform infrared spectroscopy and small-angle neutron scattering, *Biochemistry* **29**: 1377–1380.

Haris, P. I., Coke, M., and Chapman, D., 1989, Fourier-transform infrared spectroscopic investigation of rhodopsin structure and its comparison with bacteriorhodopsin, *Biochim. Biophys. Acta* **995**: 160–167.

Haris, P. I., Lee, D. C., and Chapman, D., 1986, A Fourier-transform infrared investigation of the structural differences between ribonuclease A and ribonuclease S, *Biochim. Biophys. Acta* **874**: 255–265.

Haris, P. I., Robillard, G. T., van Dijk, A. A., and Chapman, D., 1992, Potential of ^{13}C and ^{15}N labeling for studying protein-protein interactions using Fourier-transform infrared spectroscopy, *Biochemistry* **31**: 6279–6284.

He, W-Z., Newel, W. R., Haris, P. I., Chapman, D., and Barber, J., 1991, Protein secondary structure of the isolated photosystem II reaction center and conformational changes studied by Fourier-transform infrared spectroscopy, *Biochemistry* **30**: 4552–4559.

He, W-Z., Newell, W. R., Haris, P. I., Chapman, D., and Barber, J., 1991, Protein secondary structure of the isolated photosystem II reaction center and conformational changes studied by Fourier-transform infrared spectroscopy, *Biochemistry* **30**: 4552–4559.

Heise, H. M., Marbach, R., Janatsch, G., and Kruse-Jarres, J. D., 1989, Multivariate determination of glucose in whole blood by attenuated total reflection infrared spectroscopy, *Anal. Chem.* **61**: 2009–2015.

Hennessey, J. P., and Johnson, W. C., 1981, Information content in the circular dichroism of proteins, *Biochemistry* **20**: 1085–1094.

Herzyk, E., Lee, D. C., Dunn, R. C., Bruckdorfer, K. R., and Chapman, D., 1987, Changes in the secondary structure of apolipoprotein B-100 after Cu^{++} catalysed oxidation of human low-density lipoproteins monitored by Fourier-transform infrared spectroscopy, *Biochim. Biophys. Acta* **922**: 145–154.

Holloway, P. W., and Buchheit, C., 1990, Topography of the membrane-binding domain of cyto-chrome b5 in lipids by Fourier-transform infrared spectroscopy, *Biochemistry* **29**: 9631–9637.

Holloway, P. W., and Mantsch, H. H., 1989, Structure of cytochrome b_5 in solution by Fourier-transform infrared spectroscopy, *Biochemistry* **28**: 931–935.

Hunt J. F. 1988, An influence of tertiary structure on protein infrared spectra, *Biophys. J.* **53**: 97a.

Hunt, J. F., Earnest, T. N., Engelman, D. M., and Rothschild, K. J., 1988, An FTIR study of integral membrane protein folding, *Biophys. J.* **53**: 97a.

Ichikawa, T., and Terada, H., 1979, Estimation of state and amount of phenylalanine residues in proteins by second derivative spectrophotometry, *Biochim. Biophys. Acta* **580**: 120–128.

Jakobson, R. J., Brown, L. L., Hutson, T. B., Fink, D. J., and Veis, A., 1983, Intramolecular interactions in collagen self assembly as revealed by Fourier-transform infrared spectroscopy, *Science* **220**: 1288–1290.

Jakobson, R. J., Wasacz, F. M., Brasch, J. W., and Smith, K. B., 1986, The relationship of bound water to the IR amide I bandwidth of albumin, *Biopolymers* **25**: 639–654.

Janatsch, G., Kruse-Jarres, J. D., Marbach, R., and Heise, H. M., 1989, Multivariate calibration for assays in clinical chemistry using attenuatted total reflection infrared spectra of human blood plasma, *Anal. Chem.* **61**: 2016–2023.

Jap, B. K., Maestre, M. F., Hayward, S. B., and Glaeser, R. M., 1983, Peptide-chain secondary structure of bacteriorhodopsin, *Biophys. J.* **43**: 81–89.

Kabach, W., and Sander, S., 1983, *Biopolymers* **22**: 2577–2637.

Kackson, M., Haris, P. I., and Chapman, D., 1991, Fourier-transform infrared spectroscopic studies of Ca^{2+}-binding proteins, *Biochemistry* **30**: 9681–9686.

Kalnin, N. N., Baikalov, I. A., and Venyaminov, S. Y., 1990, Quantitative IR spectrophotometry of peptides compounds in water (H_2O) solutions. III. Estimation of the protein secondary structure, *Biopolymers Biopolymers* **30**: 1273–1280.

Kamegai, J., Kimura, S., and Imanishi, Y., 1986, Conformation of sequential polypeptide poly (Leu-Leu-D-Phe-Pro) and formation of ion channel across bilayer lipid membrane, *Biophys. J.* **49**: 1101–1108.

Kauppinen, J. K., Moffat, D. J., and Mantsch, H. H., 1981c, Noise in Fourier self-deconvolution, *Appl. Opt.* **20**: 1866–1880.

Kauppinen, J. K., Moffat, D. J., Mantsch, H. H., and Cameron, D. G., 1981a, Fourier self-deconvolution: A method for resolving intrinsically overlapped bands, *Appl. Spectrosc.* **35**: 271–276.

Kauppinen, J. K., Moffat, D. J., Mantsch, H. H., and Cameron, D. G., 1981b, Fourier transform in the computation of self-deconvolution and first-order derivatives spectra of overlapped band contours, *Anal. Chem.* **53**: 1454–1457.

Kauppinen, J. K., Moffat, D. J., Mantsch, H. H., and Cameron, D. G., 1982, Smoothing of spectral data in the Fourier domain, *Appl. Opt.* **21**: 1866–1872.

Kennedy, D. F., Crisman, M., Toniolo, C., and Chapman, D., 1991, Studies of peptides forming 3_{10^-} and α-helices and β-bend ribbon structures in organic solutions and in model bio-membranes by Fourier-transform infrared spectroscopy, *Biochemistry* **30**: 6541–6548.

Kennedy, D. F., Slotboom, A. J., de Haas, G. H., and Chapman, D., 1990, A Fourier-transform infrared spectroscopy (FTIR) of porine and bovine pancreatic phospholipase A_2 and their inter-actions with substrate analogues and a transition-state inhibition, *Biochim. Biophys. Acta* **1040**: 317–326.

Kleffel, B., Garavito, R. M., Baumeister, W., and Rosenbusch, J. P., 1985, Secondary structure of a channel-forming protein: porin from *E. coli* outer membranes, *EMBO J.* **4**: 1589–1592.

Krimm, S., and Dwivedi, A. M., 1982, Infrared spectrum of the purple membrane: Clue to a proton conduction mechanism? *Science* **23**: 407–408.

Lacsko, I., Hollosi, M., Ürge, L., Ugen, K. E., Weiner, D. B., Mantsch, H. H., Thurin, J., and Ötvös, Jr., 1992, Synthesis and conformational studies of N-glycosylated analogues of the HIV-1 principal neutralizing determinant, *Biochemistry* **31**: 4282–4288.

Lavialle, F., Adams, R. G., and Levin, I. W., 1982, Infrared spectroscopic study of the secondary structures of Melittin in water, 2-chloroethanol and phospholipid dispersion, *Biochemistry* **21**: 2305–2312.

Lazarev, Y. A., Grishkovsky, B. A., and Khromova, T. B., 1985, Amide I band of IR spectrum and structure of collagen and related polypeptides, *Biopolymers* **24**: 1449–1478.

Lee, D. C., Haris, P. I., Chapman, D., and Mitchell, R. C., 1990, Determination of protein secondary structure using factor analysis of infrared spectra, *Biochemistry* **29**: 9185–9193.

Lee, D. C., Hayward, J. A., Restall, C. J., and Chapman, D., 1985, Second-derivative infrared spectroscopic studies of the secondary structures of bacteriorhodopsin and Ca^{++} ATPase.

Levitt, M., and Greer, J., 1977, Secondary structure in globular proteins, *J. Mol. Biol.* **114**: 181–239.

Maddams, W. F., 1980, The scope and limitations of curve fitting, *Appl. Spectrosc.* **34**: 245–267.

Mitchell, R. C., Haris, P. I., Fallowfield, C., Keeling, D. J., and Chapman, D., 1988, Fourier-transform infrared spectroscopy studies on gastric H^+/K^+-ATPase. *Biochim. Biophys. Acta* **947**: 31–38.

Moh, P. P., Fiamingo, F. G., and Alben, J. O., 1987, Conformational sensitivity of beta-93 cysteine SH to ligation of hemoglobin observed by FT–IR spectroscopy, *Biochemistry* **26**: 6243–6249.

Muga, A., Mantsch, H. H., and Surewicz, W. K., 1991a, Apoytochrome c interaction with phospho-lipid membranes studied by Fourier-transform infrared spectroscopy, *Biochemistry* **30**: 2629–2625.

Muga, A., Mantsch, H. H., and Surewitcz, W. K., 1991b, Membrane-binding induces destabiliza-tion of cytochrome c structure, *Biochemistry* **30**: 7219–7224.

Nabedryk, E., Garavito, R. M., and Breton, J., 1988, The orientation of β-sheets in porin. A polarized Fourier-transform infrared spectoscopic investigation, *Biophys. J.* **53**: 671–676.

Naumann D., Schultz C., Görne-Tschelnokow, U., and Hucho F., 1993, Secondary structure and temperature behavior of the acetylcholine receptor by Fourier-transform infrared spectroscopy, *Biochemistry* **32**: 3162–3168.

Nevskaya, N. A., and Chirgadze, Y. N., 1976, Infrared spectra and resonance interactions of amide I and II vibrations of α-helix, *Biopolymers* **15**: 637–648.

Nikolov, S., and Kantchev, K., 1987, Deconvolution of Lorentzian broadened spectra. I. Direct deconvolution, *Nucl. Instrum. Methods Phys. Res.* A256, 161–167.

Ollinger, J. M., Hill, D. M., Jakobsen, R. S., and Broody, R. S., 1986, Fourier-transform infrared studies of ribonuclease in H_2O and ^2H_2O solutions, *Biochim. Biophys. Acta* **869**: 89–98.

Oxtoby, D. W., 1981, Vibrational relaxation in liquids, *Annu. Rev. Phys. Chem.* **32**: 77–101.

Padros, E., Dunach, M., Morros, A., Sabes, M., and Manosa, J., 1984, Fourth-derivative spectrophotometry of proteins, *Trends Bioch. Soc.* **36**: 508–511.

Perkins, S. J., Nealis, A. S., Haris, P. I., Chapman, D., Goundis, D., and Reid, K. B. M., 1989, Secondary structure in properdin of the complement cascade and related proteins: A study by Fourier-transform infred spectroscopy, *Biochemistry* **28**: 7176–7182.

Petrelski, S. J., Abrakawa, T., Kenney, W. C., and Byler, D. M., 1991c, The secondary structure of two recombinant human growth factor, platelet-derived growth factor and basic fibroblast growth factor as determined by Fourier-transform infrared spectroscopy, *Arch. Biochem. Biophys.* **285**: 111–115.

Petrelski, S. J., Byler, D. M., and Liebman, M. N., 1991a, Comparison of various molecular forms of bovine trypsin: Correlation of infrared spectra with X-ray crystal structures, *Biochemistry* **30**: 133–143.

Petrelski, S. J., Byler, D. M., and Thompson, M. P., 1991b, Infrared spectroscopy descrimination between α-and 3_{10}-helices in globular proteins: Reexamination of Amide I infrared bands of α-lactalbumin and their assignment to secondary structures, *Int. J. Protein Res.* **37**: 508–512.

Petrelski, S. J., Byler, D. M., and Thompson, M. P., 1991d, Effect of metal ion binding on the secondary structure of bovine α-lactalbumin as examined by infrared spectroscopy, *Biochemistry* **30**: 8797–8804.

Purcell, J. M., and Susi, H., 1984, Solvent denaturation of proteins as observed by resolution-enhanced Fourier-transform infrared spectroscopy, *J. Biochem. Biophys. Methods* **9**: 193–199.

Rainetau, D., Wolf, C., and Lavialle, F., 1989, Effect of calcium and calcium analogs on calmodulin: A Fourier-transform infrared and electron spin resonance investigation, *Biochim. Biophys. Acta* **1011**: 81–87.

Rao, G. R., and Zerbi, G., 1984, Factor analysis and least-squares curve-fitting of infrared spectra: an application to the study of phase transition in organic molecules, *Appl. Spectrosc.* **38**: 795–803.

Rath, P., Bouché, O., Merril, A. R., Cramer, W. A., and Rotschild, K. J., 1991, Fourier-transform infrared evidence for a predominantly α-helical structure of the membrane bound channel forming COOH-terminal peptide of colicin E1, *Biophys. J.* **59**: 516–522.

Renugopalakrishnan, V., Horowitz, P. M., and Glimcher, M. J., 1985, Structural studies of phosvitin in solution and in the solid state, *J. Mol. Biol.* **260**: 11406–11413.

Restall, C. J., Coke, M., Phillips, E., and Chapman, D., 1986, Derivative spectroscopy of tryptophan fluorescence used to study conformational transitions in the $Ca^{++}+Mg^{++}$-adenosine triphosphatase of sarcoplasmic reticulum, *Biochim. Biophys. Acta.* **874**: 305–311.

Rial, E., Muga, A., Valpuesta, J. M., Arrondo, J. L. R., and Goñi, F. M., 1990, Infrared spectroscopic studies of detergent-solubilized uncoupling protein from brown-adipose-tissue mitochondrial, *Eur. J. Biochem.* **188**: 83–89.

Rothschild, K. J., and Clark, N. A., 1979a, Polarized infrared spectroscopy of oriented purple membrane, *Biophys. J.* **25**: 473–488.

Rothschild, K. J., and Clark, N. A., 1979b, Anomalous amide I infrared absorption of purple membrane, *Science* **204**: 311–312.

Rothschild, K. J., Sanches, R., and Clark, N. A., 1982, Infrared absorption of photoreceptor and purple membranes, *Methods Enzymol.* **88**: 696–714.

Rothschild, K. J., Sanches, R., De Grip, W., 1979, Fourier-transform infrared absorption of photoreceptors membranes:I. Group assignments based on rhodopsin delipidation and regeneration, *Biochim. Biophys. Acta* **596**: 333–351.

Rothschild, K. J., Sanches, R., Hsiao, T. L., and Clark, N. A., 1980, A spectroscopic study of rhodopsin alpha-helix orientation, *Biophys. J.* **31**: 53–64.

Sarver, R. W., and Krueger, W. C., 1993, Infrared investigation of proteins deposited on poly-ethylene films, *Anal. Biochem.* 519–525.

Sarver, R. W., and Kruerger, W. C., 1991, Protein secondary structure from Fourier-transform infrared spectroscopy: A database analysis, *Anal. Biochem.* **194:** 89–100.

Savitzky, A. and Golay M. J. E. 1964, Smoothing and differentiation of data by simplified least squares procedures, *Anal. Chem.* **36:** 1627–1639.

Suga, H., Shirabe, K., Yamamoto, T., Tosumi, M., Umeda, M., Nishinmura, C., Nakazawa, A., Nakamishi, M. and Arata, Y., 1991, Structural analyses of a channel-forming fragment of colicin E1 incorporated into lipid vesicles: Fourier-transform infrared and tryptophan fluorescence studies, *J. Biol. Chem.* **266:** 13537–13543.

Surewicz, W. K., and Mantsch, H. H., 1988a, New insight into protein secondary structure from resolution-enhanced infrared spectra, *Biochim. Biophys. Acta* **952:** 115–130.

Surewicz, W. K. and Mantsch, H. H., 1988b, Solution and membrane structure of enkephalins as studied by infrared spectroscopy, *Biochem. Biophys. Res. Comm.* **150:** 245–251.

Surewicz, W. K., Leddy, J. J., and Mantsch, H. H., 1990, Structure, stability, and receptor interaction of cholera toxin as studied by Fourier-transform infrared spectroscopy, *Biochemistry* **29:** 8106–8111.

Surewicz, W. K., Mantsch, H. H., and Chapman, D., 1993, Determination of protein secondary structure by Fourier-transform infrared spectroscopy: A critical assessement, *Biochemistry* **32:** 389–394.

Surewicz, W. K., Mantsch, H. H., Stahl, G. L., and Epand, R. M., 1987c, Infrared spectroscopy evidence of conformational transitions of an atrial natriuretic peptide, *Proc. Natl. Acad. Sci. USA* **84:** 7028–7030.

Surewicz, W. K., Moscarello, M. A., and Mantsch, H. H., 1987b, Fourier transform infrared investigation of the interaction between myelin basic protein and dimyristoylphosphatidyl-glycerol bilayers, *Biochemistry* **26:** 3881–3886.

Surewicz, W. K., Moscarello, M. A., and Mantsch, H. H., 1987a, Secondary structure of the hydrophobic myelin protein in a lipid environment as determined by Fourier-transform infrared spectroscopy, *J. Biol. Chem.* **262:** 8598–8602.

Surewicz, W. K., Szabo, A. G., and Mantsch, H. H., 1987d, Conformational properties of azurin in solutions determined from resolution-enhanced Fourier-transform infrared spectra, *Eur. J. Biochem.* **167:** 519–523.

Susi, H., and Byler, D. M., 1983, Protein structure by Fourier-transform infrared spectroscopy: second derivative spectra, *Biochem. Biophys. Res. Commun.* **115:** 391–397.

Susi, H., and Byler, D. M., 1986, Resolution-enhanced Fourier-transform infrared spectroscopy of enzymes, *Methods Enzymol.* **130:** 290–311.

Susi, H., and Byler, D. M., 1987, Fourier-transform infrared study of proteins with parallel β-chains, *Arch. Biochem. Biophys.* **258:** 465–469.

Susi, H., Timaseff, S. N., and Stevens, L., 1967, Infrared spectra and protein conformations in aqueous solutions : I. The amide I band in H_2O and D_2O solution, *J. Biol. Chem.* **242:** 5460–5466.

Suzuki, E., 1967, A quantitative study of the amide vibrations in the infrared spectrum of silk fibroin, *Spectrochimica Acta* **23A:** 2303–2308.

Swedberg, S. A., Pesek, J. J., and Fink, A. L., 1990, Attenuated total reflection Fourier-transform infrared analysis of an acyl-enzyme intermediate of α-chymotrypsin, *Anal. Biochem.* **186:** 153–158.

Taillandier, E., Fort, L., Liquier, J., Couppez, M., and Sautiere, P., 1984, Role of the protein a helices in histone-DNA interactions studied by vibrational spectroscopy, *Biochemistry* **23:** 2644–2650.

Talsky, G., Mayring, L., and Kreuzer, H., 1978, High resolution, higher-order UV/VIS derivative spectrophotometry, *Angew. Chem. Int. Ed. Engl.* **17:** 785–799.

Teruel, J. A., Villalain, J., and Gomez-Fernandez, J. C., 1990, Effect of protease digestion on the secondary structure of sarcoplasmic reticulum Ca^{++}-ATPase as seen by FTIR spectroscopy, *Int. J. Biochem.* **22:** 779–783.

Thiaudière, E., Soekarjo M., Kuchinka E., Kuhn A. and Vogel H. 1993, Structural characterization of membrane insertion of M13 procoat, M13 coat, and Pf3 coat proteins, *Biochemistry* **32:** 12186–12196.

Timasheff, S. N., Susi, H., and Stevens, L., 1967, Infrared spectra and protein conformations in aqueous solutions: II. Survey of globular proteins, *J. Biol. Chem.* **242:** 5467–5473.

Tinker, D. A., Krebs, E. A., Feltham, I. C., Atta-Poku, S. K., and Ananthanarayanan, V. S., 1988, Synthetic β-turn peptides as substrates for a tyrosine protein kinase, *J. Biol. Chem.* **263:** 5024–5026.

Toniolo, C., and Benedetti, E. 1991, The polypeptide 3_{10}-helix, *TIBS* **16:** 350–353.

Toniolo, C., Bonora, G. M., Heimer, E. P., and Felix, A. M., 1987, Structure, solubility, and reactivity of peptides. A conformational study of two protected key intermediate from a large-scale synthesis of thyomosin alpha 1, *Int. J. Pept. Protein Res.* **30:** 232–239.

Torii, H., and Tasumi M., 1992, Model calculations on the amide I infrared bands of globular proteins, *J. Chem. Phys.* **96:** 3379–3387.

Trewhella, J., Liddle, W. K., Heidon, D. B., and Strymnadka, N., 1989, Calmodulin and troponin c structures studied by Fourier transform infrared spectroscopy: Effects of Ca^{++} and Mg^{++} binding, *Biochemistry* **28:** 1294–1301.

Valpuesta, J. M., Arrondo, J. L. R., Barbero, M. C., Pons, M., and Goñi, F. M., 1986, Membrane-surfactant interactions: The role of surfactant in mitochondrial complex III-phospholipid-triton X-100 mixed micelles, *J. Biol. Chem.* **261:** 6578–6584.

Van Hoek, A. N., Wiener, M., Bicknese, S., Miercke, L., Biwersi, J., and Verkman, A. S., 1993, Secondary structure analysis of purified functional CHIP28 water channels by CD and FTIR spectroscopy, *Biochemistry* **32:** 11847–11856.

Van Wart, H. E., and Sheraga, H. A., 1978, Raman and resonance Raman spectroscopy, *Methods Enzymol.* **49:** 67–149.

Venyaminov, S. Y., and Kalnin, N. N., 1991a, Quantitative IR spectrophotometry of peptides compounds in water (H$_2$O) solutions. I. Spectral parameters of amino acid residue absorption band, *Biopolymers* **30:** 1243–1257.

Venyaminov, S. Y., and Kalnin, N. N., 1991b, Quantitative IR spectrophotometry of peptides compounds in water (H$_2$O) solutions. II. Amide absorption bands of polypeptides and fibrous proteins in α-, β- and random conformations, *Biopolymers* **30:** 1259–1271.

Villalain, J., Gomez-Fernandez, J. C., Jackson, M., and Chapman, D., 1989, Fourier-transform infrared spectroscopyic studies on the secondary structure of the Ca^{++} ATPase of sarcoplasmic reticulum, *Biochim. Biophys. Acta* **978:** 305–312.

Vincent, S. S., Star, C. J., and Levin, I. W., 1984, Infrared spectroscopic study of the pH-dependent secondary structure of brain clathrin, *Biochemistry* **23:** 625–631.

Vorherr, T., James, P., Krebs, J., Enyedi, A., McCormick, D. J., Penniston, J. T., and Carafoli, E., 1990, Interaction of calmodulin with the calmodulin binding domain of the plasma membrane Ca^{2+} pump, *Biochemistry* **29:** 355–365.

Wantyghem, J., Baron, M.-H., Picquart, M., and Llavialle, F., 1990, Conformational changes of Robinia pseudoacacia lectin related to modifications of the environment: FTIR investigation, *Biochemistry* **29:** 6600–6609.

Wasacz, F. M., Olinger, J. M., and Jakobsen, R. J., 1987, Fourier-transform infrared studies of proteins using non aqueous solvents. Effects of methanol and ethylene glycol on albumin and immunoglobulin G, *Biochemistry* **26:** 1464–1470.

Wei, J., Lin, Y-Z, Zhou, J. M., and Tsou, C. L., 1991, FTIR studies of secondary structures of bovine insulin and its derivatives, *Biochim. Biophys. Acta* **1080:** 29–33.

Wilder, C. L., Friedrich, A. D., Potts, R. O., Daumy, G. O., and Francoeur, M. L., 1992,

Secondary structural analysis of two recombinant murine proteins, interluekins 1α and 1β: is infrared spectroscopy sufficient to assign structures, *Biochemistry* **31:** 27–31.

Yang, J. T., Wu, C-S., and Martinez, H. M., 1986, Calculation of protein conformation from circular dichroism, *Methods Enzymol.* **130:** 208–269.

Yang, P. W., Mantsch, H. H., Arrondo, J. L. R., Saint-Girons, I., Guillon, Y., Cohen, G. N., and Bârzu, O., 1987, Fourier-transform infrared investigation of the *Escherichia coli* methionine aporepressor, *Biochemistry* **26:** 2706–2711.

Yang, W. J., Griffiths, P. R., Byler, D. M., and Susi, H., 1985, protein conformation by infrared spectroscopy: Resolution enhancement by Fourier self-deconvolution, *Applied Spectrosc.* **39:** 282–287.

Yang, W. S., and Griffiths, P. R., 1983, Optimization of parameters for Fourier self-deconvolution. I. Minimization of noise and side-lobes without apodization, *Comput. Enhanced Spectrosc.* **1:** 157–165.

Yang, W. S., and Griffiths, P. R., 1984, Optimization of parameters for Fourier self-deconvolution. II. Band multiplits, Comput. Enhanced Spectrosc, **2:** 69–74.

Zhang, Y-P., Lewis, R. N., Hodges, R. S., and McElhaney, R. N., 1992a, FTIR spectroscopic studies of the conformation and amide hydrogen exchange of a peptide model of the hydrophobic transmembrane α-helices of membrane proteins, *Biochemistry* **31:** 11572–11578.

Zhang, Y-P., Lewis, R. N., Hodges, R. S., and McElhaney, R. N., 1992b, Interaction of a peptide model of a hydrophobic transmembrane α-helical segment of membrane protein with phosphatidylcholine bilayers: differential scanning calorimetry and FTIR spectroscopic studies, *Biochemistry* **31:** 11579–11588.

Chapter 11

X-Ray Diffraction on Biomembranes with Emphasis on Lipid Moiety

Peter Laggner

1. INTRODUCTION

The theme of the fluid mosaic membrane by Singer and Nicholson (1972), in any colored variation, is known to every student of the biosciences. Yet, there exists no complete picture of the structure and dynamics of any real biological membrane, based on a sufficient set of experimentally determined atomic coordinates and trajectories. While the paradigm of static structure as a determinant of biological function has proven extremely fruitful for nucleic acids, and to a large extent also for proteins, this seems not to be the case with biological membranes. From all we know today, the static structures and mechanistic interactions are necessary to describe the protein moiety of membranes; this could be sufficiently defined in terms of three-dimensional atomic coordinates and their dislocations on conformational changes. However, a major species of membranes constituents, the lipids, appears to be physiologically in a state of dynamic disorder within certain limits given by the overall morphology—planar, tubular, or micellar—of the membrane. This is related to the situation in liquid crystals (De

Peter Laggner   Institute of Biophysics and X-ray Structure Research, Austrian Academy of Sciences, A-8010 Graz, Austria.

Subcellular Biochemistry, Volume 23: Physicochemical Methods in the Study of Biomembranes, edited by Herwig J. Hilderson and Gregory B. Ralston. Plenum Press, New York, 1994.

Gennes, 1974), and the physical description of membrane lipids has greatly benefited from the phenomenology applied to that field.

The structural approach to membranes, in particular that of present X-ray diffraction methods, therefore focuses on separate parts of membranes. For the investigation of membrane proteins, the ultimate aim is the crystal structure analysis of isolated membrane proteins. Because of the difficulties inherent in the crystallization of amphipathic proteins (for a comprehensive treatise, see Michel, 1990), which requires the presence of detergents, this has been successfully accomplished only in very few cases so far. Another related method, although of lower resolution, is the investigation of solubilized membrane proteins by X-ray small-angle scattering in solution (for reviews on the method, see Glatter and Kratky, 1982; Feigin and Svergun, 1987; Kratky and Laggner, 1992). This method can provide information on the overall size and shape of membrane proteins and is particularly suited to investigating global conformational changes on specific reactions. Neutron small-angle scattering, which is analogous in its physical basis, can be of particular value in studies on soluble lipoproteins or detergent–protein complexes because of the fact that specific parts of the structure can be highlighted through replacement of hydrogen by deuterium (which has different scattering contrast for neutrons), either in the solvent or in the complex (Blasie et al., 1984; Atkinson and Shipley, 1984; Laggner et al., 1981, 1984).

The present review is concerned with the other major component of membranes, the lipid moiety. While the fluid lipid bilayer structure is generally taken to be the dominating feature in membranes, the spectrum of morphologies that can be attained by polar lipids under different conditions is amazingly wide. Thus, in lipid model systems there are not only a variety of bilayer phases, but also several nonbilayer structures are known to occur (Figure 1). Because of the possible functional significance in biomembranes, e.g., in fusion (for an overview, see Wilschut, 1991), and related to the general importance of this polymorphism toward the physicochemical description of lipids, X-ray structure analysis has played a central role in this field ever since the pioneering studies by Bear et al. (1941) and Palmer and Schmitt (1941). The information has contributed significantly to the enormous amount of data existing on the structural phase behavior of lipids, which cannot possibly be combined in a review of the present size. In fact, recently a computerized data base on lipid phases has been established (Caffrey et al., 1991a) which provides a useful retrieval system for those interested in specific lipid systems and the sources for further data.

Rather than attempting a complete overview on the method and structural results, for which there exist excellent older (Luzzati, 1968; Levine, 1972; Shipley, 1973; Franks and Levine, 1981; Blaurock, 1982) and more recent reviews (Lindblom and Rilfors, 1989; Seddon, 1990; Caffrey, 1989b, c; Tate et al., 1991), the emphasis in this chapter shall be on recent developments in X-ray

A CLASSES OF POLYMORPHIC LIPID PHASES

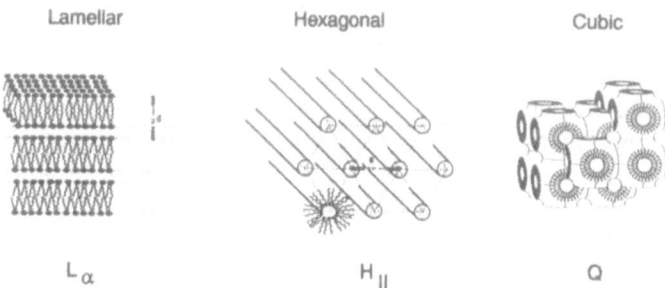

Lamellar	Hexagonal	Cubic
L_α	H_{II}	Q

B SOME FORMS OF LOW-TEMPERATURE LAMELLAR PHASES

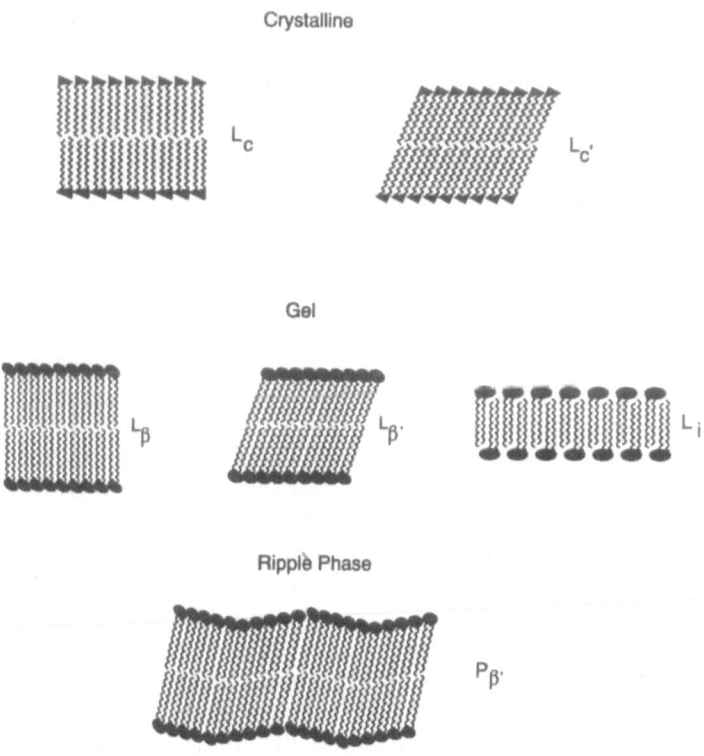

Crystalline

L_c $L_{c'}$

Gel

L_β $L_{\beta'}$ L_i

Ripple Phase

$P_{\beta'}$

FIGURE 1. Some of the characteristic lipid phase structures in schematic representation. (A) The more frequent types of polymorphic phases; note the differences in local curvature at the lipid–water interface. (B) The most frequent types of bilayer structures in lamellar lipid phases below the chain-melting transition.

diffraction techniques which have largely been the consequence of breakthroughs in the fields of fast detectors and brilliant synchrotron X-ray sources. Through these advances it has become possible to overcome the notorious limitations of conventional X-ray diffraction, which had essentially been given by the weak signals and the consequently low speed of the technique. Among other features, perhaps the most important innovations to the field of membranes are the possibilities of fast time-resolved experiments for cinematographic measurements of phase transitions, and of structural studies on single lipid monolayers. On the more conventional side of the methodological spectrum, new technologies have resulted in grossly reduced demands in time and effort to the experimenter, opening the field also to the nonspecialist laboratory.

2. CRYSTAL STRUCTURE OF MEMBRANE LIPIDS

From the static and dynamic disorder that prevails in most natural or model membrane systems, it follows that X-ray diffraction or diffuse scattering data from such systems are rather limited in resolution, and consequently the assignment of atomic coordinates is impossible. Even in the liquid state, however, the molecular packing is so dense that most local structural parameters must resemble those of the solid state in the crystal, despite the lack of long-range order. The only solid reference points for the detailed quantitative interpretation of low-resolution data are therefore the molecular packing coordinates, distances, volumes, and areas from single-crystal diffraction studies. On the other hand, it is clear that the crystal structure provides very little information on the important aspects of hydration, except for the few molecules of crystal water.

A very good basis of knowledge exists for the crystal structures of simple lipids, such as alkanes, alkanols, fatty acids, soaps, and glycerides (Malkin, 1952, 1954; Abrahamson et al., 1978; Shipley, 1986). Much greater difficulties have been encountered in the crystallization of membrane lipids, such as phospho- or sphingolipids, but several important structures have been reported over the past two decades (Pascher, 1976; Pascher et al., 1992; Hauser et al., 1981). As a representative example, Figure 2 shows the structure of 1,2-dimyristoyl-sn-glycero-3-phosphocholine dihydrate (DMPC) solved by Pearson and Pascher (1979). The molecules are in a bilayer arrangement but the crystal forces determine two different polar head group conformations alternatingly side by side, and a mutual displacement in their long molecular axis by 2.5 Å, or by one whole zigzag unit of a hydrocarbon chain. In both conformations the head groups are inclined to the plane of the bilayer, but to different angles (17 and 27°, respectively, for the P–N vectors). The two hydrocarbon chains in one molecule are nonequivalent, in that the sn-2 chain initially runs along the plane of the layer and is bent at the C-2 atom to become parallel to the sn-1 chain. Pearson and

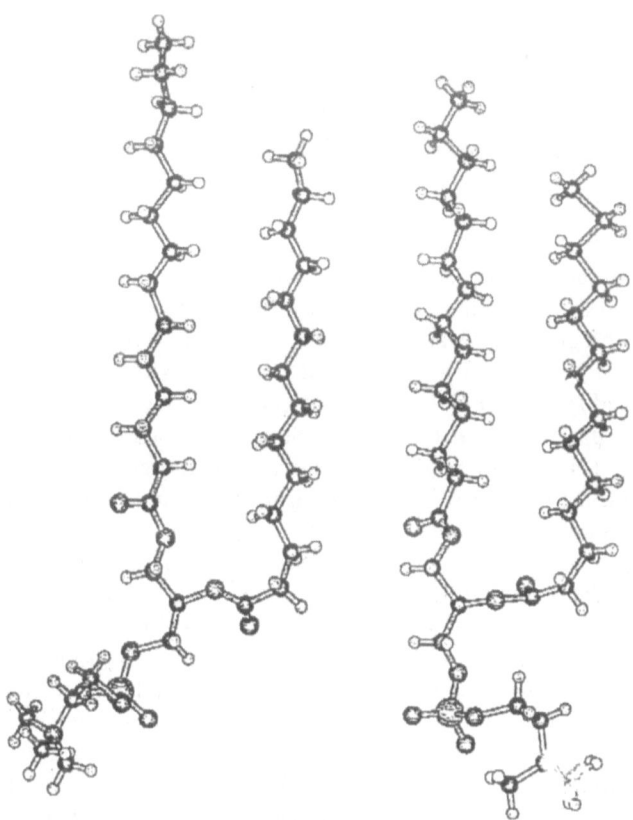

FIGURE 2. Molecular structure and packing within the unit cell of 1,2-dimyristoyl-*sn*-glycero-3-phosphocholine dihydrate (DMPC). From Pearson and Pascher (1979) with permission.

Pascher have pointed out that this accessibility of the carboxyl group of the *sn*-2 chain at the hydrophilic/hydrophobic interface of the lipid bilayer may explain its susceptibility to enzymatic splitting. As a note of caution, however, it has to be stressed that these crystals have been obtained from organic solvents (ether/ethanol/water). In deed, Fourier transform infrared (FTIR) experiments on aqueous samples have shown considerable structural discrepancies to the single-crystal X-ray results, particularly with respect to the glycerol backbone conformation (Lewis and McElhaney, 1992).

Many of the features, e.g., nonequivalence of *sn*-1 and *sn*-2 chains and bending back of the head groups toward the bilayer plane, are similar to those shown by the crystal structure of 1,2-dilauroyl-DL-phosphatidylethanolamine (Hitchcock *et al.*, 1974); however, there are distinct differences in the head group interactions and hydration which may account for the well-known differences in phase behavior of the two lipid classes.

3. DIFFRACTION FROM PARTLY ORDERED OR DISORDERED LIPID SYSTEMS

It is a widespread but mistaken belief among biologists, who are most familiar with the spectacular success of protein crystallography, that X-ray diffraction is limited to the availability of well-developed single crystals. In actual fact, with biological membranes, the main contribution by X-ray diffraction was made through its potential to obtain structural information also from random or partly ordered systems. Figure 3 shows a schematic representation of the nature of structural information that can be extracted from such studies on lipid model membranes. This information can be directly juxtaposed to the thermodynamic data defining the relative stabilities and transition enthalpies as obtained, e.g., from differential scanning calorimetry. Recently, Chung and Caffrey (1992) have reported on a combined experimental approach to obtain X-ray structural and calorimetric data simultaneously, which is certainly of high value in directly assigning certain structural changes to the respective transition enthalpies.

In the following sections, a brief recollection is given on the various approaches by diffraction techniques to study problems relevant to supramolecular lipid structures. For more in-depth information the reader is referred to comprehensive monographs and reviews on the general theory of X-ray diffraction (e.g., Hosemann and Bagchi, 1962; Guinier, 1964; Alexander, 1969; Warren, 1969) and X-ray small-angle scattering (Guinier and Fournet, 1955; Porod, 1982; Feigin and Svergun, 1987; Kratky and Laggner, 1992). For small-angle scattering from random, dilute systems the user can resort to a variety of numerical approaches with well-developed computer programs for data evaluation and interpretation (Glatter, 1977, 1980; Stuhrmann and Miller, 1978; Moore, 1980; Taupin and Luzzati, 1982; Svergun, 1991; Hansen and Skov Pedersen, 1991).

As will be shown, X-ray diffraction offers a highly convenient and informative way to study lipid polymorphism by a comparatively simple approach which, in most cases, does not require extensive procedures for data analysis. In this field, X-ray methods have become comparable in their ease of application to spectroscopic or thermodynamic techniques, and thus form an integrating component in the methodical arsenal of membrane biophysics.

3.1. Lipid Polymorphism

Polar lipids, such as phospholipids, when dispersed in water, form aggregates with a certain type of local symmetry, which can be classified crudely in

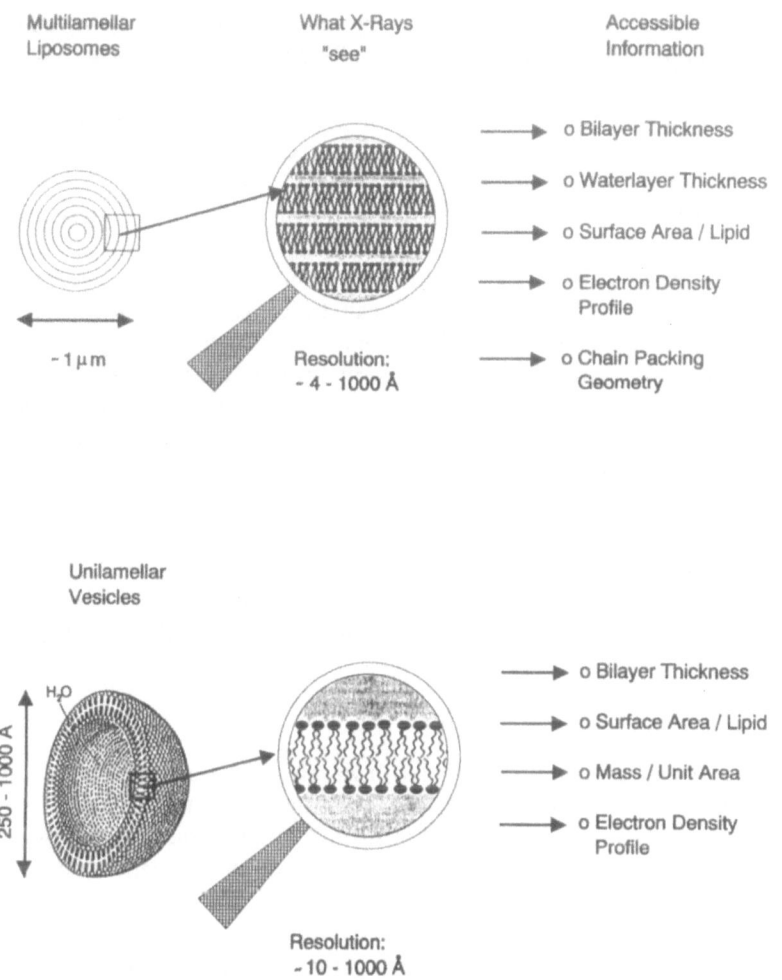

FIGURE 3. Scheme of the information available from X-ray diffraction on lipid/water dispersions in the partly ordered form of multilamellar liposomes and in the form of unilamellar vesicles.

lamellar, hexagonal, or cubic structures, depending on the nature of the lipid and the physicochemical state of the system (T, p, concentration, and nature of salts or other additives). These hydrated lipid structures display, to a good approximation, the characteristics of phases in the thermodynamic sense, i.e., they follow the Gibbs phase rule and exhibit phase transitions. With the exception of cubic phases, which can grow to large monocrystalline dimensions of the size of the sample container, these phases normally are present in microscopic crystalline domains, randomly oriented throughout the sample, so that the X-ray beam (its cross section being much larger than the individual crystallite) "sees" the situation of a crystalline powder. This has the effect that all lattice planes meet the Bragg condition simultaneously thus leading to a powder pattern of concentric rings about the direction of the primary X-ray beam. A scheme of this powder diffraction is shown in Figure 4.

If available to sufficient resolution, the powder diffraction pattern provides all necessary information to define the crystallographic unit cell symmetry and dimensions. In the general case of three-dimensional crystals the indexing of a large number of individual reflections can be an elaborate procedure. With phospholipid dispersions, except for complex cubic structures, however, the problem is grossly simplified because normally only relatively few reflections are sufficient to describe the type of symmetry (e.g., lamellar or hexagonal) and the relevant dimensions. This is related to their liquid-crystalline nature; i.e., the fact that a structural regularity is only expressed at the supramolecular level, while the individual molecules and atomic groups are not fixed in a long-range ordered crystal.

Owing to the fact that the supramolecular lattices of polymorphic phases can have characteristic distances on the order of 10 to several hundred angstroms, and the reciprocity between real-space distances d and scattering angle 2θ as expressed by Bragg's equation,

$$n \cdot \lambda = 2d \cdot \sin\theta \tag{1}$$

the corresponding diffraction signals (with wavelengths λ on the order of 1 Å) lie in the small-angle region, typically between some 10 and 100 mrad. There is, however, also important structural information to be extracted from the wide-angle region, corresponding to the dimensions of the subcell lattices of the hydrocarbon chain packing (Malkin, 1952, 1954; Abrahamson et al., 1978; Maulik et al., 1990), if they are not in a liquid disordered state. Ideally, these two regions of the diffraction pattern, which pose quite different demands on the detection system (see Section 3.2.2), should be measured simultaneously to characterize the structure of a given phase. The most frequently encountered mesomorphic phases are described below with respect to their typical diffraction features.

3.2. Lamellar Phases

3.2.1. Small-Angle Diffraction

Depending on the chemical nature of the lipid, the state variables (T, p, and degree of hydration), or the presence of co-solutes, the lipid bilayer can occur in various structural forms, such as normal or interdigitated, or partially interdigi-

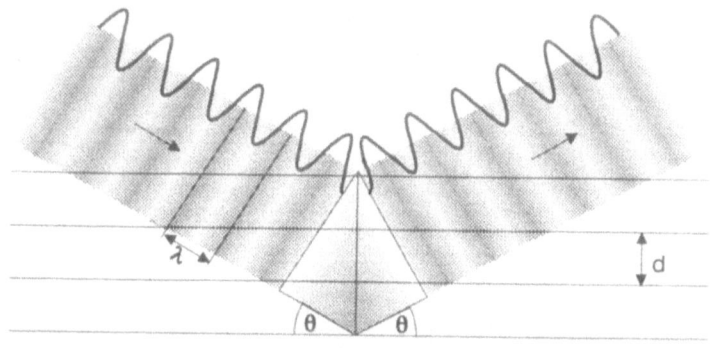

a

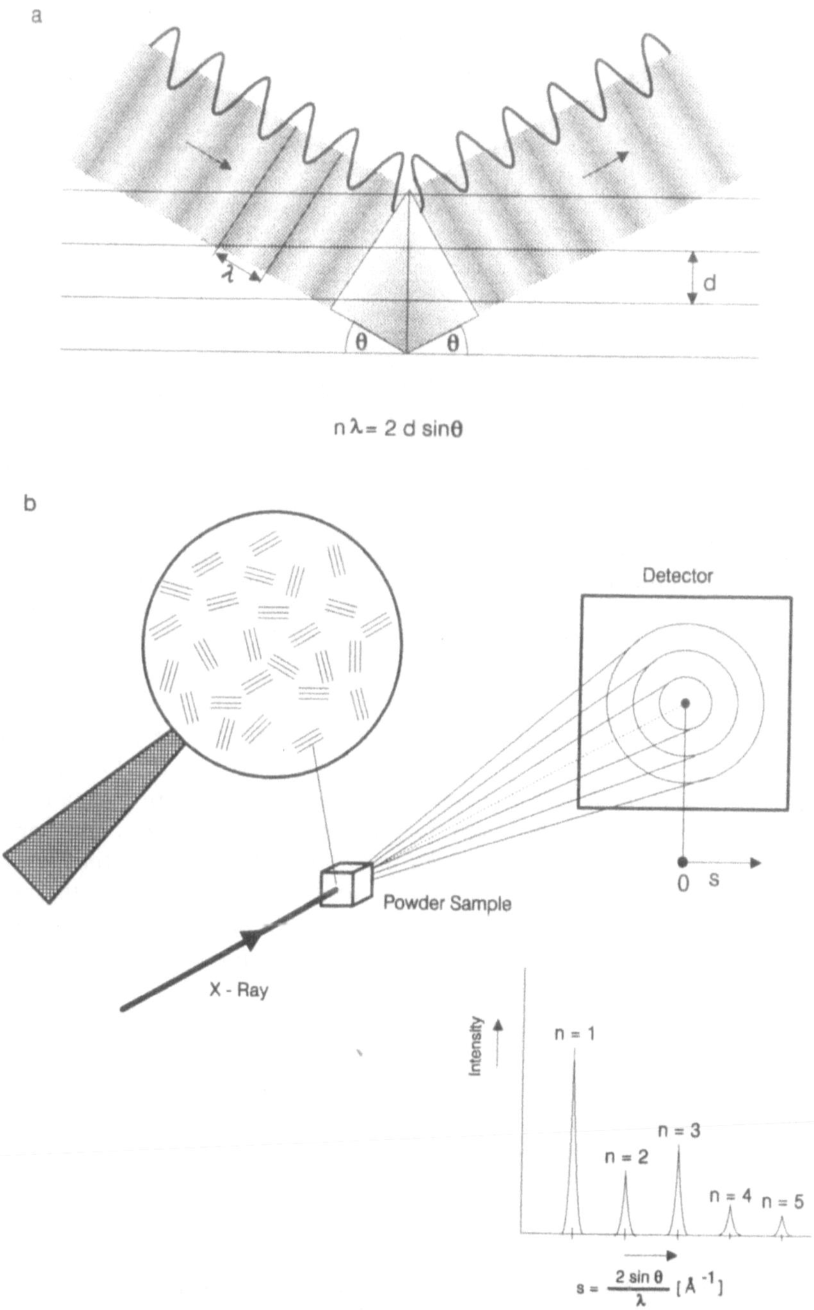

$$n\lambda = 2\ d\ \sin\theta$$

b

Detector

0 s

Powder Sample

X - Ray

Intensity

n = 1

n = 3

n = 2

n = 4 n = 5

$$s = \frac{2\sin\theta}{\lambda}\ [\text{Å}^{-1}]$$

FIGURE 4. (a) Schematic representation of diffraction of a wave on a lattice of planes (Bragg diffraction). (b) The optical arrangement in a powder diffraction experiment; the inset shows an idealized diffraction pattern, i.e., the radial intensity profile through the diffractogram at the detector.

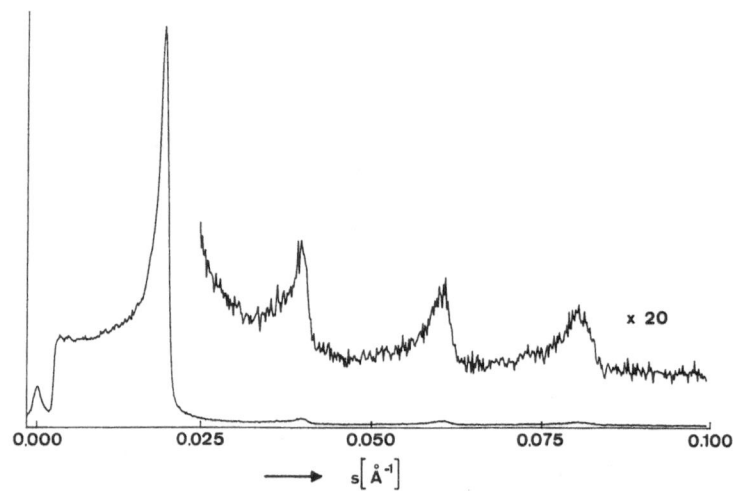

FIGURE 5. A representative experimental X-ray small-angle powder diffraction pattern for L-α-dipalmitoylphosphatidylethanolamine in aqueous dispersion, taken with a Kratky line-collimation camera and MBraun position-sensitive detector OED 50-M (MBraun Graz Optical Systems, Graz, Austria) at a laboratory X-ray generator; exposure time: 5 min.

tated, which all can be characterized by their respective small-angle diffraction patterns (Mattai *et al.*, 1987; Laggner *et al.*, 1987a; Lohner *et al.*, 1987). As an example, Figure 5 shows a typical small-angle diffraction pattern from a multilamellar phospholipid in the L_β phase. In this case, the existence of only integral multiples of the position of the first-order reflection indicates a one-dimensional lattice. The structural regularity relates only to the periodic repeat of the lipid bilayer together with the water of hydration. The repeat distance d can be simply calculated from the angular positions s_n of the reflection peaks, according to

$$d = n/s_n = \lambda /(2 \sin \theta_n) \qquad (2)$$

If the volume fraction of water f_w in this phase (not in the total sample volume) is known, e.g., from a series of measurements of d as a function of water content, the thickness of the lipid bilayer d_1 can be calculated from

$$d_1 = d \cdot (1 - f_w) \qquad (3)$$

It should be noted, however, that strictly this requires the knowledge of the specific volumes ϕ_w and ϕ_1, of water and lipid, respectively, to calculate the volume fraction. The molecular area A, in $Å^2$, of a lipid molecule at the bilayer–water interface can be obtained in a basically similar way from the formula

$$A = (2_{\phi 1} \cdot M \cdot 10^{24}) / (d_1 \cdot N_L) \quad (4)$$

where M is the molecular weight and N_L is Loschmidt's number. Care has to be taken, however, to consider the effects of the basic assumption inherent in this type of calculation, i.e., that the water and the lipid occupy discrete domains along the direction normal to the bilayer, which of course is a rather crude simplification (Murthy and Worthington, 1991). Second, the precise values for the specific volume of water (or the buffer) in the interbilayer space may dif fer significantly from its bulk value, particularly at water contents below 10% (Laggner *et al.*, 1987a; Scherer, 1987; White *et al.*, 1987). For most practical purposes, however, these errors are barely relevant. Quantitative procedures to determine the membrane-bound water by a combined approach with X-ray and neutron scattering have been reviewed by Knott and Schoenborn (1986).

In addition to these integral structural parameters, the powder diffraction patterns can be analyzed also in terms of the electron density profile, $\rho(x)$, centrosymmetrical and normal to the bilayer plane. This requires the measurement of the intensities $I(n)$ (with n the order of the diffraction peaks) and the determination of the signs of the amplitudes $\pm|\sqrt{I(n)}|$, as schematically shown in Figure 6. The relation between $\rho(x)$ and $I(n)$ is given by

$$\rho(x) = (2/d) \cdot \Sigma \pm |n \cdot \sqrt{I(n)}| \cos(2\pi xn/d) \quad (5)$$

Several procedures for the determination of the signs in this equation have been described in the literature (Franks, 1976; McIntosh and Holloway, 1987; Worthington, 1981; Wiener *et al.*, 1991; Wiener and White, 1992; for reviews, see Franks and Levine 1981; and Laggner, 1988). Particularly elegant ways of verifying the choice of signs for the amplitudes are given in neutron scattering by contrast variation with H_2O/D_2O mixtures (Worcester and Franks, 1976).

3.2.2. Wide-Angle Diffraction

At temperatures below the chain-melting transition, the hydrocarbon chains of membrane lipids adopt ordered states, in which the motions in the plane of the bilayer are frozen. This may occur in discrete stages: first, only the lateral motions are inhibited while the rotation about the long molecular axis is still fast. This can lead to a hexagonal or pseudohexagonal (2-D rectangular) packing of the chain rotation cylinders, which have a specific signature in the wide-angle diffraction patterns (Figure 6). At even lower temperatures, the chains may adopt

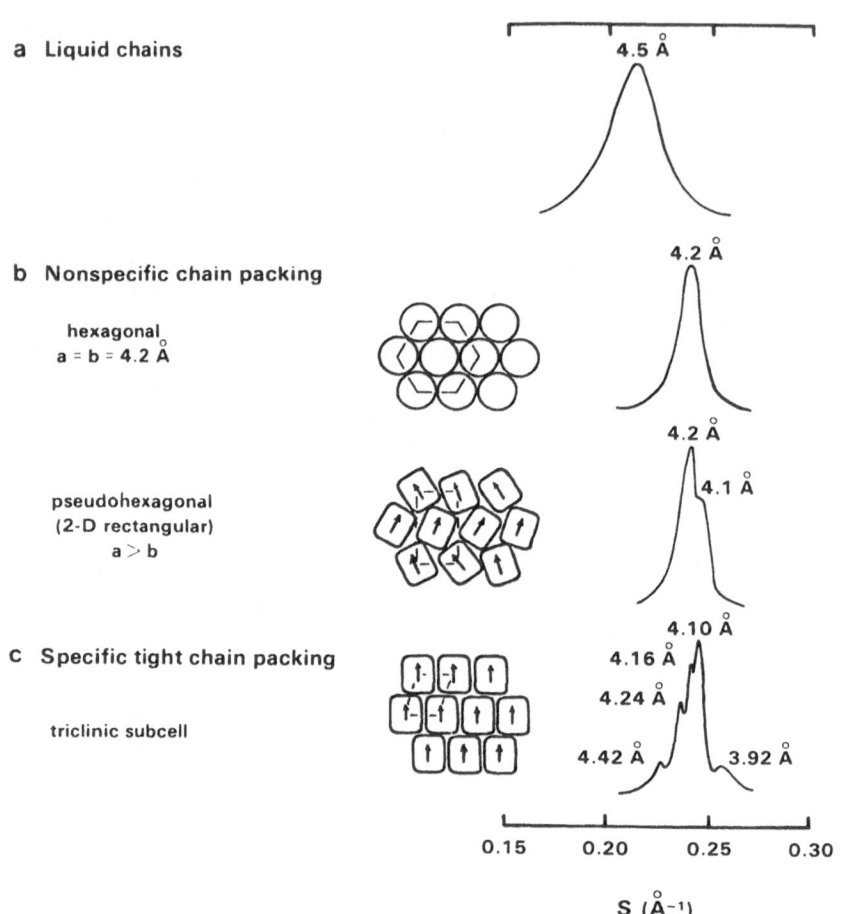

FIGURE 6. (Left) Typical examples of hydrocarbon chain packing symmetries viewed in projection onto the bilayer plane; (right) the corresponding X-ray wide-angle diffraction patterns. From Small (1986) with permission.

a crystalline configuration, with a more differentiated wide-angle diffraction pattern. Recent examples, where both small- and wide-angle diffraction were used to analyze lipid polymorphism, can be found in Laggner *et al.* (1987a) for dihexadecylphosphatidylcholine, in Kodali *et al.* (1990) for diacylglycerols, and in Sen *et al.* (1990) for glycolipids.

Until recently, this required two separate diffraction experiments since special cameras, differing from those suitable for resolving the small-angle patterns, had to be used for the measurement of these signals. Apart from being a tedious duplication of efforts, this has the serious drawback that one can never be sure whether the two experiments are indeed representative of the same state of the sample, both with respect to the state variables and to the sample history. This has been overcome by the development of a dual-detector camera, the SWAX camera (MBraun Graz Optical Systems, Graz, Austria), which allows the simultaneous detection of both small- and wide-angle diffraction patterns without compromising the resolution in either range (Laggner and Mio, 1992). Figure 7 shows this camera and an example of its performance. With this instrument, the discussion of the structural events occurring at lipid phase transitions (e.g., metastabilities or the development of defects; Gordeliy et al., 1991) as measured by calorimetry can be put on a solid basis.

3.2.3. Macroscopically Oriented Lamellar Systems

The quality of diffraction data from lamellar systems can be enhanced through attempts to macroscopically orient the hydrated lipids on a solid surface and exposure to a controlled humidity environment (Hentschel and Hosemann, 1983; Katsaras et al., 1991; McIntosh et al., 1989a, b, 1992). In addition to improving the quality of diffraction data, this approach provides experimental access to the osmotic properties and repulsive forces between lipid bilayers (see also review by Parsegian et al., 1986).

Recently, Katsaras et al. (1992) have made full use of sample orientation through a two-dimensional image plate detector system covering also the wide-angle part of the diffraction pattern. Thus, they were able to directly determine not only the angle of tilt of the hydrocarbon chains of DPPC under different relative humidity conditions, but also the direction of tilt. These results have, therefore, provided an important clue to a long-standing problem for an otherwise extensively studied lipid membrane model system.

3.3. Hexagonal Phases

Closely analogous to the one-dimensional approach outlined above for lamellar phases is the analysis of powder diffraction patterns from two-dimensional hexagonal lipid phases. The fingerprint of a hexagonal phase in powder patterns, which in fact has been proven to be more critical than the ^{31}P NMR signal (Gasset et al., 1988), is the sequence of spacings

$$s = (1, \sqrt{3}, 2, \sqrt{7}, 3, \ldots)/d \qquad (6)$$

The first d spacing (corresponding to the [1,0] Bragg peak) is related to the distance a between the water cylinders by

$$a = 2d/\sqrt{3} \qquad (7)$$

a

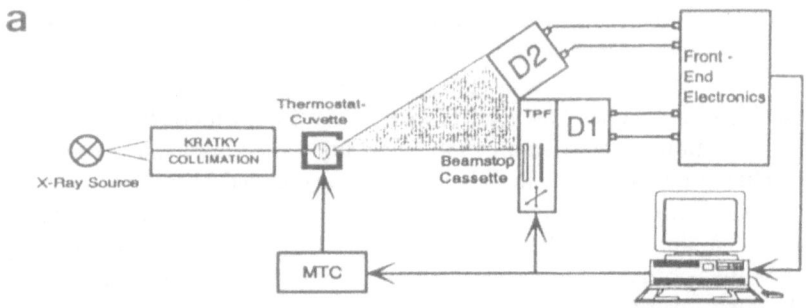

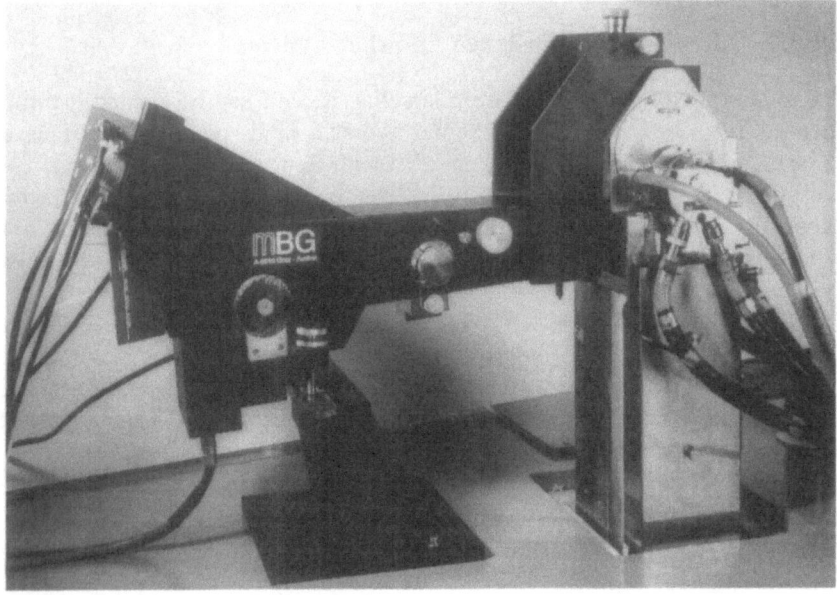

FIGURE 7. The X-ray small- and wide-angle camera SWAX (MBraun Graz Optical Systems, Graz, Austria). (a) Block scheme of the components (MTC, measurement and temperature controller; TPF, transparent beam-stop cassette; D1 and D2, small- and wide-angle position-sensitive detectors. (b) Natural view. (c) Typical data set with small- (left) and wide-angle (right) diffractograms taken simultaneously in a temperature–time protocol.

(see also Figure 8a). Again, this distance can be geometrically separated into the lipid layer thickness and the water cylinder radius by a volume fraction calculation similar to the one indicated above for the lamellar case. Similarly, the molecular surface area at the lipid/water interface can be calculated. All equations are listed and discussed in Seddon (1990). Methods to reconstruct the two-dimensional electron density relief of the hexagonal H_{II} phase in the plane

C

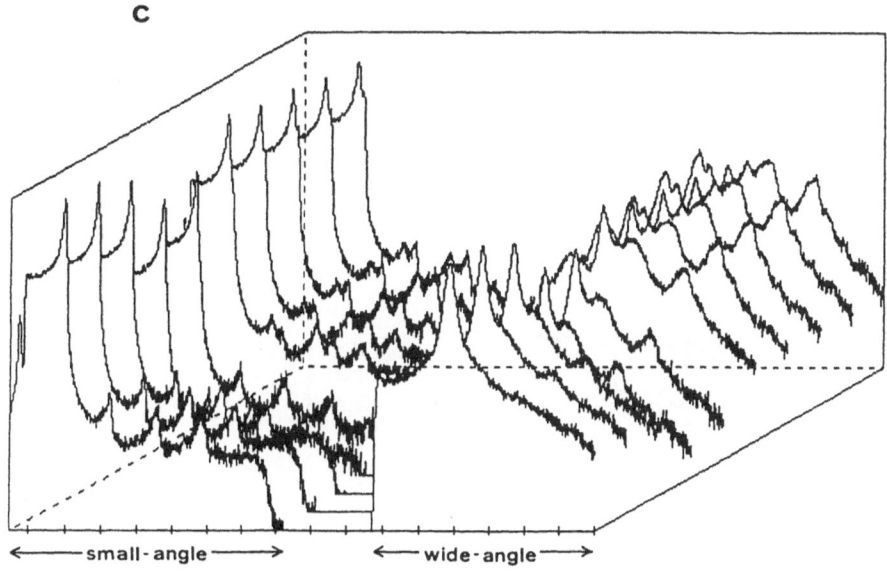

←——— small-angle ———→   ←——— wide-angle ———→

FIGURE 7. *Continued.*

vertical to the cylinder axes from the observed X-ray intensities have been described first by Caron *et al.* (1974) and Gulik *et al.* (1985, 1988), and recently refined and extended to electron-density reconstruction of a noncircular averaged shape of the water tubes by Turner and Gruner (1992) and Turner *et al.* (1992). A typical result of their work is shown in Figure 8b.

3.4. Cubic Phases

In the context of membrane fusion and, more generally, of the question of membrane curvature, the cubic phases of lipids have recently attracted increasing interest. Such three-dimensional periodic structures were first observed in bacterial lipid extracts (Pseudomonas fluorescens; Tardieu, 1972). Various crystallographic space groups for cubic phases of membrane lipids have been found by X-ray diffraction analysis and discussed (Luzzati *et al.*, 1992; Mariani *et al.*, 1988, 1990; Vargas *et al.*, 1992; Seddon *et al.*, 1990; Shyamsunder *et al.*, 1988). In all cases so far studied, these structures have been found to lie in between the Lα and H$_{II}$ phases, and it has therefore been postulated that cubic structures play a general mechanistic role in the interconversion between lamellar and hexagonal phases (Siegel, 1987; Siegel and Banschbach, 1990).

From the practical point of X-ray diffraction, both the measurement and the

a   DIMENSIONS OF THE H$_{II}$ - PHASE

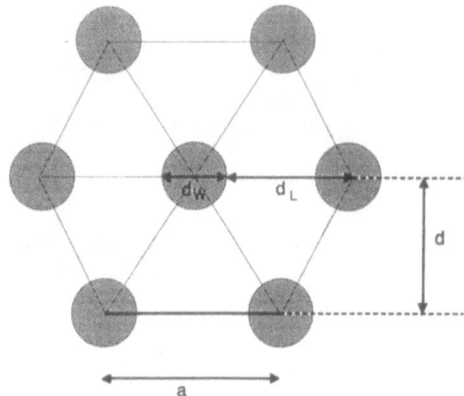

b   MOLECULAR PACKING

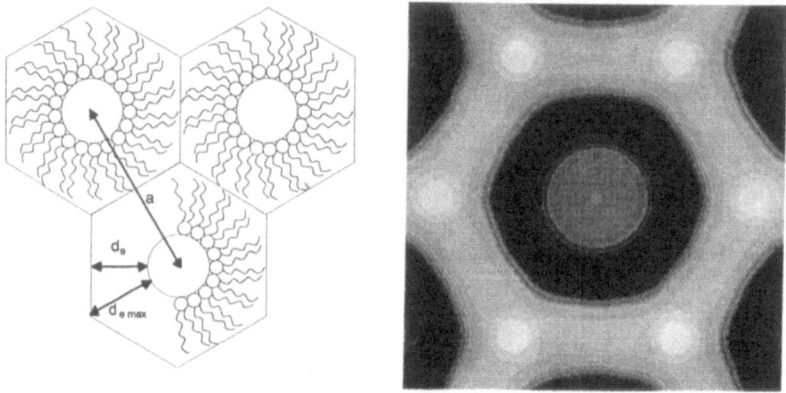

FIGURE 8. Geometric scheme and molecular packing in the hexagonal H$_{II}$ phase. Bottom right shows the electron density projection onto the plane perpendicular to the water cylinders.

analysis of cubic phases are definitely more difficult than for lamellar or hexagonal ones. Cubic phases of membrane lipids tend to form macrodomains in the samples and therefore the powder diffraction approach is often unsuccessful in picking up the characteristic reflections. They are also often metastable, their generation can be enhanced by temperature cycling, and in general their formation is poorly reproducible (Shyamsunder *et al.*, 1988); moreover, the pattern of discrete reflections may be blurred by a strong diffuse scattering background, so that an insufficient number of peaks can be indexed and assigned unambiguously. For reliable results it is therefore essential to have a precise record of the sample history and its effects on the observed diffraction pattern. Preferably, two-dimensional detectors (film or 2-D position-sensitive detectors) as well as sample rotation should be employed. A representative example of a cubic phospholipid diffraction pattern is shown in Figure 9.

3.5. Studies on Interactions

With this structural information at hand, the method of X-ray diffraction offers wide possibilities to investigate the interactions of different lipid species or of lipids with other molecular species. The literature abounds with such studies, and the majority of them are not concerned with an extensive structure analysis but rather with the dose–effect correlations in a specific type of interaction. In this sense, X-ray diffraction in membrane biophysics has become a scanning technique for membrane-active substances similar to other physical approaches, such as the various spectroscopic or thermodynamic methods. It is a definite advantage of X-ray diffraction, nevertheless, that even if used in scanning the boundaries of phase diagrams (Ruocco and Shipley, 1982; Mattai *et al.*, 1987; Laggner *et al.*, 1987a; Lohner *et al.*, 1987, 1991; Lewis *et al.*, 1990; Hing *et al.*, 1991; Hinz *et al.*, 1991; Gawrisch *et al.*, 1992; see also the review by Caffrey, 1989c), it provides at the same time quantitative structural information which lends itself to model building. Some examples from the more recent literature include studies of the effects of osmotically active co-solutes (Rand *et al.*, 1990; Klose *et al.*, 1991; Simon *et al.*, 1991; Burgess *et al.*, 1992), of salts (Tilcock *et al.* 1988; Sanderson *et al.*, 1991), of sugars and cryoprotectants (Lee *et al.*, 1989; Williams *et al.*, 1991), of cholesterol (Tilcock *et al.*, 1988; Needham *et al.*, 1988; Finean and Hutchinson, 1988; Finean, 1989; Mason *et al.*, 1992a, b; McIntosh *et al.*, 1992), of squalene (Lohner *et al.*, 1993), of tocopherols and antioxidants (Nakajima *et al.*, 1990), of antitumor (de Wolf *et al.*, 1990; Colotto *et al.*, 1992a) and antiarrhythmic drugs (Jendrasiak *et al.*, 1990; Young *et al.*, 1992), and of cannabinoids (Mavromoustakos *et al.*, 1990).

Of particular importance to an understanding of the interplay between lipids and proteins in biological membranes is the fact that proteins and peptides can specifically alter the morphology of lipid phases. This aspect has received atten-

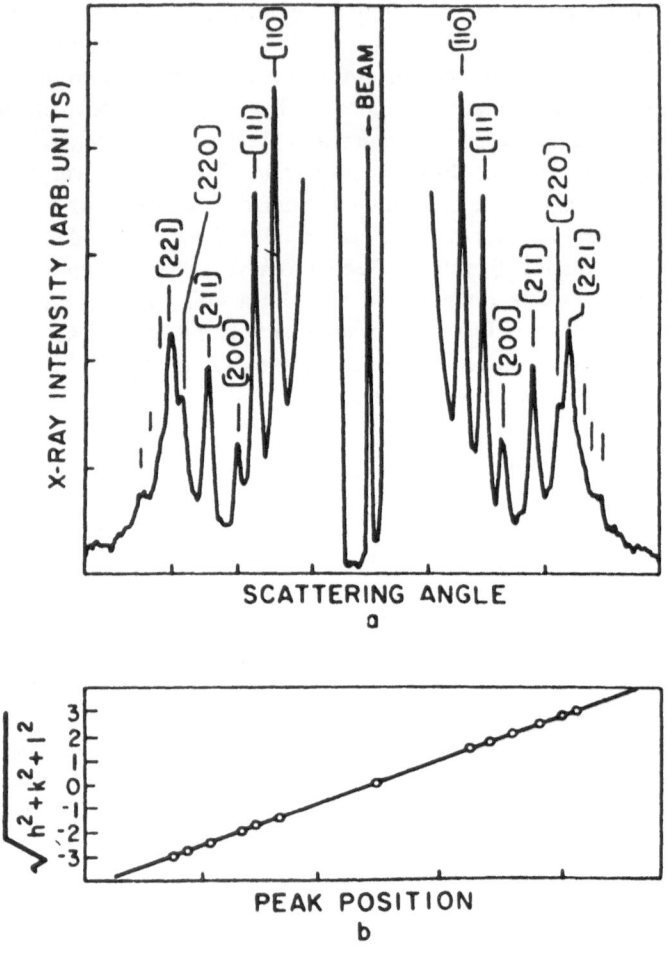

FIGURE 9. (a) X-ray diffraction of hydrated DOPE (dioleoylphosphatidylcholine) in the cubic phase, after 1200 cycles across the L_a–H_{II} transition. The marks indicate the expected peak positions for the $Pn3m/Pn3$ lattice of unit cell spacing 122 Å. (b) The function $(h^2 + k^2 + l^2)^{1/2}$ versus measured peak positions. The spacing between the lines fall in the ratios of $\sqrt{2}:\sqrt{3}:\sqrt{4}:\sqrt{6}:\sqrt{8}:\sqrt{9}$ as would be expected for a $Pn3m$ or $Pn3$ cubic space group. From Shyamsunder *et al.* (1988) with permission.

tion in the context of ion-channel formation by ionophores, such as gramicidin (Aranda *et al.*, 1987). A most notable example is that of melittin, a 26-amino-acid amphiphilic peptide from bee venom, which alters the thermotropic behavior of phospholipids even at molar ratios as low as 1:100 (peptide : lipid). Extensive studies on the structural background of this interaction (Batenburg *et al.*, 1987a, b; Colotto *et al.*, 1991, 1992b) have shown clearly that this low-dose effect of melittin is primarily one of curvature modulation of bilayers which may even lead to nonbilayer structures. It can be expected that the application of X-ray diffraction on related peptide–lipid systems will provide deeper insight into biologically relevant membrane processes, such as peptide-induced fusion and morphogenesis of cells and organelles.

3.6. Diffuse Small-Angle Scattering from Unilamellar Vesicles

Because of the very weak, diffuse scattering signal and the consequently rather long exposure times (5–10 hr on conventional X-ray sources) necessary to obtain good results, this method has not become as popular as the above-described diffraction approach to partly ordered systems. Basically, however, the information contained in the continuous scattering curves is very valuable since it reflects the cross-sectional structure of the single lipid bilayer membrane: the angular scattering function $I(h)$, where $h = (4\pi\cdot\sin\theta/\lambda$, is connected to the electron density profile ρ (r) of the bilayer by the pair of Fourier transform relations

$$\pm|\sqrt{I_t(h)}| = \pm|\sqrt{I(h)}| \cdot h = c \cdot \int_0^\infty \rho(r) \exp(-ihr) \cdot dr \qquad (8)$$

$$\rho(r) = c' \cdot \int_0^\infty \pm|\sqrt{I_t(h)}| \exp(-ihr) \cdot dh \qquad (9)$$

The direct transformation of the intensities $It(h)$ leads to the autocorrelation or Patterson function, which obviates the need to find the correct signs for the root terms in Equations 8 and 9, and provides unambiguously the membrane thickness from the value of r where it finally becomes zero (for a review on the theory, see Laggner, 1988). However, because of the limited range of h over which $I(h)$ can be measured, the resolution is also limited. Taking the canonical resolution as π/h_{max}, this will normally not extend below 5 Å; however, this does not apply to the detectability of differences, e.g., in the electron density peaks of the polar head groups or the bilayer thickness in different systems. With good quality of the scattering patterns, such differences can be reliably found to a resolution of about 3Å, or even better. A schematic view of the functional relationships is shown in Figure 10.

Despite the experimental limitations of this method, its advantages over the powder diffraction approach can be seen in the following points: (1) the structural information, e.g., of the bilayer thickness, refers to the individual membrane, not affected by interbilayer interactions, and possible errors arising from the

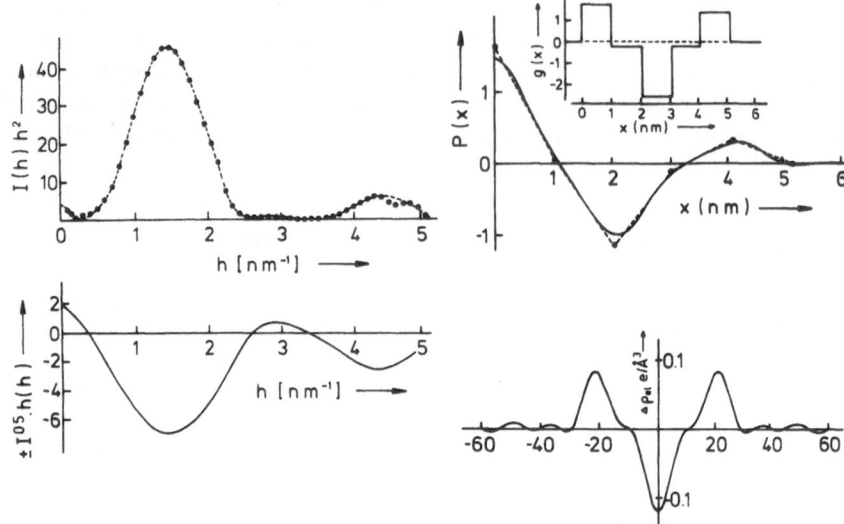

FIGURE 10. Small-angle scattering from a dilute, random dispersion of lipid vesicles. (a) Thickness factor obtained from the experimental intensity distribution $I(h)$ by multiplication with h_2. (b) Structure factor (amplitude function) with arbitrarily chosen signs $(+,-,+,-)$. (c) Autocorrelation function of the electron density (x) profile obtained by cosine transformation of $I_t(h)$; the inset shows the profile obtained by deconvolution. (d) Centrosymmetric electron density profile obtained by Fourier transformation of $F_t(h)$. From Laggner (1988) with permission.

unknown local specific volume terms (see Section 3.2.7) are avoided; (2) structural changes induced by the addition of co-solutes, such as salts or proteins, can be studied without interference of osmotic or steric effects, which may be different in multilamellar preparations; (3) lipids which spontaneously tend to form vesicles and are not forming regular mesophases can be structurally investigated.

In fact, this approach has provided crucial evidence for the notion that the lipid bilayer is the basic structural motif of biological membranes (Wilkins *et al.*, 1971). Lewis and Engelman (1983) have used this method to study systematically the bilayer thickness for a homologous series of diacylphosphatidylcholines with different carbon numbers, and found that the thickness varies linearly, while the surface area remained essentially constant. This supported the view that the phospholipid head groups at the bilayer interface predominate in determining the equilibrium configuration in fluid bilayers.

The above-indicated point, that lattice constraints in multilamellar preparations may lead to conclusions which cannot be extrapolated to the situation of the isolated membrane in a vesicle, is illustrated by the case of dihexadecylphosphatidylcholine (DHPC), which is the di-ether analogue of dipalmitoylphosphatidyl-

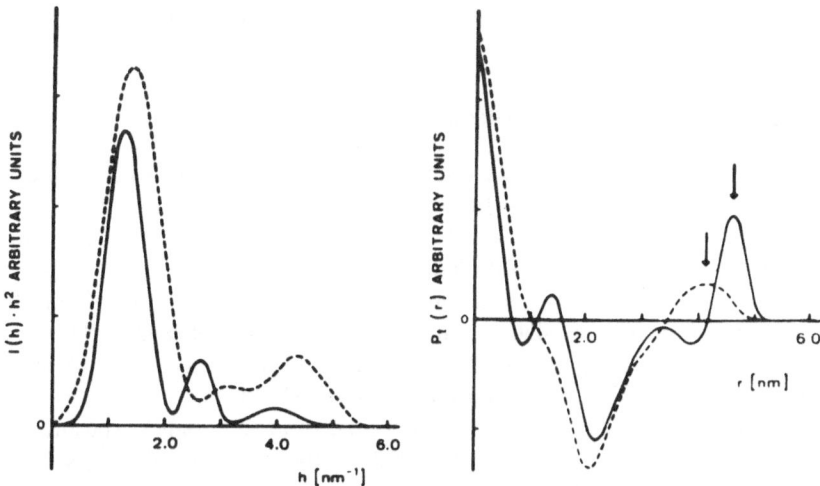

FIGURE 11. X-ray small-angle scattering from DHPC small unilamellar vesicles at 20°C (solid line) and 50°C (dashed line). (Left) The Lorentz corrected scattering curves. (Right) Electron pair correlation function obtained by Fourier transformation of $I(h) \cdot h^2$. Arrows indicate the approximate distances between the phosphate head groups across the bilayer normal. These distances clearly show that in single-shelled vesicles of DHPC, no interdigitation occurs. From Laggner et al. (1967a) with permission.

choline (DPPC), and forms interdigitated bilayers in the multilamellar dispersion (Laggner et al., 1987a). In contrast, this is not the case in small unilamellar vesicles, as shown by diffuse small-angle scattering experiments (Figure 11).

It may be pointed out that the present knowledge on the structure of soluble plasma lipoproteins, which bear many relationships to membrane structure, is strongly based on diffuse X-ray and neutron small-angle scattering (for reviews, see Laggner and Müller, 1978; Kostner and Laggner, 1989).

4. X-RAY DIFFRACTION WITH SYNCHROTRON RADIATION

Synchrotron radiation from electron or positron storage rings (Hodgson and Doniach, 1978; Stuhrmann, 1978; Margaritondo, 1988; Brefeld and Gürtler, 1992) offers several excellent features for performing X-ray diffraction studies, such as extremely high X-ray flux, spectral tunability, linear polarization, and pulsed time structure. Especially the high X-ray flux—gain factors of about 10^5 as compared with conventional laboratory X-ray sources can presently be obtained—has led to new and exciting applications for X-ray diffraction, also in the field of membrane structural studies (for reviews, see Laggner, 1986, 1988;

Gruner, 1987; Caffrey, 1989b, c; Lis and Quinn, 1991), which shall be outlined in the following.

4.1.   Time-Resolved X-ray Diffraction on Lipid Phase Transitions

The transitions between mesomorphic lipid phases pose some interesting and hitherto unsolved problems. In simple terms, these can be summarized in the following two questions: what are the kinetics (one or more relaxation components), and how can they be mechanistically interpreted in terms of structural rearrangements between the initial and final phase structures? In this field, there is presently still much speculation that needs to be further substantiated both with respect to the underlying physics and to their possible biological significance in membrane processes (Laggner and Kriechbaum, 1991).

While with conventional X-ray sources the shortest possible exposure times to obtain a powder diffraction pattern are on the order of 10–100 sec, synchrotron radiation experiments can presently reach a time resolution of tenths of milliseconds (Laggner et al., 1989; Kriechbaum et al., 1990). This opens the field for cinematographic structural investigations on membrane processes.

At this point, a brief technical remark is necessary to illustrate the focal points of the present development in time-resolved X-ray diffraction. The possibility opened by synchrotron radiation sources to take millisecond exposures is but one prerequisite for cinematographic X-ray studies. Another often neglected point concerns the timing and triggering of the phase transitions. In most of the studies so far, temperature has been chosen as the variable for triggering the transitions. Originally, simply external heating by thermostat fluid (Laggner, 1986) or heating by a heat gun (Caffrey, 1985) has been applied. With such methods, however, the problems of slow heat conduction within the aqueous samples pose severe limits: the inevitably occurring internal temperature gradients lead to the fact that at any instant the X-ray beam "sees" a broad distribution of states. This limitation cannot be overcome: the faster the external heating, the larger will be the internal gradient, and consequently no clear-cut structural information can be drawn from such experiments, unless the structural processes are much slower than the time scale of T-equilibration within the sample, in which case the use of synchrotron radiation is obsolete. What is really required are defined jumps in the thermodynamic variables (T, p, or concentration of an additive) so that the entire system under investigation crosses the thermodynamic phase boundary instantaneously.

For temperature as the variable there is presently only the choice between two alternatives: infrared laser or microwaves. Of these two, only the IR-laser option has so far proven to provide the potential to obtain T-jumps of sufficient height in the millisecond time range (on the order of 10°C/msec; Laggner et al., 1989; Kriechbaum et al., 1990). A schematic view of this method is given in

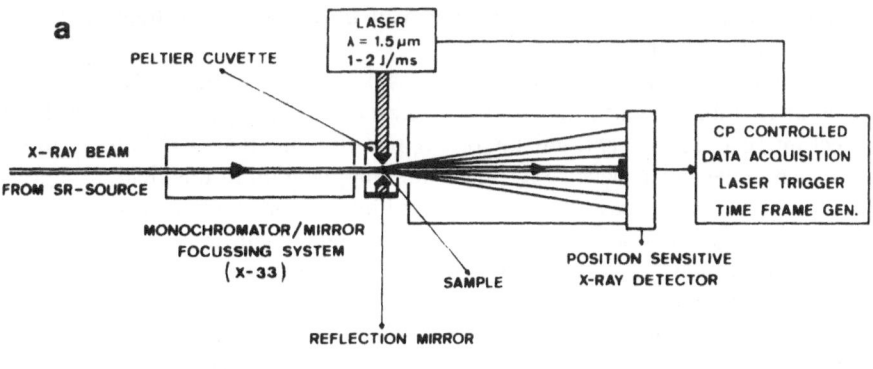

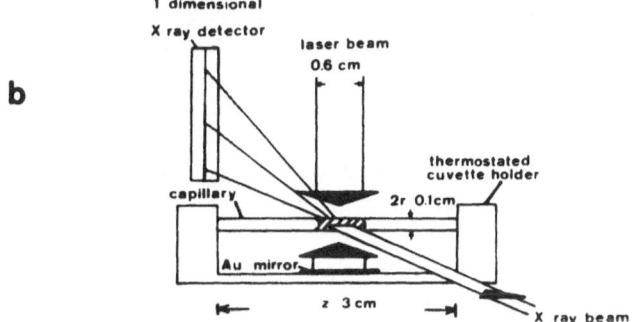

FIGURE 12. Schematic view of the millisecond time-resolved X-ray diffraction with synchrotron radiation and IR-laser temperature jump at camera X-33 of the EMBL Hamburg Outstation at the storage ring DORIS II. (a) Overall scheme. (b) Close-up view of the sample geometry in the cross-fire position of X-ray and laser beams.

Figure 12. The microwave heating approach has been taken by Caffrey *et al.* (1990) in time-resolved studies on the L_α–H_{II} transition with jump rates of up to 29°C/sec.

Pressure jumps, as the thermodynamic alternative to T-jumps, have only recently been applied for the first time in the field of time-resolved X-ray studies. This approach, though more difficult in practice, has the advantage that jumps in either direction through a phase transition can be performed, and that pressure is homogeneous at any time throughout the sample; first results on jumps of up to 2 kbar/10 msec on lipid systems were reported by Kriechbaum *et al.* (1993). Mencke and Caffrey (1991) and Caffrey *et al.* (1991b) have reported also on first pressure jump experiments, as well as on a repetitive technique involving oscillating pressure (Mencke *et al.*, 1993), which has great potential in gaining further insight into the kinetics and mechanism of lipid phase transitions (Van Osdol et al., 1989).

Beyond T and p as variables in time-resolved studies, the rapid admixture of salts or membrane-active substances (detergents, drugs, peptides) would be of great interest. There, rapid-mixing or stopped-flow methods would be required, but such experiments have so far found little application in this field, except for one report on the process of detergent–lipid interaction (Laggner et al., 1987b). A lyotrope-gradient technique which might lend itself to studies related to composition-induced phase changes under diffusion control has been proposed by Caffrey (1989a).

In the following, a summary of the hitherto performed time-resolved X-ray diffraction studies by IR-laser T-jump shall be given. In the discussion of structural phase transitions we have found it convenient to introduce an operational classification based on the geometric nature of the long-range rearrangements involved. Thus, we use the term *homologous* for transitions between affine lattices, i.e., where only the lattice parameters change, while the symmetry type remains unchanged (e.g., lamellar–lamellar). *Heterologous* transitions, on the other hand, are those for which the symmetry type changes (e.g., lamellar–hexagonal). Despite its simplicity and lack of theoretical foundation, it will be seen that this system allows for certain predictions regarding the transition mechanism.

4.2. Homologous Transitions: The Concept of Martensitic "Umklapp" Transformations

A typical result of a T-jump experiment for a transition between two lamellar lattices, lamellar gel to lamellar liquid-crystalline (L_α–L_β) of an ethanolamine phospholipid (1-palmitoyl-2-oleoyl-*sn*-glycero-3-phosphoethanolamine) in water, is shown in Figure 13a. On the time scale of the experiment, this transition is immediate. The disappearance of the parent and the appearance of the nascent diffraction pattern coincide in time, and no intermediate product, ordered or disordered, can be detected. This leads to the conclusion that the transition is, at least at this time scale and with respect to the long-range structural characteristics, a one-step, two-state process.

In an attempt to rationalize these findings, in particular the notion of coexistence of the two lamellar structures with different repeat distances without intermediate disorder, we have postulated (Kriechbaum et al., 1990) that the transition occurs by cooperative folding about a disclination plane, as indicated in Figure 13b, which propagates rapidly through the system. The ratio of the repeat distances is simply related to the cosine of the disclination angle. This mechanism accounts for the conservation in lattice order and for a minimum in defect

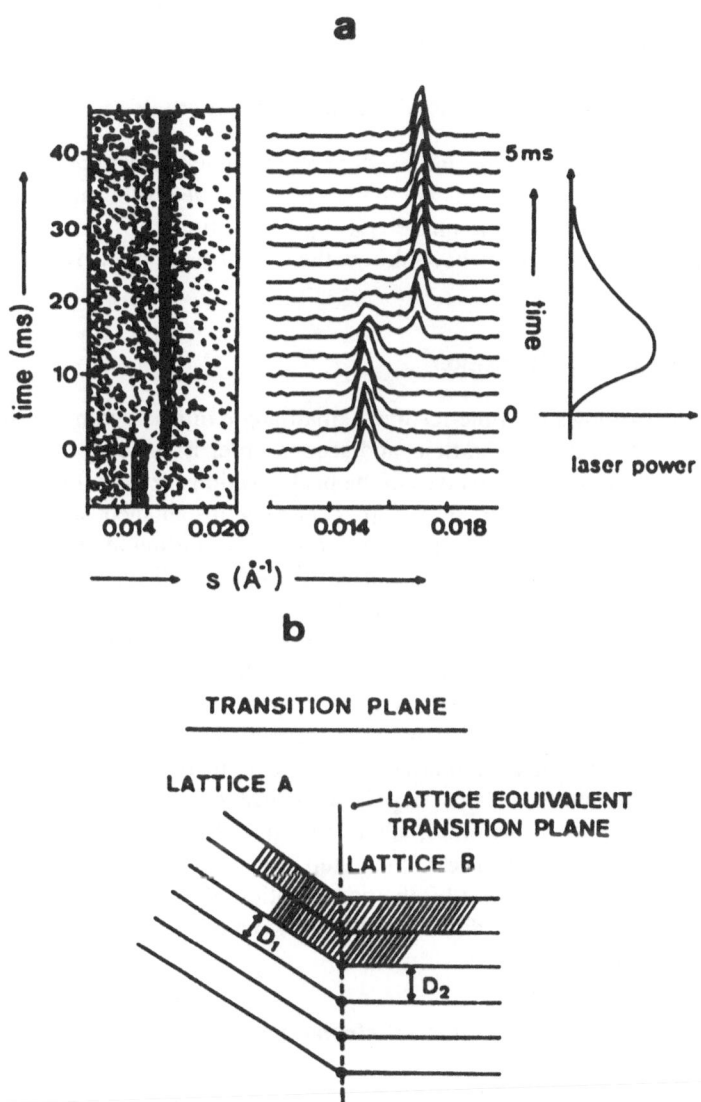

FIGURE 13. (a) Time development of the first-order lamellar repeat peak of the powder pattern for the POPE/water system in a T-jump from 25°C to 32°C (L_β–L_α transition), with a nominal resolution of 250 μsec per detection frame. The physically relevant resolution is given by the duration of the laser pulse of 1–2 msec. The inset shows the intensity contour plot. (b) Scheme of a martensitic "Umklapp" transition; the transition plane moves rapidly through the system.

formation. Transitions of this type are known to metallurgists as martensitic "Umklapp" transitions (Nishiyama, 1978).

4.3. Heterologous Transitions: Structural Intermediates

In these transitions, the lattice dimensions and the symmetry types are changed. The best studied cases are the lamellar-to-inverted-hexagonal (L_α–H_{II}) and the lamellar-gel-to-ripple-phase (L_β–$P_{\beta'}$) transitions, respectively. In both cases, the temperature jumps have shown the existence of short-lived intermediates.

4.3.1. The L_α–H_{II} Transition

Figure 14 shows the time-resolved small-angle diffractogram of an L_α–H_{II} transition in the region of the first- and second-order Bragg reflections together with the time course of the changes in the peak positions. Evidently there are two steps to be distinguished: a first step, occurring instantaneously, in which a thinner lamellar lattice is formed, and a second step starting after a lag phase of some 20 msec in which the hexagonal lattice, initially distorted and in coexistence with the thin lamellar intermediate, is formed slowly with a completion time on the order of seconds. A scheme of the structural interpretation is also shown in Figure 14. The first, rapid step is discussed again in terms of a martensitic Umklapp process. This leads to the necessary close approach of adjacent head group regions, and must be associated with the rapid loss of some of the interlamellar water, or alternatively, with an increase in molecular area at the lipid–water interface. This step is also visible in slow temperature scan experiments, and it is, therefore, likely that this thin lamellar intermediate is thermodynamically stable. The second step is a relatively slow nucleation and growth process, characterized by the conservation of overall regularity as indicated by the low amount of diffuse small-angle scattering and the relative sharpness of the peaks.

4.3.2. The $L_{\beta'}$–$P_{\beta'}$ Transition (Pretransition)

In saturated diacylphosphatidylcholine lipids, there exists a stable ripple-phase ($P_{\beta'}$) between the lamellar gel ($L_{\beta'}$) and the lamellar liquid-crystalline phase (L_α) (Janiak et al., 1976). Our time-resolved small-angle diffraction experiments (Laggner et al., 1990) have shown the following sequence of events (Figure 15). In a first step, occurring instantaneously on the experimental time scale, a lamellar lattice is formed with a smaller repeat distance, which coexists for various lengths of time (depending on salt conditions and the chain length of the lipids), with the parent lattice. These two lattices, as manifested by their distinct Bragg

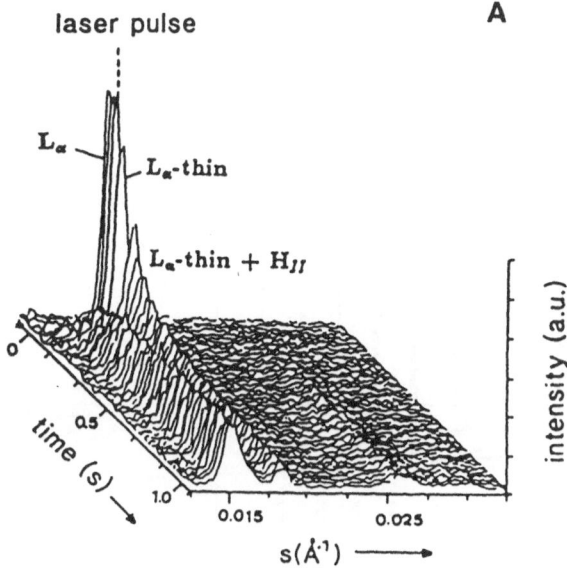

laser pulse

A

L_α

L_α-thin

L_α-thin + H_{II}

intensity (a.u.)

0

time (s)

0.5

1.0

0.015   0.025

$s(\text{Å}^{-1})$ ⟶

BIPHASIC TIME-PATTERN OF $L_\alpha \rightarrow H_{II}$ TRANSITION

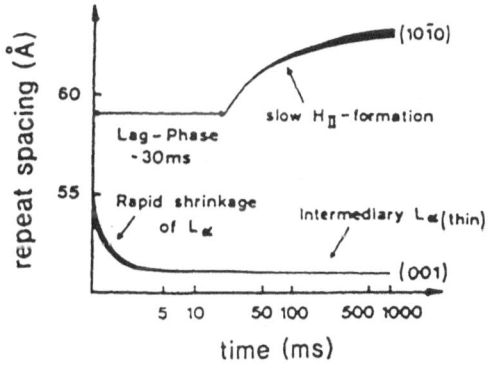

repeat spacing (Å)

60

$(10\bar{1}0)$

slow H_{II}-formation

Lag-Phase ~30ms

55

Rapid shrinkage of L_α

Intermediary L_α(thin)

(001)

5 10   50 100   500 1000

time (ms)

B

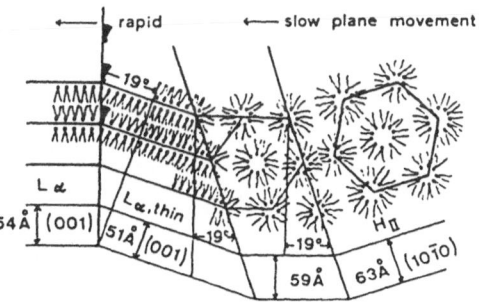

⟵ rapid   ⟵ slow plane movement

-19°

L_α

$L_{\alpha, thin}$

H_{II}

54Å (001)

51Å (001)

-19°

-19°

59Å   63Å $(10\bar{1}0)$

FIGURE 14. (A) Time-resolved small-angle diffractogram of the POPE/water system, and time development of the peak positions. (B) Structural model for the L_α–H_{II} transition mechanism.

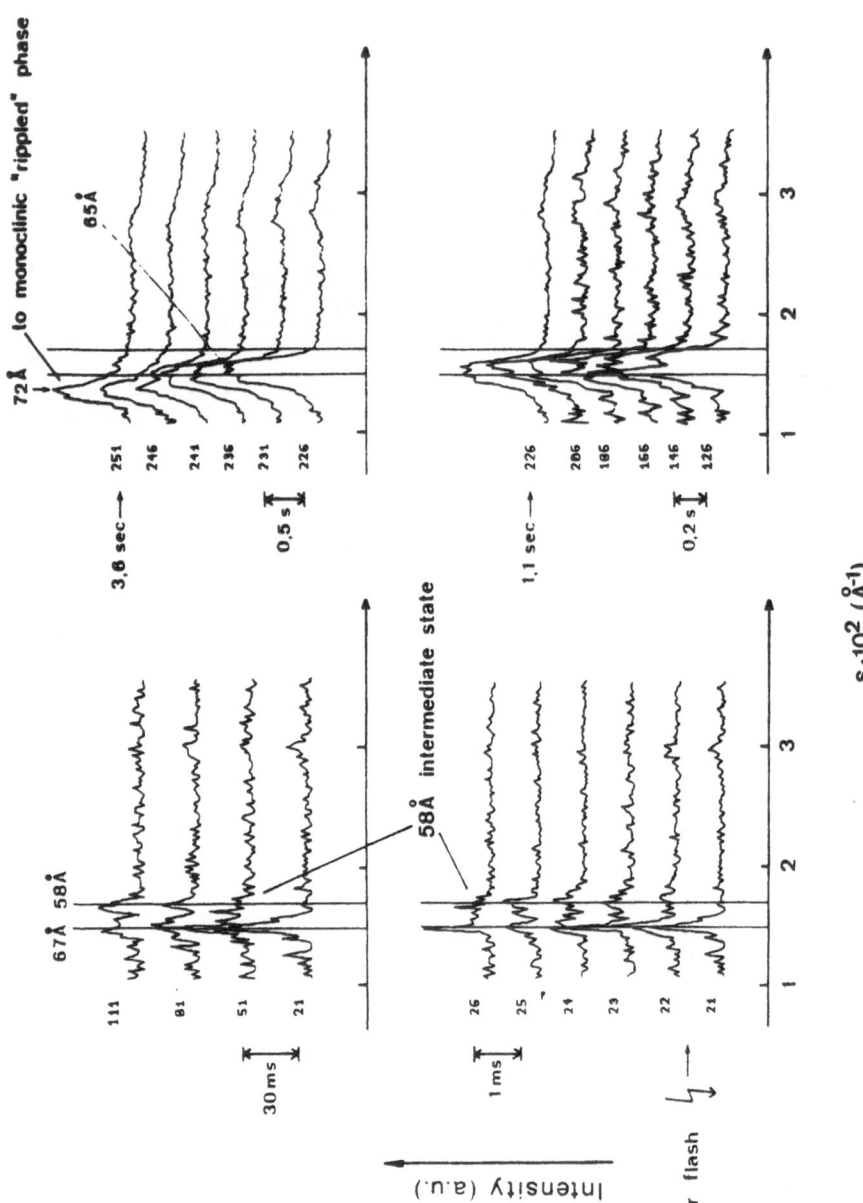

FIGURE 15. Time-resolved small-angle diffractogram of the pretransition of DPPC/water (1,2-dipalmitoyl-*sn*-glycero-3-phosphocholine) after a laser-induced *T*-jump.

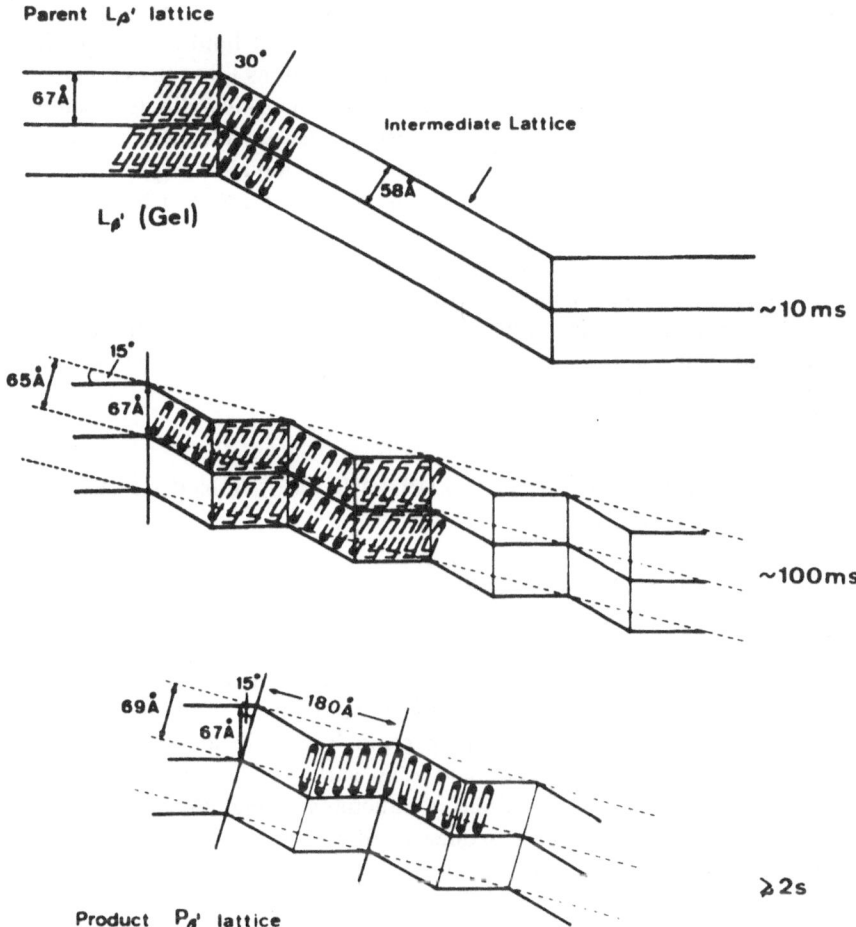

FIGURE 16. Structural model for the mechanism of the DPPC pretransition of DPPC under *T*-jump conditions. The first, rapid step is the formation of a thin (58 Å) intermediate lattice, by a martensitic "Umklapp" mechanism. Coexistence of many such zones with the parent lattice leads to a transient frustration, which lasts for several hundred milliseconds, and finally anneals slowly in times of minutes to hours into the equilibrium ripple phase.

reflections at 64 and 58 Å, respectively, merge slowly over a time of several seconds to give way to the final lattice with a relatively broad first-order peak at about 72 Å. A tentative mechanistic model for this process is shown in Figure 16. There remain several important questions, such as the time course of the chain-packing transition and the fate of the interbilayer water during the existence of the intermediate, thin lattice, which will have to be solved by future investigations. However, it is worth noting that this is the first verification of a dissipative intermediate structure in phase transitions of lipids, since this intermediate cannot be detected under equilibrium conditions (Tenchov et al., 1989; Matuoka et al., 1990). Thus, structural transitions can follow different pathways depending on the conditions and it is, therefore, important in the discussion of transition mechanisms to consider not just the energy but also the power involved in triggering the transition. It is also tempting to speculate that such intermediate structures could play a biological role in signal reception and noise-filtering since they provide a mechanism for discerning high-power signals from random noise. Further support for this notion has recently come from T-jump experiments within the temperature range of the L_α-Phase (Laggner, Kriechbaum, and Rapp, unpublished), which also showed a transient, intermediate 58 Å structure, that is, a thin lamellar liquid-crystalline lattice.

5. DIFFRACTION FROM THIN LIPID FILMS: MONOLAYERS AND LANGMUIR–BLODGETT FILMS

Although the spreading properties of membrane lipids have been known since the early 1930s and indeed have led to early insight into the basic feature of lipid bilayer arrangements in membranes, it was only during the last decade that this area received strongly renewed interest. This had its origin not so much in the biological implications, but rather in the recognition of highly interesting, potential applications in molecular engineering. The basic notion is that through subsequent pickup of monolayers onto solid surfaces it should be possible to artificially build up ordered multilayer structures (Langmuir–Blodgett films, Figure 17 with purpose-designed sequences of physical or chemical properties, and thus produce supramolecular architectures for specific technological applications. This has become a thriving field of interdisciplinary research (for a recent overview, see Ulman, 1991). X-ray diffraction techniques serve as important tools to analyze the structure of such thin organic films (Huang et al., 1991).

For diffraction on monolayers at the air–water interface, the advent of high-X-ray-flux synchrotron sources has triggered a breakthrough, since they provide the necessary intensity to measure the obviously very weak signal from a monomolecular film. Special optical geometries are required for reflection and diffraction at grazing incidence, close to the angle of total reflection. For a comprehensive review, the reader is referred to Als-Nielsen and Möhwald (1991). In its basic formalism, the information content of such experiments is similar to that outlined above for layered, planar systems (Section 3.7.5). Information is available on the electron density profile vertical to the monolayer surface and on the lateral packing of the hydrocarbon chains. The special advantage of such experiments stems from the fact that monolayers can be studied under a variety of

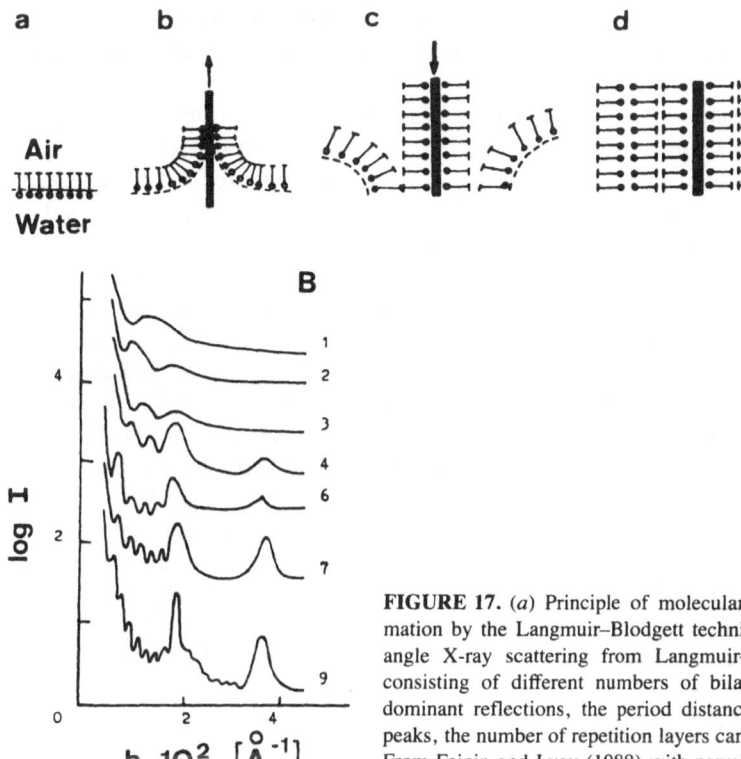

FIGURE 17. (a) Principle of molecular oligolayer formation by the Langmuir–Blodgett technique. (b) Small-angle X-ray scattering from Langmuir–Blodgett films consisting of different numbers of bilayers. From the dominant reflections, the period distance, and the side peaks, the number of repetition layers can be determined. From Feigin and Lvov (1988) with permission.

conditions (Helm et al., 1991), e.g., temperature, position at the pressure–area isotherm, and composition of the aqueous subphase.

For L-B films, informative X-ray diffraction patterns can be obtained even with conventional X-ray sources, and the data can provide important insight into the structure and degree of ordering that has been achieved by a given preparation technique (Figure 17; Feigin and Lvov, 1988). It is also noteworthy that L-B films can produce such highly interesting X-ray optical phenomena as standing waves (Bedzyk et al., 1988) and resonance enhancement between refracted and reflected X rays (Wang et al., 1991), which may find useful applications in and beyond membrane biophysics.

6. RADIATION DAMAGE

With conventional laboratory X-ray sources, aspects of radiation damage have not received much attention. However, with the 1000-fold or even higher increase in X-ray flux density on the samples offered by modern synchrotron radiation sources, this can become a real problem and needs to be considered seriously.

Of the manyfold reactions that are possible between the ionizing X rays and a lipid–water system (von Sonntag, 1989), the production of OH radicals from interaction with water seems to be the major potential source for damage. Secondary interaction with phospholipids can result in hydrogen abstraction, and the ensuing radicals may combine leading to cross-linking. Lipid peroxidation may also occur which results in alkenals as the final products (Esterbauer, 1982). It is clear that these local chemical and structural changes can induce point or line defects and alter the physical properties of the lipid phase (Caffrey, 1984; Ianzini et al., 1984; Albertini et al., 1987). As such effects can be observed even with the relatively stable, fully saturated phospholipid DPPC, it is likely that the problem of radiation damage may be worse with other less chemically stable lipids. In certain delicate cases this may even lead to unwanted phase transitions. It is known that short-chain hydrocarbons stabilize the inverted hexagonal H_{II} phase; if such hydrocarbons are formed as a consequence of radiation damage in a lamellar L_{α} phase, close to the L_{α}–H_{II} transition temperature, a concentration can be reached where the transition occurs isothermally (Chung and Caffrey, 1992; Cheng et al., 1993). It is therefore highly advisable to control the integrity of the samples not only before but also after exposure to X rays by suitable analytical methods.

Another problem, not directly related to radiation damage, may arise from the fact that the absorbed X-ray energy will eventually lead to a heating effect within the sample. At modern synchrotron sources, X-ray power densities on the order of 1 mW/mm^2 can be estimated, which can lead to heating of an aqueous sample, with the typical thickness of 1 mm, by some tenths of a degree Celsius per second. This may not seem dramatic, but with even higher X-ray fluxes as are to be expected with future sources, this effect will have to be taken seriously into account. In this respect, the use of shorter wavelengths than usually employed (CuK_{α} radiation, 1.54 Å) is considered as a way to effectively reduce radiation damage (Helliwell, 1992).

Finally, however, it is justified to make a positive note in favor of synchrotron sources: radiation damage is primarily a dose-determined effect, which means that the same amount of primary damage is caused by a given dose, irrespective of whether this is obtained by low- or high-flux radiation. Thus, if no damage occurs during an exposure over hours at conventional sources, a corresponding exposure at a synchrotron source, which takes fractions of a second, will also be without damage. Indeed, the shorter exposure times may even be favorable, since all secondary reactions are diffusion controlled and therefore a quick exposure may even lead to valid data before the secondary destruction processes are complete.

ACKNOWLEDGMENTS. The original works by the author cited herein have been supported by the Österreichischer Fonds zur Förderung der Wissenschaftlichen Forschung under grant S4614. Stimulating discussions with Drs. M. Kriechbaum, K. Lohner, and G. Rapp are gratefully acknowledged. Thanks are also due to Ms. M. Zechner for valuable help with preparing the graphics.

7. REFERENCES

Abrahamson, S., Dahlen, B., Löfgren, H., and Pascher, I., 1978, Lateral packing of hydrocarbon chains, *Prog. Chem. Fats Other Lipids* **16**:125–143.
Albertini, G., Fanelli, E., Guidoni, L., Ianzizi, F., Mariani, P., Masella, R., Rustichelli, F., and Viti, V., 1987, Studies of structural modifications induced by gamma-irradiation in distearoylphosphatidylcholine liposomes, *Int. J. Radiat. Biol. Relat. Stud. Phys. Chem. Med.* **52**:145–156.
Alexander, L. E., 1969, *X-ray Diffraction Methods in Polymer Science*, Wiley–Interscience, New York.
Als-Nielsen, J., and Möhwald, H., 1991, Synchrotron X-ray scattering studies of Langmuir films, in: *Handbook of Synchrotron Radiation*, Vol 4 (S. Ebashi, M. Koch, and E. Rubenstein, eds.), pp. 1–53, North-Holland, Amsterdam.
Aranda, F. J., Killian, J. A., and de Kruijff, B., 1987, Importance of gramicidin for its lipid structure modulating activity in lysophosphatidylcholine and phosphatidylethanolamine model membranes. A comparative study employing gramicidin analogs and a synthetic alpha-helical hydrophobic polypeptide, *Biochim. Biophys. Acta* **901**:217–228.
Atkinson, D., And Shipley, G. G., 1984, Structural studies of plasma lipoproteins, in *Neutrons in Biology* (B. Schoenborn, ed.), pp. 211–226, Plenum Press, New York.
Batenburg, A. M., Hibbeln, J. C., Verkleij, A. J., and de Kruijff, B., 1987a, Melittin induces HII phase formation in cardiolipin model membranes, *Biochim. Biophys. Acta* **903**:142–154.
Batenburg, A. M., van Esch, J. H., Leunissen-Bijvelt, J., Verkleij, A. J., and de Kruijff, B., 1987b, Interaction of melittin with negatively charged phospholipids: Consequences for lipid organization, *FEBS Lett.* **223**:148–154.
Dear, R. S., Palmer, K. J., and Schmitt, F. O., 1941, X-ray diffraction studies of nerve lipides, *J. Cell. Comp. Physiol.* **17**:355–367.
Bedzyk, M. J., Bilderback, D. H., Bommarito, G. M., Caffrey, M., and Schildkraut, J. S., 1988, X-ray standing waves: A molecular yardstick for biological membranes, *Science* **241**:1788–1791.
Blasie, J. K., Pachence, J. M., and Herbette, L. G., 1984, Neutron diffraction and the decomposition of membrane scattering profiles into the scattering profiles of their molecular components, in: *Neutrons in Biology* (B. Schoenborn, ed.), pp. 201–210, Plenum Press, New York.
Blaurock, A. E., 1982, Evidence of bilayer structure and of membrane interactions from x-ray diffraction analysis, *Biochim. Biophys. Acta* **650**:167–207.
Brefeld, W., and Gürtler, P., 1992, Synchrotron radiation sources, in: *Handbook on Synchrotron Radiation* (S. Ebashi, M. Koch, and E. Rubenstein, eds.), Vol. 4, pp. 269–296, North-Holland, Amsterdam.
Burgess, S. W., McIntosh, T. J., and Lentz, B. R., 1992, Modulation of poly(ethylene glycol)-induced fusion by membrane hydration: Importance of interbilayer separation, *Biochemistry* **31**:2653–2661.
Caffrey, M., 1984, X-radiation damage of hydrated lecithin membranes detected by real-time X-ray

diffraction using Wiggler-enhanced synchrotron radiation as the ionizing radiation source, *Nucl. Instrum. Methods Phys. Res.* **222**:329–338.

Caffrey, M., 1985, Kinetics and mechanism of the lamellar gel/lamellar liquid-crystal/inverted hexagonal phase transition in phosphatidylethanolamine: A real-time x-ray diffraction study using synchrotron radiation, *Biochemistry* **24**:4826–4844.

Caffrey, M., 1989a, A lyotrope gradient method for liquid crystal temperature–composition–mesomorph diagram construction using time-resolved x-ray diffraction, *Biophys. J.* **55**:47–52.

Caffrey, M., 1989b, Structural, mesomorphic and time-resolved studies of biological liquid crystals and lipid membranes using synchrotron x-radiation, in: *Synchrotron Radiation in Chemistry and Biology III* (E. Mandelkow, ed.), pp. 75–109, Springer, Berlin.

Caffrey, M., 1989c, The study of lipid phase transition kinetics by time-resolved X-ray diffraction, *Annu. Rev. Biophys. Biophys. Chem.* **18**:159–186.

Caffrey, M., Magin, R. L., Hummel, B., and Zhang, J., 1990, Kinetics of the lamellar and hexagonal phase transitions in phosphatidylethanolamine. Time-resolved x-ray diffraction study using a microwave-induced temperatue jump, *Biophys. J.* **58**:21–29.

Caffrey, M., Moynihan, D., and Hogan, J., 1991a, A database of lipid phase transition temperatures and enthalpy changes, *Chem. Phys. Lipids* **57**:275–292.

Caffrey, M., Hogan, J., and Mencke, A., 1991b, Kinetics of the barotropic ripple ($P_{\beta'}$)/lamellar liquid crystal (L_α) phase transition in fully hydrated dimyristoylphosphatidylcholine (DMPC) monitored by time-resolved x-ray diffraction, *Biophys. J.* **60**:456–466.

Caron, F., Mateu, L., Rigny, P., and Azerad, R., 1974, Chain motions in lipid–water and protein–lipid–water phases: A spin-label and X-ray diffraction study, *J. Mol. Biol.* **85**:279–300.

Cheng, A.-C., Hogan, J. L., and Caffrey, M., 1993, X-rays destroy the lamellar structure of model membranes, *J. Mol. Biol.* **229**:291–294.

Chung, H., and Caffrey, M., 1992, Direct correlation of structure changes and thermal events in hydrated lipid established by simultaneous calorimetry and time-resolved x-ray diffraction, *Biophys. J.* **63**:438–447.

Colotto, A., Lohner, K., Nd Laggner, P., 1991, Small-angle x-ray diffraction studies on the effects of melittin on lipid bilayer assemblies, *J. Appl. Crystallogr.* **24**:847–851.

Colotto, A., Mariani, P., Ponzi-Bossi, M. G., Rustichelli, F., Albertini, G., and Amaral, L. Q., 1992a, Lipid–drug interaction: A structural analysis of pindolol effects on model membranes, *Biochim. Biophys. Acta* **1107**:165–174.

Colotto, A., Lohner, K., and Laggner, P., 1992b, Low-dose effects of melittin on phospholipid structure. Differences between diester and diether phospholipids, *Prog. Colloid Polym. Sci.* **89**:334.

De Gennes, P. G., 1974, *The Physics of Liquid Crystals,* Oxford University Press, London.

de Wolf, F. A., Maliepaard, M., van Dorsten, F., Berghuis, I., Nicolay, K., and de Kruijff, B., 1990, Comparable interaction of doxorubicin with various acidic phospholipids results in changes of lipid order and dynamics, *Biochim. Biophys. Acta* **1096**:67–80.

Esterbauer, H., 1982, Aldehydic products of lipid peroxidation, in: *Free Radicals, Lipid Peroxidation and Cancer* (D. H. C. McBrien and T. F. Slater, eds.), pp. 101–128, Academic Press, New York.

Feigin, L., and Lvov, Y., 1988, Structure studies of Langmuir–Blodgett films, *Makromol. Chem. Makromol. Symp.* **15**:259–274.

Feigin, L., and Svergun, D. I., 1987, *Structure Analysis by Small-Angle X-ray and Neutron Scattering,* Plenum Press, New York.

Finean, J. B., 1989, X-ray diffraction studies of lipid phase transitions in hydrated mixtures of cholesterol and diacylphosphatidylcholines and their relevance to the structure of biological membranes, *Chem. Phys. Lipids* **49**:265–269.

Finean, J. B., and Hutchinson, A. L., 1988 X-ray diffraction studies of lipid phase transitions in cholesterol-rich membranes at sub-zero temperatures, *Chem. Phys. Lipids* **46**:63–71.

Franks, N. P., 1976, Structural analysis of hydrated egg lecithin and cholesterol bilayers. I. X-ray diffraction, *J. Mol. Biol.* **100**:345–358.

Franks, N. P., and Levine Y. K., 1981, Low-angle x-ray diffraction, in: *Membrane Spectroscopy* (E. Grell, ed.), pp. 437–492, Springer, Berlin.

Gasset, M., Killian, J. A., Tournois, H., and de Kruijff, B., 1988, Influence of cholesterol on gramicidin-induced HII phase formation in phosphatidylcholine model membranes, *Biochim. Biophys. Acta* **939**:79–88.

Gawrisch, K., Ruston, D., Zimmerberg, J., Parsegian, V. A., Rand, R. P., and Fuller, N., 1992, Membrane dipole potentials, hydration forces, and the ordering of water at membrane surfaces, *Biophys. J.* **61**:1213–1223.

Glatter, O., 1977, A new method for the evaluation of small-angle scattering data, *J. Appl. Crystallogr.* **10**:415–421.

Glatter, O., 1980, Evaluation of small-angle scattering data from lamellar and cylindrical particles by the indirect transformation method, *J. Appl. Crystallogr.* **13**:577–584.

Glatter, O., and Kratky, O., 1982, *Small Angle X-ray Scattering* (O. Glatter and O. Kratky, eds.), Academic Press, New York.

Gordeliy, V. I., Ivkov, V. G., Ostanievich, Y. M., and Yaguzhinskij, L. S., 1991, Detection of structural defects in phosphatidylcholine membranes by small-angle neutron scattering, *Biochim. Biophys. Acta* **1061**:39–48.

Gruner, S. M., 1987, Time-resolved x-ray diffraction of biological materials, *Science* **238**:305–312.

Guinier, A., 1964 *Theorie et Technique de la Radiocrystallographie,* Dunod, Paris.

Guinier, A., and Fournet, G., 1955, *Small-Angle Scattering of X-Rays,* Wiley, New York.

Gulik, A., Luzzati, V., DeRosa, M., and Gambacorta, A., 1985, Structure and polymorphism of bipolar isoprenyl ether lipids from archebacteria, *J. Mol. Biol.* **182**:131–149.

Gulik, A., Luzzati, V., DeRosa, M., and Gambacorta, A., 1988, Tetraether lipid components from a thermoacidophilic archebacterium. Chemical structure and physical polymorphism, *J. Mol. Biol.* **210**:429–435.

Hansen, S., and Skov Pedersen, J., 1991, A comparison of three different methods for analysing small-angle scattering data, *J. Appl. Crystallogr.* **24**:541–548.

Hauser, H., Pascher, I., Pearson, R. H., and Sundell, S., 1981, Preferred conformation and molecular packing of phosphatidylethanolamine and phosphatidylcholine, *Biochim. Biophys. Acta* **650**:21–51.

Helliwell, J. R., 1992, *Macromolecular Crystallography with Synchrotron Radiation,* Cambridge University Press, London.

Helm, C. A., Tippmann-Krayer, P., Möhwald, H., Als-Nielsen, J., and Kjaer, K., 1991, Phases of phosphatidyl ethanolamine monolayers studied by synchrotron x-ray scattering, *Biophys. J.* **60**:1457–1476.

Hentschel, M., and Hosemann, R., 1983, Small- and wide angle x-ray scattering of oriented lecithin multilayers, *Mol. Cryst. Liq. Cryst.* **94**:291–316.

Hing, F. S., Maulik, P. R., and Shipley, G. G., 1991, Structure and interactions of ether- and ester-linked phosphatidylethanolamines, *Biochemistry* **30**:9007–9015.

Hinz, H. J., Kuttenreich, H., Meyer, R., Renner, M., Frund, R., Koynova, R., Boyanov, A. I., and Tenchov, B. G., 1991, Stereochemistry and size of sugar head groups determine structure and phase behavior of glycolipid membranes: Densitometric, calorimetric, and x-ray studies, *Biochemistry* **30**:5125–5138.

Hitchcock, P. B., Mason, R., Thomas, K. M., and Shipley, G. G., 1974, Structural chemistry of 1,2 dilauroyl-DL-phosphatidylethanolamine: Molecular conformation and intermolecular packing of phospholipids, *Proc. Natl. Acad. Sci. USA* **195**:3036–3049.

Hodgson, K. O., and Doniach, S., 1978, Synchrotron radiation, a new tool for chemical and structural studies, *Chem. Eng. News* **21**:26–37.

Hosemann, R., and Bagchi, S. N., 1962, *Direct Analysis of Diffraction by Matter*, North-Holland, Amsterdam.

Huang, T. C., Cohen, P. J., and Eaglesham, D. J., 1991, *Advances in Surface and Thin Film Diffraction*, Materials Research Society Symposium Proceedings, Vol. 208, Materials Research Society, Pittsburgh.

Ianzini, F., Guidoni, L., Indovina, P. L., Viti, V., Erriu, G., Onnis, S., and Randaccio, P., 1984, Gamma-irradiation effects on phosphatidylcholine multilayer liposomes: Calorimetric, NMR, and spectrofluorimetric studies, *Radiat. Res.* **98:**154–166.

Janiak, M. J., Small, D. M., and Shipley, G. G., 1976, Nature of the thermal pretransition of synthetic phospholipids: Dimyristoyl- and dipalmitoyllecithin, *Biochemistry* **15:**4575–4580.

Jendrasiak, G. L., McIntosh, T. L., Ribeiro, A., and Porter, R. S., 1990, Amiodarone–liposome interaction: A multinuclear NMR and X-ray diffraction study, *Biochim. Biophys. Acta* **1024:**19–31.

Katsaras, J., Stinson, R. H., Davis, J. H., and Kendall, E. J., 1991, Location of two antioxidants in oriented model membranes. Small-angle x-ray diffraction study, *Biophys. J.* **59:**645–653.

Katsaras, J., Yang, D. S.-C., and Epand, R. M., 1992, Fatty-acid chain tilt angles and directions in dipalmitoyl phosphatidylcholine bilayers, *Biophys. J.* **63:**1170–1175.

Klose, G., König, B., Gordeliy, V. I., and Schulze, G., 1991, Incorporation of phosphonic acid diesters into lipid model membranes. Part II. X-ray and neutron diffraction studies, *Chem. Phys. Lipids* **59:**137–149.

Knott, R. B., and Schoenborn, B. P., 1986, Quantitation of water in membranes by neutron diffraction and x-ray techniques, *Methods Enzymol.* **127:**217–229.

Kodali, D. R., Fahey, D. A., and Small, D. M., 1990, Structure and polymorphism of saturated monoacid 1,2-diacyl-sn-glycerols, *Biochemistry* **29:**10771–10779.

Kostner, G., and Laggner, P., 1989, Chemical and physical properties of lipoproteins, in: *Human Plasma Lipoproteins* (J. C. Fruchart and J. Shepherd, eds.), pp. 22–54, de Gruyter, Berlin.

Kratky, O., and Laggner, P., 1992, X-ray small-angle scattering, in: *Encyclopedia of Physical Science and Technology* (R. A. Meyers, ed.), Vol. 17, pp. 727–781, Academic Press, New York.

Kriechbaum, M., Laggner, P., and Rapp, G., 1990, Fast time-resolved X-ray diffraction for studying laser T-jump induced phase transitions, *Nucl. Instrum. Methods Phys. Res.* **A291:**41–45.

Kriechbaum, M., Osterberg, M. W., Tate, M. W., Shyamsunder, E., Polcyn, A. D., So, P. C. T., and Gruner, S. M., 1993, Time-resolved pressure-jump studies on membrane lipids by synchrotron x-ray diffraction, *Biophys. J.* **64:**A296.

Laggner, P., 1986, Time-resolved diffraction studies on thermotropic phase transitions of phospholipids, in: *New Methods in X-ray Absorption, Scattering and Diffraction* (B. Chance and H. D. Bartunik, eds.), pp. 171–182, Academic Press, New York.

Laggner, P., 1988, X-ray studies on biological membranes using synchrotron radiation, in: *Synchrotron Radiation in Chemistry and Biology I* (E. Mandelkow, ed.), 171–202, Springer, Berlin.

Laggner, P., and Kriechbaum, M., 1991, Phospholipid phase transitions: Kinetics and structural mechanisms, in: *Phospholipid Phase Transitions* (P. Kinnunen and P. Laggner, eds.), Special Issue of *Chem. Phys. Lipids* **57:**121–145, Elsevier, Amsterdam.

Laggner, P., and Mio, H., 1992, SWAX—a dual-detector camera for simultaneous small- and wide-angle diffraction in polymer and liquid crystal research, *Nucl. Instrum. Methods Phys. Res.* **A323:**86–90.

Laggner, P., and Müller, K., 1978, The structure of serum lipoproteins as analysed by x-ray small-angle scattering, *Q. Rev. Biophys.* **11:**371–425.

Laggner, P., Kostner, G. M., Rakusch, U., and Worcester, D., 1981, Neutron small angle scattering on selectively deuterated human plasma low density lipoproteins. The location of polar phospholipid headgroups, *J. Biol. Chem.* **256:**11832–11839.

Laggner, P., Kostner, G. M., Degovics, G., and Worcester, D. L., 1984, Structure of the cholesteryl ester core of human plasma low density lipoproteins: Selective deuteration and neutron scattering, *Proc. Natl. Acad. Sci. USA* **81**:4389–4393.

Laggner, P., Lohner, K., Degovics, G., Mueller, K., and Schuster, A., 1987a, Structure and thermodynamics of the dihexadecyl–phosphatidylcholine–water system, *Chem. Phys. Lipids* **44**:31–60.

Laggner, P., Lohner, K., and Müller, K., 1987b, X-ray cinematography of phospholipid phase transformations with synchrotron radiation, *Mol. Cryst. Liq. Cryst.* **151**:373–388.

Laggner, P., Kriechbaum, M., Hermetter, A., Paltauf, F., Hendrix, J., and Rapp, G., 1989, Laser-induced temperature jump and time-resolved x-ray powder diffraction on phospholipid phase transitions, *Prog. Colloid Polym. Sci.* **79**:33–37.

Laggner, P., Kriechbaum, M., Rapp, G., and Hendrix, J., 1990, Structural pathways and short-lived intermediates in phospholipid phase transitions. Millisecond synchrotron x-ray diffraction studies, in: *2nd European Conference on Progress in X-ray Synchrotron Radiation Research* (A. Balerna, E. Bernieri, and S. Mobilio, eds.), pp. 995–998, SIF Editrice Compositori, Bologna.

Lee, C. W., Das Gupta, S. K., Mattai, J., Shipley, G. G., Abdel-Mageed, O. H., Makriyannis, A., and Griffin, R. G., 1989, Characterization of the L lambda phase in trehalose-stabilized dry membranes by solid-state NMR and X-ray diffraction, *Biochemistry* **28**:5000–5009.

Levine, Y. K., 1972, Physical studies of membrane structure, *Prog. Biophys. Mol. Biol.* **24**:1–74.

Lewis, B. A., and Engelman, D. M., 1983, Lipid bilayer thickness varies linearly with acyl chain length in fluid phosphatidylcholine vesicles, *J. Mol. Biol.* **166**:211–217.

Lewis, R. N., and McElhaney, R. N., 1992, Structures of the subgel phases of n-saturated diacyl phosphatidylcholine bilayers: FTIR spectroscopic studies of $^{13}C = 0$ and ^{2}H labeled lipids, *Biophys. J.* **61**:63–77.

Lewis, R. N., Yue, A. W., McElhaney, R. N., Turner, D. C., and Gruner, S. M., 1990, Thermotropic characterization of the 2-0-acyl, polyprenyl a-D-glucopyranoside isolated from palmitate-enriched *Acholeplasma laidlawii* B membranes, *Biochim. Biophys. Acta* **1026**:21–28.

Lindblom, G., Nd Rilfors, L., 1989, Cubic phases and isotropic structures formed by membrane lipids—Possible biological relevance, *Biochim. Biophys. Acta* **998**:221–256.

Lis, L., and Quinn, P., 1991, The application of synchrotron x-radiation for the study of phase transitions in lipid model membrane systems, *J. Appl. Crystallogr.* **24**:48–60.

Lohner, K., Schuster, A., Degovics, G., Mueller, K., and Laggner, P., 1987, Thermal phase behavior and structure of hydrated mixtures between dipalmitoyl- and dihexadecylphosphatidylcholine, *Chem. Phys. Lipids* **44**:61–70.

Lohner, K., Balgavy, P., Hermetter, A., Paltauf, F., and Laggner, P., 1991, Stabilization of non-bilayer structures by the etherlipid ethanolamine plasmalogen, *Biochim. Biophys. Acta* **1061**:132–140.

Lohner, K., Degovics, G., Laggner, P., Gnamusch, E., and Paltauf, F., 1993, Squalene promotes the formation of non-bilayer structures in phospholipid model membranes, *Biochim. Biophys. Acta* **1152**:63–77.

Luzzati, V., 1968, X-ray diffraction studies of lipid–water systems, in: *Biological Membranes* (D. Chapman, ed.), pp. 71–123, Academic Press, New York.

Luzzati, V., Vargas, R., Gulik, A., Mariani, P., Seddon, J. M., and Rivas, E., 1992, Lipid polymorphism: A correction. The structure of the cubic phase of extinction symbol Fd— consists of two types of disjointed reverse micelles imbedded in a three-dimensional hydrocarbon matrix, *Biochemistry* **31**:279–285.

McIntosh, T. J., and Holloway, P. W., 1987, Determination of the depth of bromine atoms in bilayers formed from bromolipid probes, *Biochemistry* **26**:1783–1788.

McIntosh, T. J., Magid, A. D., and Simon, S. A., 1989a, Range of the solvation pressure between lipid membranes: Dependence on the packing density of solvent molecules, *Biochemistry* **28**:7904–7912.

McIntosh, T. J., Magid, A. D., and Simon, S. A., 1989b, Cholesterol modifies the short-range repulsive interactions between phosphatidylcholine membranes, *Biochemistry* **28**:17–25.

McIntosh, T. J., Simon, S. S., Needham, D., and Huang, C.-H., 1992, Structure and cohesive properties of sphingomyelin/cholesterol bilayers, *Biochemistry* **31**:2012–2020.

Malkin, T., 1952, The molecular structure and polymorphism of fatty acids and their derivatives, *Prog. Chem. Fats Other Lipids* **2**:1–17.

Malkin, T., 1954, The polymorphism of glycerides, *Prog. Chem. Fats Other Lipids* **2**:1–24.

Margaritondo, G., 1988, *Introduction to Synchrotron Radiation*, Oxford University Press, London.

Mariani, P., Luzzati, V., and Delacroix, H., 1988, Cubic phases of lipid-containing systems. Structure analysis and biological implications, *J. Mol. Biol.* **204**:165–189.

Mariani, P., Rivas, E., Luzzati, V., and Delacroix, H., 1990, Polymorphism of lipid extract from *Pseudomonas fluorescens:* Structure analysis of a hexagonal phase and of a novel cubic phase of extinction symbol Fd—, *Biochemistry* **29**:6799–6810.

Mason, R. P., Shoemaker, W. J., Shajenko, L., Chambers, T. E., and Herbette, L. G., 1992a, Evidence for changes in the Alzheimer's disease brain cortical membrane structure mediated by cholesterol, *Neurobiol. Aging* **13**:413–419.

Mason, R. P., Moisey, D. M., and Shajenko, L., 1992b, Cholesterol alters the binding of Ca^{2+} channel blockers to the membrane lipid bilayer, *Mol. Pharmacol.* **41**:315–321.

Mateu, L., Luzzati, V., Vargas, R., Vonasek, E., and Borgo, M., 1990, Order—disorder phenomena in myelinated nerve sheaths. II. The structure of myelin in native and swollen rat sciatic nerves and in the course of myelinogenesis, *J. Mol. Biol.* **215**:385–402.

Mattai, J., Witzke, N. M., Bittman, R., and Shipley, G. G., 1987, Structure and thermotropic properties of hydrated 1-eicosyl-2-dodecly-rac-glycero-3-phosphocholine and 1-dodecyl-2-eicosyl-rac-glycero-3-phosphocholine bilayer membranes, *Biochemistry* **26**:623–633.

Matuoka, S., Kato, S., Akiyama, M., Amemiya, Y., and Hatta I., 1990, Temperature dependence of the ripple structure in dimyristoylphosphatidylcholine studied by synchrotron x-ray small-angle diffraction, *Biochim. Biophys. Acta* **1028**:103–109.

Maulik, P. R., Ruocco, M. J., and Shipley, G. G., 1990, Hydrocarbon chain packing modes in lipids; effect of altered sub-cell dimensions and chain rotation, *Chem. Phys. Lipids* **56**:123–133.

Mavromoustakos, T., Yang, D. P., Charalambous, A., Herbette, L. G., and Makriyannis, A., 1990, Study of the topography of cannabinoids in model membranes using X-ray diffraction, *Biochim. Biophys. Acta* **1024**:336–344.

Mencke, A. P., and Caffrey, M., 1991, Kinetics and mechanism of the pressure-induced lamellar order/disorder transition in phosphatidylethanolamine: A time-resolved X-ray diffraction study, *Biochemistry* **30**:2453–2463.

Mencke, A., Cheng, A., and Caffrey, M., 1993, A simple apparatus for time-resolved x-ray diffraction biostructure studies using static and oscillating pressures and pressure-jumps, *Rev. Sci. Instrum.* **64**:383–389.

Michel, H., 1990, General and practical aspects of membrane protein crystallization, in: *Crystallization of Membrane Proteins* (H. Michel, ed.), pp. 72–88, CRC Press, Boca Raton, FL.

Moore, P. B., 1980, Small-angle scattering. Information content and error analysis, *J. Appl. Crystallogr.* **13**:168–175.

Murthy, N. S., and Worthington, C. R., 1991, X-ray diffraction evidence for the presence of discrete water layers on the surface of membranes, *Biochim. Biophys. Acta* **1062**:172–176.

Nakajima, K., Utsumi, H., Kazama, M., and Hamada, A., 1990, Alpha-tocopherol-induced hexagonal HII phase formation in egg yolk phosphatidylcholine membranes, *Chem. Pharm. Bull.* **38**:1–4.

Needham, D., McIntosh, T. J., and Evans, E., 1988, Thermomechanical and transition properties of dimyristoylphosphatidylcholine/cholesterol bilayers, *Biochemistry* **27**:4668–4673.

Nishiyama, Y., 1978, *Martensitic Transformation*, Academic Press, New York.

Palmer, K. J., and Schmitt, F. O., 1941, X-ray diffraction studies of lipid emulsions, *J. Cell. Comp. Physiol.* **17**:385–394.

Parsegian, V. A., Rand, R. P., Fuller, N. L., and Rau, D. C., 1986, Osmotic stress for the direct measurement of intermolecular forces, *Methods Enzymol.* **127**:400–416.

Pascher, I., 1976, Molecular arrangements in sphingolipids. Conformation and hydrogen bonding of ceramide and their implication on membrane stability and permeability, *Biochim. Biophys. Acta* **455**:433–451.

Pascher, I., Lundmark, M., Nydham, P.-G., and Sundell, S., 1992, Crystal structure of membrane lipids, *Biochim. Biophys. Acta* **1113**:339–373.

Pearson, R. H., and Pascher, I., 1979, The molecular structure of lecithin dihydrate, *Nature* **281**:499–501.

Porod, G., 1982, General theory, in: *Small Angle X-Ray Scattering* (O. Glatter and O. Kratky, eds.), pp. 17–53, Academic Press, New York.

Rand, R. P., Fuller, N. L., Gruner, S. M., and Parsegian, V. A., 1990, Membrane curvature, lipid segregation, and structural transitions for phospholipids under dual-solvent stress, *Biochemistry* **29**:76–87.

Ruocco, M. J., and Shipley, G. G., 1982, Characterization of the sub-transition of hydrated dipalmitoylphosphatidylcholine bilayers. Kinetic, hydration and structural study, *Biochim. Biophys. Acta* **691**:309–320.

Sanderson, P. W., Lis, L. J., Quinn, P. J., and Williams, W. P., 1991, The Hofmeister effect in relation to membrane lipid phase stability, *Biochim. Biophys. Acta* **1067**:43–50.

Scherer, J. R., 1987, The partial molar volume of water in biological membranes, *Proc. Natl. Acad. Sci. USA* **84**:7938–7942.

Seddon, J. M., 1990, Structure of the inverted hexagonal (H_{II}) phase, and non-lamellar phase transitions of lipids, *Biochim. Biophys. Acta* **1031**:1–69.

Seddon, J. M., Hogan, J. L., Warrender, N. A., and Pebay-Peyroula, E., 1990, Structural studies of phospholipid cubic phases, *Prog. Colloid Polym. Sci.* **81**:189–197.

Sen, A., Hui, S. W., Mannock, D. A., Lewis, R. N., and McElhaney, R. N., 1990, Physical properties of glycosyl diacylglycerols. 2. X-ray diffraction studies of a homologous series of 1,2-di-0-acyl-3-0-(alpha-D-glucopyranosyl)-sn-glycerols, *Biochemistry* **29**:7799–7804.

Shipley, G. G., 1973, Recent X-ray diffraction studies of biological membranes and membrane components, in: *Biological Membranes* (D. Chapman and D. F. H. Wallach, eds.), pp. 1–89, Academic Press, New York.

Shipley, G. G., 1986, X-ray crystallographic studies of aliphatic lipids, in: *Handbook of Lipid Research,* Vol. 4 (D. M. Small, ed.), pp. 97–147, Plenum Press, New York.

Shyamsunder, E., Gruner, S. M., Tate, M. W., Turner, D. C., So, P. T., and Tilcock, C. P., 1988, Observation of inverted cubic phase in hydrated dioleoylphosphatidylethanolamine membranes, *Biochemistry* **27**:2332–2336.

Siegel, D. P., 1987, Inverted micellar intermediates and the transitions between lamellar, cubic, and inverted hexagonal amphiphile phases. III. Isotropic and inverted cubic state formation via intermediates in transitions between Lα and HII phases, *Chem. Phys. Lipids* **42**:279–301.

Siegel, D. P., and Banschbach, J. L., 1990, Lamellar/inverted cubic (Lα–QII) phase transition in N-methylated dioleoylphosphatidylethanolamine, *Biochemistry* **29**:5975–5981.

Simon, S. A., Fink, C. A., Kenworthy, A. K., and McIntosh, T. J., 1991, The hydration pressure between lipid bilayers. Comparison of measurements using x-ray diffraction and calorimetry, *Biophys. J.* **59**:538–546.

Singer, S. J., and Nicholson, G. L., 1972, The fluid mosaic model of the structure of cell membranes, *Science* **175**:720–724.

Small, D. M., 1986, The physical states of lipids: Solids, mesomorphic states, and liquids, in: *Handbook of Lipid Research*, Vol. 4 (D. M. Small, ed.), pp. 43–87, Plenum Press, New York.

Stuhrmann, H. B., 1978, The use of X-ray synchrotron radiation for structural research in biology, *Q. Rev. Biophys.* **11**:71–98.

Stuhrmann, H., and Miller A., 1978, Small-angle scattering of biological structures, *J. Appl. Crystallogr.* **11**:325–345.

Svergun, D. I., 1991, Mathematical methods in small-angle scattering data analysis, *J. Appl. Crystallogr.* **24**:485–492.

Tardieu, A., 1972, Etude cristallographique de systemes lipides-eau, Thesis, Universite de Paris-Sud.

Tate, M. W., Eikenberry, E. F., Turner, D. C., Shyamsunder, E., and Gruner, S., 1991, Nonbilayer phases of membrane lipids, *Chem. Phys. Lipids* **57**:147–164.

Taupin, D., and Luzzati, V., 1982, Information content and retrieval in solution scattering studies. 1. Degrees of freedom and data reduction, *J. Appl. Crystallogr.* **15**:289–300.

Tenchov, B. G., Yao, H., and Hatta, I., 1989, Time-resolved x-ray diffraction and calorimetric studies at low scan rates. I. Fully hydrated DPPC and DPPC/water/ethanol mixtures, *Biophys. J.* **56**:757–768.

Tilcock, C. P., Cullis, P. R., and Gruner, S. M., 1988, Calcium-induced phase separation phenomena in multicomponent unsaturated lipid mixtures, *Biochemistry* **27**:1415–1420.

Turner, D. C., and Gruner, S. M., 1992, X-ray diffraction reconstruction of the inverted hexagonal (HII) phase in lipid–water systems, *Biochemistry* **31**:1340–1355.

Turner, D. C., Gruner, S. M., and Huang, J. S., 1992, Distribution of decane within the unit cell of the inverted hexagonal (H_{II}) phase of lipid–water–decane systems determined by neutron diffraction, *Biochemistry* **31**:1356–1363.

Ulman, A., 1991, *An Introduction to Ultrathin Organic Films. From Langmuir–Blodgett to Self-Assembly*, Academic Press, New York.

Van Osdol, W. W., Biltonen, R. L., and Johnson, M. L., 1989, Measuring the kinetics of membrane phase transitions, *J. Biochem. Biophys. Methods* **20**:1–46.

Vargas, R., Mariani, P., Gulik, A., and Luzzati, V., 1992, Cubic phases of lipid-containing systems. The structure of phase Q223 (space group Pm3n). An X-ray scattering study, *J. Mol. Biol.* **225**:137–145.

von Sonntag, C., 1989, *The Chemical Basis of Radiation Biology*, Taylor & Francis, London.

Wang, J., Bedzyk, M. J., Penner, T. L., and Caffrey, M., 1991, Structural studies of membrane and surface layers up to 1000 Å thick using x-ray standing waves, *Nature* **354**:377–380.

Warren, B. E., 1969, *X-Ray Diffraction*, Addison–Wesley, Reading, MA.

White, S. H., Jacobs, R. E., and King, G. I., 1987, Partial specific volumes of lipid and water in mixtures of egg lecithin and water, *Biophys. J.* **52**:663–665.

Wiener, M. C., and White, S. H., 1992, Structure of a fluid dioleoylphosphatidylcholine bilayer determined by joint refinement of X-ray and neutron diffraction data. III. Complete structure, *Biophys. J.* **61**:437–447.

Wiener, M. C., King, G. I., and White, S. H., 1991, Structure of a fluid dioleoylphosphatidylcholine bilayer determined by joint refinement of x-ray and neutron diffraction data. I. Scaling of neutron data and the distribution of double bonds and water, *Biophys. J.* **60**:568–576.

Wilkins, M. H. F., Blaurock, A. E., and Engelman D. M., 1971, Bilayer structure in membranes, *Nature New Biol.* **230**:72–76.

Williams, W. P., Quinn, P. J., Tsonev, L. I., and Koynova, R. D., 1991, The effects of glycerol on the phase behaviour of hydrated distearoylphosphatidylethanolamine and its possible relation to the mode of action of cryoprotectants, *Biochim. Biophys. Acta* **1062**:123–132.

X-Ray Diffraction on Biomembranes 491

Wilschut, J., 1991, Membrane fusion in lipid vesicle systems, an overview, in: *Membrane Fusion* (J. Wilschut and D. Hoekstra, eds.), pp. 89–126, Dekker, New York.
Worcester, D. L., and Franks, N. P., 1976, Structural analysis of hydrated egg lecithin and cholesterol bilayers. II. Neutron diffraction, *J. Mol. Biol.* **100:**359–378.
Worthington, C. R., 1981, The determination of the first-order phase in membrane diffraction using electron density strip models, *J. Appl. Crystallogr.* **14:**387–391.
Young, H. S., Skita, V., Mason, R. P., and Herbette, L. G., 1992, Molecular basis for the inhibition of 1,4-dihydropyridine calcium channel drugs binding to their receptors by a nonspecific site interaction mechanism, *Biophys. J.* **61:**1244–1255.

Index